BREEDING BIOLOGY OF BIRDS

*PROCEEDINGS OF A SYMPOSIUM
ON BREEDING BEHAVIOR AND
REPRODUCTIVE PHYSIOLOGY IN BIRDS*

Donald S. Farner, Editor

*DENVER, COLORADO
FEBRUARY 1972*

NATIONAL ACADEMY OF SCIENCES
WASHINGTON, D.C.   1973

NOTICE: The symposium reported herein was undertaken under the aegis of the National Research Council with the express approval of its Governing Board. Such approval indicated that the Board considered the problem to be of national significance, that its elucidation required scientific or technical competence, and that the resources of NRC were particularly suitable to the conduct of the project. The institutional responsibilities of the NRC were then discharged in the following manner:

The participants in the symposium were selected for their individual scholarly competence and judgment with due consideration for the balance and breadth of disciplines. Responsibility for all aspects of this report rests with them and with the organizing committee, to whom sincere appreciation is expressed.

Although proceedings of symposia are not submitted for approval to the Academy membership or to the Council, each is reviewed according to procedures established and monitored by the Academy's Report Review Committee. Such reviews are intended to determine *inter alia*, whether the major questions and relevant points of view have been addressed and whether the reported findings and conclusions arose from the available data and information. Distribution is approved, by the President, only after satisfactory completion of this review process.

This symposium was supported by the National Science Foundation, the U.S. Department of the Interior, and the National Wildlife Federation.

Available from
Printing and Publishing Office, National Academy of Sciences
2101 Constitution Avenue, N.W., Washington, D.C. 20418

LIBRARY OF CONGRESS CATALOGING IN PUBLICATION DATA

Symposium on Breeding Behavior and Reproductive Physiology in Birds, Denver, 1972.
   Breeding biology of birds.

   "Undertaken under the aegis of the National Research Council."
   Includes bibliographies.
   1. Birds–Physiology–Congresses. 2. Birds–Behavior–Congresses. 3. Reproduction–Congresses.
I. Farner, Donald Stanley, 1915–   ed.
II. National Research Council. III. Title.
QL698.S93 1972   598.2′5′6   73-86
ISBN 0-309-02109-X

Preface

During the past two decades experimental investigations have resulted
in highly significant advances in the ethology and physiology of reproduction in birds. During this same period important game species
have come under increased hunting pressure while their habitats have
been subjected to the severe impact of changing practices in land use;
the problems of "endangered species" have become more acute; and
problems of economically important "pest" species have persisted.
Wildlife biologists, who must contend with the problems, have conducted extensive field research, the results of which provide a more
precise empirical definition of problems and some solutions. For
many reasons, however, there has been a serious failure in communication between wildlife biologists, on the one hand, and ethologists
and physiologists, on the other. There are also significant philosophic
differences and consequent barriers to communication between ethologists and physiologists. Nevertheless, it seems clear that research by
ornithologists, avian ethologists, avian physiologists, and wildlife
biologists has reached the point where highly useful interactions are
now possible.

In this symposium the Division of Biology and Agriculture func-
tioned in its not infrequent role of organizer and catalyst in an area

that falls in the realm of no single professional or scientific organization. The symposium was, in fact, a response to a joint recommendation by the presidents of Wildlife Society and the American Ornithologists' Union and the Chairman, Division of Comparative Endocrinology of the American Society of Zoologists.

The primary function of this symposium was to present neither the latest products of research nor detailed technical résumés of restricted areas of research. Rather, its goal was to attain, by adjustment of communication to, or at least toward, a "common language"— a fruitful exchange of ideas among the disciplines involved. Although it seems clear that the symposium attained its anticipated function, the true extent to which it achieved its goal can only be assessed with time. It is hoped that both its successful aspects and its shortcomings will serve as guidelines for the organizers of similar ventures in the future.

Contents

Introduction

DONALD S. FARNER

Reproduction in birds, as in most higher animals, is discontinuous and frequently periodic. This is a consequence, in the first instance, of the discontinuous nature of the fundamental events of oviposition, incubation, and care of young. The phenomena of discontinuity and periodicity in reproduction have been enhanced, in evolution, by the selection of control systems that cause reproduction to occur when environmental conditions maximize the probability of survival of the young, and minimize hazards to the reproductive adults.

Although reproductive control systems, in varying degrees, contain endogenous periodic components, these must be responsive, both in period and in phase, to environmental information of predictive value to the individual. By predictive value I mean that the environmental information is of such nature that its effect on the control system will be to further the occurrence of reproduction at the appropriate time.

I use environmental information here in a very broad sense, including information derived from the behavior of other individuals of the species whenever it is a component of the battery of information used in the timing of reproductive function. Thus members of a pair can be, in an alternating series of events, mutual sources of information [see Hinde (1970) for examples].

Some comments on the general features of reproductive control systems in birds are in order.

The diagrams that follow show, in a very general way, some of the established and presumed neuroendocrine relationships among the components of the control systems. They are, of necessity, inadequate since they do not show rate relationships and lack information on changes in relationships among components through the course of the reproductive cycle (Assenmacher, 1970; Farner, 1967; Farner *et al.*, 1967).

Recognizing full well my neuroendocrinologic bias, it nevertheless seems useful to recognize, albeit simplistically, three primary categories of functional relationships within the control systems.

1. Efferent information is transmitted from the central nervous system via

    a. the nervous system per se or

    b. the neuroendocrine apparatus to effector, with hormones as the mode of information.

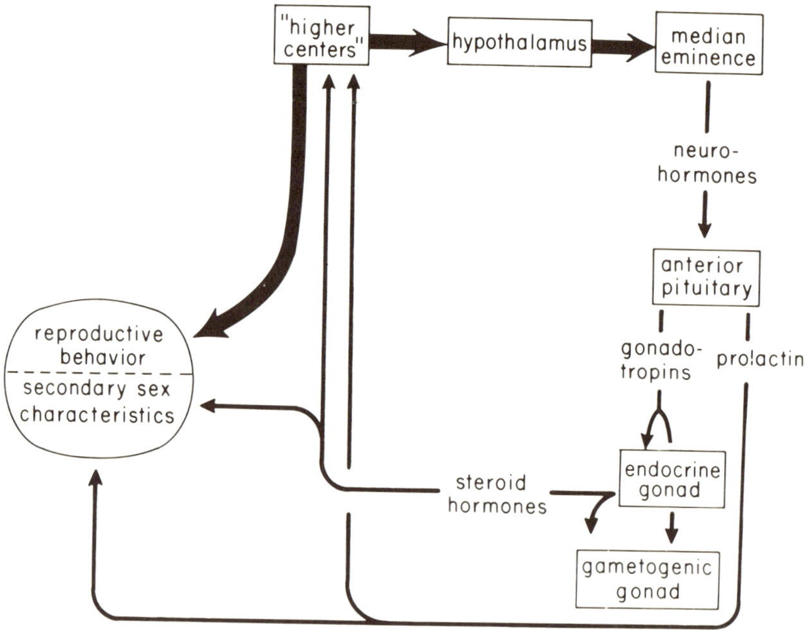

FIGURE 1 The general pathways of information in development and control of gonadal function, secondary sex characteristics, and reproductive behavior. Negative feedbacks are omitted. From Farner and Lewis (1971) courtesy of Academic Press, New York.

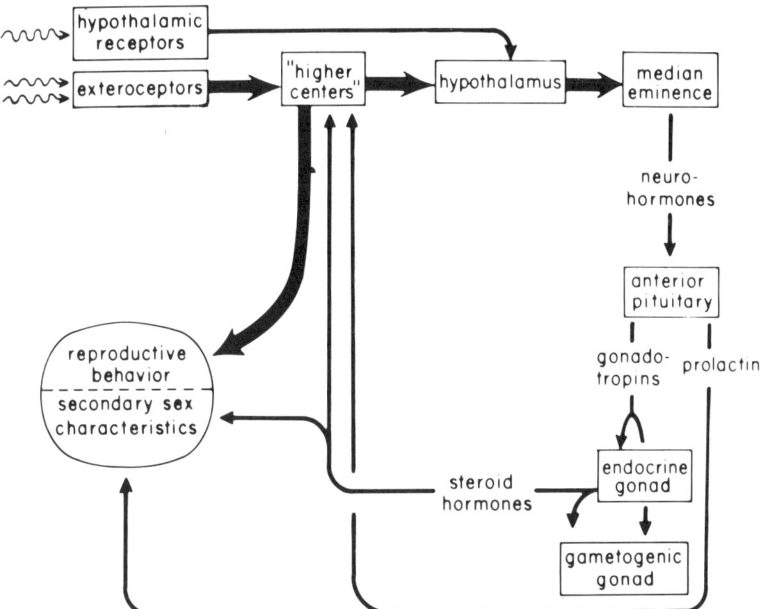

FIGURE 2 The general role of the hypothalamus in the use of external information in the development and control of gonadal function, secondary sex characteristics, and reproductive behavior. From Farner and Lewis (1971) courtesy of Academic Press, New York.

Both, of course, feed back on the central nervous system and thereby affect its subsequent output. Figure 1 shows only a part of the hormonal feedbacks and none of the neural feedbacks. The hormonal feedback here is not the typical negative feedback with its stabilizing effect but it is rather intended to show the use of sex hormones, in a general sense, as information for the development and maintenance of sexual behavior. The ethologists can rightfully condemn the hopeless simplicity of this diagram. It reflects the inadequacy of our current knowledge of the physiologic basis of behavior within a conventional neurologic framework. Indirectly, at least, it seems to recognize the oft suggested possibility that behavior cannot be rationalized entirely on classical neurophysiologic and neuro-endocrinologic bases.

2. External information to the central nervous system (Figure 2) is transduced either in conventional sense organs or isolated receptors— *or* at least in the case of photoperiodic information, by an unidentified receptor and transducer system deep in the hypothalamus—into information in the form of nerve impulses.

3. Internal information is transmitted to or is available to the central nervous system (Figure 3). Included here are

   a. Information coded in DNA (in the central nervous system) that fixes response of the system to information presented to it,

   b. Endogenous periodicities normally entrained to natural periodicities, and

   c. Neural and hormonal feedback from peripheral to central components.

Figure 3 obviously depicts a photoperiodic species (Farner and Lewis, 1971). No implication is intended with respect to the generality of photoperiodic controls. Rather, it is presented as an example of the interplay of external and internal information in the control of reproductive function.

I wish that it were possible to predict optimistically that this symposium will generate a rational synthesis of our knowledge of the physiology and ethology of reproduction in birds. This will not happen, however, since our knowledge of both areas is far too imperfect. What can be anticipated, and indeed with substantial confidence, is a

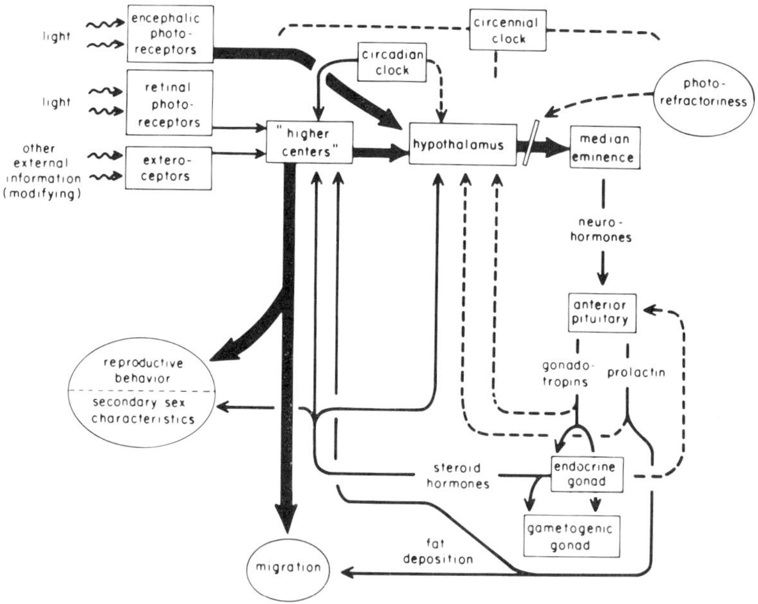

FIGURE 3 Principal pathways of information, including negative feedback, in a species in which the annual cycle is, at least in part, photoperiodically controlled. Modified from Farner and Lewis (1971) courtesy of Academic Press, New York.

useful convergence of these disciplines in a manner that involves a minimum of semantic entanglements.

I also wish that it were possible to predict optimistically that the symposium will generate ideas that can be translated immediately into effective new management practices by the wildlife biologists. This also will not occur in such a spectacular sense. Nevertheless, if we can express our discussions and ideas in a reasonably mutually comprehensive language with a minimum of semantic shackles, I am confident that the symposium will generate ideas that lead to novel experiments in management practice. But, equally important, it is hoped that the experience and observations of the wildlife biologists will lead both to evaluation of hypotheses based on experimental approaches and to challenging ideas that can be examined experimentally.

These, indeed, were the motives of the Wildlife Society, the American Society of Zoologists, and the Division of Comparative Endocrinology of the American Society of Zoologists in proposing that this symposium be organized.

REFERENCES

Assenmacher, I. 1970. Importance de la liaison hypothalamo-préhypophysaire dans le contrôle nerveux de la reproduction chez les oiseaux, p. 167–191. *In* J. Benoit and I. Assenmacher [ed.] La photorégulation de la reproduction chez les oiseaux et les mammifères. Editions du C.N.R.S., Paris.

Farner, D. S. 1967. The control of avian reproductive cycles, p. 107–133. *In* D. W. Snow [ed.] Proceedings of the XIV International Ornithological Congress. Blackwell's, Oxford.

Farner, D. S., and R. A. Lewis. 1971. Photoperiodism and reproductive cycles in birds, p. 325–370. *In* A. C. Giese [ed.] Photophysiology. Vol. 6. Academic Press, New York.

Farner, D. S., F. E. Wilson, and A. Oksche. 1967. Avian neuroendocrine mechanisms, p. 529–582. *In* L. Martini and W. F. Ganong [ed.] Neuroendocrinology. Vol. 2. Academic Press, New York.

Hinde, R. A. 1970. Animal behaviour. 2nd ed. McGraw-Hill, New York. 876 p.

Ecoethological Aspects of Reproduction

FRANK McKINNEY

The past decade saw new and exciting developments in areas of research in which evolution, ecology, and ethology overlap. Studies of avian reproduction played a key role in promoting new ways of thinking (Cullen, 1957; von Haartman, 1957; Tinbergen, 1959; Orians, 1961; Immelmann, 1961, 1962; Crook, 1964). These studies evolved through a blending of interests and ideas derived mainly from the long-term studies of Lack (1954, 1966, 1968, 1971) on avian ecology, especially population regulation, and those of Tinbergen (1953, 1954, 1959, 1963, 1967) on the evolution and adaptive significance of behavior.

Investigators using these new approaches have referred to their field as "ecoethology" or "behavioral ecology". Although there are still somewhat different emphases among investigators, depending on whether their interests are primarily ecological or behavioral, it is obvious that an important field of biology, focusing on the study of adaptations, has developed rapidly.

This new field had its roots in ornithology and already occupies a dominant place in the thinking of avian biologists. There have been parallel developments among field biologists of all types. In particular, it has been stimulated, among mammalogists, by the rapid growth of field studies on primates (e.g., Jay, 1968; Altmann, 1971; Crook, 1970). These developments have had profound influences on the orientation of research programs with ultimate objectives bearing on conservation, control, or management; this is especially apparent in Britain (Murton, 1971).

The blend of interests, stemming mainly from the research programs of Lack and Tinbergen, is illustrated by current attention to such topics as feeding ecology and behavior (e.g., Watson, 1970), spacing behavior (Watson and Moss, 1970; Brown and Orians, 1970), social systems (Crook, 1965; Orians, 1969), flocking behavior (Crook, 1961; Moynihan, 1962), habitat selection (Hildén, 1965), and communication (Crook, 1964; Marler, 1968). Many of these areas of research have been reviewed recently, primarily from an ecological aspect, by Orians (1971), and since other participants in this symposium will discuss important ecological and physiological aspects of reproduction, I will take an ethological viewpoint. Since my task is neither to present the latest products of research nor to provide a detailed review, I will concentrate on an attempt to illustrate the ecoethological approach.

THE HISTORY OF ECOETHOLOGY

Ethology had its roots in natural history and zoology, stemming from the early work of such biologists as Charles Darwin, C. O. Whitman, Oskar Heinroth, Wallace Craig, and Julian Huxley. It became a recognized branch of biology largely under the influence of Konrad Lorenz and Niko Tinbergen. Since the interests and discoveries of these two zoologists are well known through their many books and papers, I will not attempt to give a résumé of their research. It is essential to note, however, that both have applied concepts of organic evolution to the study of animal behavior. A concern with how behavior has evolved and how it is adaptive has been central to their ideas and research. To this end, they applied to behavior the methods of comparative anatomy, especially the concepts of homology, adaptive radiation, and convergent evolution. Both Lorenz and Tinbergen have consistently stressed the need for careful observation and description of natural behavior and, because of their training as systematists, they tend to think of behavior patterns as though they were "organs" possessed by a species.

This zoological approach was, and still very largely is, completely different from the way that most psychologists think about behavior. As many reviewers (e.g., Klopfer and Hailman, 1967) have pointed out, there are fundamental differences in interests. In general, for example, psychologists tend to have little interest in the evolution and adaptive significance of behavior.

Lorenz and Tinbergen laid great stress on the desirability of asking all four types of biological questions (causation, development, evolution, and function) about behavior and, for many years, attention was focused on their ideas about causation and development, the very areas where their interests overlapped with those of psychologists and physiologists. This produced endless controversy over such matters as the distinction between instinct and learning and the reality and nature of "drives." From this interaction there emerged a flourishing interdisciplinary field (sometimes called "psychobiology"), centered around such problems as motivation and imprinting, to which both ethologists and psychologists were attracted. These fields are brilliantly reviewed by Hinde (1970), whose research program has had a major catalytic effect in bridging these two disciplines.

Meanwhile, many ethologists slanted their research toward problems of evolution and adaptive significance. They followed the classic ethological orientation toward behavior as an attribute of a species, and it is understandable that they found their closest ties with evolutionists, ornithologists, and ecologists, rather than with neurophysiologists and psychologists. Splitting within the ethological ranks has been resisted vigorously, largely intuitively I think, because of the basic conviction of zoologists that all aspects of biology are interrelated. But specialization has been inevitable, as methods and techniques have become more sophisticated and as the volume of research has increased steeply. The basic ecoethological methods—field observation, description and analysis of natural behavior, field and laboratory experimentation (e.g., with models and playback of tape recordings), and comparisons between species—require prolonged, specialized training and distinctive attitudes. Most researchers in this field cannot handle with equal competence the laboratory techniques used by neurophysiologists, endocrinologists, and learning psychologists. Fortunately, interdisciplinary contact between these fields is possible through symposia such as this, and there are increasing signs of fruitful team approaches being instituted.

In the broadest sense, everyone uses comparative methods, but there have been important differences in the way that these methods have been applied by psychologists and ethologists. Comparative psychologists and physiologists frequently compare representatives from different animal groups, as they do, for example, in studying learning abilities in fish, frogs, chickens, cats, and monkeys. Such studies usually focus on a fundamental process, rather than on the particular species selected as a class representative. Ethologists, in contrast, have come to rely heavily on two specialized ways of comparing different species: They study *adaptive radiation* in a group of closely related species, or they search for *convergence* among distantly related forms living under similar environmental conditions. Both of these methods are basic to all evolutionary studies and are certainly of fundamental importance in ecoethology.

Classic examples of the approach via adaptive radiation are those of Lorenz (1941) on the comparison of duck displays and Tinbergen (1959) on gulls. The use of the convergence approach is well illustrated in von Haartman's (1957) study of hole-nesting adaptations in different groups of birds.

The investigations of Lorenz on ducks showed that courtship displays can be homologized from species to species and that many are useful taxonomic characters. This discovery caused considerable excitement among systematists, and there has been widespread application of the method in other groups of animals. Many of these studies were carried out on captive animals and interesting information was gathered on the distribution of homologous displays, often providing insights as to how these signal movements have evolved from such everyday actions as feeding and preening movements. The view of Lorenz, however, that the signal repertoires are "mere conventions," their characteristics being largely accidental in any one species, was generally accepted. This mode of thinking was reinforced by strong emphasis by evolutionary biologists on the role played by courtship displays in species isolation.

This point of view was changed radically by Cullen (1957) in a study of the Black-legged Kittiwake (*Rissa tridactyla*). Her ideas on the influence of environmental factors, in this case cliff-nesting, on all aspects of social behavior developed for two reasons: (a) the behavior of many gull species had been described and the descriptions were available to provide a contrast with the Kittiwake, and (b) these comparative studies of gulls were carried out in the field, where ideas on the adaptive significance of behavior developed naturally as selective forces were observed in action.

The ideas of Cullen were quickly confirmed and extended by studies on other bird groups, especially by Crook (1964) in extensive comparative investigations on weaver birds (Ploceinae). This has led to broader statements (e.g., Crook, 1965; Lack, 1968; Orians, 1971) of relationships between seasonal and spatial distribution of food, feeding methods, characteristics of the breeding habitat and the nest-site, vulnerability to predators, mating systems, characteristics of the pair-bond, and communication methods. It is now recognized that the behavior of each species has evolved as a compromise between many competing selection pressures and that morphological, physiological, ecological, and behavioral adaptations must be thought of as co-adapted systems.

One is tempted to use the term "breakthrough" in considering these advances in thinking about adaptive aspects of avian reproduction; the effects have indeed been far-reaching. The fundamental importance of observation and description of the behavior of each species (the building of "ethograms") is now easy to justify. The need to study birds in the field in their natural habitats can be argued logically. The comparative method need no longer be used tentatively and intuitively, as it was for many years, but can now be applied deliberately with specific objectives in mind. In short, the study of species-specific behavior, as pioneered by species-oriented ornithologists and ethologists, seems to have finally come of age and has taken a key place in adaptation research.

A COMPARISON OF STELLER'S EIDER AND COMMON EIDER

Studies of the behavior of the Common Eider (*Somateria mollissima*) in England and Alaska (McKinney, 1961) and of Steller's Eider (*Polysticta stelleri*) in Alaska (McKinney, 1965b) disclosed many differences (Table 1). On the Alaska Peninsula in spring, the small Steller's Eider feeds mostly in shallow water, apparently on small molluscs and crustacea. Some feeding is done along shorelines but mostly the birds dive to take food off the bottom. It is an extremely sociable species at this time of year, living in very large, densely packed flocks. When diving for food, there is a marked tendency for birds to dive synchronously, and at times a whole flock, including several thousand birds, will submerge almost simultaneously. The birds are very easily flushed by the appearance of such birds of prey as eagles and gyrfalcons, flocks taking to the air when the predator is still a speck on the horizon. In summary, the Steller's Eider shows strong tendencies to be sociable and to take wing readily, both characteristics probably being primarily antipredator adaptations.

In contrast, the large Common Eider feeds mainly by diving in

TABLE 1   Major Behavioral Differences between Steller's Eider and Common Eider

Characteristic	Steller's Eider	Common Eider
Body size	Small	Large
Feeding habitat	Shallow water	Deep water
Food	Small invertebrates	Large invertebrates
Diving flocks	Large, densely packed; synchronize dives	Small not densely packed; dives not synchronized
Response to birds of prey	Flocks fly up readily	Fly less readily; may dive to escape
Precopulatory behavior	Brief; single shake; final display invariable	Up to several minutes; conspicuous wing-flap; many shakes; some calls by male; final display variable
Postcopulatory behavior	Pair flies back to flock; male displays silent	Male calls
Courtship behavior	Preflight movements; short flights; aerial pursuits; displays rapid; males silent	Underwater pursuits; displays slower; some compound displays; males give cooing calls

deep water, often bringing large food items (e.g., crabs, razor-shells) to the surface to swallow. Feeding flocks are usually quite small, less densely-packed, and there is much less pronounced synchronous diving than in Steller's Eider. In general, they seem less likely to fly in response to birds of prey and there are reports that they will dive to escape. Thus, these large birds are less markedly sociable, less ready to fly, and, in general, their preference for deep-water feeding probably makes them less vulnerable to surprise attacks from predators.

These differences seem to have had repercussions on the characteristics of their social behavior and signaling methods, notably the behavior associated with pair formation and copulation (Figure 1). In courting groups, Steller's Eider males frequently adopt erect preflight postures, they perform ritualized "short flights," and aerial pursuit flights are common. Furthermore, the courtship displays are rapid and males give no calls. In Common Eider, preflight and aerial components are absent, and courting parties indulge in underwater pursuits instead. The male displays are slower and more elaborate and many are accompanied by cooing calls.

To copulate, a pair of Steller's Eider flies out several hundred meters from the flock, quickly performs precopulatory displays, mounting occurs, and almost immediately they fly back to the flock.

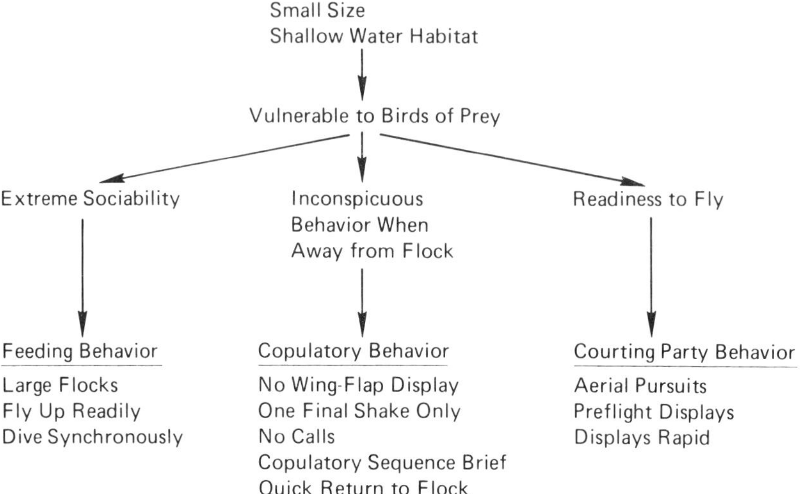

FIGURE 1   Suggested relationships between body size, habitat characteristics, vulnerability to predators, and various characteristics of social behavior in Steller's Eider.

Presumably, while away from the flock, the pair is much more vulnerable to attack by birds of prey, and their uneasiness is often apparent. Selection has apparently favored the use of inconspicuous displays in this situation, and the evolutionary changes that have occurred in the precopulatory displays are nicely shown by comparing them to the displays of the Common Eider. In the latter species, Shakes are frequent, and Bathing movements are usually followed by a conspicuous ritualized Wing-flap. In Steller's Eider, only a single Shake occurs, immediately before mounting, and while Bathe is very frequent, a ritualized Wing-flap is absent.

Thus there is good evidence to suggest that the small size and shallow-water feeding habits of Steller's Eider make this species more vulnerable to predators. This has favored extreme sociability and a general readiness to take wing. These tendencies are reflected in the kinds of display used and selection has favored the use of inconspicuous displays when birds are away from the safety of the flock.

BREEDING BEHAVIOR OF THE SHOVELER AND PINTAIL

Territorial behavior and aerial pursuit activity in dabbling ducks on the breeding grounds have been topics of controversy for many years.

There have been strong disagreements on whether ducks do defend territories, whether chasing activity disperses breeding pairs, and whether the motivation of chasing males is primarily aggressive or sexual. Strong observational evidence in favor of territorial behavior was presented by Hochbaum (1944), Sowls (1955), Dzubin (1955), and others, but it is now clear that generalizations about such behavior in ducks, or even in dabblers and divers, are very dangerous. There are important specific differences and, apparently, each species has evolved somewhat different breeding strategies in response to selection pressures of several kinds (McKinney, 1965a).

It has been recognized for some time, for example, that the average size of the "home range" of a breeding pair varies from species to species (Dzubin, 1955), but progress has been very slow in discovering *why* each species behaves differently. It now seems likely that specific differences in such factors as food requirements and feeding methods, seasonal and spatial distribution of food, nesting cover and brood rearing habitat, physical characteristics of the habitat, and vulnerability of nests to predation have placed different demands on each species, resulting in a variety of breeding strategies. The complexity and subtlety of the relationships between such factors can be illustrated by comparing two extreme types in the dabbling duck group, the Shoveler (*Anas clypeata*) and the Pintail (*A. acuta*) (Table 2).

The Shoveler, with its large spatulate bill, specializes in feeding by straining plankton at the surface. This appears to be a time-consuming feeding method and breeding pairs prefer prairie habitats, where there is a rich supply of suitable food available within a restricted area throughout the breeding season. Also it must be especially important for females during the prelaying, laying, and incubation periods to have assured feeding areas, close to the nest site, where they are free from harassment by conspecific individuals. Observations on *A. clypeata*, both in the wild (McKinney, 1965a; Poston, 1969) and in captivity (McKinney, 1967), and on the closely related Cape Shoveler (*A. smithi*) (McKinney, 1970) have shown that these requirements are met by a classical territorial arrangement. Breeding-pair ranges are small, well defended by the male, and there is usually little overlap between neighboring territories. Associated with this system, Shovelers have long-lasting pair-bonds, the male often behaving aggressively on his territory until the time of hatching.

These pair-bond and territorial characteristics have widespread repercussions on all aspects of breeding behavior in Shovelers

TABLE 2 Major Differences in Ecology and Breeding Behavior between the Shoveler and the Pintail

Behavior	Shoveler	Pintail
Habitat	Permanent water areas rich in plankton	Temporary pools and widely dispersed permanent water areas
Nest site	Concealed in dense cover near water	In open or sparse cover often far from water
Feeding methods	Mainly strains plankton	Up-ends frequently
Breeding home range	Small discrete territories with little overlap	Large with much overlap
Brood movements	Short distances	Can move long distances
Distant threat signals	Head-pumping and calls; wing noise	—
Hostility between paired males	Highly developed ritualized circular fighting	Rare
Strength of pair-bond	Strong	Weak
Duration of pair-bond	Throughout incubation	Breaks early in incubation
Aerial chases	Brief and short-range; male hostility common	Prolonged and wide-ranging
Rape attempts	Rare	Common
Copulation frequency in pairs	Daily before and during laying	Rarely observed
Courtship displays	Ritualized feeding movements; Jump Flight	Grunt-whistle; Head-up–tail-up

(McKinney, 1970) (Figure 2). Pair-bond copulations occur daily, promiscuous tendencies of males are weak, and rape of strange females is rare. Aerial chases involving a territorial male pursuing an intruding pair (three-bird flights) are frequent but brief and of short range; the pursuer quickly returns to his territory and mate. Males threaten intruders at a distance with conspicuous Hostile Pumping movements accompanied by calls. Their wings make a characteristic, loud buzzing sound, especially at takeoff; this also appears to have a threat function, discouraging intrusions. At territorial boundaries, disputes between males are resolved through highly ritualized circular fighting.

The evolution of Hostile Pumping, wing noise, and the time-consuming feeding method appear to have influenced the courtship displays. Thus the form of the female's Inciting movement is quite different from that of other *Anas* species, including a mixture of head-pumping and head-low pointing movements, clearly influenced by Hostile Pumping. Males have conspicuous Wing-flap and Jump-flight displays, reinforced by wing noise. The Shoveler, and its relatives in the blue-winged duck group, have used three feeding movements as the basis for the courtship displays: Lateral Dabbling, Head-dip, and Up-end. Presumably, the evolution of these signals was facilitated by the possibility they offered of continuing to feed while signaling, and this may still be the case in Lateral Dabbling.

The Pintail strategy is completely different, though as yet it is not so well understood. The following account is based on Smith's (1968) important study, although Figure 3 has been worked out in collaboration with Mr. Scott Derrickson who is currently studying this species.

Pintails seem best adapted to open prairie or tundra habitats. Although they are early spring migrants and make use of temporary pools, they are able to breed in areas with sparse, widely scattered permanent water areas. Mobility is their specialty and pairs range widely, using locally abundant food sources as they become available throughout the breeding season. The broods are mobile and the cryptically colored ducklings often make long overland journeys from nest to water. The long neck is probably an adaptation for feeding by up-ending, which they do a great deal, and it may also be an important aid to vision in grassland habitats. The diet is probably more vegetarian than that of the Shoveler.

In contrast to the Shoveler, discrete defended territories are impossible and the home ranges of many pairs overlap. Males are strik-

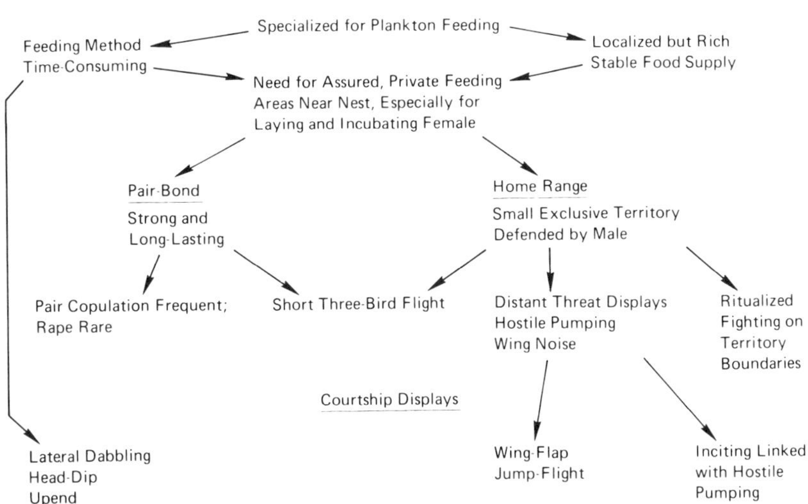

FIGURE 2 Suggested relationships between feeding strategy and social behavior in the Shoveler.

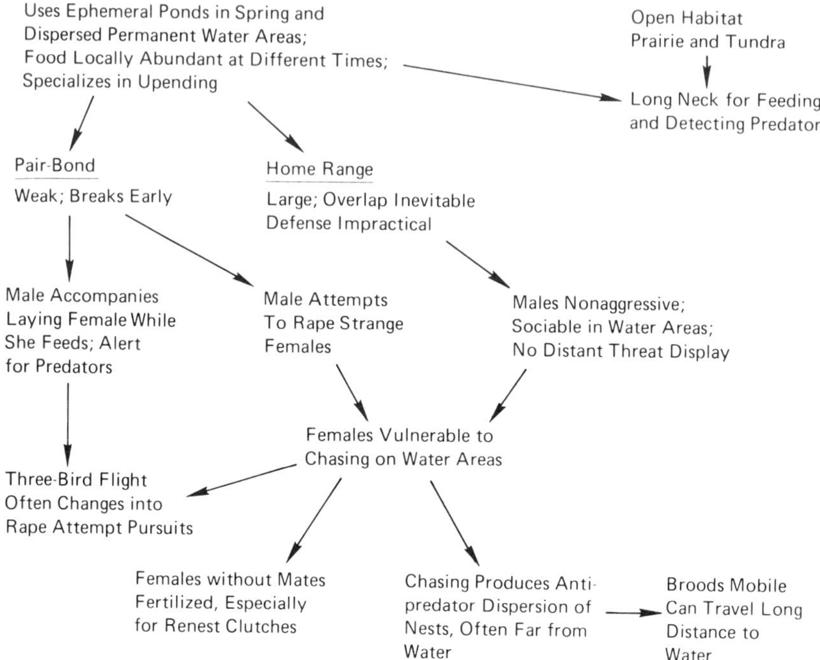

FIGURE 3   Suggested relationships between species ecology and breeding strategy in the Pintail.

ingly nonaggressive and are sociable at the water areas. There are no distant threat displays, and fighting between males is rare.

The pair-bond is much "weaker" than in the Shoveler, and it breaks early in incubation. Males do accompany their mates much of the time during the prelaying and laying periods, when they presumably afford protection both from harassing males and from predators. But paired males also spend much time pursuing the females of other pairs. These aerial pursuits may begin as three-bird flights, but they are so prolonged that they often attract other males and change into "rape-attempt flights" that may end, after great distances are covered, in rape of the female by several males when the party lands.

Because of their localized nature, the same water areas are used by many males and females, and the latter are very vulnerable to being chased when they come to feed. Because the vigorous, persistent chasing seems to result in many females moving far from water to lay, nests are dispersed. This must be especially important for the Pintail since the site is often very open (e.g., they will use burns and ploughed fields).

Raping is likely important in fertilizing eggs, especially in the case of females without a mate who are preparing to lay renest clutches.

RESEARCH STRATEGIES

There is evidence from these comparative studies of eiders and dabbling ducks to suggest that differences in feeding ecology, habitat characteristics, and vulnerability to predators have had wide-ranging repercussions on territorial behavior, characteristics of the pair-bond, and other aspects of social behavior. The picture is still only dimly outlined for any species of waterfowl, but the possibility of really "understanding how everything fits" is tantalizing. Progress so far has come mainly from the strategy that I have outlined and has stemmed from an interest in the evolution of displays. Now, however, the need is apparent for much more detailed information on waterfowl foods and feeding methods, on predation, on time and energy budgets, on home ranges and their features, on the physical characteristics of different habitats for communication by various methods, on the process of pair formation and the duration of pair-bonds, on the functions of individual displays, and so on. Any one of these topics, however, could occupy a team of investigators for years, even if they devoted all of their energy to a single species.

The worth of intensive research on single species over many years has been well illustrated by the studies of Watson and Moss (1970), Murton (1965) and Tinbergen (1967) and the need for experimental approaches that involve models, playback of tape-recorded vocalizations, and habitat manipulations is obvious. But insights on adaptation seem to come mainly from comparative study, and there are strong tendencies these days to neglect this approach in favor of detailed analytical and experimental work on single species. Individual choice will no doubt lead different workers in different directions, and some ecoethologists will certainly continue to search for broad correlations through comparative methods. As Scott (1968) has expressed it, "the game of assigning adaptive significance to either structure or behavior is one of the rare areas still open to the free imaginations of scientists" and it is likely that many will continue to be attracted to this pursuit.

There are practical difficulties in deciding which groups should be studied, how much detail is needed in behavioral descriptions for each species, how many species should be examined, and in what sequence. Should ecological information be collected simultaneously and, if so, what parameters must be measured, and how precisely? After preliminary comparative studies of behavior and ecology have been made, should attention then be given to (a) analytical, (b) experimental, or (c) further comparative study? How far do we need to go, for example, with preliminary motivational analyses on the social behavior of each species?

Not the least obligation is that of keeping abreast of the literature. On birds alone, the literature relating to adaptations is widely scattered in a multitude of journals. Comparative, descriptive papers tend to be long and filled with details that are of interest mainly to people working on the same or closely related species.

The answer is obvious, I think, i.e., team approach will be essential. All techniques must be applied in such a way that they are complementary; we can afford to neglect none. Studies of captive birds have an important role in many groups, but field studies of each species must also be made. Long-term intensive investigations, along many analytical and experimental lines, are needed on single species, but comparative studies are essential to provide the ideas. Morphological and physiological studies on each species are necessary, as well as ecological and ethological investigations. None of these studies will ever be truly "complete." The price we pay for being interested in evolutionary problems is that every study we make will be "preliminary." The best we can hope for is to set the stage for deeper understanding.

IMPLICATIONS FOR WILDLIFE MANAGEMENT

For many years, the comparative study of displays seemed far removed from the practical tasks of monitoring population levels; predicting production; contolling depredations; preserving, manipulating, and creating habitat; regulating harvests; and saving species from extinction. This held true as long as ethologists were primarily interested in taxonomic questions. With broadening of interests to include the relationships between signaling methods and adaptive strategies of each species, however, this is no longer the case. Ethologists have had to blend their interests with those of ecologists and, conversely, ecologists have found that many problems lead them to study behavior. It is obvious now that we cannot hope to understand what makes a species "tick" without knowing how its ecology, behavior, morphology, and physiology are interrelated. This is surely the only firm basis for sound conservation, control, and management of its populations.

For example, the problems involved in censusing breeding pairs, interpreting the significance of male aggregations, pursuit flights and raping activity, and home range characteristics must be quite different for the Shoveler and the Pintail. In spite of much research, we do not yet seem to have a clear understanding of the relationships between breeding behavior and the ecological requirements of such important game species as the Mallard (*Anas platyrhynchos*), Baldpate (*A. americana*), Gadwall (*A. strepera*), Green-winged Teal (*A. crecca*), or Wood Duck (*Aix sponsa*). Perhaps the clues to understanding the breeding strategies of these species will come from comparing and contrasting their behavior with that of the Shoveler and Pintail.

CONCLUSION

By combining intensive studies of ecology and behavior of single species with comparative studies of closely related species, new ideas can emerge on the adaptive radiation of ecoethological strategies in the study group. Long-term field studies on each species are essential, and they should be deliberately directed toward questions of adaptive significance. Beyond the early exploratory stages, the breadth of this approach will usually require team effort, combining the talents of specialists in such fields as feeding ecology, nutrition, and reproductive physiology. Comparative ethologists can play a key role in such teams by revealing specific differences that require explanation. The adaptive characteristics of each species are interrelated in complex and often unexpected ways, and overemphasis on one source of selection, or on one research strategy, is likely to lead to an unbalanced picture. Ecoethological research is time-consuming, and long-term programs are essential; but there seems to be no alternative if we wish to understand why each species behaves as it does.

ACKNOWLEDGMENT

The research reported here was in part supported by the Atomic Energy Commission through Grant COO-1332-77.

REFERENCES

Altmann, S. A. 1971. Baboon ecology. University of Chicago Press, Chicago.

Brown, J. L., and G. H. Orians. 1970. Spacing patterns in mobile animals. Ann. Rev. Ecol. Syst. 1:239–262.

Crook, J. H. 1961. The basis of flock organization in birds, p. 125–149. *In* W. H. Thorpe and O. L. Zangwill [ed.] Current problems in animal behaviour. Cambridge University Press, London.

Crook, J. H. 1964. The evolution of social organisation and visual communication in the weaver birds (Ploceinae). Behaviour Suppl. 10.

Crook, J. H. 1965. The adaptive significance of avian social organizations. Symp. Zool. Soc. London 14:181–218.

Crook, J. H. 1970. The socio-ecology of primates, p. 103–166. *In* J. H. Crook [ed.] Social behaviour in birds and mammals. Academic Press, New York.

Cullen, E. 1957. Adaptations in the Kittiwake to cliff-nesting. Ibis 99:275–302.

Dzubin, A. 1955. Some evidences of home range in waterfowl, p. 278–298. *In* Transactions of the 20th North American Wildlife Conference. Wildlife Management Institute, Washington, D.C.

Hildén, O. 1965. Habitat selection in birds. Ann. Zool. Fenn. 2:53–75.

Hinde, R. A. 1970. Animal behaviour: A synthesis of ethology and comparative psychology. McGraw-Hill, New York. 876 p.

Hochbaum, H. A. 1944. The Canvasback on a prairie marsh. American Wildlife Institute, Washington, D.C. 187 p.

Immelmann, K. 1961. Beiträge zur Biologie und Ethologie australischer Honigfresser (Meliphagidae). J. Ornithol. 102:164–207.

Immelmann, K. 1962. Beiträge zu einer vergleichenden Biologie australischer Prachtfinken (Spermestidae). Zool. Jahr. Abt. Syst. Oekol. Geogr. Tiere 90:1–96.

Jay, P. [ed.]. 1968. Primates: Studies in adaptation and variability. Holt, Rinehart and Winston, New York. 529 p.

Klopfer, P., and J. P. Hailman. 1965. Habitat selection in birds. Adv. Study Behav. 1:279–303.

Klopfer, P., and J. P. Hailman. 1967. Ethology's first century. Prentice-Hall, Englewood Cliffs, New Jersey. 297 p.

Lack, D. 1954. The natural regulation of animal numbers. Oxford University Press, London. 343 p.

Lack, D. 1966. Population studies in birds. Oxford University Press, London. 341 p.

Lack, D. 1968. Ecological adaptations for breeding in birds. Methuen, London. 409 p.

Lack, D. 1971. Ecological isolation in birds. Blackwell's, Oxford. 404 p.

Lorenz, K. 1941. Vergleichende Bewegungsstudien an Anatinen. J. Ornithol. Suppl. 89:194–294.

Marler, P. 1968. Visual systems, p. 103–126. *In* T. A. Sebeok [ed.] Animal communication. University of Indiana Press, Bloomington.

McKinney, F. 1961. An analysis of the displays of the European Eider *Somateria mollissima mollissima* (Linnaeus) and the Pacific Eider *Somateria mollissima v. nigra* Bonaparte. Behaviour Suppl. 7:1–24.

McKinney, F. 1965a. Spacing and chasing in breeding ducks, p. 92–106. *In* Wildfowl Trust 16th Annual Report. Country Life Ltd, London.

McKinney, F. 1965b. The spring behavior of wild Steller's Eiders. Condor 67:273–290.

McKinney, F. 1967. Breeding behaviour of captive Shovelers, p. 108–121. *In* Wildfowl Trust 18th Annual Report. The Wildfowl Trust, Slimbridge, England.

McKinney, F. 1970. Displays of four species of blue-winged ducks. Living Bird 9:29–64.

Moynihan, M. 1962. The organization and probable evolution of some mixed species flocks of neotropical birds. Smithsonian Misc. Collect. 143(7):1–140.

Murton, R. K. 1965. The Wood Pigeon. New Naturalist Monograph. Collins, London. 256 p.

Murton, R. K. 1971. Man and birds. Collins, London. 364 p.

Orians, G. H. 1961. The ecology of blackbird (*Agelaius*) social systems. Ecol. Monogr. 31:285–312.

Orians, G. H. 1969. On the evolution of mating systems in birds and mammals. Am. Nat. 103:589–603.

Orians, G. H. 1971. Ecological aspects of behavior, p. 513–546. *In* D. S. Farner and J. R. King [ed.] Avian biology. Vol. 1. Academic Press, New York.

Poston, H. J. 1969. Relationships between the Shoveler and its breeding habitat at Strathmore, Alberta, p. 132–137. *In* Saskatoon wetlands seminar. Rep. Ser. No. 6. Canadian Wildlife Service, Ottawa, Canada.

Scott, J. P. 1968. Observation, p. 17–30. *In* T. A. Sebeok [ed.] Animal communication. University of Indiana Press, Bloomington.

Smith, R. I. 1968. The social aspects of reproductive behavior in the Pintail. Auk 85:381–396.

Sowls, L. K. 1955. Prairie ducks. Stackpole, Harrisburg, Pennsylvania. 193 p.

Tinbergen, N. 1953. The Herring Gull's world. Collins, London. 255 p.

Tinbergen, N. 1954. The origin and evolution of courtship and threat display, p. 233–250. *In* J. S. Huxley *et al.* [ed.] Evolution as a process. Allen & Unwin, London.

Tinbergen, N. 1959. Comparative studies of the behaviour of gulls (Laridae): A progress report. Behaviour 15:1–70.

Tinbergen, N. 1963. On adaptive radiation in gulls (Tribe Larini). Zool. Mededelingen 39:209–223.

Tinbergen, N. 1967. Adaptive features of the Black-headed Gull *Larus ridibundus* L., p. 43–59. *In* Proceedings of the XIV International Ornithological Congress. Almquist & Wiksells, Uppsala.

von Haartman, L. 1957. Adaptation in hole-nesting birds. Evolution 11:339–347.

Watson, A. [ed.]. 1970. Animal populations in relation to their food resources. Blackwell's, Oxford. 477 p.

Watson, A., and R. Moss. 1970. Dominance, spacing behaviour and aggression in relation to population limitation in vertebrates, p. 167–220. *In* A. Watson [ed.] Animal populations in relation to their food resources. Blackwell's, Oxford.

DISCUSSION

Jack P. Hailman

I wish to focus on a few additional aspects of the ecoethology of avian reproduction that I believe are worthy of further attention.

DEVELOPMENT OF YOUNG AND PARENTAL CARE

I would like to emphasize the need to include the development of the young and parental care as an important topic of avian reproduction. Parental care is, from a behavioral viewpoint at any rate, perhaps more important in the life of birds than it is in any other animal, except for some mammals. Ornithology has for too long been wanting a new synthetic viewpoint and classification of the diversity of parental strategies utilized in avian reproduction.

Most contemporary students of avian reproduction use only the dichotomy "precocial–altricial" to characterize young birds—and generalize about parental care accordingly—despite the fact that there are available two other sets of useful terms utilized thus far by only a few avian biologists. Venn diagrams can be used to show the relations concisely.*

Figure 1a shows the three characteristics considered of prime importance: feathering, motility, and feeding of the young. Since all birds that leave the nest are downy, and all that feed themselves soon after hatching leave the nest, four of the logic cells are empty in Figure 1b, 1c, and 1d. The diagram shows that all precocial birds are nidifugous (and ptilopaedic) and that all nidifugous birds are ptilopaedic.

It is an almost unbelievable fact that the three sets of characteristics in Figure 1a specify $\sum_{r=1}^{8} \binom{8}{r} = 256$ possible ornithological "categories" (single cells and combinations of cells); in other words, there are 256 distinct ways to shade the diagram, of which Figures 1b through 1d show six. It is a tribute to the intuitive logic of ornitholo-

*Thanks to J. E. Donmoyer, C. S. Furchner, T. C. Grubb, Jr., M. P. Heironimus, M. A. Novak, D. L. Schramm, R. B. Waide, and G. E. Wright for helping to clarify the material that follows.

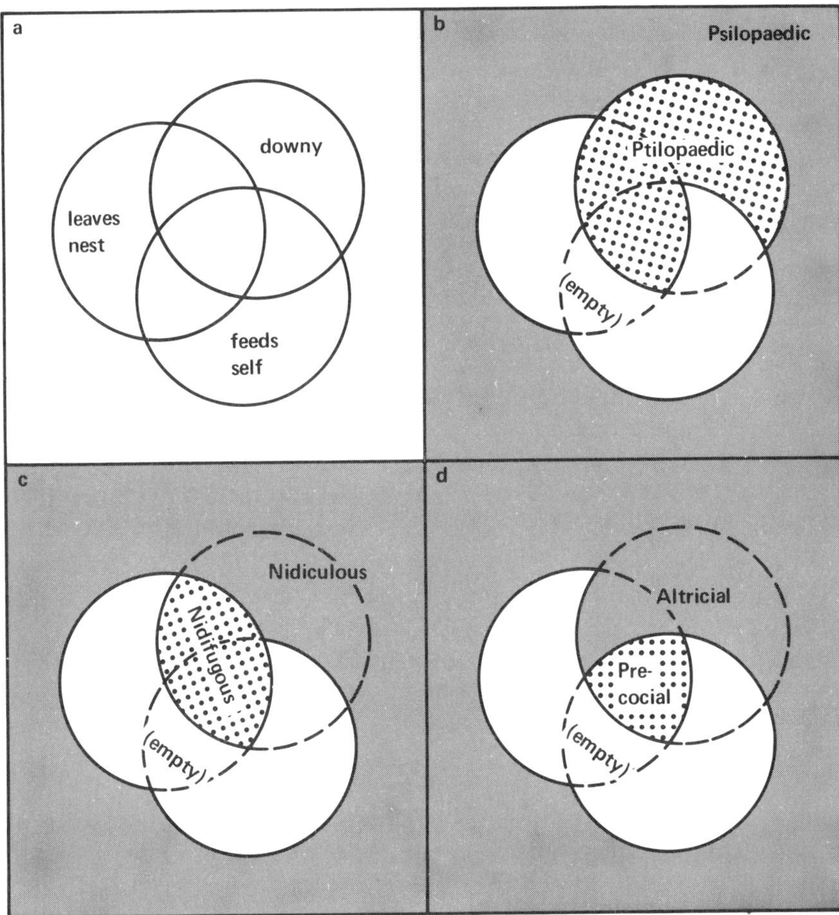

FIGURE 1   (a) Each circle represents the set of young birds having the labeled characteristic; birds outside each circle lack the characteristic of that circle. (b) The ornithological
dichotomy ptilopaedic–psilopaedic is represented by two kinds of shading, while the un-
shaded portion is empty (no bird species have young with such combinations of characters).
(c) The dichotomy nidifugous–nidiculous is similarly shown by shading. (d) The dichotomy
precocial–altricial is shown by shading, although some authors use altricial to mean "having
none of the attributes of precocial" (i.e., psilopaedic) rather than the complement of precocial. One or two terms from the three dichotomies are sufficient to specify any one of the
four nonempty cells or 11 combinations of nonempty cells; this efficiency of terminology is
a tribute to ornithological intuition.

gists that this number has been collapsed through clever terminology, such that the 16 remaining relevant (nonempty) possible categories can be described with only one or two terms, each taken from the three ornithological dichotomies. For instance, ducklings (*Anas*) are precocial, whereas Oystercatcher (*Haematopus*) young are nidifugous and altricial.*

Judicious use of the three ornithological dichotomies can eliminate such lengthy descriptions as "semiprecocial chicks that stay at or near the nest, though able to walk" (Nice, 1962), where, presumably, "semiprecocial" refers, at least in part, to the feathering and parental feeding in the gull (*Larus*) chicks described. How much simpler to say merely ptilopaedic and altricial! The scheme of Figure 1 allows not only specifications of particular cells for particular species, but also allows useful generalizations about such categories as "all ptilopaedic birds."

The point of this modest example is that even more complex schemes of classifying young and parental care are capable of succinct expression, through the elimination of irrelevant combinations of factors and through clever linguistic coding. The ultimate scheme will probably have at least two characteristics—(a) the transformation of variables from nominal dichotomies to spectra of intermediate types (e.g., from "leaves nest" or "stays" to a scale based on when the nest is deserted) and (b) an increase in the number of categories of spectra used (e.g., whether one or both parents attend the young, how many siblings there are in the nest, and synchrony of hatching). The point to be made here—aside from pointing to the underused system already available and the desirability for an even more complete system— is that we should not be dissuaded from complex schemes, since there may be logical methods for reducing the needed terminology to a surprisingly simple language for expressing precisely a very complex range of possibilities.

BEHAVIOR AND RELATIONSHIPS OF INDIVIDUAL BIRDS

The study of individual animals is going to have to play an increasingly important role in solving the problems of avian reproduction. The highly developed social interactions among birds implies a high degree

---

*Some authors use altricial to mean not having *any* of the defining characteristics of precocial, instead of "not precocial," but such usage destroys the logical usefulness of the terminology and defeats its purpose. If altricial is thus used as a

of individual recognition, yet there are precious few studies of this phenomenon that are anything more than superficial. Colin Beer's (1970) fine study, showing that young gulls learn to recognize the individual voices of their own parents, is only a beginning. Niko Tinbergen (1953) found that parent gulls learn to recognize their own chicks within a few days after hatching, but Mr. Edward Burtt in my laboratory has found no evidence, in careful experimentation, that swallow parents learn to recognize their own altricial young prior to flight from the nest. More studies of this sort are clearly needed.

Primate behaviorists have recognized that variations among individuals are not merely random perturbations of species-typical behavior, but are the result of predictable factors of social interactions that can be discovered only by studying known individuals and genealogies. Thus, the study of individual animals can not only make sense of what at first glance appears to be unadaptive variation in reproductive behavior, but can moreover become the critical foundation for the study of traditional behavior and its transmission among individuals. Faithfulness of mated Canada Geese (*Branta canadensis*) and traditional migratory patterns in waterfowl (Hochbaum, 1955) are but two examples of relevant topics requiring studies of individual birds and their relationships.

COORDINATION AND TIMING IN REPRODUCTION

Something needs to be said of the role played by behavior in coordinating the reproductive effort with the environment and between the mates. Following the earlier studies of A. J. Marshall (1954), Lehrman (1959) demonstrated the interplay between reproductive displays and endocrinological events, and in his subsequent studies on reproductive behavior in Ring Doves (*Streptopelia risoria*) Lehrman (1964) provides us with a clear example of how displays induce endocrinological events in the mate, whose changed hormonal state, in turn, produces new behavior affecting the first bird. This interactive play keeps the mates in step through the reproductive cycle, while other, non-display stimuli from the environment keep them both in step with environmental variables crucial to successful reproduction. The full understanding of these behavior–physiology interactions must be set

---

synonym for psilopaedic, one can still utilize the complement of precocial, by describing the Oystercatcher, for instance, as nidifugous and "not precocial."

in the context of the species' ecology if the timing mechanisms are to
be properly understood. Laboratory studies are not going to answer
all the questions, as indicated by a recent study of reproductive timing in gulls (Emlen and Miller, 1969).

COMMUNICATION COMPLEXITIES IN REPRODUCTION

The fourth and last point I wish to raise is exemplified by the incredible signal diversity involved in avian reproduction. The functional
significance of certain classes of signals is known in many birds. For
instance, it is clear that there are signals that lead to and synchronize
copulation in the pair, that switch the pair from courtship to nest-
building, that bring about incubation behavior; and still further signals that facilitate the transition to parental behavior. These, however,
are largely categories, not specific signals. For instance, no one has
shown to my satisfaction why the Mallard (*Anas platyrhynchos*) and
other ducks should possess such an incredible repertoire of premating
displays, such as the Grunt-whistle, Head-up-Tail-up, Down-up, Head-
shake, Tail-wag, and others so carefully charted by Konrad Lorenz
(1941), Uli Weidmann (1956), Paul Johnsgard (1965), and others.
McKinney (1965) has shown that to this list should probably be
added a huge number of comfort movements or maintenance activities that—in their basic form or slightly altered forms—take on signal
value as reproductive displays.

The very existence of such a variety of signals suggests that for us
to lump them under the rubric of "courtship displays" is no more
than the crudest analysis of the functional significance of the detailed events apparently necessary for successful reproduction. While
the effects of pesticides on reproduction of birds have been traced
with some accuracy to the induction of hepatic enzymes that, in
turn, disrupt calcium deposition in eggshells, there are suggestions
that the courtship behavior of some species is also being affected
(Risebrough, 1968). Yet, we cannot possibly pinpoint such effects
unless we understand in detail the delicate interactions of male and
female in reproductive activities—and we seem to be a long way from
that goal.

Whereas Dr. McKinney spoke positively in delivering a tight summary of achievements, I have purposely raised the importance of
some remaining issues. These four examples—development of the
young and parental care, individual differences, timing coordinating

roles of behavior, and signal diversity—open Pandora's box. If we believe there is merit in bringing together experts in a field of study, as witnessed by our participation in this symposium, it is as much to decide what we do not know as it is to summarize what we do.

REFERENCES

Beer, C. G. 1970. On the responses of Laughing Gull chicks (*Larus atricilla*) to calls of adults. Anim. Behav. 18:652–677.

Emlen, J. T., and D. E. Miller. 1969. Pace-setting mechanisms of the nesting cycle in the Ring-billed Gull. Behaviour 33:237–261.

Hochbaum, H. A. 1955. Travels and traditions of waterfowl. University of Minnesota Press, Minneapolis, 301 p.

Johnsgard, P. A. 1965. Handbook of waterfowl behavior. Cornell University Press (Comstock), Ithaca, New York. 378 p.

Lehrman, D. S. 1959. Hormonal responses to external stimuli in birds. Ibis 101:478–496.

Lehrman, D. S. 1964. The reproductive behavior of Ring Doves. Sci. Am. 211(5): 48–54.

Lorenz, K. 1941. Vergleichende Bewegungsstudien an Anatinen. J. Ornithol. 89:194–294.

Marshall, A. J. 1954. Bower birds: Their displays and breeding cycles. Oxford University Press, Oxford. 208 p.

McKinney, F. 1965. The comfort movements of Anatidae. Behaviour 25:120–220.

Nice, M. M. 1962. Development of behavior in precocial birds. Trans. Linnaean Soc. N.Y. 8:1–211.

Risebrough, R. W. 1968. Pollution, wildlife and science. Can. Field-Nat. 82:241–243.

Tinbergen, N. 1953. The Herring Gull's world. Collins, London. 255 p.

Weidmann, U. 1956. Verhaltensstudien an der Stockente (*Anas platyrhynchos* L.). Z. Tierpsychol. 13:208–271.

DISCUSSION

Gordon H. Orians

The results discussed by Frank McKinney, which are, in fact, the backbone of wildlife-management studies, have come from intensive long-term studies of individual species. The monographs, the classic studies in wildlife management, are extensive studies of the ecology and behavior of particular species. On the basis of these studies models are erected that contain values and parameters that pertain to that particular species and that may serve as the basis for a management program for that particular species. We all recognize the values and benefits that have come from this kind of effort.

But these studies are extremely time-consuming. They take tremendous numbers of man-hours, and they have very little potential for generalizing once completed. As McKinney rightly points out, once you find out what the Pintail does, you don't necessarily know what any other duck does. Each species must be studied in detail to learn the details of its biology. It would be useful therefore if we could use ecological–ethological theory in wildlife management and thereby avoid having to do a 20-year study on each individual species. Can the particular characteristics of a species be seen as representative of certain broad patterns that have evolved in nature? Can we establish models of these patterns that will yield results faster than long-term studies of individual species? The empirical generalizations from wildlife studies are very highly species specific and have a great degree of reality, but no generality. One procedure in modeling would be to sacrifice precision and reality in favor of generality (Levins, 1966). Our most general biological model is natural selection. If we ask why the Shoveler has that particular shape of bill, the answer would be that it makes him more fit. And why does the Pintail have its mating pattern? Obviously because it is more fit! The theory of natural selection justifies everything, but clearly we have paid an enormous price in terms of precision. No one will contradict the explanation that I have just given that these traits make their bearers more fit. But I really have not said anything that is either very interesting or enlightening. Nevertheless, there are some intermediate grounds whereon we could use the theory of natural selection to develop models of intermediate specificity that combine some generality with the comparative approach to yield some "quick and dirty" results of use to the manager.

I propose to look specifically at habitat selection and the response to space, as was illustrated by cases of the Shoveler and the Pintail. I will discuss a general model that can perhaps treat these species as examples of more widespread patterns and suggest some research efforts that might quickly yield interesting results. Figure 1 presents a model of the habitat selection process based on a paper by Fretwell

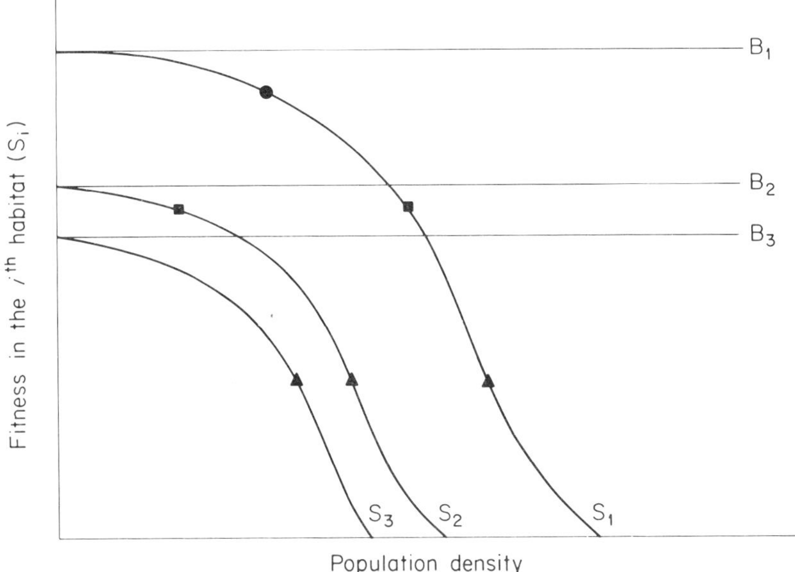

FIGURE 1　Model for habitat selection in a patchy environment. After Fretwell and Lucas (1969). The three different habitats S_1, S_2, and S_3 are of decreasing value to the species being considered such that the maximum fitness in each is B_1, B_2, and B_3, respectively.

and Lucas (1969). Consider the problem faced by an organism choosing a habitat from among environments of different quality, and assume three different environments (S_1, S_2, and S_3) in decreasing order of value to a particular species. Assume also that, as population density increases, conditions for each individual in the population deteriorate. Hence the expected fitness in each of these habitats drops off at increasing densities until finally, at a certain density, there is no further possibility of success. How ought organisms settle in this environment if they "wish" to maximize their fitness? Clearly, the organisms settling first in the environment should pick a place where they have the best chance of success and therefore they should start establishing themselves only in S_1 until that curve crosses the B_2 line, which is the maximum fitness that can be achieved in S_2. At that point, it becomes equally beneficial to settle in the first environment or to go to the second environment, which is not intrinsically so good, but is empty. We then expect additional individuals to settle those two environments until the densities cross the B_3 line, at which point it becomes equally fit to add on to any one of these environments.

When all the individuals have been accommodated, there would be different densities in the three habitats, so adjusted as to exactly equalize fitnesses. Therefore, if we measured reproductive success in those three environments, we would find no significant differences. This is called the "ideal free distribution." It is ideal because each individual has maximized its fitness and because we have assumed that each individual has perfect knowledge of these habitats and their intrinsic qualities and the abundance of conspecifics in all of them. It is free because there are no constraints. Such a distribution is not expected to occur in nature because perfect knowledge and freedom from constraints are probably never realized. For example, the organisms may have difficulty finding the environments and assessing densities. It may be costly in time and energy to move from area to area. There may be critical time limits on the searching period. Some individuals may return to familiar areas or may use other criteria that result in a piling up in certain environments and settling at low densities elsewhere. If we know something about the dispersal characteristics of the species and the ease with which it can assess various environments, we may be able to determine how far it will depart from the ideal free distribution for these reasons.

Another reason for deviations from the ideal free distribution is that there may be behavioral constraints on settling. Each individual that settles makes the environment worse for those individuals already there. Hence, there is an advantage to preventing other individuals from coming in and, if exclusion can be achieved with a modest amount of energy expenditure, an evolution of space-holding behavior patterns can be expected. A way of approaching this problem is to consider the costs and benefits of holding an exclusive area (Figure 2). The larger the area, the greater is the cost of holding it, because there will be a larger perimeter to patrol and greater distances between edges. Also, if some individuals are being excluded by the behavior, the larger the area, the greater is the pressure from these individuals. Therefore, we expect a rising cost with increasing size of territory, the exact function of which depending on a number of factors.

What benefits might accrue from holding an area? One possibility is that the resources are reserved for exclusive use. If this is true, an area that is too small has only enough resource to sustain the adults and nothing at all left over for reproduction. Therefore the curve of energy for reproduction must fall to zero at some minimal size. Conversely, as territories increase in size, a point is again reached when the individual spends all of its time engaged in what Hutchinson called "ag-

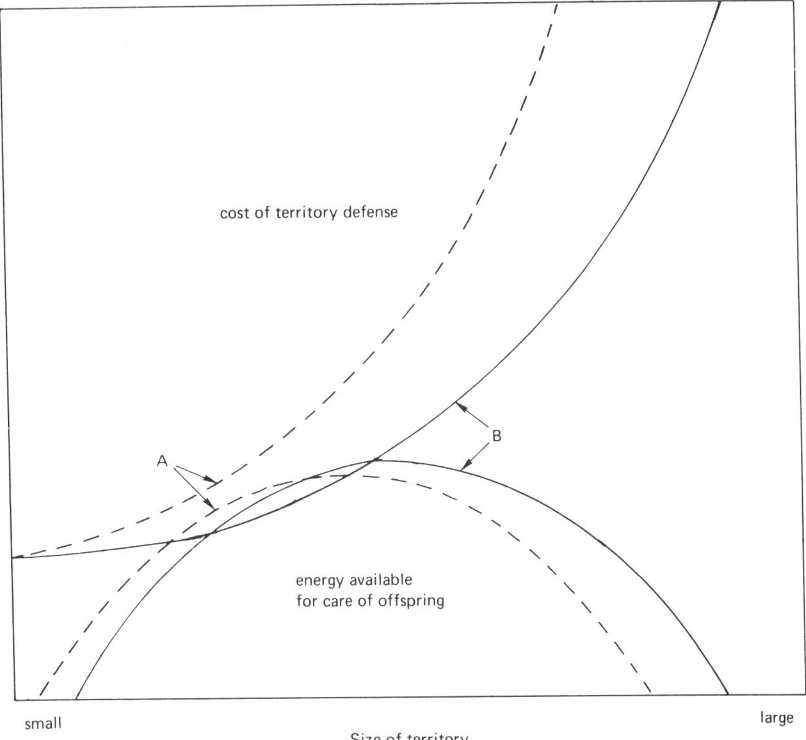

FIGURE 2 Costs and benefits of holding a territory as a function of territory size. A = a productive environment; B = a less productive environment. After Orians (1969).

gressive neglect." The defense of the area requires all of the time and energy of the individual. Since organisms *do* reproduce, we assume that the curve rises above zero in the middle and some function similar to that shown in Figure 2 seems reasonable. If so, there is an optimal intermediate size at which the cost and the benefit of holding the area are such that the maximum amount of energy is left over for reproduction. If the area is highly productive, the energy curve will become zero farther to the left and the optimal territory should be smaller. Conversely, if the area is one of low productivity, optimal size should be larger. If the costs of defense rise steeply, smaller territories are favored. If the costs rise slowly, larger territories are expected. Some of these variations are also shown in Figure 2.

One of the important things that will determine the energy returns and the cost of defense is the reliability and predictability of resources.

From general theoretical considerations, we know that it is more difficult to exploit an unpredictable, uncertain environment (Brown, 1964; Brown and Orians, 1970; Horn, 1968). The organism must spend more time simply searching if the resources are here one day and there the next. Temporarily available resources are harder to find and may be missed. A moving resource pattern inevitably requires more traveling time. Thus, for a variety of reasons, uncertainty of resource distribution and abundance increases the size of the area that has to be exploited and the effort of finding the resources.

Resource uncertainty as it relates to holding space is shown in Figure 3. We assume here that, as the resources change from being highly predictable to being highly unpredictable, the size of the territory will have to increase. Therefore, the cost of holding a territory ought to be lowest for spacially and temporarily reliable resources and increase with greater heterogeneity of the resource. In Brown's (1964)

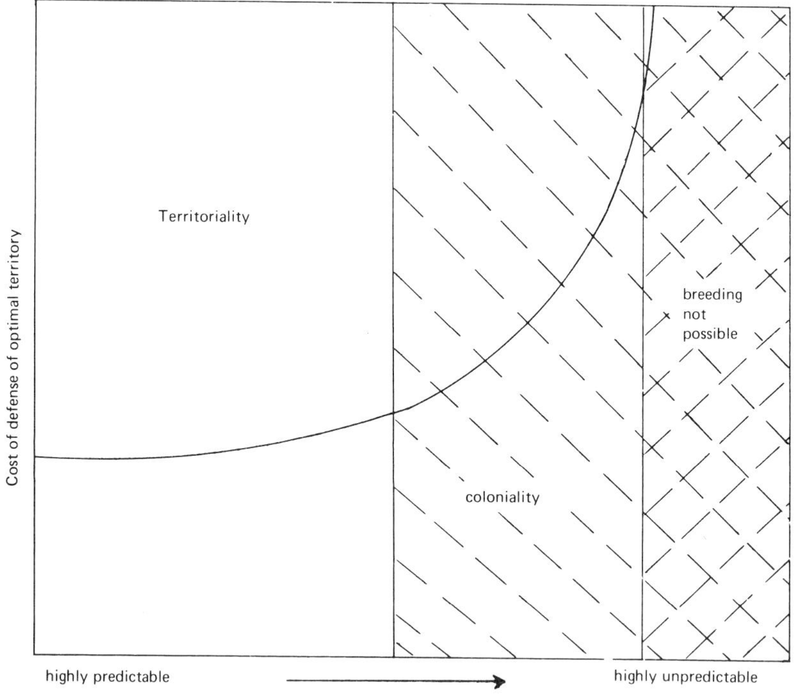

FIGURE 3   The effect of resource predictability on animal spacing patterns. After Orians (1969).

terminology, the resource becomes less economically defensible.

Consider now the waterfowl discussed by McKinney. The Shoveler—which has a reliable, spacially fairly predictable food supply in the form of plankton, which can be reasonably defended at a cost less than the advantages—is strongly territorial. Apparently, in the Pintail habitat—where the marshes vary in depth and size depending on the rains—there is great variability in the resources from year to year, and there may be enormous changes in time during the season as marshes dry up or expand. This is a situation in which it is very difficult to predict the site and time of availability of resources. Accordingly, the cost of being territorial will be great. Not surprisingly, these ducks clump in areas close to the center of gravity of the expected resource distribution and then shift in foraging areas over long distances during the breeding season.

Marshes are environments with extremely variable spacial and temporal distributions of resources for breeding birds. I can illustrate this with some of the data that we have gathered on the emergence of aquatic insects in marshes in Washington. We have sampled emergence with wire mesh traps that are open on the bottom. One type rests on stakes driven into the mud such that emerging aquatic insects can enter from below and metamorphose inside, but cannot get out. It has a heavy wire base over which a light screen is stretched. This strength is necessary because muskrats like to sit on the traps while they eat their meals. A second type of trap, made of similar materials, rests on the shore and intercepts insects coming onto land to emerge.

A large number of these traps, placed on 10 different lakes and in different positions have yielded data on the enormous variability of this resource. Figure 4 shows the seasonal pattern of damselfly (*Odonata*) emergence from several lakes. This seasonal pattern is modified from day to day, especially by weather conditions; there are enormous differences among lakes. Willow Pond, which has a very low emergence, is a lake that dries out or nearly dries out in the late summer. Therefore damselflies, which constitute the large spring emergences and which overwinter in late instars, are eliminated and the damselflies that do live in these lakes are ones that overwinter as eggs that drop into the water in spring. Thus the drying up and flooding characteristics will have an enormous impact on the resources available in these marshes over a season.

Many of these aquatic resources vary in availability during the day. Figure 5 presents data from three different lakes in eastern Washington; again, most of the emerging aquatic insects are damselflies.

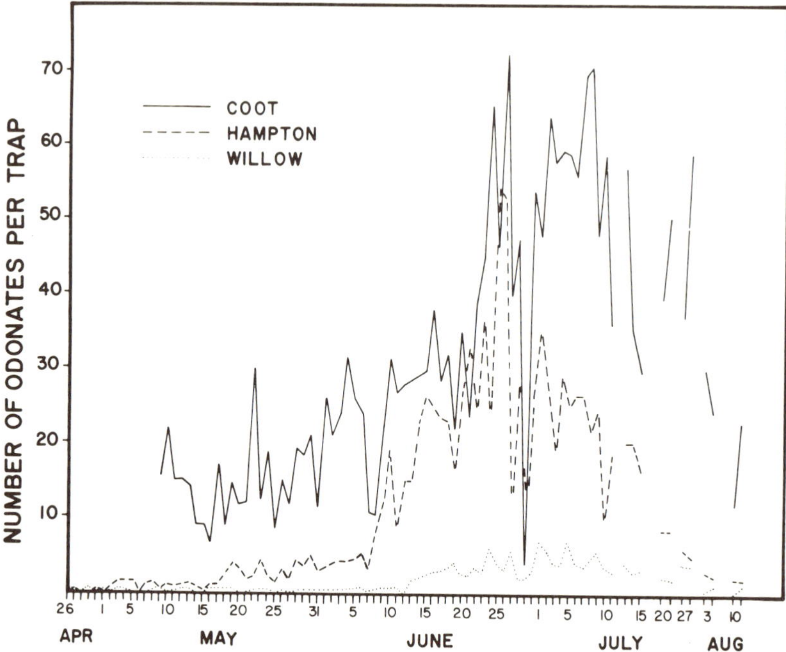

FIGURE 4   Seasonal pattern of damselfly emergence on three Washington lakes.

Notice that emergence peaks during the late morning hours, while during the early morning and late hours in the afternoon the resource availability is very low. Breeding blackbirds are very, very sensitive to this. Their foraging locations and the prey they take vary enormously with respect to time of the day (Figure 6). I don't know the resource requirements for most of the duck species, but I suspect that some of their resources are also highly variable in space and time.

In addition to this sort of temporal pattern, there is significant spatial variation that may be of great significance to waterfowl (Table 1). The data reveal the importance of the location of shore traps relative to beds of cattails and other emergent vegetation. We believe that this is because the aquatic insects, which are distributed throughout the lake, must come to the shore to emerge and that they crawl out on the first vegetation they encounter. If it is a cattail patch, they come out on the outer edge; few of them penetrate very far into it, so that within and behind the patch is a poor place to find emerging insects. Dense cattail beds also intercept most of the light. Consequently, the productivity of the system goes into the cattails and insects that eat cattails. Little remains to enter the water and the aquatic food chain, which is one of the reasons why wildlife managers discovered intuitively that good marsh management requires breaking up the cattail patches. Dense stands of cattails are poor duck and blackbird habitat. On the basis of this knowledge, one can manipulate the edge situations and geometry of cattails and exert a great influence on productivity patterns for certain kinds of aquatic organisms.

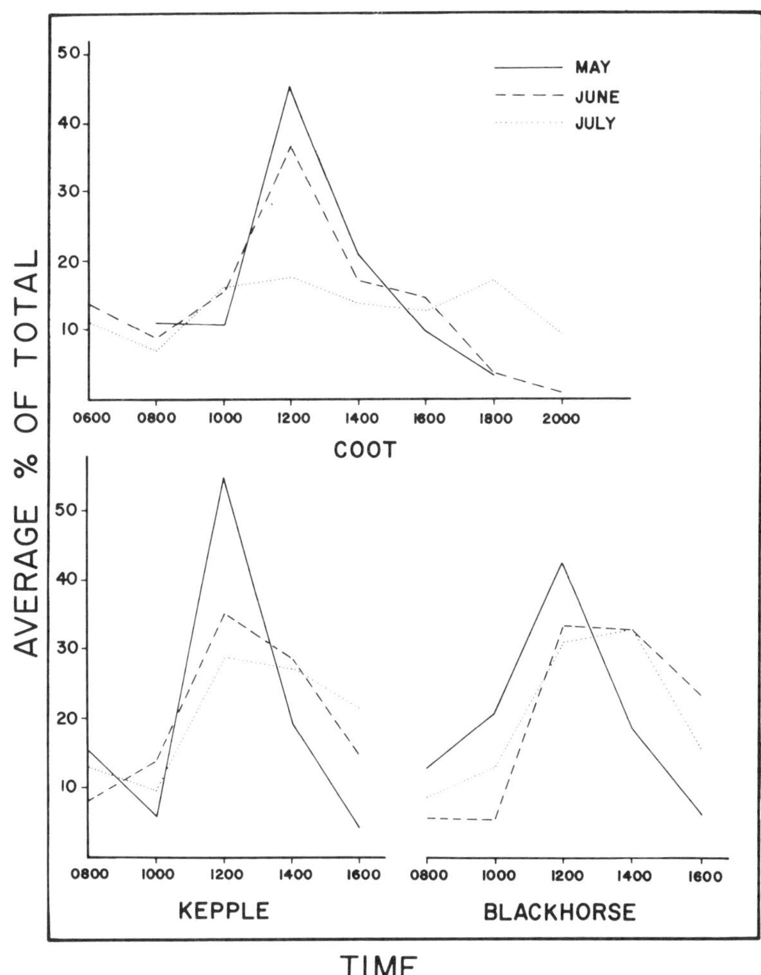

FIGURE 5   Diurnal pattern of emergence of insects from three Washington lakes during the spring and summer.

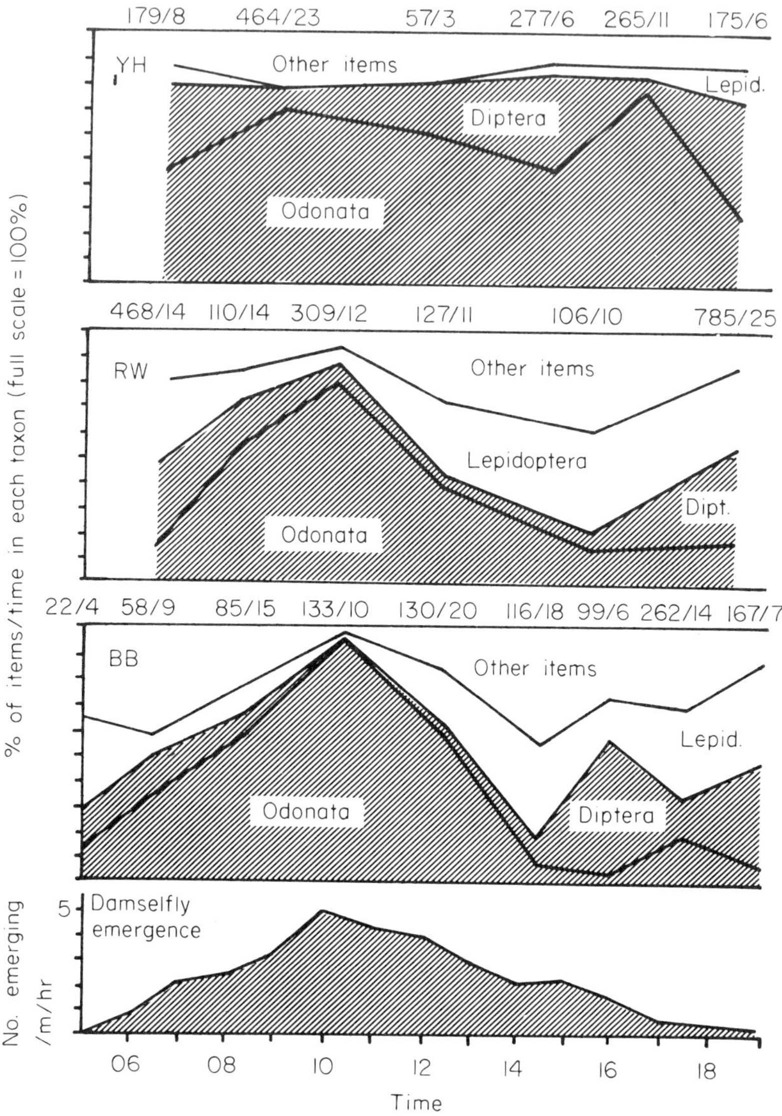

FIGURE 6   Daily pattern of prey delivered to nestling Yellow-headed Blackbird (YH) (*Xanthocephalus xanthocephalus*), Red-winged Blackbird (RW) (*Agelaius phoenicius*), and Brewer's Blackbird (BB) (*Euphagus cyanocephalus*) in the potholes of central Washington. From Orians and Horn (1969).

TABLE 1   Effect of Trap Location on the Quantity of Emergence of Odonata in Washington, 1968

A. Shore Traps

		Behind Dense Cattail Bed			Behind Open Cattail Bed			Exposed to Open Water		
Area[a]	Lake	No. Traps	Total Odonata	Odonata/ Trap/ Week	No. Traps	Total Odonata	Odonata/ Trap/ Week	No. Traps	Total Odonata	Odonata/ Trap/ Week
CNWR	Hampton	2	222	10	2	688	31	1	1657	151
TNWR	Kepple[b]	2	1098	49	—	—	—	3	8952	271

B. Floating Traps

		Outer Edge of Emergent Vegetation			Within Beds of Emergent Vegetation		
Area[a]	Lake	No. Traps	Total Odonata	Odonata/ Trap/ Week	No. Traps	Total Odonata	Odonata/ Trap/ Week
CNWR	Hampton	3	2925	122	3	964	40
TNWR	Kepple[c]	3	2577	107	2	2119	132

[a] CNWR = Columbia National Wildlife Refuge; TNWR = Turnbull National Wildlife Refuge.
[b] Traps operated through July 16.
[c] Traps operated through June 25.

37

Once we realize that the temporal and spacial characteristics of the resource used by an organism influence extensively the kind of social system, spacing system, and exploitation patterns that a species will evolve, we can examine many species in a general way to discover relationships between spacing systems and reliability of resources. For example, a quick check on many species of ducks might yield some pattern like this. Then, on the basis of such correlations, one might be able to make some very significant management decisions in terms of how to manipulate resource geometry to increase its availability for the species in which we are particularly interested. This should yield results faster than a 20- or 30-year study of each individual species.

The most important recent development in ecological–ethological theory is probably the recognition of the importance of the heterogeneity of natural environments. We generally refer to this heterogeneity as the "grain" of the environment (Levins, 1968). Marshes are perhaps the grainiest of natural environments; there are substantial differences within marshes and even greater differences among marshes. I have attempted to classify the graininess in space and time according to scale and to identify the attributes of breeding birds that would be expected to evolve in response to these different scales of graininess. Space scales are shown in Table 2 and time scales in Table 3. They were derived primarily from my work with blackbirds, but the ideas should be applicable, with slight modifications, to breeding waterfowl as well.

These developing ideas about adaptations related to the grain of the environment in space and time, and how various responses to spatial distributions of resources might be expected to evolve, offer

TABLE 2   Effects of Spatial Grain of Marshes on the Attributes of Breeding Birds

Type of Grain	Scale	Attributes of Breeding Birds Affected
Fine	Smaller than foraging area of individual birds	Individual foraging decisions: patch selection, prey discrimination; choice of nest site
Coarse	Larger than individual foraging areas or territories	Territory size (males); selection of mate and territory (females); colony site location; phenotypic variability of populations
Geographical	Regional differences in properties of marshes and uplands	Clines in morphology; clines in clutch size; clines in foraging behavior; clines in nest-site selection; termination of range; locations of territories

TABLE 3 Effects of Temporal Grain of Marshes on the Attributes of Breeding Birds

Type of Grain	Scale	Types of Pattern	Attributes of Breeding Birds Affected
Fine	Diurnal	Emergence patterns; temperature; light	Foraging areas and prey taken
Coarse	Seasonal	Weather; emergence patterns; vegetative growth; water levels	Breeding season limits; clutch size; nest locations; movements of broods during season
		Predator activity	
Rough	Yearly	Vegetative succession	Phenotypic variance—niche breadth
		Changes in water levels; differences in quantity of emergence of aquatic insects	Territory locations and size

hope of being generalized to many species of birds. They do not contain specific details about the forms of the courtship of any species, exact predictions about individual foraging behavior, or food selections. As with all generalizations, there is a loss of information that accompanies the shift away from the specific. Nevertheless, I think that the gain in general insight that comes from this type of theory may pay enormous dividends in management studies and management practices. It is clearly worth a try.

REFERENCES

Brown, J. L. 1964. The evolution of diversity in avian territorial systems. Wilson Bull. 76:160–169.

Brown, J. L., and G. H. Orians. 1970. Spacing patterns in mobile animals. Ann. Rev. Ecol. Syst. 1:239–262.

Fretwell, S. D., and H. L. Lucas. 1969. On territorial behavior and other factors influencing distribution in birds. I. Theoretical development. Acta Biotheor. 19:1–20.

Horn, H. S. 1968. The adaptive significance of colonial nesting in the Brewer's Blackbird (*Euphagus cyanocephalus*). Ecology 49:682–694.

Levins, R. 1966. Strategy of model building in population biology. Am. Sci. 54:421–431.

Levins, R. 1968. Evolution in changing environments. Princeton University Press, Princeton. 120 p.

Orians, G. H. 1969. The study of life. Allyn and Bacon, Inc., Boston. 841 p.

Orians, G. H., and H. S. Horn. 1969. Overlap in foods of four species of blackbirds in the potholes of central Washington. Ecology 50:930–938.

INFORMAL DISCUSSION

McKINNEY: I agree almost completely with what has been said by the two discussants.

I deliberately took a species-oriented approach and would like to point out that we need both fine-grain and coarse-grain analyses. I would also stress that many of the problems in the management, control, and conservation of animals come down to the individual species and that we must know a great deal about a species in order to accomplish these manipulations properly.

In respect to the size of display repertoires in ducks, Moynihan* has recently surveyed the numbers of displays in different kinds of animals and has come up with rather interesting data. His is a pretty coarse-grain analysis, because it is very difficult to give numbers to this kind of thing, but he finds that in a large range of different kinds of animals there are somewhere between 15 and 35 displays. He poses the question: Why are there not more different kinds of displays? Apparently, animals encounter a number of situations that they have to resolve by using these special signal movements or vocalizations. In the case of the ducks, I think we know enough to go quite a long way in saying why it is necessary, for example, for a Mallard to have 25 or so different displays. I think considerable progress is being made along this line.

KRAPU: I would like to comment briefly on a point made by Doctor McKinney concerning the feeding ecology of Pintails, since it may help explain certain aspects of their breeding behavior. Past studies of food habits conducted in North America have shown that the Pintail is primarily a vegetarian. However, research recently completed in North Dakota indicated that animal foods accounted for nearly four fifths of the diet of hens collected during the laying phase of the reproductive cycle.

This latter study also indicated that in April breeding Pintail hens feed largely on invertebrates that become temporarily available after flooding of shallow wetlands. For example, earthworms present in the dried bottoms of wetlands became an important food item for a short period after flooding. Fairy shrimp hatched after flooding and became a major item in the diet for a few weeks, but disappeared in late April and early May. I stress the relationship between habitat instability and fluctuating invertebrate supply in view of the apparent need laying hens have for animal foods. In this situation, a large home range, with overlap among pairs and a general lack of hostility among males, may help to ensure an adequate food supply for hens during the period of egg formation.

*Moynihan, M. 1970. Control, suppression, decay, disappearance and replacement of displays. J. Theor. Biol. 29:85–112.

FARNER: I would like to ask a relatively simple question of any one of the three speakers. I gather that you are implicitly saying that the development of behavior pattern somehow follows or is dependent on preceding morphological changes. I was reminded of this particularly in Frank McKinney's comparison of the eiders, in which he pointed out that smaller size had implications so far as flock behavior is concerned.

My general question is why do we assume that morphological or even physiological changes precede behavioral changes? Does not the development of behavior provide the opportunity for the survival of genetic patterns that alter morphology?

HAILMAN: The sequence is not always the same. I want to cite just one counterexample to show how in my own work we were able to approach this chicken-egg question in the evolution of signal patterns for begging by the gull chick. The preliminary evidence is that the behavior of the chick and the stimulus to which it responds is fairly fixed through various species. As some of the species evolved into larger forms, the adults, in a sense, had to change bill coloration and shape in order to provide the optimum stimulus for releasing begging response by the chick. In this case, the morphology actually followed the evolution of the behavior or at least we seem to have some basis to say that. But I would say this is probably one of the knottiest problems in behavioral ecology—resolving the chicken-egg question as to which are the causal factors in these correlations. I don't know of a really good method that allows us to solve this problem except by taking the usual evolutionary strategy of comparative studies—to look at a large number of species. Then, if ones that are judged to be related on other grounds show the same kind of behavioral pattern, one says it is nonadaptive; whereas if quite unrelated species show it, one gets some idea of what the cause and effect is.

ORIANS: I certainly can't speak for ethologists in general, but I see no particular reason to favor either one as having primacy. Where selection normally falls is when an individual acts: Either the act is appropriate or inappropriate; it is effective or ineffective. Hence, it is not during contemplation or anything else when the selection takes place. Presumably, selection can be acting on behavioral acts, and thereby lead to a morphology that influences the acts, or there can be certain physiological preconditions that a species brings into a situation, which then influence the effectiveness or ineffectiveness of given behavior patterns. I see no particular theoretical reason to assume that it is predominantly either one way or the other.

CHIASSON: We encountered, some time ago, an anatomical problem along this line that seems very interesting. We discovered that many doves have iridophores in the iris and that these produce structural colors. In some cases, the color most often produced is a yellow. In the one other I know about, the African Green Fruit Pigeon, the structural color is blue. These iridophores are covered over in some, but not all, doves or pigeons by blood vessels, and the amount of blood in these vessels can be controlled. When the vessels on the surface of the iris are engorged with blood, the iris is red. When there is less blood in these vessels, then the color of the iris is orange. Inca Doves have a bright red color when hand held; apparently, this color also is exhibited when there is another male in their territory. I know that the same color changes will occur in the Mexican Ground Dove, but I don't know whether this has any behavioral attributes or not.

Some of the doves have the ability to change eye color, some do not. Some are brown-eyed and have no structural pigments in the iris. It has become a very interesting evolutionary problem; one that I can't explain on the basis of behavioral adaptations by the entire order or even by a genus. It does not seem to be directly related to behavior, but it certainly involves behavior.

FARNER: Are there other comments or questions?

LINDER: One of you mentioned pesticides and their effect on some behavioral aspects. Some of us have been working in the area of pesticides and have been trying to make measurements of this kind. I, personally, have hit a stone wall after 4 or 5 years.

My questions to the ethologists is how do we go about measuring the effects of pesticides on some of the behavioral aspects during the reproductive season, especially on the displays? How do you suggest that we make these measurements?

HAILMAN: I don't know, simply because the animals that we are looking at now are perhaps the most difficult in which to study the real interactions of courtship. If you wish to go out and look at Marsh Hawks (*Circus cyaneus*), one of the species in which pesticides may be having some effect on the courtship, you must do a lot of looking up at their displaying activities high in the sky. They may do something at one point, and then be 2 miles away in a couple of minutes doing something else. It is therefore very difficult even to make a complete index of the kinds of signals that they use. But listing the kinds of signals, I think, is only a beginning, because we want to know in some detailed way what normal reproduction is, the sequence of displays and events, and how they are used. This was my contention: We don't understand this well enough, even in thoroughly studied species.

I want to qualify this by reiterating that what I said was that, to my satisfaction, they are not well enough understood. It may be that my standards are too high. I am really amazed that, for instance, if you watch a Mallard, it goes through a whole sequence of distinct behavior patterns in a short space of time. The question is, why is that necessary? Why does he have to do the Head-up/Tail-up? The Down-up? The Grunt-whistle? And then 20 minutes later do the same three or four behavior patterns? It certainly does not seem to be sequential. I think that we need to develop more hypotheses about courtship and consider that it may be the relative frequency of various displays that really signals the readiness of the animal. In other words, a given display, per se, may not necessarily be meaningful. I think that the "single-meaning" notion is an older concept of ethology that we have too long considered fact. That is, we expect that there is a warning call, that there is a precopulatory display, and so forth. I rather doubt that that is the way bird communication is set up. Consider displays A, B, and C. If A is twice as frequent as B, it means something different from B being twice as frequent as A. Or the order A, B, C, means one thing; whereas the order C, B, A, means something quite different.

McKINNEY: I will try to respond to both questions simultaneously. For some years now, we have been doing detailed analyses of Green-winged Teal (*Anas crecca*) courtship in flight pens, where we are able to film long sequences of behavior. As a result of frame-by-frame analysis, we find that it is possible to predict the situations in which a certain display will occur or is likely to occur. For example, the distance between the male and female, say, when a Grunt-whistle is performed tends to be about 5 feet. In the case of the Down-up display, the eliciting situation is very different: There must be a second male present as well as the female and the distances among the three birds are much shorter than in the case of the Grunt-whistle.

While I would agree that it is not always easy to see an immediate response to a display in the case of social courtship in ducks, I think there is evidence that the displays serve various functions. Some displays between males clearly have a threat function. This is not always obvious until many sequences are analyzed, but there is often a relationship between certain displays and overt hostility. For example, in the Green-winged Teal (*Anas erecca*), the Bill-up posture is very frequent in male-male hostile encounters, and the Down-up is much more closely associated with male-male hostility than is the Grunt-whistle.

We do know a good bit about situations that evoke each of these displays, and I think that rather specific messages are being transmitted in many cases. Also, I think the whole process of social courtship may have several general functions. There may be psychosomatic effects similar to those shown by Lehrman in Ring Doves. There may also be a general canalization of hostility. During social courtship, males are competing for a female and conflicts have to be resolved. There must be aggression, and dominance relationships between the males seem to be very important in determining which male pairs with the female. These relationships between males must be worked out at the same time as males are indicating their preference for a specific female. The orientation of the male's body as he gives displays is extremely important—most are performed with the body lateral to the female.

Courtship displays seem to be serving several functions, but I think the overall function of social courtship may be to give the female time to make a choice. In some cases, one can see that the performance of a male display is followed by a response from the female, but often one cannot. In many cases, I think there is an attraction function involved. The message, so to speak, is: "I am interested in pairing with you." The male is not only indicating his readiness to pair but he is also specifying which female interests him. This must be very important when birds are pairing in large flocks.

As for what behavior patterns might reflect pesticide effects, I think these displays provide examples. We have been studying individual variation and duration of displays and have attempted to detect radiation effects on these courtship patterns. Our results are negative, but I think these displays have special advantages in that they can be easily measured from movie films. Of course, one can go further by testing relative success of treated and untreated males in competitive pair-formation experiments.

RAVELING: Perhaps we have to ask what it is we don't know, but I seriously question the philosophy of interpreting behavior by saying "Why don't they do something?" or "Why don't they look some way?" This is endless. "Why aren't they purple? Why aren't they gold? Why don't they flap their wings twenty times or two times?"

I think it is much more important to look at the problem in terms of specific knowledge of individual behavior, as Frank McKinney just outlined, to obtain the quality of information needed, so that we can make valid interpretations. That we cannot specify cause and effect for a particular behavioral response does not mean that there is not a function.

I agree that there could be pleiotropic effects, particularly in plumage, that could be carried on and reflected in display, which we look at and conclude there must be a function. Yet there may not be.

Finally, I am interested in the question that was raised about pesticides and their effects on behavior. I thought that the statement was made during the presentation that there was evidence that pesticides were affecting courtship behavior. I would like to know the evidence, in fact, as to what species show that courtship behavior is affected by pesticides.

HAILMAN: Risebrough* cites effects upon the courtship flights of Marsh Hawks, but I am not conversant with the original literature. There are numerous statements in the literature, though, mainly about raptorial birds. I don't know of anything, say, on ducks, although Frank McKinney might.

McKINNEY: Perhaps someone from Fort Collins can tell us about the studies that have been going on there.

FARNER: Is there someone who can comment?

MULLER: We have done some work on reproductive success with low levels of pesticide.† Basically, we found debility in reproductive success, mostly in depressed embryonic survival, not in the shell-thinning syndrome that has been reported.

---

*Risebrough, R. W. 1968. Pollution, wildlife and science. Can. Field-Nat. 82:241–243.

†Weill, D. D., H. D. Muller, and J. V. Shutze. 1971. Pesticide effects on the fecundity of the Gray Partridge. Bull. Environ. Contam. Toxicol. 6:546–551; Muller, H. D., and D. C. Lockman. 1972. Fecundity and progeny growth following subacute insecticide ingestion by the Mallard. Poultry Sci. 51:239–241.

FARNER: Doctor Follett, were you asking for the floor?

FOLLETT: There seems to be some belief that pesticides alter behavior, but not much hard evidence. What strikes me, as an endocrinologist, is that, if these studies are undertaken in the future, it might be useful to try to distinguish whether the pesticides act directly on the central nervous system to alter behavior or act indirectly via the endocrine system.

SCOTT: This was the same vein in which I also wanted to make a suggestion. Both DDT and PCB's cause induction of enzymes in the liver. These enzymes, which are cytochrome P-450's, have the ability to hydroxylate many compounds, including DDT's, PCB's, and estrogens. It seems to me that using pesticides experimentally might be a way of finding out to what extent alterations in the endocrine balance in these animals might change their patterns.

I had one other comment about Farner's question on which comes first, the morphology or the associated behavior patterns. I think for an answer to this we might go to studies with human beings. It is well known that most of our children are taller than we are. Studies on calcium metabolism indicate that American young men require about 800 mg of calcium for calcium balance, because they drink milk from the time they are born right up to adulthood and have had a lot of calcium. Generations are growing taller and taller.

In Peru, where people are short, the young people get only about 200 mg of calcium and are in calcium balance at this level. One might hypothesize that this is why they are smaller.

In the Philippines, we attempted to ascertain how much calcium Filipinos were getting. With the whole fish that they eat, they get just about enough calcium to become the size they are. So I wonder about feeding patterns? The feeding patterns of these birds may have something to do with keeping them the size they are.

Nutrition in Reproduction – Direct Effects and Predictive Functions

M. L. SCOTT

Nutrition furnishes the cells of the animal with materials needed for optimum functioning of the many metabolic chemical reactions involved in growth, maintenance, work, production, and reproduction. Qualitatively, nutrition is very simple. Approximately 40 nutrients are needed by all animals. In birds, these include 13 essential amino acids, 13 essential inorganic elements, 13 vitamins, linoleic (or arachidonic) acid, adequate nitrogen for synthesis of the nonessential amino acids, and an adequate supply of energy. The science of nutrition encompasses procurement, ingestion, digestion, and absorption of the chemical elements that serve as food. In addition, it includes the transport of these elements to cells in physical and chemical forms suitable for assimilation and use. Thus, even though we may have good information concerning the exact nutrient requirements of a particular species, and may be able to formulate a diet containing all of the nutrients in proper amounts, the diet is of no use if the animal refuses to eat it or cannot digest it.

Fortunately, most birds, especially the game birds, readily consume formulated mixed feeds. In some instances, especially in the case of wild ducks and wild geese, feed consumption is much improved if the feeds are pelleted rather than used as mash mixtures.

Pellets are especially desirable if the feeds are to be scattered on the ground or on the surface of icy ponds.

Research with breeding pheasants at the Ithaca Game Farm showed, however, that it is possible to devise a complete feed having characteristics similar to those encountered in the wilds by coating a mixture of coarse cracked corn, wheat, and oats with molasses and then mixing the protein, vitamin, and mineral supplement in such a way that the molasses holds the supplement as a sticky crumble on the surface of the grain particles.

Optimum nutrition is a relative term. Usually, the concept of optimum nutrition depends on the goals set by the nutritionist. Thus diets can be scientifically formulated to produce maximum body growth rate, maximum ability to resist disease and other stresses, maximum egg production and reproduction, and good feathering. Other diets can be formulated that will produce good strong bodies and good feathers but, due to a very low energy content, will not produce a maximum rate of growth. However, retardation of growth that occurs as a consequence of a deficiency of an essential amino acid, an essential mineral, or a required vitamin usually results in poor body development, poor resistance to disease and stresses, poor feathering, and poor hatchability.

Nutrition in the wilds depends on the food supply and the competition for that food. In areas where food is plentiful, wild birds obviously are able to choose diets that support good body development, good feathering, and good reproduction.

What is known concerning the quantitative nutritional requirements of birds? What happens when birds cannot find enough food, or the right kinds of foods, to meet these nutritional requirements?

Because the domestic chicken is economically important, much effort has been expended to determine its critical nutritional needs. Furthermore, because the young chick is an excellent experimental animal for fundamental nutrition studies, it has been used by many laboratories. This combination of circumstances has produced more information concerning the nutrition of the chicken than of any other species and has placed the formulation of poultry feeds on a sound scientific basis.

With information on poultry as the starting point, a series of experiments has been continued at Cornell University in cooperation with the Ithaca Game Farm of the New York State Conservation Department on the quantitative nutritional requirements of pheasants, Bobwhite Quail (*Colinus virginianus*), wild duck, wild geese,

Hungarian Partridge (*Perdix perdix*), and Wild Turkeys (*Meleagri gallopavo*). These studies have succeeded in providing information on the exact nutritional requirements of pheasants and quail for most of the essential nutrients. Other studies have provided good evidence concerning nutrient requirements for satisfactory results in the other species.

The nutrient requirements that are of importance in the formulation of adequate diets for pheasants, Bobwhite Quail, wild ducks, wild geese, and Wild Turkeys are shown in Table 1. These data could be used in assessing the nutritional value of any specific dietary regimen for any of these species, or in predicting probable effects on one of these species of the elimination of one or more of the food items presently being consumed in the wild as an important part of the total diet.

Examination of the data in Table 1 shows that the nutritional requirements for reproduction in birds are very similar to the requirements of the young birds for survival and optimum growth for the period from hatching through the critical starting period of approximately 3–6 weeks. The dam must consume sufficient amounts of all of the essential amino acids, vitamins, and minerals to produce an egg and to supply to that egg all of the nutrients needed by the embryo throughout its entire incubation period. For optimum survival and growth of the young, the dam also must provide sufficient reserves in the spare yolk to supply nutrients to the newly hatched bird.

The nutritional requirements during the starting period are higher than for any other period throughout the life of the bird. This is the time when a bird grows at a very rapid rate and does not yet demand a large amount of food for maintenance. Thus, during the early stages, most of the food that the chick consumes is used for growth. As the chick grows larger, more and more of the food is used for maintenance and relatively less for growth. Thus the percentages of the various amino acids and minerals and the levels of the vitamins needed in the food become less and less as the animal grows older. When the bird becomes mature, its only nutritional needs are for maintenance until the time arrives for egg production, reproduction, and feather replacement in molt.

More is known concerning the nutrition of pheasants than of any other wild bird. However, the requirements for reproduction appear to be very similar for all the species of birds studied thus far. The carcass composition, and the nutritional need to replace wear and tear of body tissues appear to be very similar for all adult birds.

TABLE 1 Nutrient Requirements[a] of Pheasants, Bobwhite Quail, Wild Ducks and Geese, and Wild Turkeys

	Pheasants (Chinese, Mongolian, Korean, and Green)			Bobwhite Quail			Wild Ducks and Geese			Wild Turkeys		
	Starting	Growing	Breeding	Starting	Growing	Breeding	Starting	Growing	Breeding	Starting	Growing	Breeding
Energy, Kcal ME/kg	3,000	3,000	2,900	3,000	3,000	2,900	2,750	2,750	2,900	2,900	3,000	2,900
Protein, %	30[b]	24	19	27	24	19	22.5	18.5	19	30	26	19
Methionine, % of protein	2.0	2.0	2.0	2.0	2.0	2.0	2.0	2.0	2.0	2.0	2.0	2.0
Methionine + cystine, % of protein	3.7	3.5	3.5	3.5	3.5	3.5	3.5	3.5	3.5	3.5	3.5	3.5
Lysine, % of protein	5.0	4.5	4.0	5.0	4.5	4.0	5.0	4.5	4.0	5.0	5.0	4.0
Calcium, %	1.0	0.7	2.7	0.9	0.7	2.7	1.1	0.9	2.7	1.2	0.5	2.7
Phosphorus, available, %	0.5	0.35	0.5	0.45	0.35	0.5	0.55	0.45	0.5	0.6	0.45	0.5
Sodium, %	0.1	0.1	0.1	0.1	0.1	0.1	0.1	0.1	0.1	0.1	0.1	0.1
Chloride, %	0.15	0.15	0.15	0.15	0.15	0.15	0.15	0.15	0.15	0.15	0.15	0.15
Potassium, %	0.6	0.6	0.6	0.6	0.6	0.6	0.6	0.6	0.6	0.6	0.6	0.6
Manganese, ppm	95	70	70	95	70	70	75	60	60	95	70	70
Zinc, ppm	62	60	60	62	60	60	65	60	60	70	60	60
Magnesium, ppm	500	400	350	500	400	350	500	400	350	500	400	350

Continued overleaf

49

TABLE 1 (Continued)

	Pheasants (Chinese, Mongolian, Korean, and Green)			Bobwhite Quail			Wild Ducks and Geese			Wild Turkeys		
	Starting	Growing	Breeding	Starting	Growing	Breeding	Starting	Growing	Breeding	Starting	Growing	Breeding
Iron, ppm	130	90	130	130	90	130	130	90	130	130	90	130
Copper, ppm	13	9	13	13	9	13	13	9	13	13	9	13
Iodide, ppm	0.3	0.3	0.3	0.3	0.3	0.3	0.3	0.3	0.3	0.3	0.3	0.3
Selenium, ppm	0.1	0.1	0.1	0.1	0.1	0.1	0.1	0.1	0.1	0.2	0.2	0.2
Vitamin A, IU/kg	11,000[c]	2,000	11,000	13,000	10,000	13,000	11,000	2,000	11,000	11,000	4,000	11,000
Vitamin D_3, IU/kg	1,500	1,500	1,500	1,500	1,500	1,500	1,500	1,500	1,500	1,500	1,500	1,500
Vitamin E, IU/kg	15	10	25	15	10	25	20	10	25	20	10	25
Vitamin K, mg/kg[d]	1.0	0.5	0.5	1.0	0.5	0.5	1.0	0.5	0.5	1.0	0.5	0.5
Riboflavin, mg/kg	4.0	3.5	3.5	4.0	3.5	3.5	4.0	3.5	3.5	4.0	3.5	3.5
Nicotinic acid, mg/kg	70	60	60	70	60	60	70	60	60	70	60	60
Pantothenic acid, mg/kg	10	8	15	10	8	15	11	8	15	11	8	16
Vitamin B_{12}, µg/kg	5	3	5	5	3	5	5	3	5	5	3	5
Choline, mg/kg	1,500	1,100	1,100	1,500	1,100	1,100	1,500	1,100	1,100	1,980	1,500	1,500
Linoleic acid, %	1.5	0.8	1.5	1.5	0.8	1.5	1.5	0.8	1.5	1.5	0.8	1.5

[a]SOURCES: Scott and Reynolds (1949), Scott *et al.* (1954, 1957, 1958a, b, 1959, 1960, 1961, 1963, 1964, 1969).
[b]This level of protein is used to improve ability of young pheasant to adapt to stresses. Minimum level for growth and feathering is 26.5 percent.
[c]These levels of vitamin A recommended for improved resistance to stresses and for adequate carry-over of the vitamin through the egg to the young.
[d]Vitamin K should be increased fivefold in diets containing sulfaquinoxaline.

50

TABLE 2 Protein as Percentage of Fresh Carcass[a] of Domestic and Game Birds

| Age | Chickens | | Ring-necked Pheasants | Bobwhite Quail | Turkeys | |
	Leghorns	Broilers			Domestic	Wild
8 weeks						
Male	18.6	18.5	21.1	23.7	20.5	21.8
Female	18.7	18.3	20.5	22.8	20.9	20.9
Maturity						
Male	20.0	19.6	20.5	[b]	20.4	20.7
Female	17.3	18.5	19.4	[b]	18.4	19.8

[a]Includes all tissues (except feathers) washed free of digesta.
[b]Bobwhite Quail are mature size at 8 weeks of age.

We have conducted extensive studies on the nutrient composition of the carcasses of pheasants, Bobwhite Quail, partridge, wild waterfowl, and Wild Turkeys in terms of protein, fat, moisture, calcium, phosphorus, and essential amino acid composition. Some of these results are shown in Tables 2, 3, 4, 5, and 6. This information, coupled with measures of growth rate in these various species (Figures 1, 2, and 3), and information on feathering (Figure 4) allows one to predict nutritional requirements for growth and development. Similar data on the composition of eggs make possible a prediction of the nutrient needs for egg production and hatchability.

Birds, as do all other animals, tend to consume a certain amount of energy each day, depending on body size, stage of development, and environmental temperature. It is possible, therefore, to deter-

TABLE 3 Fat as Percentage of Fresh Carcass[a] of Domestic and Game Birds

| Age | Chickens | | Ring-necked Pheasants | Bobwhite Quail | Turkeys | |
	Leghorns	Broilers			Domestic	Wild
8 weeks						
Male	9.8	11.8	3.1	2.3	4.2	1.3
Female	10.3	12.3	3.5	1.8	4.7	4.0
Maturity						
Male	11.1	11.2	10.8	[b]	17.1	9.0
Female	20.0	21.4	14.6	[b]	20.2	13.4

[a]Includes all tissues (except feathers) washed free of digesta.
[b]Bobwhite Quail are mature size at 8 weeks of age.

TABLE 4   Ash Content as Percentage of Fresh of Carcass of Domestic and Game Birds

| | Chickens | | Ring-necked Pheasants | Bobwhite Quail | Turkeys | |
Age	Leghorns	Broilers			Domestic	Wild
8 weeks						
Male	3.6	3.7	3.7	3.8	4.0	4.1
Female	3.9	3.7	3.9	4.5	4.1	4.7
Maturity						
Male	4.5	3.6	3.4	[a]	3.4	4.1
Female	3.4	3.5	3.2	[a]	2.8	3.8

[a]Bobwhite Quail are mature size at 8 weeks of age.

mine the nutrient levels in the diet being consumed and to decide whether or not these levels meet the nutrient needs of the animal.

Is it always necessary that the diet produce maximum results? What happens when the dietary level of one or more of the nutrients is increased or decreased?

In terms of reproduction, limitations in nutrient intake can have several effects upon egg production and hatchability, as follows:

1. *Energy.* Provided all other nutrients are consumed in adequate amounts, birds have the ability to adjust their food consumption over a fairly wide range in order to obtain sufficient energy for egg production.

TABLE 5   Moisture as Percentage of Fresh Carcass[a] of Domestic and Game Birds

| | Chickens | | Ring-necked Pheasants | Bobwhite Quail | Turkeys | |
Age	Leghorns	Broilers			Domestic	Wild
8 weeks						
Male	64.8	66.3	72.8	72.0	71.0	73.4
Female	66.8	67.1	72.2	71.2	70.3	72.1
Maturity						
Male	64.5	65.2	65.4	[b]	60.0	64.8
Female	60.2	57.3	63.1	[b]	60.3	63.0

[a]Includes all tissues, except feathers, washed free of digesta.
[b]Bobwhite Quail are mature size at 8 weeks of age.

TABLE 6 Essential and Nonessential Amino Acid Compositon[a] of Domestic and Game Birds

	Chickens				Turkeys	
Amino Acid	Leghorns	Broilers	Ring-necked Pheasants	Bobwhite Quail	Domestic	Wild
Essential						
Arginine	7.8	6.8	7.8	7.4	7.6	7.0
Cystine	2.6	2.4	2.8	2.8	2.4	2.6
Histidine	3.9	4.1	4.5	3.9	3.9	3.7
Isoleucine	3.9	3.9	4.0	3.9	4.2	4.7
Leucine	6.4	6.5	6.3	6.7	6.7	7.4
Lysine	9.4	9.9	10.1	10.0	10.2	8.9
Methionine	1.8	1.9	1.8	2.0	1.9	2.1
Phenylalanine	3.5	3.6	3.5	3.5	3.6	3.6
Threonine	3.4	3.4	3.4	3.5	3.5	3.7
Tryptophan	[1.0][b]	[1.0]	[1.0]	[1.0]	[1.0]	[1.0]
Tyrosine	2.9	3.1	3.1	3.0	2.9	2.9
Valine	4.4	4.4	4.3	4.3	4.3	4.3
Nonessential						
Alanine	6.1	6.0	5.7	5.9	6.0	6.6
Aspartic acid	7.1	7.1	7.4	7.2	7.7	8.4
Glutamic acid	11.4	11.2	12.0	11.7	12.2	13.2
Glycine	9.0	8.1	7.0	7.9	7.3	8.6
Proline	5.6	5.1	5.0	6.4	5.9	5.3
Serine	3.4	3.2	3.4	3.7	3.3	3.5

[a] Percentages of total carcass proteins.
[b] Estimated values given in parentheses.

Thus, the energy content of the diet can range between about 2,500 Kcal and 3,300 Kcal of metabolizable energy per kilogram. Below 2,500 Kcal/kg, the diet becomes so bulky that the birds simply cannot pack enough of it into the digestive tract to obtain the daily energy needs.

2. *Protein and Amino Acids.* To produce an egg, the female bird must have a dependable supply of total protein and essential amino acids to produce a certain sized follicle or yolk in the ovary. During passage of the yolk down the oviduct, the bird must have sufficient available amino acids and proteins to quickly secrete the albumen portion of the egg. Although she can borrow to some extent from her body tissues when the diet is not adequate in protein or essential amino acids, this ability to borrow is very limited and will allow her to produce only one or two eggs—and these eggs will be much smaller than normal. Since the egg has an exact amino acid composition, if

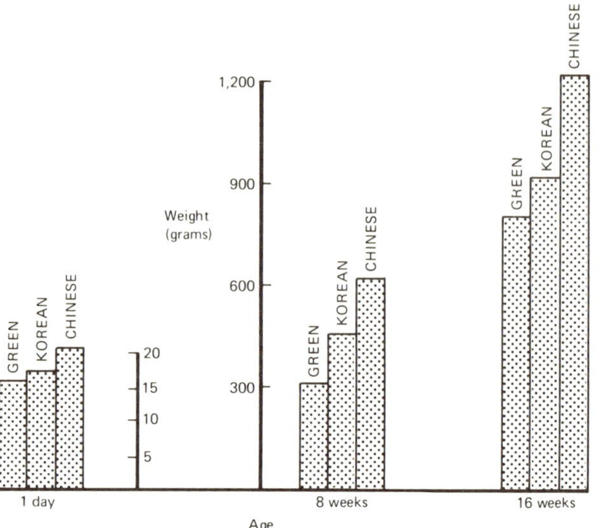

FIGURE 1   Weight as a function of age in pheasants on normal protein diet.

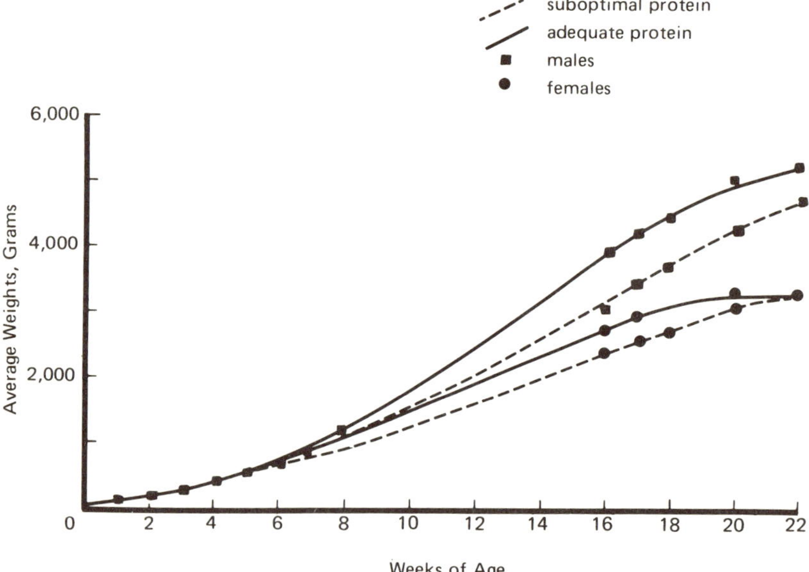

FIGURE 2   Growth of Wild Turkeys (*Meleagris gallopavo*) of both sexes fed two levels of protein.

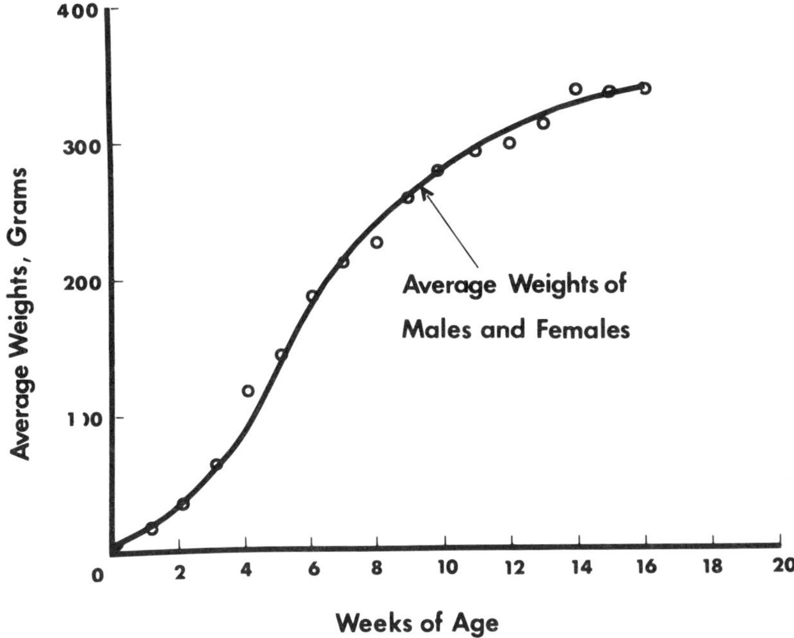

FIGURE 3 Growth curve of Hungarian Partridge (*Perdix perdix*).

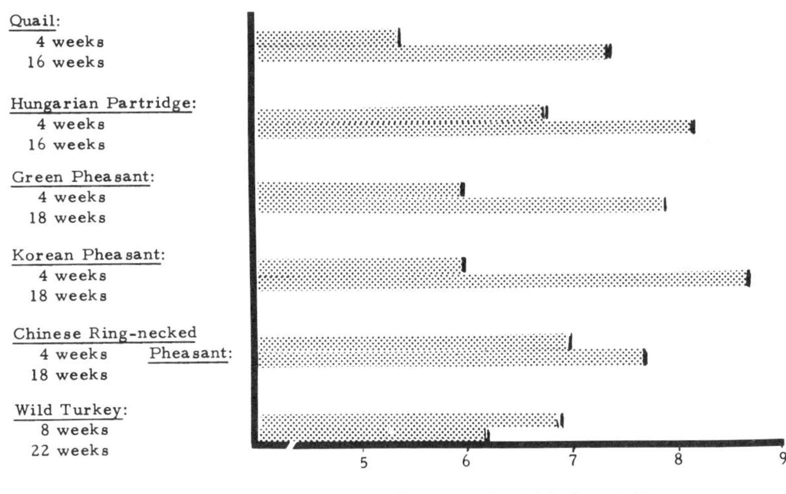

FIGURE 4 Feathering of various game birds as percent of body weight.

even one essential amino acid is missing, no egg can be produced.

3. *Linoleic Acid.* Linoleic acid apparently is required for formation of vital phospholipids necessary for egg production. Thus a total deficiency of this fatty acid eliminates egg production completely. The size of the yolk, and therefore of the egg, also is determined to some extent by the level of linoleic acid in the diet.

4. *Vitamins.* Although some of the vitamins are required for egg production, these requirements are very low in comparison to the levels of the vitamins needed for reproduction. The levels needed for reproduction are shown in Table 1.

5. *Minerals.* Calcium is needed to form the eggshell. In birds laying large numbers of eggs, the dietary calcium requirement ranges from about 2.5 to 3.5 percent. Whether or not those birds that lay only a few eggs require special high-calcium feeds during the egg laying season remains to be determined.

The calcium in the medullary bone is sufficient for two or three eggs. Once this is used up, the blood calcium level drops and a feedback mechanism calls a halt to the production of the follicle-stimulating hormone, thereby causing a cessation of egg production. This may be one of nature's mechanisms for limiting the number of eggs that various species of birds lay before incubation.

6. *Other Minerals.* Phosphorus, sodium, potassium, and certain of the other inorganic elements also are needed at low levels for egg production and reproduction as shown in Table 1. Many vitamin and trace mineral deficiencies produce characteristic deficiency signs and death in developing embryos. Among the most important are vitamin A, vitamin E, riboflavin, pantothenic acid, vitamin B_{12}, manganese, and zinc.

REQUIREMENTS FOR MAXIMUM RESISTANCE TO DISEASES AND OTHER STRESSES

The theories of Selye (1950) regarding stress and the "adaptation syndrome" are well known. Ershoff (1953) studied the relationship between nutritional deficiencies in laboratory animals and their ability to adapt to a number of different stresses. He found that certain deficiencies markedly reduced the ability of animals to adapt to and survive forced swimming in cold water or exposure to sublethal x irradiation. An unidentified nutrient was found in liver that could increase resistance of the animals to these stresses.

The relationship between vitamin A nutrition and coccidiosis was studied by Erasmus *et al.* (1960) to determine the effects of increased dietary levels of the vitamin upon growth, feed consumption, and storage of the vitamin in livers of chicks experimentally infected with oocysts. After infection, chicks receiving 17,600 IU of vitamin A per kg of diet regained their appetites and grew faster than those receiving 1,760 IU/kg, an amount shown previously to be adequate under normal environmental conditions. The infected chicks had lower liver stores of vitamin A than were found in comparable lots of uninfected chicks receiving the same levels of the vitamin and restricted to the same feed intake. An interrelationship between vitamin A nutrition and resistance to coccidiosis also has been shown in Germany by Schoop *et al.* (1954) and by Gerriets (1966), who found that administration of 60 IU per chick per day during a severe attack of coccidiosis almost completely prevented mortality, whereas 100 percent mortality occurred with the vitamin A-deficient diet.

At Texas A&M University, an interrelationship was found between the severity of coccidiosis and the requirement for vitamin K. As much as 8 mg of vitamin K per kg of diet were needed at times for maximum growth and improvement of efficiency of feed utilization. Under similar conditions, however, Harms *et al.* (1962) found the vitamin K requirement to be no more than 1.2 mg per kg of diet.

The effects of diet on the ability of young pheasant chicks to withstand stress of a cold, drenching rain were studied at Cornell. Diets fully adequate for normal growth in young pheasants were not adequate for greatest resistance to the stress of the rain applied experimentally in the laboratory. Even less resistance to this stress was observed when desiccated defatted liver was substituted for fish meal in the practical diet, resulting in poor growth and feathering. A definite improvement in resisting the cold rain was shown as the protein content of the diet was increased from 28 percent to 34 percent by addition of either liver or casein. Whether this improvement was due to increased protein per se or to unknown factors in the supplementary protein sources could not be determined (Scott *et al.*, 1955).

On the basis of Selye's hypothesis, an experiment was conducted wherein a group of 200 pheasant chicks in the brooder house were sprayed each day with a fine mist of cold water beginning at one week of age. The spraying time was gradually increased until by the time the chicks were 3 weeks old they were being thoroughly drenched each day.

It was found that when these chicks were placed on range they were able to withstand a severe overnight exposure to a cold rain without mortality, whereas the control unsprayed birds succumbed after about 2 hours' exposure to the cold rain.

Spraying the chicks in the brooder house markedly increased the severity of coccidiosis. Rather than add a coccidiostat to their daily diet, it was decided to try high levels of vitamin A and vitamin K. These, together with a weekly feeding of a mash containing 0.1 percent sulfaquinoxaline, prevented mortality and allowed the young pheasants to develop an immunity to coccidiosis.

REFERENCES

Erasmus, J., M. L. Scott, and P. P. Levine. 1960. A relationship between coccidiosis and vitamin A nutrition in chickens. Poultry Sci. 39:565–572.

Ershoff, B. H. 1953. Comparative effects of pantothenic acid deficiency and inanition on resistance to cold stress in the rat. J. Nutr. 49:373–385.

Gerriets, E. 1966. Die coccidiostaticafreie Massenaufzucht von Junghennen unter Vitamin A-Schutz. Berliner Münchener Tierärztl. Wochenschr. 79:271.

Harms, R. H., P. W. Waldroup, and D. D. Cox. 1962. Activity of vitamin K_1 and menadione sodium bisulfite complex when measured by mortality of chicks with cecal coccidiosis. Poultry Sci. 41:1836–1839.

Schoop, G., K. H. Wagner, and P. Minners. 1954. Der Vitamin A-mangel als prädisponierender Faktor für Kükenkokzidiose. Monatsh. Tierheilk. 6:154–166.

Scott, M. L., and R. E. Reynolds. 1949. Studies on the nutrition of pheasant chicks. Poultry Sci. 28:392–397.

Scott, M. L., E. R. Holm, and R. E. Reynolds. 1954. Studies on pheasant nutrition. 2. Protein and fiber levels in diets for young pheasants. Poultry Sci. 33:1237–1265.

Scott, M. L., E. R. Holm, and R. E. Reynolds. 1955. Effect of diet on the ability of young pheasant chicks to withstand the stress of cold, drenching rain. Poultry Sci. 34:949–956.

Scott, M. L., F. W. Hill, L. C. Norris, G. F. Heuser, R. E. Reynolds, E. H. Parsons, Jr., and H. E. Butters. 1957. New information on the vitamin A requirements of chickens, ducks and pheasants, p. 132–136. *In* Proceedings of the 1957 Cornell Nutrition Conference for Feed Manufacturers. Mann Library, Cornell University, Ithica, New York.

Scott, M. L., E. R. Holm, and R. E. Reynolds. 1958a. The calcium, phosphorus and vitamin D requirement of young pheasants. Poultry Sci. 37:1419–1425.

Scott, M. L., E. R. Holm, and R. E. Reynolds. 1958b. A study of the phosphorus requirements of young Bobwhite Quail. Poultry Sci. 37:1425–1428.

Scott, M. L., E. R. Holm, and R. E. Reynolds. 1959. Studies on the niacin, riboflavin, choline, manganese and zinc requirements of young Ring-necked Pheasants for growth, feathering and prevention of leg disorder. Poultry Sci. 38:1342–1350.

Scott, M. L., A. Van Tienhoven, E. R. Holm, and R. E. Reynolds. 1960. Studies on the sodium, chlorine and iodine requirements of young pheasants and quail. J. Nutr. 71:282–288.

Scott, M. L., E. R. Holm, and R. E. Reynolds. 1961. Studies on the vitamin K requirements of young pheasants and quail. Poultry Sci. 40:1593–1597.

Scott, M. L., E. R. Holm, and R. E. Reynolds. 1963. Studies on the protein and methionine requirements of young Bobwhite Quail and young Ring-necked Pheasants. Poultry Sci. 42:676–680.

Scott, M. L., E. R. Holm, and R. E. Reynolds. 1964. Studies on the pantothenic acid and unidentified factor requirements of young Ring-necked Pheasants and Bobwhite Quail. Poultry Sci. 43:1534–1539.

Scott, M. L., M. C. Nesheim, and R. J. Young. 1969. Nutrition of the chicken. M. L. Scott and Associates, Ithaca, New York. 511 p.

Selye, H. 1950. The physiology and pathology of exposure to stress. Acta, Inc., Montreal. 1025 p.

DISCUSSION

Adam Watson

I shall comment first on the importance and relevance of laboratory research on nutrition to the field ecologist working with wild populations and, second, on some examples from my own experience in which field work and laboratory work have proved useful when they go hand in hand. My background is research, with a small team in Scotland, on what controls such population processes as reproduction and recruitment in wild populations of Red Grouse (*Lagopus lagopus scoticus*) and Rock Ptarmigan (*L. mutus*), and what controls variations in social status and aggressive behavior, which are the immediate mechanisms for limiting breeding stocks of these two species. At the same time, we use laboratory studies and captive birds for experimental work, but only in close integration with the field studies.

Laboratory work on nutrition has been useful for people who are working mainly with wild populations by providing a general theoretical framework within which the field studies can be placed and useful methods and "knowhow." But it can provide no more than this; if it is allowed to substitute completely for work in the field—a frequent temptation—it is liable to provide misleading and, indeed,, sometimes even ludicrous results. Faced with a problem about populations, behavior, or nutrition of wild animals in the field, a worker who tackles it entirely in the laboratory and then extrapolates to the field without checking in the field is on slippery ground. A similar case is that wherein an ecologist studying population regulation of birds in the field may find the theoretical development of climatology or of plant physiology useful, but he may well have to plough his own new furrow in such fields when he tries to find how field populations are affected by climate or by changes in plants. Figure 1 is a simplified diagram illustrating the differences in approach of most animal nutritionists who have worked with domestic animals and of ecologists who become interested in nutritional problems. The main difference is that the ecological problem of nutrition in wild animals involves a complex network with important feedback effects. Indeed, to emphasize how simple nutrition is and how well nutritionists understand the exact levels needed for growth, reproduction, etc. (e.g., Blaxter, 1970; Scott, this volume) is likely to be misleading if applied to the much more complex wild situation.

Before I begin an appraisal of the relevance of laboratory research on nutrition to field problems, I must be fair and admit that the field ecologist has often appeared a derisory figure to animal nutritionists. In past decades, workers studying wild animal populations tended to consider food merely as the total amount of vegetation or other material—an emphasis still obvious even in the 1950's in textbooks on

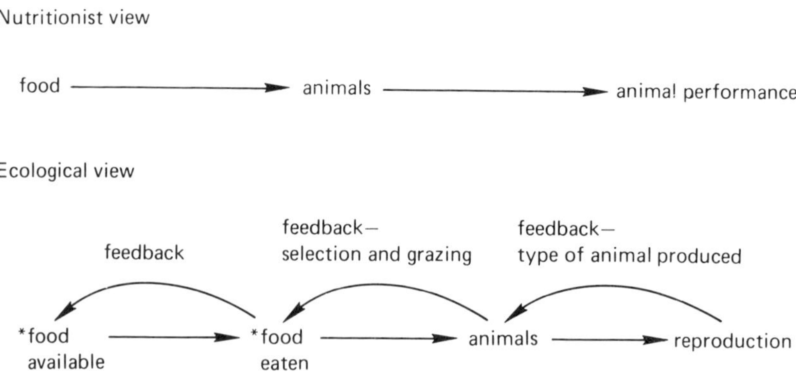

FIGURE 1   The main difference in approach of most animal nutritionists working with domestic animals and of ecologists interested in nutritional problems.
Asterisk (*) means including species composition, amount, and nutritive value.

population regulation. Until the 1960's, field ecologists largely ignored nutrient content, quality of food, digestibility, or the varying nutrient needs of the animal, and some workers came to ludicrous conclusions as a result. To avoid any charge of bias, I will quote from Watson and Moss (1971), "One of the difficulties is that food quality in biochemical or nutrient terms, as distinct from mere quantity of material, has been greatly neglected in ecology, despite its well-established importance for the survival, weight, physiology, reproduction and behavior of domestic animals and man." This neglect was surprising, as a few studies with wild populations (e.g., deer) had already shown how important these aspects were. Fortunately, this situation of neglect has changed radically in the last few years, as recent symposia and papers in the literature amply demonstrate.

Most of our knowledge of nutrition is based on laboratory work with domestic animals, which can be misleading if applied to the field without further qualification. Much of this modern work is aimed at maximizing production of meat, milk, or eggs by maximizing efficiency of food conversion. This is done by varying the constituents of the diet so as to provide a higher output by the animal, much as one might vary the fuel to get better output from a machine. This is a rather far cry from the kind of problem that the field ecologists working with wild populations face. The trend in laboratory work also aids this process of maximizing production by breeding for a more uniform animal that can make a bigger output from a given input, giving a more intensive system of management so as to lower the costs of human labor. There is a great individual variation in a wild stock, and it is unwise to ignore this heterogeneity when considering nutrition in relation to reproduction and other population processes (see Birch, 1971). Ayala (1968) found that different strains of *Drosophila* showed different efficiency at using food and perhaps as a result of this they reached different population ceilings in the same experimental environment. Kolb (1971), working with us, found that Red Grouse hens kept in the same cages and given the same food performed very differently in number of eggs laid, timing of egg laying, survival of chicks, and early growth rate of chicks. Another example wherein it would be unwise to regard all individuals as equivalent is that, in Red Grouse, the lengths of the cecum and small intestine vary between individuals, thus probably affecting how well they make use of a given diet (Moss, 1972a).

If one now considers laboratory work on the nutrition of captive animals taken from wild stock, the dangers of extrapolating to the field are probably even greater, as the pitfalls are less obvious. Much work has been done on finding how fast various species of North American and European deer calves grow on a high plane of nutrition in the laboratory and how well they perform later on, compared to "poor" growth and performance in the wild (e.g., Nordan *et al.*, 1968). This work shows that they grow faster in the laboratory, perhaps a fairly obvious result, but it is not very relevant to an understanding of the field situation. There, the important factors are how the animals can adapt and have viable populations with variable animal types over a wide variety of environments and levels of nutrition. Laboratory work, to be of most use for the field ecologist, must be directed toward such questions.

An example, in birds, of a misleading conclusion from laboratory work is the study by Hill *et al.* (1968) on Ruffed Grouse (*Bonasa umbellus*). Gullion (1970) had stated earlier, as a result of prolonged study in the field in Minnesota, that their main food in winter was buds of male aspen trees. Hill *et al.* gathered branches from male aspens and gave an excess of them to captive Ruffed Grouse that were excluded from other food. The grouse lost weight, which led the authors to conclude that aspen buds could not be their main winter diet. However, Gullion had already stated that the wild birds were very selective feeders between and within trees; clearly, captive birds are denied this selection and forced to feed very heavily on the small sample of branches given to them. Moreover, the captive birds previously had been fed on pellets. They were probably, unlike wild birds, adapted to the higher plane of nutrition afforded by the pellets and so may have been unable to quickly adapt their gut and their gut flora to the aspen diet. The likelihood that the conclusion was incorrect is confirmed by the work of our team concerning Red Grouse and their main food, heather (*Calluna vulgaris*). Red Grouse are highly selective feeders, often eating heather that is much higher in nutritive value than what is generally available to them(Moss, 1972b). When captive Red Grouse are fed on pellets and the diet is then switched suddenly so that they are on a heather sward where they can pick their own heather, the birds lose weight. They regain weight as soon as some pellets are provided. These results with captive Ruffed Grouse and Red Grouse are probably partly due to a prevention of normal feeding selection. But in Red Grouse, and probably in Ruffed Grouse also, birds kept on a pellet diet on a high plane of nutrition have shorter ceca and shorter intestines than do wild birds (Moss, 1972a, b). Some individuals that lose weight less rapidly than others are probably better adapted to deal with the sudden change to a more fibrous and poorer diet.

Another pitfall here is that a captive stock of wild animals often rapidly changes in various ways, due to the change in selective mortality inherent in captivity and to further changes in selective breeding imposed consciously or unconsciously by their keepers. As soon as wild birds are put in captivity, selection for a new type ensues. In captive Red Grouse, Moss (1972a) found that the ceca and small intestines became shorter with each generation removed from the wild. This inherited selective change had reached reductions of 50 percent in the length of the ceca and 72 percent in the short intestine over a few years. Thus, conclusions about nutrition and its effects on performance are likely to vary according to how artificial one's captive stock has become.

The work done by Scott and his colleagues—just described—on the spraying of young pheasants (*Phasianus colchicus*) with cold water might be relevant to the question of whether inclement weather after the eggs hatch reduces the survival of wild young pheasants. This cannot be assumed, and, on present evidence, there is no way of knowing. Scott found that young pheasants fed varying diets varied in their susceptibility to mortality after drenching. This is not a sufficient basis for understanding the field situation, unless it can be shown that the diets of young wild pheasants that suffer high mortality in rain are nutritionally deficient.

Extrapolation from data on a species kept in captivity to the same species in the wild is sometimes misleading for other reasons. Dr. Scott's paper states that "we may have good information concerning the exact nutrient requirements of a particular species, and may be able to formulate a diet containing all of the nutrients in proper amounts"; obviously, however, this information is mainly of use in keeping captives, not in understanding the wild situation. His Table 1 is of great value for birds in captivity but not for those in the field. It does not consider the huge individual variation in wild birds or the interactions of different nutrients and other dietary constituents when, for example, one is reduced and thus has an influence on other constituents. Even in captivity, different constituents of a diet interact.

Dr. Scott's title contains the words "predictive functions" and his paper suggests that Table 1 can be used in "predicting probable effects upon one of these species [of bird] of the elimination of one or more of the food items presently being consumed in the wilds as an important part of the total diet." However, this prediction is unlikely to be simple, as it does not take into account the importance of feeding selection, or the availability of the nutrients in different plants even though one knows the nutrient content (i.e., one has also to know digestibility and other factors), or of adaptation by the birds when one or more of the plants is absent.

A wide adaptation is possible, and good examples are afforded by the work of Gardarsson and Moss (1970) on feeding selection and food availability in Icelandic Rock Ptarmigan and the largely grass diet of a local population of pheasants near Seattle (Rieck *et al.*, 1971). Also, Rock Ptarmigan with longer ceca occur on areas where the diet has lower *in vitro* digestibility (Gardarsson, 1971). Moreover, individual Red Grouse having a small food intake digest a higher proportion of the heather than do those with a large intake; with a small intake, this involves a relatively high rate of digestion of lignin and holocellulose (Moss, 1972b). This digestion of lignin and cellulose is facultative and is not used, or is little used, when intake is very high.

In dealing with wild populations, it is useful above all to study in depth in the field. Questions that are difficult or impossible to tackle in the field can then be considered for the laboratory. The important thing is that they are first asked in the field and are brought to light by the field work. This means a close dialogue between field worker and laboratory worker; ideally, a worker ought to be both. Under this approach, field and laboratory complement one another and each gives rise to new questions for the other. In the small team on which I serve, our nutritionist Robert Moss started as a biochemist but he has increasingly become an ecologist and an ethologist, with wider research interests.

Field work on food and nutrition usually begins with exploratory surveys of vegetation and of the amount and chemical composition of food in relation to underlying soils and rock type, weather, animal stocks, performance of different animal populations, and human exploitation. Much of the research and monitoring work on wildlife management in North America takes this slant. My own work has also involved a lot of this exploratory surveying. For example, it led to correlations between the age, cover, and performance of heather and differences in numbers of Red Grouse within areas and between areas (e.g., Miller *et al.*, 1966; Watson *et al.*, 1966). Another finding was that areas over richer rocks, where the vegetation is of higher nutritive value, support higher breeding stocks of Red Grouse and Rock Ptarmigan, which also rear bigger broods, than do areas over poorer rocks (Watson and Moss, 1972). This even applies to the invertebrate-eating Dotterels (*Eudromias morinellus*), which also rear more young over rich rocks on the Scottish mountain tops than over poor rocks (Watson *et al.*, 1970).

Such exploratory work and correlations are useful, but only as first steps that, if possible, should lead to the framing of hypotheses for testing by experiment. In the Red Grouse, we have done a number of experiments that were suggested by earlier surveys and correlations. These involved manipulating the environment by burning the heather, by fertilizing it, and by excluding the grazing sheep and cattle that are potential competitors and then measuring effects on the behavior and population processes of the grouse (Miller *et al.*, 1970; Watson and O'Hare, 1970).

Another useful approach is to compare results on birds in captivity, where certain conditions can be kept standard, with results in the wild where the same conditions are highly variable. For example, variations in the survival of young Red Grouse depend mainly on pre-hatching factors in the egg, not on weather after hatching. We discovered this by taking eggs from wild Red Grouse, hatching and rearing in captivity, and then predicting and comparing with the survival of wild chicks in the stock from which the sample of eggs was taken. The two were closely correlated (Jenkins *et al.*, 1965), thus showing that prehatching factors account for much of the variability in reproductive success. This early finding has since been confirmed by more intensive experiments with Red Grouse and Rock Ptarmigan. Also, recent unpublished work shows that this prehatching factor is connected with variations in the nutrition of the parent before egg-laying. Poultry researchers have paid less attention to nutritional effects on chick viability than to egg hatchability. This is perhaps not surprising, as domestic chicks tend to survive well after hatching. But this trait may have been highly selected for domestic conditions for hundreds of years. Thus, many factors that are of little consequence in the domestic chicken might be important in wild populations.

In another instance of a field problem giving rise to laboratory work, I noticed that in years when the heather started growing earlier in spring, the Red Grouse later bred more successfully than in years when the heather was late. This finding gave rise to experiments from which we found (Moss *et al.*, 1971) that captive Red Grouse hens lay more eggs when given a small supplement of growing heather in spring than when the supplement is dormant winter heather or when no heather is added to the pelleted diet.

Puzzling results from field experiments may sometimes be cleared up by laboratory work. For example, we found that after strip burning of the heather, Red Grouse increased their breeding stocks via a reduction in territory size 3 years after the first fires (Miller *et al.*, 1970). This was puzzling since regeneration was good in the first year and the nutritive value appeared to be highest then. Captive birds were presented with heather of different ages. They preferred 3- and 4-year-old heather to 1- and 2-year-old (Moss *et al.*, 1972). This preference explains the lag in response by the population after the experiment. This laboratory check of a process that was first suggested by field work on populations has since led to a detailed check in the field, where Savory (in preparation) has measured the feeding rate of wild grouse directly and confirms their preference for 3- and 4-year-old heather. This is a good example of the supporting dialogue between field and laboratory when the two go hand in hand.

Field studies of nutrition can be very useful if one is trying to measure food selection; clearly, selection for quality may be important in subsequent reproductive performance. In Red Grouse and Rock Ptarmigan, we have measured feeding selection by comparing the nutritive value of the food generally available at the feeding site with that in the bird's crop (Moss, 1972b). This work shows that birds tend to be more selective for nitrogen and phosphorus if less of these nutrients is available. Hens are particularly selective in spring before they lay their eggs. But birds are not always as selective as possible, which suggests that the given nutrient or constituent is probably then not in short supply. For example, there is good evidence that Red Grouse do not select food for its energy content in spring.

In recent years, much of the ecological work done in the field and in the laboratory on the nutrition of birds has involved energetics and energy flow. It is worth remembering that other factors are much more important in some species. For instance, comparison of populations on different areas indicates that the secondary production of Red Grouse, Rock Ptarmigan, and mountain hares is not related to the production of energy but to the nutrient content of the food. Thus, studies of energy flow or of plant production are largely irrelevant in providing a useful understanding of what controls such animal populations, and this may well happen more commonly than is generally realized.

REFERENCES

Ayala, F. J. 1968. Genotype, environment, and population numbers. Science 162:1453–1459.

Birch, L. C. 1971. The role of environmental heterogeneity and genetical heterogeneity in determining distribution and abundance, p. 109–126. *In* P. J. den
Boer and G. R. Gradwell [ed.] Dynamics of populations. Centre for Agricultural Publishing and Documentation, Wageningen, Netherlands. 611 p.

Blaxter, K. L. 1970. Chairman's remarks. Symp. Brit. Ecol. Soc. 10:5–6.

Gardarsson, A. 1971. Food ecology and spacing behavior of Rock Ptarmigan
(*Lagopus mutus*) in Iceland. PhD thesis. University of California, Berkeley.
382 p.

Gardarsson, A., and R. Moss. 1970. Selection of food by Icelandic Ptarmigan in
relation to its availability and nutritive value. Symp. Brit. Ecol. Soc. 10:47–69.

Gullion, G. W. 1970. Factors affecting Ruffed Grouse populations in the boreal
forests of northern Minnesota, USA. Trans. VIII Int. Congr. Game Biol. 8:103–
117.

Hill, D. C., E. V. Evans, and H. G. Lumsden. 1968. Metabolizable energy of
aspen flower buds for captive Ruffed Grouse. J. Wildlife Manag. 32:854–858.

Jenkins, D., A. Watson, and N. Picozzi. 1965. Red Grouse chick survival in captivity and in the wild. Trans. VI Int. Congr. Game Biol. 6:63–70.

Kolb, H. H. 1971. Development and aggressive behavior in Red Grouse (*L. lagopus
scoticus*, Lath.) in captivity. PhD thesis, University of Aberdeen, Scotland.
176 p.

Miller, G. R., D. Jenkins, and A. Watson. 1966. Heather performance and Red
Grouse populations. I. Visual estimates of heather performance. J. Appl. Ecol.
3:313–326.

Miller, G. R., A. Watson, and D. Jenkins. 1970. Responses of Red Grouse populations to experimental improvement of their food. Symp. Brit. Ecol. Soc.
10:323–334.

Moss, R. 1972a. Effects of captivity on gut lengths in Red Grouse. J. Wildlife
Manage. 36:99–104.

Moss, R. 1972b. Food selection by Red Grouse (*Lagopus lagopus scoticus*) in
relation to chemical composition. J. Anim. Ecol. 41:411–428.

Moss, R., A. Watson, R. Parr, and W. Glennie. 1971. Effects of dietary supple-
ments of newly growing heather on the breeding of captive Red Grouse. Brit.
J. Nutr. 25:135–143.

Moss, R., G. R., Miller, and S. E. Allen. 1972. Selection of heather by captive
Red Grouse in relation to the age of the plant. J. Appl. Ecol. 9:837–847.

Nordan, H. C., I. MeT. Cowan, and A. J. Wood. 1968. Nutritional requirements
and growth of black-tailed deer, *Odocoileus hemionus columbianus*, in captivity. Symp. Zool. Soc. London 21:89–96.

Rieck, C. R., E. S. Dziedzic, and R. G. Jeffrey. 1971. A high density of pheasants
at Seattle, Washington. Murrelet 52:7–9.

Watson, A., and P. J. O'Hare. 1970. Bringing back the grouse. The Irish Times,
3 September, p. 10.

Watson, A., and R. Moss. 1971. Spacing as affected by territorial behavior,
habitat and nutrition in Red Grouse (*Lagopus l. scoticus*), p. 92–111. *In*
Behavior and environment. Plenum Press, New York. 411 p.

Watson, A., and R. Moss. 1972. A current model of population dynamics in
Red Grouse. Proc. 15th Int. Ornithol. Cong. 15:110–125.

Watson, A., G. R. Miller, and F. H. W. Green. 1966. Winter browning of heather
(*Calluna vulgaris*) and other moorland plants. Trans. Bot. Soc. Edinburgh
40:195–203.

Watson, A., N. Bayfield, and S. M. Moyes. 1970. Research on human pressures on Scottish mountain tundra, soils, and animals, p. 256–266. *In* W. A. Fuller and P. G. Kevan [ed.] Productivity and conservation in northern circumpolar lands. IUCN Publ., New Series No. 16. International Union for Conservation of Nature and Natural Resources, Morges, Switzerland. 344 p.

DISCUSSION

Ivan Assenmacher

Although I am not by any means a nutritionist, I would like to add to the report of Dr. Scott two comments and to ask two questions. All are related and concern the possible role of food availability and food consumption as a clue to the regulation of the avian reproduction cycle.

In Professor Marshall's (1970) last scientific report it was demonstrated that, excluding photoperiodicity, at least eight other external factors may be influential in the control of sexual cycles in birds and animals. Among these was food supply. In fact, food supply can be considered from two different standpoints: seasonal fluctuations in food availability and food consumption.

For birds living under natural conditions, the available food is obviously subject to marked seasonal variations. Several observations have pointed to a possible causal link between this particular environmental factor and the breeding cycle. Whether the natural cycle of food supply will act, for a given species, as a main or as a secondary factor in the regulation of the breeding cycle, depends on the relative amplitude of fluctuations of the various external factors, to which the bird is submitted, and the relative sensitivity of the bird to those factors; the latter point is of major importance in ecological adaptation. A clear example of adaptation to food supply, as a main regulatory factor in reproduction, has been studied by Wells (1966), who was unable to detect any relationship between the reproductive cycle of tropical Munias living at 3°8′N, with classical clues such as day length and rainfall, but noticed a close correlation with food abundance associated with agricultural cycles.

Now food supply has a quantitative and a qualitative aspect.

Since the pioneer work of Grandis (1889) on the pigeon, the deleterious action of caloric restriction on gonads has been many times verified. Figure 1 shows such observations in the domestic male Mallard. In a first experiment several groups of sexually reactive ducks were submitted to continuous light (LL), together with various levels of food supply. Caloric restriction to 50 percent of *ad libitum* fed controls had no effect on gonadal growth, and a decrease to 18 percent of the control food intake induced only partial inhibition of the gonadal stimulation, which still led to full spermatogenesis. Even 17 days of complete starvation was inadequate to block completely the photo–gonadal response. However two additional points can be made: Under natural conditions, where light-induced gonadal stimulation extends over a much longer time span than in this experiment, starvation has more drastic effects on the gonads and leads to complete testicular atrophy (Figure 2). Even under forced gonadotropic stimu-

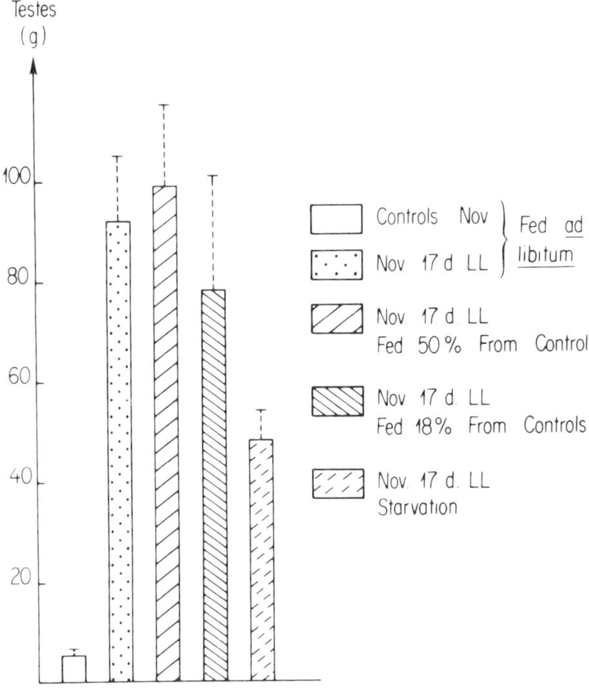

FIGURE 1 Effects of malnutrition on the light-induced gonadal development in the domestic (Mallard) drake. LL: continuous light.

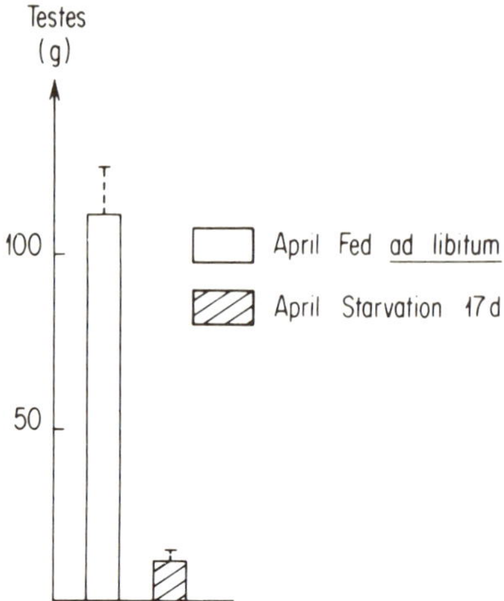

FIGURE 2   Effects of starvation on the gonads in the domestic (Mallard) drake, during the breeding season.

lation by artificial "long days" (18L–6D), the endocrine function of the testis is more sensitive to starvation than is spermatogenesis (Table 1). The plasma testosterone level is as low as in sexually quiescent controls; since the metabolic clearance rate (MCR) of the hormone is unaltered, this low plasma content must reflect blockade of testicular secretion.

In the chick, caloric restriction to 60 percent of the controls induced testicular atrophy, with a drastic decrease of sperm production (Parker and Ascott, 1964), and reduced egg production (Card and Nesheim, 1966).

As could be expected, restricted food intake has more drastic gonadal effects in passerine birds. In his classical observations on House Sparrows (*Passer domesticus*), Vaugien (1960) has shown that a 25 percent food restriction reduced the testis weight response to light stimulation (LL) to about 20 percent of the controls fed *ad libitum* and inhibited full spermatogenesis.

However, it must be kept in mind, that, although severe malnutrition usually leads to almost complete gonadal atrophy, the general endocrine balance induced by starvation cannot be considered as one of pseudohypophysectomy, as was made clear by the classical work of Mulinos and Pomerantz (1940) in rats. In the duck experiment shown in Figure 2, in which 17 days of starvation in spring resulted in gonadal atrophy, the adrenal cortical function was rather stimulated, and, for the thyroid, the radioiodine uptake and the peripheral metabolism of thyroxine were normal; only the protein-bound iodine (PB [131] I) was decreased (Table 2). In the case of severe lack in available food, the reproductive function obviously appears from an individual standpoint as a luxury and is therefore cut off, whereas other endocrine functions, which regulate the general metabolism, are recruited and oriented in such a way as to offer the bird a maximal chance of survival.

Turning toward qualitative requirements of the gonadal function, a number of studies have established the essentiality of protein, or better of specific amino acids intake, as well as of the essential fatty acids, minerals, and vitamins (see Van Tienhoven, 1968). These requirements, which have been analyzed by Dr. Scott, apply mainly to egg production. I would like to know from Dr. Scott if, among the various nutritional components that were listed in his studies, a tentative classification may be drawn for the relative importance of all these nutrients as limiting factors for testicular and for ovarian development.

My second comment concerns the other aspect of food supply as a limiting factor of reproduction, namely, food consumption. The seasonal hyperphagia of migratory bird species, which contributes to their premigratory storage of energy, is well known (King, 1970).

TABLE 1 Comparative Effects of Starvation on Light-Induced (18L–6D) Testis Development and Testosterone Secretion in the Domestic (Mallard) Drake[a]

Treatment	Testes (g) m ± SD	Plasma Testosterone (ng/10ml) m ± SD	Metabolic Clearance Rate Testosterone (ml/min/kg) m ± SD	Secretion Rate Testosterone (ng/min/kg)
Controls (Nov.–Dec.)	1.6 ± 0.2	4.6 ± 1.7	104 ± 31	47.8
Nov.–Dec. 21d.: 18L–6D	72.6 ± 10.0	29.0 ± 8.7	96.5 ± 43	280.0
Nov.–Dec. 21d.: 18L–6D +17d. Starvation	16.8 ± 6	2.0 ± 1.2	117 ± 20	23.4

[a] From M. Jallageas and I. Assenmacher (unpublished data).

TABLE 2   Physiological State of Adrenal Cortex and Thyroid after 17 Days
Starvation Inducing Testicular Atrophy in Domestic (Mallard) Drakes

Adrenal	Plasma Corticosterone (μg/100 ml)	$t^{1/2}$ (min)[b] of ^{3}H-Corticosterone
Controls	12.28 ± 0.42	10.69 ± 0.25
17d.–Starvation	19.75 ± 1.32	12.00 ± 0.52

Thyroid[a]	Thyroid ^{131}I Uptake (% i.d. at 24 hr)	PB ^{131}I (% i.d. at 24 hr)	$t^{1/2}$ (hr)[b] of ^{125}I-Thyroxine
Controls	14.61 ±1.89	48.34 ±3.06	3.52 ±0.27
17d.–Feed restriction to 50%	18.54 ±0.10	32.22 ±3.17	3.80 ±0.20
17d.–Starvation	18.50 ±3.38	8.27 ±4.97	3.62 ±0.25

[a] From H. Astier and I. Assenmacher (unpublished data).
[b] $t^{1/2}$, biological half-life.

However, even nonmigratory species display, even under optimal *ad libitum* feeding conditions, significant fluctuations in their feeding behavior. In ducks reared under field conditions, the daily individual food intake shows a significant 9 percent decrease during the breeding season and a 17 percent fall during light-induced (LL) gonadostimulation (Assenmacher *et al.*, 1965). In the House Sparrow, Vaugien (1960) observed even higher seasonal variations in the food intake, since sexually active birds exhibited a 25 percent decrease of their daily food intake versus sexually quiescent controls. The normally occurring seasonal restriction of food consumption thus attained the threshhold that has been shown to induce testicular atrophy in this species. The physiological origin of the behavioral restriction of food consumption during the breeding season is unclear. The extent of the resulting malnutrition, however, could contribute in the most sensitive bird species to induction of sexual regression (Vaugien, 1960). This leads me to my second question: What is actually known as to the selective drive of birds toward a peculiar diet during the breeding cycle, apart from the well-known search for calcium in laying females?

REFERENCES

Assenmacher, I., A. Tixier-Vidal, and H. Astier. 1965. Effets de la sous-alimentation et du jeûne sur la gonadostimulation du canard. Ann. Endocrinol. (Paris) 26:1–26.

Card, L. E., and M. C. Nesheim. 1966. Poultry production. Lea and Febiger, Philadelphia. 850 p.

Grandis, T. 1889. La spermatogenèse durant l'inanition. Arch. Italiennes Biol. 12:215–222.

King, J. R. 1970. Photoregulation of food intake and fat metabolism in relation to avian sexual cycles, p. 365–385. *In* J. Benoit and I. Assenmacher [ed.] La photorégulation de la reproduction chez les oiseaux et les mammifères. Editions du C.N.R.S., Paris. 588 p.

Marshall, A. J. 1970. Environmental factors other than light involved in the control of sexual cycles in birds and mammals, p. 53–69. *In* J. Benoit and I. Assenmacher [ed.] La photorégulation de la reproduction chez les oiseaux et les mammifères. Editions du C.N.R.S., Paris, 588 p.

Mulinos, M. G., and L. Pomerantz. 1940. Pseudohypophysectomy. A condition resembling hypophysectomy produced by malnutrition. J. Nutr. 19:493–504.

Parker, J. E., and G. H. Arscott. 1964. Energy intake and fertility in male chickens. J. Nutr. 82:183–187.

Van Tienhoven, A. 1968. Reproductive physiology of vertebrates. W. B. Saunders Co., Philadelphia. 498 p.

Vaugien, L. 1960. L'appétit du Moineau domestique s'abaisse graduellement durant le cycle sexuel jusqu'à une valeur néfaste à la gonadostimulation expérimentale par illumination. C. R. Acad. Sci. (Paris) 251:1570–1572.

Wells, D. R. 1966. Breeding season, gonad cycles and moult in three Malayan Munias (Aves=Estrildidae). PhD thesis, University of Malaya.

INFORMAL DISCUSSION

SCOTT: Doctor Assenmacher has asked about nutrients other than calcium, but that is the only one I know to have an effect upon feed selection. As for calcium itself, I have already mentioned the deposition of medullary bone. And Taylor, in England, has shown in chickens that, if mature hens are laying, and calcium is then removed from the diet, most of them will very quickly stop laying eggs. If these hens are then given follicle-stimulating hormone, they will continue to lay 16–17 eggs in succession and will use the calcium from the cortical bone until they go completely off their legs. But this is not the normal situation—normally, as soon as the medullary bone is used up, if there is no calcium coming from the food, the calcium level of the blood drops sufficiently to stop the production of the follicle-stimulating hormone, and egg production ceases.

So it seems to me that those birds that do not have access to very much calcium in their diet are limited to the production of one to two eggs or, at most, about three. This is the number of eggs that hens can produce if they have a maximum amount of medullary bone, but do not get any extra calcium in their diet at onset of egg production. This factor may be one of those that limit egg production in certain birds.

Calcium deficiency cannot account for the fact that fish-eating birds, which receive abundant calcium from the bones of fish, also may lay only three or four eggs. But it may be a limiting factor in some species.

As for protein, we have found that a severely deficient hen will lay a few eggs and, in so doing, will lose 200 g of body weight. Thereupon egg production will drop off. Thus protein can also be a limiting factor in the number of eggs that will be laid, even though a hen on a diet very low in protein will lay a few.

Concerning the effects of nutrition on testicular development, the only experiment I can remember was conducted by Dutcher* many years ago. He showed that, with a vitamin A-deficient diet, there was very poor testicular development. Because he used alfalfa meal as his source of vitamin A in those days, we don't know for sure that he was studying only vitamin A. Other nutrients also could have been provided by the alfalfa meal. When he altered the treatments it was possible to show that cockerels with poor testicular development, after receiving vitamin A from alfalfa, showed an increase in testicular development.

FARNER: Concerning predictive functions, I would like to exercise a prerogative of the Chairman to propose, cautiously, a hypothesis that, in effect, Professor Assenmacher has already proposed. I think that it is clear, largely from anecdotal evidence but also from the very good data of Professor Assenmacher, that a state of inadequate food availability and intake constitutes a source of information to the bird in the developmental phase of the reproductive cycle that, in essence, says it is better to use the energy for its own survival than to run the chance of spending it unsuccessfully in reproduction.

We have tested this with White-crowned Sparrows (*Zonotrichia leucophrys gambelii*) and obtained completely negative results, but I think in a way they are consistent with the hypothesis. What we did was to photostimulate White-crowned Sparrows to initiate gonadal growth and to induce migratory state and then reduced the food intake to the point where they were almost at the level of death by starvation. This did not affect gonadal development at all.

---

*Dutcher, R. A., and S. D. Wilkins. 1921. Vitamin Studies, VIII. The influence of fresh alfalfa upon the weight of testes in single comb leghorn cockerels. Am. J. Physiol. 57:437.

But, I need only remind you that what the White-crowned Sparrow must recognize is that a deficiency of food supply at some place along the migratory route is not an indicator of a deficient food supply in the breeding area. So I think that perhaps these negative results are still consistent with this hypothesis and, certainly, open the way for further experiments.

NORTON: I have a question for Doctor Scott. You mentioned protein denial to, I believe, Wild Turkeys, and the fact that they could catch up with their non-deprived counterparts or controls. Have you any evidence on the effect of a general denial of an adequately composed diet, and whether, following general starvation, growing birds of one kind or another can catch up to controls?

SCOTT: Yes, there has been quite a lot of work done on restricting complete diets to a low level. The restricted birds do catch up. This is one of the ways, by restricting the complete diet, that commercial poultrymen try to hold pullets back, i.e., to delay sexual maturity, so that the pullets will be a little bit older when they begin to lay. It is more effective than providing a diet with some kind of an odd deficiency.

When the hens are returned to a complete diet, they do catch up; they weigh at the end of the year about the same as those that were not restricted and lay just about as many eggs.

MURTON: I have three points that are relevant to this theme. The first concerns the effect of photoperiod on the nutritional state of the animal. A bird held on an 8-hr day must eat for its immediate requirements and must also accumulate sufficient reserves, in the form of crop stores or diurnal fat deposits, for the long night; if not, it must break down its protein reserves overnight. A bird kept on a 16-hr day would not have to prepare for a long night and might well use any energy bonus for gonad growth. Recently,* we have found that Collared Turtle Doves (*Streptopelia decaocto*) kept on light schedules of 6L–16D ½L–1½D eat significantly more than do others that are kept on a schedule of 6L–2D ½L–15½D and also exhibit more gonad development. Obviously, from such data it cannot be argued that gonad growth depends on the photoperiod anymore than it does on the extra nutrition. It transpires, however, that birds on a schedule of 6L–8D ½L–9½D achieve still more gonad development, but eat the same amount as do the short-day subjects. Since various circadian schedules stimulate endocrine mechanisms that variously cause gonad growth, fat deposition, body moult, and other phenomena, a more fully controlled photoperiod experiment is needed to clarify the causal factors involved.

My second point concerns the Wood Pigeon (*Columba palumbus*), which has a photoperiodic mechanism that results in gonad recrudescence and the onset of breeding in spring.† At this season, the birds possess fairly large lipid

---

*Murton, R. K., N. J. Westwood, and R. J. P. Thearle. 1973. Food supply and stress factors in photo-period experiments. Ibis 115:132–134.

†Lofts, B., R. K. Murton, and N. J. Westwood. 1967. Photoresponse of the Wood Pigeon *Columba palumbus* in relation to the breeding season. Ibis 109:338–358.

reserves and it seems that these can be used partly for egg production, as a result of which the birds produce their heaviest eggs at this time of the year. Egg weight declines as the season progresses and is relatively low in July and August. But this is just when most Wood Pigeons breed in the agricultural portions of Britain and when breeding success is highest, because now the birds have access to the ripe cereal crops. Although the birds have modified their breeding season to suit a changed environmental food supply, they have not yet evolved modified physiological mechanisms—these mechanisms are still adapted to the conditions under which the species evolved. It follows that experimental subjects could manifest rather a different physiological state as between the beginning and the end of the season and we must be cognizant of such factors in designing experiments.

The third point again concerns the Wood Pigeon, which invariably lays two eggs to a clutch. Total clutch weight may vary by as much as 40 percent, yet there is less than a 5 percent variation in egg weight within clutches. Heavy eggs are more likely to hatch than are light eggs, and those weighing less than 15 g virtually never succeed. What prevents a female, with the potential of laying a clutch of two light eggs, from producing instead one good egg only, or one good and one very poor egg? The reason seems to be that two quite distinct selective factors are involved: Clutch size is related to the brood size that is on average most productive,[*] whereas the capacity to produce an egg of given size depends partly on stored reserves and partly on what the bird obtains during the immediate period of egg formation.

W. O. WILSON: Some interesting work on the effect of vitamin D_3 deficiency on reproduction in pullets has been reported, showing that a deficiency of vitamin D results in a marked decrease in egg production. The reproductive organs do not atrophy, even though egg production is decreased.[†] Turk and McGinnis suggested that vitamin D is involved in regulating ovulation and/or follicular growth in the laying hen.

Concerning specific nutritional deficiencies in Japanese Quail (*Coturnix coturnix japonica*), Robinson *et al.*[‡] studied the effect of protein deprivation at different ages. They examined the type of growth that resulted to determine whether it was a function of increased population of cells (hyperplasia) or an increase in the size of the individual cells (hypertrophy). At hatch, *Coturnix* chicks were fed a poult starter (28 percent protein). Beginning at 6 days of age, four similar groups were separated from the original population and fed a complete ration, except that it contained only 9 percent protein. At 13 days, the four groups were returned to the 28 percent protein

*Lack, D. 1954. The natural regulation of animal numbers. Clarendon Press, Oxford. 343 p.

†Turk, J. L., and J. McGinnis. 1964. Influence of vitamin D on various aspects of the reproductive process in mature hens. Poultry Sci. 43:539–546.

‡Robinson, D. W., P. Vohra, D. W. Peterson, and K. Hazelwood. 1971. Responses to nutritional deprivation in the Japanese Quail (*Coturnix coturnix japonica*). Brit. Vet. J. 127: 384–393.

ration in consecutive weeks. In this work it was observed that, under conditions of deprivation, all of the tissues exhibit cellular hyperplasia and are susceptible to nutritional deficiencies that restrict their growth. When a balanced diet is again fed, the tissues resume growth by cellular hyperplasia, except for the brain, which was irreversibly stunted. Other tissues studied included liver, kidney, heart, and pectoral muscles.

A second report* indicated that sexual maturity, as measured by 50 percent lay, was delayed by approximately the length of time that the birds were kept on the early growth-restricting regimens.

RUSSELL: Doctor Scott gave evidence that when pheasant chicks were maintained on a low-protein diet, feathers developed poorly. Subsequently, when these chicks were wetted, substantial numbers in the poorly feathered groups died. There is indirect evidence that, in nesting and rearing seasons of below normal rainfall, pheasant reproductive success is below normal. If, by chance, protein availability is in some way related to the amount of rainfall during a preceding period—perhaps through an influence on insect population densities—it might well be that, in abnormally dry nesting and rearing seasons (hence, abnormally low protein availability during rearing), a significant portion of the pheasant chick population is poorly feathered. Then, when a heavy rainfall does occur, chick mortality rates might be abnormally high, which is seen as a year of below normal reproductive success.

*Morse, K., and P. Vohra. 1971. The effect of early growth retardation of *Coturnix* (Japanese Quail) on their sexual maturity. Poultry Sci. 50:283–284.

Energetics of
Reproduction
in Birds

JAMES R. KING

Ecologically or demographically significant reproduction culminates
in self-sufficient offspring of adult size that are themselves capable
of reproduction if appropriately stimulated. The energy costs of
reproduction stem not only from gonadal events but also from the
accompanying behavioral patterns. Although behavioral aspects of
reproduction will be considered elsewhere in this symposium, the
energy requirements they impose cannot be satisfactorily evaluated
at present. The following account, therefore, necessarily focuses attention only on the ways in which absolute food shortage or relative
food shortage (as during periods of thermal stress, for instance) may
curtail or stop gonadal function and incubation.

The scientific literature is replete with reports of correlations between apparent food shortage and reduced reproductive activity or
success [for recent examples among birds see Miller *et al.* (1970),
Southern (1970), Bengston and Ulfstrand (1971)]. Most contemporary theories of the evolution of reproductive potential in birds,
in fact, identify food supply during the breeding season as a major
selection pressure (Lack, 1954, 1968; Immelmann, 1971; Cody,
1971), although certainly not the only one (von Haartman, 1971).

No free-living avian species has been adequately investigated with respect to its complete energy budget during the reproductive period. Extensive data are available only from domestic birds and from the few wild or semidomesticated species that can be induced to breed in captivity. Extrapolations from these data are the basis of many of the estimates of energy costs presented here. It therefore is apparent at the outset that these estimates can serve, at best, as a fragmentary model that will help to identify patterns and problems, to be modified and replaced as soon as better data from wild species become available.

CORRELATIONS OF GONADAL FUNCTION AND THE PLANE OF NUTRITION

The energy costs of gonadal function include the costs of growing the tissues themselves and of producing semen and eggs. Investigations of birds that were starved or fed a ration below the maintenance level show that nutritional deficiency strongly affects the endocrine stimulus of gonadal growth, as well as the level of productivity. Except in very carefully designed investigations, severe caloric restriction may be accompanied by debilitating restriction in essential nutrients. Although the term "caloric restriction" is used below, it should be understood that the investigations mentioned cannot discriminate between the effects of caloric deficiency and the potential effects of deficiency in essential nutrients.

THE MALE

The postnatal growth of testes in chickens is significantly retarded by a diet containing only about 50 percent of the *ad libitum* caloric supply (Breneman, 1940). In adult chickens, semen volume, sperm count, and fertilizing capacity are progressively reduced during the feeding of rations affording 42–72 percent (usually about 65 percent) of *ad libitum* intake (Parker and McSpadden, 1943). The effects of food shortage in these experiments were less prominent in spring than at other times of the year (cf Assenmacher *et al.*, 1965). Parker and Arscott (1964) reported similar results in groups of chickens fed 81 percent and 62 percent of *ad libitum* intake. Although there was a steady decrease of semen volume and fertilizing capacity, there were no significant differences in the hatchability of eggs fertilized by semen obtained from any group. This indicates that normal sperm were produced, although in smaller than normal amounts. The experimental birds became essentially infertile when the loss of body weight reached 30 percent of initial body weight.

In domestic pigeons, Marrian and Parkes (1928) found that testicular weight decreased to about one third of the control level (birds feeding *ad libitum*) and that spermatogenesis ceased, within 11 days after the onset of fasting in birds fed only water and 1 g of yeast extract daily. The body weights of the fasting birds had decreased by this time by about 35 percent of the initial weight.

Domestic ducks supplied with water but no food for 17 days in spring, when the testes were of maximum weight, showed testicular involution comparable to that following hypophysectomy (Assenmacher *et al.*, 1965). Testicular involution was accompanied by inactivation of the follicle-stimulating-hormone(FSH-), luteinizing-hormone(LH-), and prolactin-producing cells of the hypophysis and by neurosecretory stasis in the neurohemal interface of the median eminence. The authors interpret the result to indicate that the testicular effects of inanition are exerted primarily by reducing or blocking the output of hypothalamic gonadotropic-releasing factors. Additional experiments in this series were conducted in autumn, when testicular function is normally minimal in ducks but is sensitive to photostimulation. Testicular and hypothalamo–hypophysial functions were studied in controls (natural photoperiod) and in groups of ducks photostimulated by continuous illumination and fed *ad libitum* (photostimulated controls), at levels of 50 percent and 18 percent of *ad libitum* intake, and at zero caloric intake. The ducks on 50 percent and 18 percent of *ad libitum* intake showed no significant depression of testicular function at the end of 17 days, while those on zero caloric intake showed testicular weights significantly below those of photostimulated controls (48.1 g versus 78.9 g, or 0.0213 versus 0.0305 percent of final body weight) but significantly above those of natural controls (48.1 g versus 5.43 g, or 0.0213 versus 0.00128 percent of final body weight). Hypothalamo–hypophysial function was suppressed noticeably only in the ducks kept on zero caloric intake. All groups of photostimulated birds produced spermatozoa. The ducks on zero caloric intake had lost an average of 24 percent of initial body weight at the end of 17 days.

Practically no data are available on the effects of caloric restrictions on testicular function in wild species of birds. Vaugien (1959a) asserted that the testicular response to photostimulation in Linnets

(*Carduelis cannabina*) was reduced if food intake was also reduced (by an unspecified amount). These and additional experiments with House Sparrows (*Passer domesticus*), involving peculiar feeding regimes that depressed or enhanced testicular function (Vaugien, 1959b, 1960a,b), are of considerable interest, but the effects are not necessarily related directly to the energy supply. It is possible that they result from the entrainment of synergistic neuroendocrine cycles in the manner shown by Meier *et al.* (1971) in other passerine species. Kendeigh (1941) suggested that the caloric supply rarely, if ever, limits testicular function of House Sparrows under natural conditions.

Environmental factors that potentially affect energy balance may slightly advance or delay the seasonal advent of spermatogenesis in wild birds. Low air temperature is known to slow testicular growth slightly in both experimental (Farner and Wilson, 1957; Engels and Jenner, 1956) and natural situations [Marshall, 1949; Lofts and Murton, 1966; for review, see Immelmann (1971)]. In view of the slight energy costs of testicular growth, it seems probable that the modifying effect of air temperature is exerted through the neuro-endocrine system and modulations in the output of gonadotropins.

In summary, the evidence available from a few species of birds indicates that testicular function is highly resistant to caloric deficiency. In already functioning testes, depression or failure of spermatogenesis occurs rather quickly in response to complete starvation, but develops relatively slowly in the face of even severe caloric deficiency and persists until the loss of body weight approaches 30 percent of normal weight.

THE FEMALE

Several studies of domestic fowl that were restricted during postnatal growth to 70–80 percent of the ration necessary for maximal growth rate show that sexual maturity (onset of egg laying) is typically delayed by about 2 weeks. Subsequent egg production, however, is not affected as to number and size of eggs if the hens are returned to full feed during the laying season (Singsen *et al.*, 1954; Davis and Watt, 1955; Schneider *et al.*, 1955; Milby and Sherwood, 1956; MacIntyre and Aitken, 1959; Gowe *et al.*, 1960). Caloric restriction to 88 percent of full feed during the laying period caused a slight decrease of egg production (206 versus 233 eggs/hen-year) but caused no decrease of egg weight (Walter and Aitken, 1961). More severe caloric restriction in domestic fowl may have more prominent effects. Restriction

of hens to 70 percent of full feed reduced production significantly
from 83 to 55 eggs/hen per 17 weeks in one experiment and from 71
to 44 eggs/hen per 17 weeks in a second experiment, but had no effect on egg weight (Heywang, 1940). In contrast, Scott and Payne
(1941) found in domestic turkeys that restriction of hens to 61 percent of full feed during the laying period had no significant effect on
either productivity or egg size, but significantly reduced the hatchability of the eggs (from 65.4 to 54.1 percent). Restriction to 78
percent of full feed in a second experiment produced no statistically
significant effect on any index of reproductive performance.

Caloric shortage in Ring-necked Pheasants (*Phasianus colchicus*)
had a more drastic effect than did comparable levels of restriction in
chickens (Breitenbach *et al.*, 1963). Hen pheasants were kept experimentally on a maintenance level of about 72 kcal/day, or about
60–75 percent of the *ad libitum* intake during the laying season
(Figure 1). The seasonal onset of egg-laying was delayed by about
1 month in the experimental birds, compared to controls, and laying
was terminated earlier. Egg production in the limited-intake birds was
only 9 percent of the number of eggs layed by the control birds,
averaging 7.3 eggs/hen during the season. The eggs that were laid by
the experimental birds weighed slightly but not significantly less than
those of the control birds, reflecting the same pattern of conservation
of size at the expense of numbers that is reported for domesticated
species confronted with caloric deficit.

Ovarian degeneration begins, and ovulation ceases, within 2 or 3
days after the onset of complete caloric deprivation in domestic fowl,
and within a week the follicles become atretic. Injection of gonadotropins prevents these changes (Hosada *et al.*, 1955) and may even
maintain regular ovulation for 8 or more days after the withdrawal
of food (Morris and Nalbandov, 1961). This indicates, as in the case
of testicular involution (Assenmacher *et al.*, 1965), that the primary
effect of food shortage is at the level of the hypothalamo–hypophysial
system.

Only indirect evidence is available concerning the effects of caloric
deficiency on ovarian function in wild species of birds under natural
conditions. Investigations of several species have shown that there are
correlations between the onset of the egg-laying season and the air
temperature during the 3 or 4 previous days (e.g., Kendeigh, 1941,
1963; Krüger, 1965; Tanner, 1966; Pinowski, 1968). A return of cold
weather after the onset of egg laying often is correlated with arrest of
follicular development and with interruption of egg-laying (e.g., Nice,

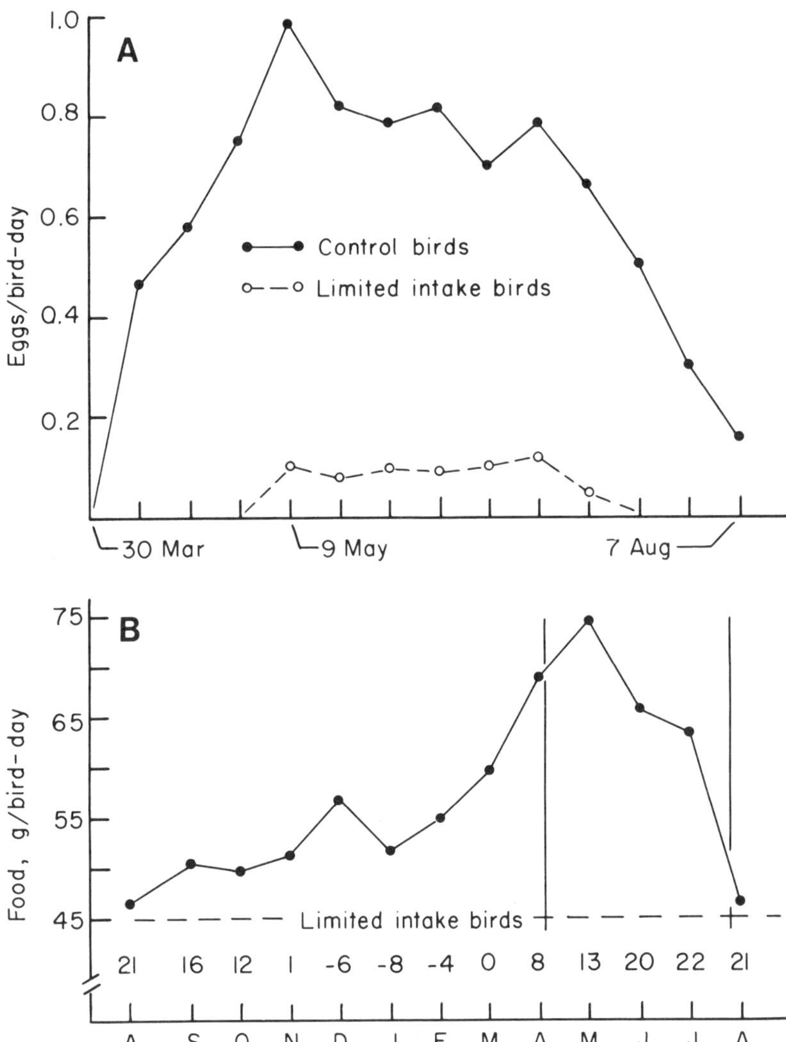

FIGURE 1 (A) The relationship between egg production and food consumption in caged Ring-necked Pheasants in Wisconsin. The "limited intake birds" were restricted to 45 g of feed/bird/day. The row of numerals above the abcissa in graph B shows the mean monthly air temperature. The vertical lines in (B) demark the egg-laying season, shown above. Modified from Breitenbach *et al.* (1963).

1937; Myres, 1955). It seems most plausible that these responses result from reduction of gonadotropin output in reaction to thermoregulatory stress, rather than to direct limitations of caloric supply in growing tissues.

In both male and female birds, the reproductive consequences of caloric deficit appear to follow the pattern that is typical of mammals (Samuels, 1946; Moustgaard, 1959): Food shortage in young animals delays sexual maturity and suppresses gonadal function. It reduces fertility and productivity in adults and may stop gonadal function entirely if the deficit is severe and prolonged. Adult females are affected more promptly and more drastically than are adult males. Most of the available evidence indicates that the hypothalamo–hypophysial axis is the primary site of functional impairment. Gonadal function will continue, supplied from endogenous energy sources, for an unknown period of time during starvation if supported by hormone therapy. Refeeding at a normal level of caloric intake following restriction restores normal gonadal function. There is no evidence of permanent impairment of function. At the risk of oversimplification, it can be said that the neuroendocrine system monitors the level of caloric intake in relation to expenditure and, in the face of caloric deficit, reduces or suspends somatic activities that are not essential to the survival of the individual. In effect, adequate energy resources must be available from the environment or from endogenous stores if reproduction is to occur at a normal level.

THE ENERGY COSTS OF GONADAL FUNCTION

Data now available are inadequate to calculate the exact caloric costs of gonadal growth, maintenance, and production. Nevertheless, it is possible to make some estimates that are accurate enough to reveal the general magnitude of these expenditures and to aid in a preliminary analysis of the problems confronting birds in coping with the caloric costs of reproduction. In the following estimates, it is assumed that the caloric content of gonadal tissue is 2 kcal/g wet weight [extrapolated from data of Ricklefs (1967), Brisbin (1968), Myrcha and Pinowski (1970)] and that the partial (net) efficiency of synthesis is 70 percent (Brody, 1945; Kleiber, 1961). Additional assumptions will be identified as encountered. When necessary, the wet weights of the fresh reproductive organs were estimated from the weights of the histologically fixed organs, using the equation: $y = 0.364 + 1.388x$, where y and x are the weights (mg) of the fresh and fixed organs, respectively ($n = 80$, $r = 0.997$; Kern, 1970).

THE MALE SEX ORGANS

Published data on the seasonal increase of testicular weight have been treated according to the assumptions mentioned above; the results are shown in Table 1. The estimated basal metabolic rate (BMR) is also shown as a caloric reference. The ratio of the total caloric costs of testicular growth to the BMR indicates, in general, the relative cost among species, but provides no information about the daily fraction of BMR required by testicular growth. The daily rate at which testicular tissue is synthesized changes markedly through the prebreeding season (Figure 2) and the daily caloric costs change in parallel. Even at the maximum rate, however, these costs are negligible. It can be estimated, for instance, from data for California Quail (Anthony, 1970) that the maximum growth rate of the testes in this species is about 14 mg/day, or equivalent to about 0.04 kcal/day. This is about 0.2 percent of the concurrent BMR. In Ring-necked Pheasants (Greeley and Meyer, 1953), the maximum rate of testicular growth

TABLE 1 Estimates of the Caloric Cost of Testicular Growth between Resting and Maximal Size[a]

Species	Body Weight (g)	Δ Weight Testes (g)	Δ kcal Testes	BMR[b] (kcal/day)	$\dfrac{\Delta \text{kcal}}{\text{BMR}}$
Mallard					
(*Anas platyrhynchos*)	1,200	11.2	32.2	84.0	0.38
Ring-necked Pheasant					
(*Phasianus colchicus*)	900	7.90	22.6	68.0	0.33
Chukar					
(*Alectoris graeca*)	650	2.04	5.85	53.6	0.11
Wood Pigeon					
(*Columba palumbus*)	500	2.33	6.53	44.2	0.15
California Quail					
(*Lophortyx californicus*)	180	1.04	2.98	20.9	0.14
White-crowned Sparrow					
(*Zonotrichia leucophrys*)	27	0.61	1.75	8.34	0.21
House Finch					
(*Carpodacus mexicanus*)	20	0.22	0.64	6.71	0.095

[a] Data from Johnson (1961), Greeley and Meyer (1953), Mackie and Buechner (1963), Ljunggren (1969), Anthony (1970), King *et al.* (1966), and Hamner (1968).
[b] Basal metabolic rate, estimated from appropriate equation (birds at rest) in Aschoff and Pohl (1970).

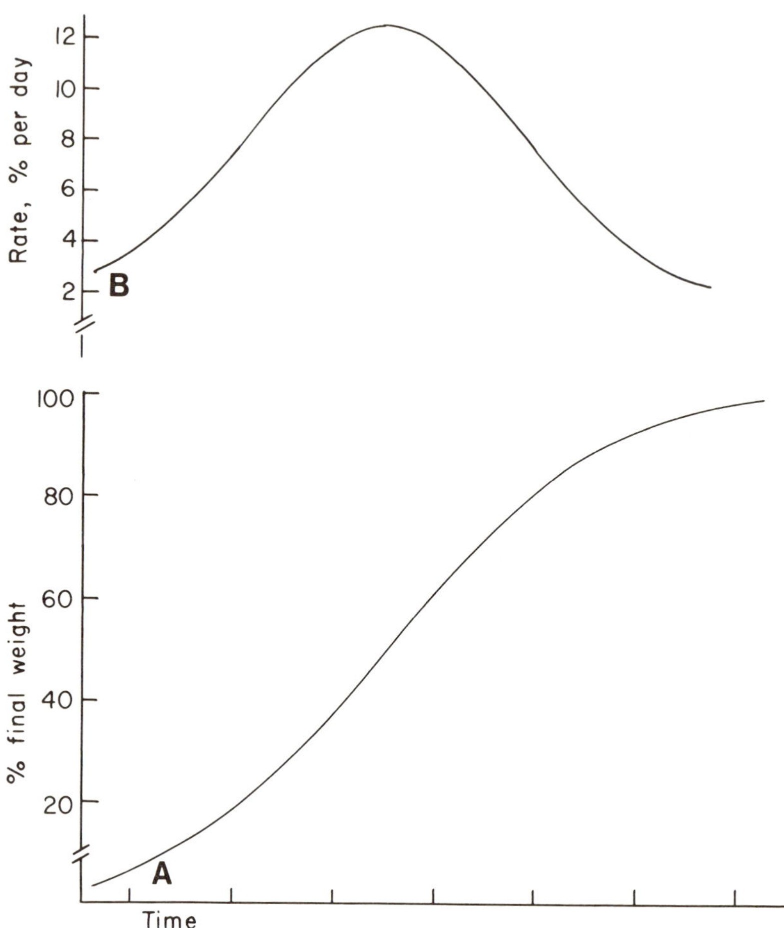

FIGURE 2   The temporal course of cumulative growth (A) in relation to instantaneous growth rate (B), assuming that testicular (or ovarian plus oviducal) growth follows a logistic curve. The growth rate reaches a maximum equivalent to 12.5 percent (per day) of the asymptotic organ weight.

is equivalent to about 0.23 kcal/day, or about 0.3 percent of the concurrent BMR. Analogous estimates can be made for several other species in Table 1, all showing that the caloric cost of testicular synthesis is, at a maximum, less than 1 percent of the BMR. It therefore is clear that no significant caloric saving can be effected by a reduction in the rate of testicular growth in times of stress.

THE FEMALE SEX ORGANS

Estimates of the caloric costs of synthesizing a fully functional oviduct, plus a functional but nonovulating ovary, are shown in Table 2 for selected species. The cost of developing a functional brood-patch cannot be estimated at present, but is undoubtedly slight. As may be expected, the seasonal growth of the ovary and oviduct is much more costly than the growth of the testes. The daily cost at the time of maximal growth rate in the combined organs in the California Quail (data of Anthony, 1970) is about 1.48 kcal/day (about 520 mg/day). This is about 7.1 percent of the concurrent BMR. The available data are not adequate to make analogous estimates from other species, but it appears that the maximal rate of caloric expenditure does not exceed 10 percent of the BMR.

It would be greatly interesting to relate the caloric costs of gonadogenesis to body size, clutch size, egg size, and other variables, but the available data are not adequate to do so. Additional data are needed on tissue weights and rates of growth. Nevertheless, it is evident that the caloric costs of ovarian and oviducal recrudescence can be an appreciable fraction of the female's energy budget during the period of maximal growth rate. In the late winter and early spring,

TABLE 2 Estimates of the Caloric Cost of Combined Ovarian and Oviducal Growth between Resting and Maximal Size[a]

Species	Body Weight (g)	Δ Weight Organs (g)	Δ kcal Organs	BMR[b] (kcal/day)	$\dfrac{\Delta \text{kcal}}{\text{BMR}}$
Mallard					
(*Anas platyrhynchos*)	1,100	61.0	174.3	78.8	2.21
Ring-necked Pheasant					
(*Phasianus colchicus*)	900	51.0	145.7	68.0	2.14
Chukar					
(*Alectoris graeca*)	650	37.1	106.0	53.6	1.98
California Quail					
(*Lophortyx californicus*)	180	12.2	34.9	20.9	1.67
Gambel Quail					
(*Lophortyx gambelii*)	170	10.4	29.8	20.0	1.49
White-crowned Sparrow					
(*Zonotrichia leucophrys*)	27	1.8	5.5	8.34	0.66

[a] Functional but nonovulating ovary. Data from Johnson (1961), Breitenbach *et al.* (1963), Anthony (1970), Lewin (1963), Raitt and Ohmart (1966), and King *et al.* (1966).
[b] Basal metabolic rate, estimated from appropriate equation (birds at rest) in Aschoff and Pohl (1970).

however, when the growth rates are less than half maximum, it is questionable that sexual activation is a significant caloric burden. Breitenbach *et al.* (1963), in fact, found that restriction of Ring-necked Pheasants to a ration just adequate to maintain normal body weight and vigor suppressed ovarian and oviducal weights only slightly (not statistically significant), compared to the organ weights in hens feeding *ad libitum*. Similar experiments are needed with additional species.

EGG PRODUCTION

The foregoing estimates of the caloric costs of seasonal gonadogenesis are fairly simple and are based on methods applicable to all species of birds. The estimation of the caloric cost of egg production is a much more complex problem, owing not only to the dearth of empirical data, but also to interspecific variability in the rate of ovogenesis, laying interval, clutch size, egg size, and so on. One must rely heavily on extrapolations from data on egg production in poultry in developing some crude models concerning game birds.

The basic empirical data are as follows. The mean energy content of a fresh egg is affected by several variables (Romanoff and Romanoff, 1949), but most strongly by the state of the chick at the time of hatching. The eggs of species producing precocial young contain, proportionately, about twice as much yolk as do those of species producing altricial young (Romanoff and Romanoff, 1949; p. 114); the available data suggest that the caloric content per gram whole egg varies accordingly. Among altricial birds, House Sparrow (*Passer domesticus*) eggs average 1.099 kcal/g (Tangl, 1903), Tree Sparrow (*Passer montanus*) eggs average 1.02 kcal/g (Mackowicz *et al.*, 1970), and Long-billed Marsh Wren (*Cistothorus palustris*) eggs average 1.06 kcal/g (Kale, 1965) in direct calorimetric measurements. Among precocial birds, eggs of galliform species average about 1.6 kcal/g (Romanoff and Romanoff, 1949, p. 599), which is also the standard value in poultry husbandry (Brody, 1945). Eggs of anseriforms appear to average somewhat higher (1.73–2.02 kcal/g) in domestic ducks and geese (Romanoff and Romanoff, 1949, p. 599) in correlation with their relatively high fat content. Eggs of the Herring Gull (*Larus argentatus*) average 1.677 kcal/g (Drent, 1967). For purposes of this analysis, a value of 1.05 kcal/g is used for eggs of species producing altricial young, 1.6 kcal/g for galliform species, and 1.8 kcal/g for anseriform species.

The partial (net) efficiency of egg production in chickens in extensive experiments averaged 77 percent (Brody, 1945). Indirect computations with other data for egg production yielded a partial efficiency of 54 percent. The partial efficiency of milk formation in cattle—a production process similar to that of egg formation—averages about 64 percent. For the present analysis, a partial efficiency of 70 percent is used.

The assumptions regarding the caloric content of an egg and the efficiency of synthesis, combined with data on average egg weight and clutch size, make it possible to estimate the total energy requirement for production of an egg or a clutch of eggs. There is no simple way, however, to estimate the temporal course of energy expenditure during the laying of the clutch, or the peak energy requirement when several ovarian follicles are growing concurrently. The time course for energy expenditure in the synthesis of each egg is unknown, yolk deposition occurs in several ova simultaneously, and birds of different sizes and species ovulate at different rates. A crude model will serve, nevertheless, to establish some limits and to identify problems worthy of future investigation.

The phase of rapid yolk synthesis begins in the ovum of large birds 7–10 days before ovulation and in small birds 3–4 days before ovulation (Table 3). Including transit time in the oviduct, the period for synthesis and deposition of a single egg thus varies from about 4 days to about 11 days in species (laying about one egg per day on the average) in the size range shown in Table 3. The growth rate of the follicle is approximately sigmoid (e.g., Romanoff, 1943; Anthony, 1970), and the energy requirement thus follows a bell-shaped curve in time (cf Figure 2). The energy requirement for albumin secretion, shell formation, and oviducal transport constitutes a second peak of unknown form. If it is assumed, as a limiting, simplified case, that the energy requirement for producing a single egg describes a sine wave in time and that the area beneath one cycle (Figure 3) is proportional to total energy costs, then it can be shown by simple calculations that the peak (daily) energy expenditure for one egg is given by the equation

$$\text{kcal}_{\text{peak}} = 2A/p,$$

where A = kcal/egg and p = the period of the cycle in days. The peak energy cost per day rises sharply with decreasing period of egg formation (Figure 4), and it appears from the few data available (Table 3)

TABLE 3   Estimates of Time Required for Rapid Phase Growth (Yolk Deposition) in Avian Ovarian Follicles[a]

Species	Body Weight (g)	Egg Weight (g)	Follicular Growth (days)
Domestic duck	3,000	80	6–7
Domestic chicken	2,500	58	7–8
Herring Gull			
(*Larus argentatus*)	895	82	9–10
Domestic pigeon	350	17	5–8
Chukar			
(*Alectoris graeca*)	650	30	∼6
Jackdaw			
(*Corvus monedula*)	250	15.2	∼5
California Quail			
(*Lophortyx californicus*)	180	10.3	6–7
Ring Dove			
(*Streptopelia* sp.)	150	9.2	5–7
Tricolored Blackbird			
(*Agelaius tricolor*)	46	3.7	3–4
White-crowned Sparrow			
(*Zonotrichia leucophrys*)	27	2.4	∼4
Song Sparrow			
(*Melospiza melodia*)	21	2.3	∼4
Great Tit			
(*Parus major*)	18	1.75	3–4

[a]Sources of data: Riddle (1916), Marza and Marza (1935), Warren and Conrad (1939), Nalbandov and James (1949), Romanoff and Romanoff (1949), Paludan (1951), Riddle (1911), Bartelmez (1912), Mackie and Buechner (1963), Stieve (1919), Lewin (1963), Anthony (1970), Cuthbert (1945), Payne (1969), Kern (1970), Nice (1937), Kluijver (1951).

that birds tend to function near the point at which the peak energy increases sharply at the left in Figure 4. This probably represents an evolutionary compromise between high daily energy costs of egg production and the total (gross) efficiency of production. As the period for producing the egg lengthens, the total efficiency diminishes and is diluted by the continuing expenses of maintenance. It therefore is advantageous to the bird to operate as far to the left as possible in Figure 4, but not so far that the daily costs become difficult to meet.

The peak energy expenditure in egg production during the laying of a clutch is determined by the amount of overlap among cycles for individual eggs and, hence, by the period of synthesis for a single egg, the unit rate of production (or laying), and the clutch size (Figure 3). It can be shown from the sine model that the daily energy cost of several ova growing concurrently reaches a peak equivalent to

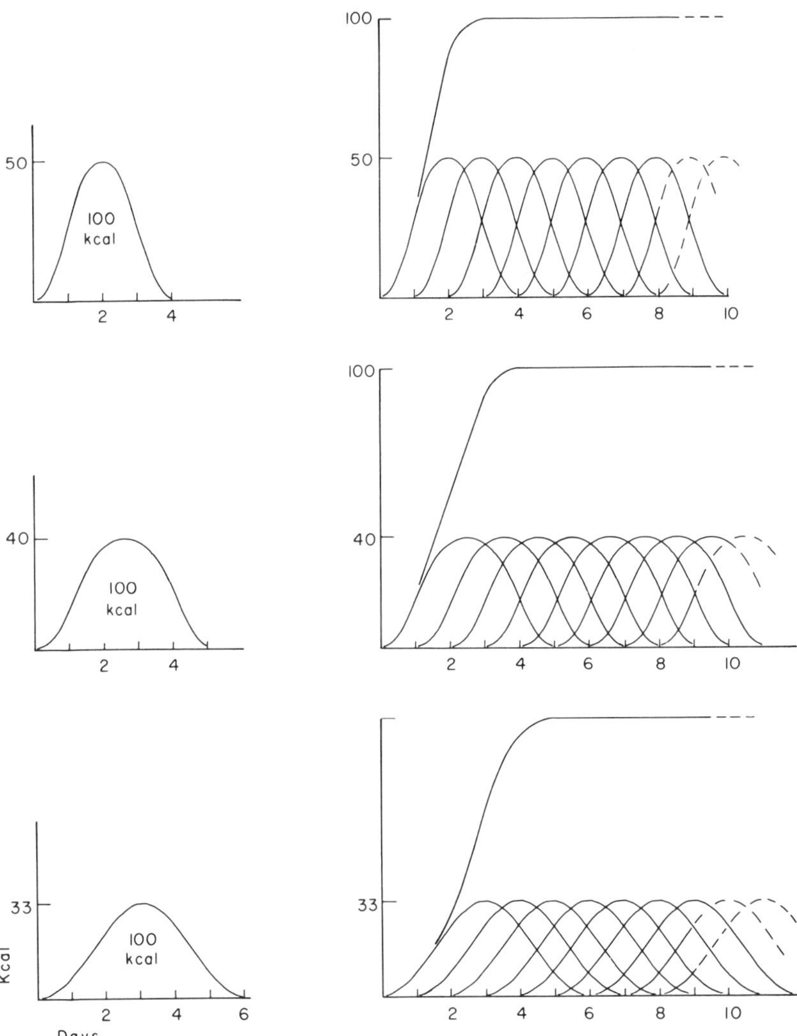

FIGURE 3   A model, assuming that the energy requirement for egg formation follows a sine function in time, illustrating the effects of concurrent growth of several ova on the total energy requirement. At the left: Interactions of period of formation (4, 5, or 6 days) and peak energy requirement in a hypothetical 100-kcal egg. At the right: Total energy cost (upper curve in each panel) of egg formation assuming that the egg-laying interval is one day. Note that the peak energy requirement for the clutch is determined by, and equivalent to, the energy content of a single egg and is independent of the period of formation or the peak energy for a single egg.

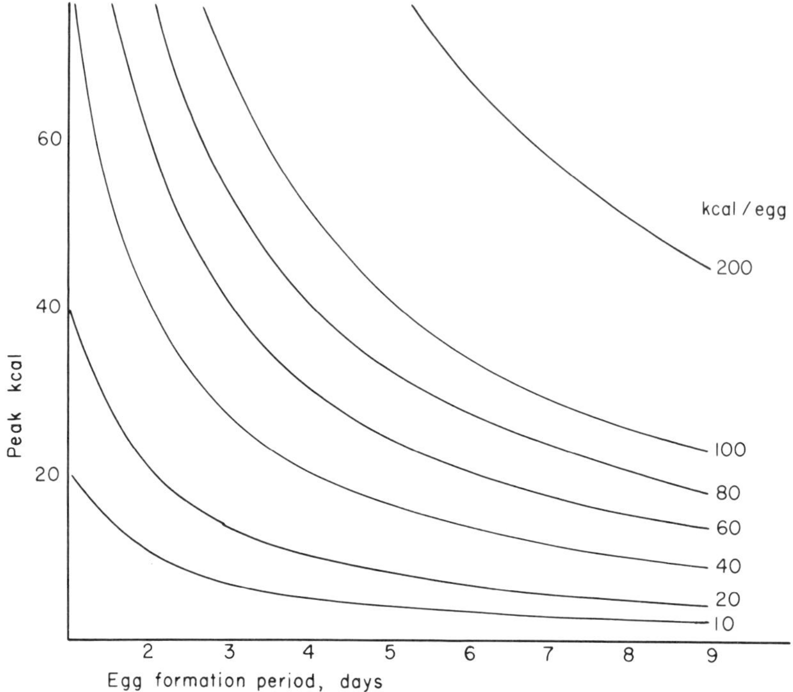

FIGURE 4   Relationship among the period of egg formation (abcissa), the estimated caloric cost of formation (individual curves, scale at right), and peak energy requirement (ordinate) for synthesis of a single egg.

the total energy cost of a single egg on $p - 1$ days after the onset of rapid yolk deposition in the first egg, where p equals the period in days required in the synthesis and laying of the egg (and assuming that the unit rate of production is one egg per day). For instance, if the period is 8 days (as in the chicken, approximately), then the daily energy expense of egg production reaches a peak on the seventh day after the activation of the first follicle. If the clutch is 7 eggs or less, then the theoretical maximum is not attained.

Estimates assembled in Table 4 show that the daily peak energy requirement of egg production is a substantial fraction of the daily energy budget, equaling 0.7–1.1 times the BMR in galliform birds and 1.6–2.4 times BMR in a selection of anseriform birds. The other species fall within these ranges.

Although the BMR provides a convenient standard for comparison of energy costs, free-living birds do not operate at the basal level for significant·periods of time. Other bases of comparison are therefore more appropriate. One of these is the maintenance level of metabolism, which includes the BMR plus the heat increment of feeding plus the energy costs of movement, feeding, and self-maintenance under the usual conditions of husbandry in domestic animals of stable body weight whenever not producing eggs or milk. The energy cost of maintenance in chickens is about 1.7 times BMR (calculated from appropriate regression equations) (Figure 5). The average energy cost of sustained egg production in chickens is about 62 percent (increment/maintenance) greater than the maintenance level of metabolism, regardless of body weight (Figure 5). Data for Ring-necked Pheasants in captivity, when corrected for seasonal changes of body weight, conform closely ($\sim$ 65 percent) at peak production with the increment found in chickens (data of Breitenbach *et al.*, 1963). In contrast, the production increment in captive Willow Ptarmigan in Alaska averaged only 21.3 percent of the apparent maintenance metabolism. The hens were apparently losing weight during the laying period, however, and the caloric cost of egg formation was therefore supported, in part, from endogenous energy sources, resulting in a reduced food increment. Additional quantitative data are needed from other species.

Free-living birds, however, do not function at a maintenance level, and it is necessary to guess the relative plane of daily energy requirement during the laying period. It is very unlikely that free-living animals function at average levels below 1.5 times or (for appreciable periods of time except during migration) above 4 times the maintenance plane. If it is assumed, as an illustrative case, that free-living birds function at about twice the maintenance plane (or about 3.4 times the basal plane), then it can be estimated that the daily maximum cost of egg production is 21–30 percent of the daily energy intake for normal function without gain or loss of body weight in galliforms, 52-70 percent in anseriforms, and 13–16 percent in passeriforms. If birds operate at a lower level of active metabolism, then the increment required for egg production is relatively greater. Conversely if they operate at a higher level, the increment is proportionately less. In any case, these estimates are adequate to indicate that the peak energy requirement for egg formation in typical game birds is at least a quarter of the daily energy budget during egg-laying. It is therefore to be expected, as indicated by the empirical data (e.g., Figure 1), that egg production is very sensitive to nutritional deficiency.

TABLE 4 Estimates of the Caloric Cost of Producing Eggs and of the Peak Caloric Cost per Day during the Laying Period[a]

Species	Body Weight (g)	Mean Clutch	Mean Egg Weight (g)	kcal/ Egg[b]	Eggs/ Day	kcal/ Clutch	BMR[c]	$\dfrac{\text{kcal/Egg}}{\text{BMR}}$
Wild Turkey								
(*Meleagris gallopavo*)	4,200	11.1	65.2	149.0	1.0	1,654	210.8	0.71
Sage Grouse								
(*Centrocercus urophasianus*)	1,530	7.4	40.5	92.6	0.77	685	100.4	0.92
Ring-necked Pheasant								
(*Phasianus colchicus*)	900	11.5	30.0	68.6	0.67	789	68.0	1.01
Ruffed Grouse								
(*Bonasa umbellus*)	540	11.5	17.8	40.7	0.67	468	46.8	0.87
Willow Ptarmigan								
(*Lagopus lagopus*)	530	11.0	?	37.9[d]	0.59	417	46.1	0.82
California Quail								
(*Lophortyx californicus*)	180	14.0	10.3	23.5	0.72	330	20.9	1.13
Bobwhite Quail								
(*Colinus virginianus*)	173	14.4	8.6	19.4	1.0	283	20.3	0.97

Canada Goose							
(*Branta c. canadensis*)	5,000	6.0	163	372.6	1.0	2,235	239.5	1.56
Mallard							
(*Anas platyrhynchos*)	1,100	11.0	54	138.8	1.0	1,527	78.8	1.76
Lesser Scaup							
(*Aythya affinis*)	650	9.5	44	113.1	1.0	1,075	53.6	2.11
Cinnamon Teal							
(*Anas cyanoptera*)	355	10.0	32	82.3	1.0	823	34.4	2.39
White-winged Dove							
(*Zenaida asiatica*)	170	2.0	7.7	11.6	0.5	23.1	20.0	0.58
White-crowned Sparrow							
(*Zonotrichia leucophrys*)	27	4.7	2.5	3.75	1.0	17.6	8.34	0.45
House Wren							
(*Troglodytes aedon*)	11.5	6.4	1.45	2.11	0.67	13.5	4.33	0.49

[a] Data from Hewitt (1967), Mosby and Handley (1943), Patterson (1952), Lack (1968), Bump *et al.* (1947), West (1968), Lewin (1963), Stoddard (1931), Cottam and Trefethen (1968), King *et al.* (1966), Kendeigh (1963), Kendeigh *et al.* (1956).
[b] Equivalent to maximum daily cost during the laying cycle (see text).
[c] Basal metabolic rate, estimated from the appropriate equation (birds at rest) in Aschoff and Pohl (1970).
[d] Direct measurement (West, 1969).

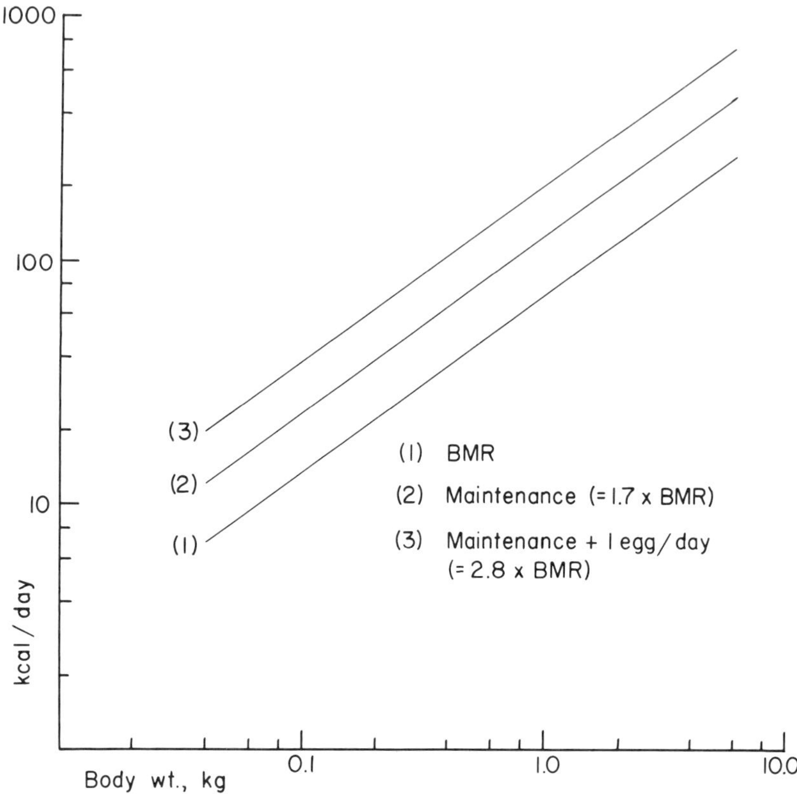

FIGURE 5   Estimates of energy requirements in relation to basal metabolic rate (BMR) in captive galliform birds maintaining stable body weight and composition. The figure illustrates that the relative caloric increment for egg production is independent of body weight. The BMR follows the equation: kcal/day = 73.5 $W^{0.734}$, where W = body weight in kg (Aschoff and Pohl, 1970); maintenance metabolism follows the equation: TDN (g/day) = 0.273 $W^{0.73}$, where W = body weight in grams and TDN = total digestible nutrients equivalent to 3 kcal/g (Brody, 1945; Eq. 23.6). The caloric cost of egg production follows the equation: TDN (g/day) = 0.688 (0.255 $W^{0.74}$), where the factor in parentheses is an empirical regression of egg weight (g) on body weight (g) for galliforms (range of weight, 40–3,000 g) computed by J. R. King; the rest of the equation is from Brody (1945; Eq. 23.6). The regression of galliform egg weight on body weight does not differ significantly from the analogous equation of Brody (1945, p. 482) for birds ranging in weight from 2 g to 100 kg.

THE CALORIC COSTS OF INCUBATION

The heat required for the incubation of an egg is derived partly from the parent or parents (with the exception of the megapodes), partly from the metabolism of the embryo, and in some situations partly from the nest and its surroundings. There have been no empirical investigations of the complete energy budgets of eggs and parents during natural incubation, and our information about the caloric expenditure of the parent birds during incubation is based mainly on indirect estimates and extrapolations. These may be pieced together to furnish the following generalizations.

Under typical circumstances, the parent supplies essentially all of the heat required to maintain normal egg temperature during the early stages of incubation. This contribution progressively diminishes during the later stages of incubation as the heat production of the embryo increases [for review, see Drent (1967)]. The only complete account of the heat balance of avian eggs is that of Kashkin (1961) for the eggs of domestic ducks incubated artificially. If it is assumed as a first approximation that a parent bird functions thermally like a mechanical incubator, then it is clear from Kashkin's data that heat production by the embryo can balance, or almost balance, heat loss within the range of normal incubation temperatures (Figure 6) during a large fraction of the incubation period. The eggs of other species presumably show a qualitatively similar pattern, although the time course of embryonic heat production no doubt varies with the diversity of developmental patterns. Major functions of the adult during the late stages of incubation seem to be to supply insulation, to protect the eggs from predators, to rewarm the eggs if attentiveness is not continuous, and to protect the eggs from lethal heating in very hot environments.

Since it is likely that the heat loss from eggs in nature is greater than it is in a mechanical incubator, it is also likely that the parent bird must supply a small amount of heat continuously even after the eggs have reached a normal incubation temperature. The data in Figure 6 nevertheless emphasize that the heat production of the eggs themselves cannot be ignored during that part of the incubation period in which this source of heat becomes significant in relation to the rate of heat loss (Drent, 1967, p. 103). In addition to its role in supplying insulation and maintaining heat flow while sitting on the eggs, the parent bird also must provide a large fraction of the heat required to rewarm the eggs following a period of inattentiveness. Con-

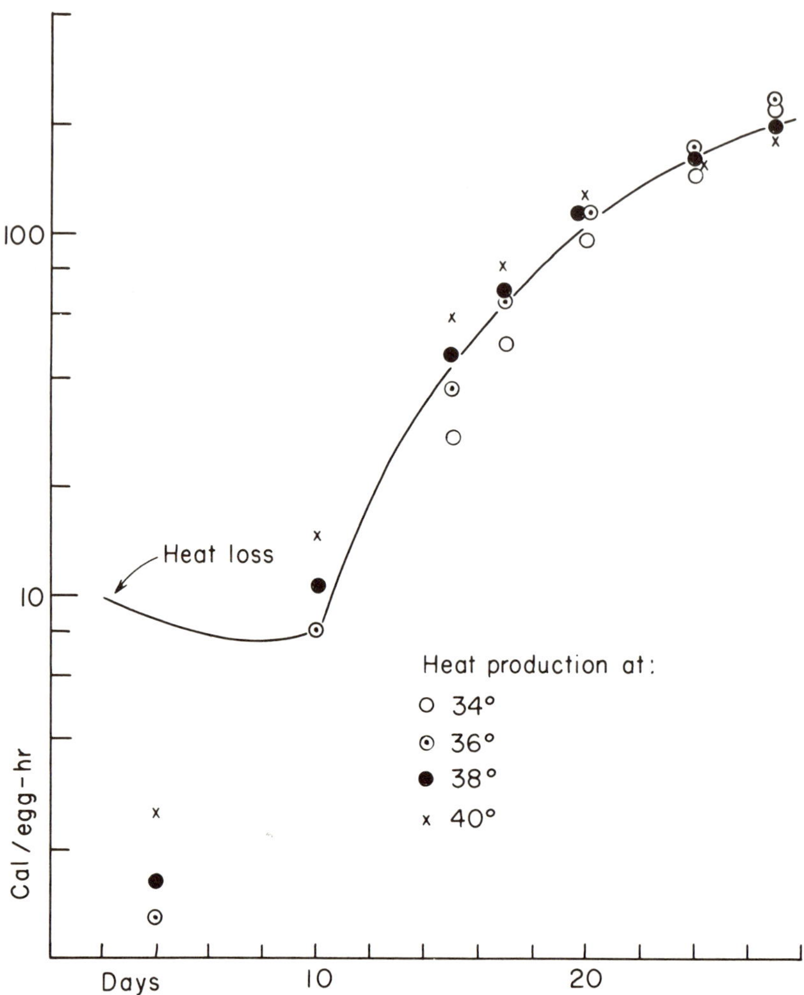

FIGURE 6   The relationship of heat loss (solid line) from domestic duck eggs (mean weight = 78 g) artificially incubated at 37.5 °C and heat production (assuming that 1 cm^3 O$_2$ = 4.8 cal) at various egg temperatures (individual points). Note that heat production after about 10 days of incubation approximately balances heat loss if egg temperature is 35–36 °C. Data from Kashkin (1961).

tact with eggs cooled well below the normal temperature elicits overt thermoregulatory responses (shivering, ptiloerection) from the incubating bird (Baerends *et al.*, 1960; Franks, 1967). Less heat is required from the parents if the eggs are not allowed to cool. The behavioral aspects of incubation (participation by the sexes, percentage attentiveness during the day, covering the clutch with down when inattentive, and so on) are thus intimately connected with the thermal physiology of the clutch and its requirements from its parents (Kendeigh, 1952).

A major question in the context of the present account is whether or not incubation requires heat in addition to that released as a by-product of the energy metabolism of the resting parent at prevailing air temperature. The data currently available do not allow a completely satisfactory examination of this question, but do assist in establishing some tentative conclusions.

In Ring Doves and domestic pigeons in captivity, there is no increase of voluntary food consumption during the incubation period, as compared to periods free of reproductive activity (Riddle and Braucher, 1933; Brisbin, 1969). If incubation requires increased heat production by the parents (both of which share in incubation), then it is clear that the cost is compensated by reduction in other routes of expenditure. These columbiforms, however, lay clutches of only two eggs and develop only a minimal incubation patch, at least in the case of the pigeon (Jones, 1971). They probably should not be regarded as representative of birds in general.

In Zebra Finches (*Poephila guttata=Taeniopgypia castanotis*), metabolizable energy intake per pair (the sexes share about equally in incubation) during the incubation period is significantly greater than control values (no reproductive activity) at air temperatures of 20.9 and 24.4 °C, but not at 29.1 and 34.4 °C (El-Wailly, 1966). As in the investigations of doves and pigeons just noted, however, it is not possible to discern whether the increased energy expenditure at the two lower temperatures was associated with the requirements of incubation or with a change in some other route of expenditure, such as locomotor activity. The empirical data thus far available do not allow rigorous conclusions, and it becomes necessary to depend temporarily on indirect estimates.

Kendeigh (1963) has developed a theoretical equation for estimating the amount of heat required (disregarding embryonic heat production) to maintain normal incubation temperature in a clutch of eggs. Drent (1967) has shown empirically that the equation satisfactorily predicts of heat requirements of Herring Gull eggs if a correction is made for the embryonic heat production. The Kendeigh equation thus somewhat overestimates the heat requirements of a clutch, particularly in the later stages of incubation, but nevertheless is useful in indicating the maximum heat transfer required from an incubating bird.

El-Wailly (1966) estimated, by means of the Kendeigh equation, that the heat requirement of Zebra Finch clutches ranged from 4.78 to 1.24 kcal/pair per day at air temperatures ranging from 14.5 to 29.1 °C. The metabolic rates of a pair of Zebra Finches at rest at these air temperatures range from about 22.1 to 11.1 kcal/pair per day (Calder, 1964). The estimated heat required for incubation as a fraction of the metabolic rate at rest thus ranges from 21.6 to 11.1 percent. It should be recalled that these are maxima that do not include the heat production of the embryos. In view of the fact that the incubation patch of a Zebra Finch covers at least a third of the surface area of the torso, it seems reasonable to suggest that a maximum of 21.6 percent of the bird's heat production could be transferred to the eggs.

Similar methods of estimation can be applied to the House Wren, a hole-nesting species in which only the female incubates the eggs. Kendeigh (1963) has estimated the heat required for incubation in this species and also has measured the metabolic rates of females at rest in various air temperatures (Kendeigh, 1939). The heat required for incubation (estimated by the Kendeigh equation) ranges from maxima of about 22 percent of the female's heat production at rest at an air temperature of 17 °C to about 10 percent at 22 °C. If as much as 22 percent (at a maximum) of the female's obligatory heat production can be transferred to the eggs, which seems reasonable, then they can be maintained at normal incubation temperature without an increase of the female's energy requirements above the maintenance level.

A final example can be derived from the elegant investigation of Herring Gulls by Drent (1967). The resting heat production of the Herring Gull evidently has not been measured, but can be estimated as 83.3 kcal/day for an 895-g gull during the daytime (Aschoff and Pohl, 1970). The peak heat requirement for incubation as estimated by the Kendeigh equation, corrected for embryonic heat production (Drent, 1967, p. 103), is equivalent to about 27 kcal/day (days 15–20 of incubation), or about 32 percent of the adult's heat production at rest; the minimum estimated heat requirement for the eggs (day 30) is about 10.6 kcal/day, or about 13 percent of the adult's heat production. Whether or not as much as 32 percent of an adult's heat output can be transferred to a clutch of eggs must remain a moot point until the situation is investigated further. Nevertheless, it is clear from the three cases examined that the metabolic rate of a bird at rest can supply a large fraction, if not all, of the heat required for incubation. Estimates that assert that incubation requires as much as 50 percent of a bird's "productive energy" (i.e., up to 50 percent of its maximum potential for energy expenditure above the maintenance level) appear to disregard the contribution of the bird's obligatory heat production at rest (West, 1960; Kendeigh, 1963; El-Wailly, 1966; Drent, 1967). This contribution cannot be disregarded, and it remains an open question as to whether or not a bird must significantly increase its energy expenditure for the specific purpose of incubating its eggs.

SUMMARY

The reproductive potential of male birds is relatively resistant to caloric deficiency. The energy requirement of the seasonal growth of the testes is a negligible fraction of the male's energy budget. Sperm production and vitality are not imparied, in those species that have been investigated thus far, until the loss of body weight during starvation reaches 25–30 percent of the initial weight.

The reproductive potential of female birds is much more susceptible than is that of males to caloric deficiency. The energy cost of the seasonal growth of the ovary and oviduct may reach a peak equivalent to about 7–10 percent of the female's daily basal energy requirement, or about 3.5–5 percent of the maintenance requirement. It appears doubtful that caloric restriction, short of complete starvation, significantly affects the seasonal growth of the sex organs to the preovulatory stage.

Egg production requires relatively great amounts of energy, which can be estimated provisionally as 20–30 percent above the daily energy requirement in a sampling of galliforms, 50–70 percent in anseriforms, and 13–16 percent in small passeriforms. Egg production, in particular, is sensitive to caloric deficiency. Ovulation stops within 48 hr or less in response to complete starvation, and the ovary involutes. In response to less severe caloric shortage, egg formation is slowed and production is reduced, although the eggs that are laid are usually of normal size and hatchability.

In both male and female birds, as in mammals, the primary effect of caloric shortage is exerted on the output of gonadotropic hormones. It appears that the hypothalamo–hypophysial system is a sensitive monitor of the bird's caloric status and modulates reproductive activity accordingly. Restoration of normal caloric intake following a period of restriction also restores normal reproductive function.

With respect to the physiologic aspects of reproduction, it is evident from the foregoing that the reproductive potential of a species is most sensitive to caloric shortage at the time of egg-laying. Reproductive processes at other times of the breeding season or prebreeding season are relatively resistant to caloric shortage.

Essentially nothing is known, quantitatively, about the effects of caloric shortage on the behavioral aspects of reproduction. Although these aspects must be ignored at present, owing to the absence of suitable data, it is to be expected that caloric shortage can have serious effects on the efficiency of territorial defense, courtship, nest-building, and so on. Although nothing has been said here about the feeding and parental care of the young, it is obvious that this phase is a second period of the breeding season during which caloric shortage exerts a major effect on reproductive potential.

REFERENCES

Anthony, R. 1970. Ecology and reproduction of California Quail in southeastern Washington. Condor 72:267–287.

Aschoff, J., and H. Pohl. 1970. Der Ruheumsatz von Vögeln als Funktion der Tageszeit und der Körpergrösse. J. Ornithol. 111:38–47.

Assenmacher, I., A. Tixier-Vidal, and H. Astier. 1965. Effets de la sous-alimentation et de jeûne sur la gonadostimulation du canard. Ann. Endocrinol. 26:1–26.

Baerends, G. P., K. H. Posthuma, and T. Joustra. 1960. Die Reaktion der brütenden Silvermöwe auf die Temperatur des Eies. Arch. Néerl. Zool. 13:586–588.

Bartelmez, G. W. 1912. The bilaterity of the pigeon's egg. A study in egg organization from first growth of the oocyte to the beginning of cleavage. J. Morphol. 23:269–328.

Bengston, S.-A., and S. Ulfstrand. 1971. Food resources and breeding frequency of the Harlequin Duck *Histrionicus histrionicus* in Iceland. Oikos 22:235–239.

Breitenbach, R. P., C. L. Nagra, and R. K. Meyer. 1963. Effect of limited food intake on cyclic annual changes in Ring-necked Pheasant hens. J. Wildlife Manage. 27:24–36.

Breneman, W. R. 1940. Limitation of food consumption as a factor influencing endocrine relations in the chick. Endocrinology 26:1091–1098.

Brisbin, I. L., Jr. 1968. A determination of the caloric density and major body components of large birds. Ecology 49:792–794.

Brisbin, I. L., Jr. 1969. Bioenergetics of the breeding cycle of the Ring Dove. Auk 86:54–74.

Brody, S. 1945. Bioenergetics and growth. Reinhold, New York. 1023 p.

Bump, G., R. W. Darrow, F. C. Edminster, and W. F. Crissey. 1947. The Ruffed Grouse. New York State Conservation Department, Albany. 915 p.

Calder, W. A. 1964. Gaseous metabolism and water relations of the Zebra Finch, *Taeniopygia castanotis*. Physiol. Zool. 37:400–413.

Cody, M. L. 1971. Ecological aspects of reproduction, p. 461–512. *In* D. S. Farner and J. R. King [ed.] Avian biology. Vol. 1. Academic Press, New York.

Cottam, C., and J. B. Trefethen [ed.]. 1968. Whitewings: The life history, status, and management of the White-winged Dove. Van Nostrand, New York. 348 p.

Cuthbert, N. L. 1945. The ovarian cycle of the Ring Dove (*Streptopelia risoria* L.). J. Morphol. 77:351–377.

Davis, W. M., and A. B. Watt. 1955. The effect of ration treatments during various stages of the chicken's life upon subsequent performance of laying hens. Poultry Sci. 34:1231–1232.

Drent, R. H. 1967. Functional aspects of incubation in the Herring Gull (*Larus argentatus* Pont.). Behav. Suppl. 17, 132 p.

El-Wailly, A. J. 1966. Energy requirements for egg-laying and incubation in the Zebra Finch, *Taeniopygia castanotis*. Condor 68:582–594.

Engels, W. L., and C. E. Jenner. 1956. The effect of temperature on testicular recrudescence in Juncos at different photoperiods. Biol. Bull. 110:129–137.

Farner, D. S., and A. C. Wilson. 1957. A quantitative examination of testicular growth in the White-crowned Sparrow. Biol. Bull. 113:254–267.

Franks, E. C. 1967. The responses of incubating Ringed Turtle Doves (*Streptopelia risoria*) to manipulated egg temperatures. Condor 69:268–276.

Gowe, R. S., A. S. Johnson, R. D. Crawford, J. H. Downs, A. T. Hill, W. T. Mountain, J. R. Pelletier, and J. H. Strain. 1960. Restricted versus full-feeding during the growing period for egg production stock. Brit. Poultry Sci. 1:37–56.

Greeley, F., and R. K. Meyer. 1953. Seasonal variation in testis stimulating activity of male pheasant pituitary glands. Auk 70:350–358.

Hamner, W. M. 1968. The photorefractory period of the House Finch. Ecology 49:211–228.

Hewitt, O. H. [ed.]. 1967. The Wild Turkey and its management. The Wildlife Society, Washington, D.C. 589 p.

Heywang, B. W. 1940. The effect of restricted feed intake on egg weight, egg production, and body weight. Poultry Sci. 19:29–34.

Hosada, T., T. Kaneko, K. Mogi, and T. Abe. 1955. Effect of gonadotropic hormone on ovarian follicles and serum vitellin of fasting hens. Proc. Soc. Exp. Biol. Med. 88:502–504.

Immelmann, K. 1971. Ecological aspects of periodic reproduction, p. 341–389. *In* D. S. Farner and J. R. King [ed.] Avian biology. Vol. 1. Academic Press, New York.

Johnson, O. W. 1961. Reproductive cycle of the Mallard duck. Condor 63:351–364.

Jones, R. E. 1971. The incubation patch of birds. Biol. Rev. 46:315–339.

Kale, H. W., II. 1965. Ecology and bioenergetics of the Long-billed Marsh Wren in Georgia salt marshes. Publications of the Nuttall Ornithological Club, No. 5. Nuttall Ornithological Club, Cambridge, Mass. 142 p.

Kashkin, V. V. 1961. Heat exchange of bird eggs during incubation. Biophysics 6:57–63.

Kendeigh, S. C. 1939. The relation of metabolism to the development of temperature regulation in birds. J. Exp. Zool. 82:419–438.

Kendeigh, S. C. 1941. Length of day and energy requirements for gonad development and egg-laying in birds. Ecology 22:237–248.

Kendeigh, S. C. 1952. Parental care and its evolution in birds. Ill. Biol. Monogr. 22:1–356.

Kendeigh, S. C. 1963. Thermodynamics of incubation in the House Wren, *Troglodytes aedon*, p. 884–904. *In* C. G. Sibley, J. J. Hickey, and M. B. Hickey [ed.] Proceedings of the XIIIth International Ornithological Congress. Vol. 2. American Ornithologists' Union, Washington, D.C.

Kendeigh, S. C., T. C. Kramer, and F. Hamerstrom. 1956. Variations in the egg characteristics of the House Wren. Auk 73:42–65.

Kern, M. D. 1970. Annual and steroid-induced changes in the reproductive system of the female White-crowned Sparrow, *Zonotrichia leucophrys gambelii*. PhD dissertation, Washington State University, Pullman.

King, J. R., B. K. Follett, D. S. Farner, and M. L. Morton. 1966. Annual gonadal cycles and pituitary gonadotropins in *Zonotrichia leucophrys gambelii*. Condor 68:476–487.

Kleiber, M. 1961. The fire of life. Wiley, New York. 453 p.

Kluijver, H. N. 1951. The population ecology of the Great Tit, *Parus m. major* L. Ardea 39:1–135.

Krüger, P. 1965. Über die Einwirkung des Temperatur auf das Brutgeschäft und das Eierlegen des Rebhuhnes (*Perdix perdix* L.). Acta Zool. Fenn. 112:1–64.

Lack, D. 1954. The natural regulation of animal numbers. Oxford University Press, London. 343 p.

Lack, D. 1968. Ecological adaptations for breeding in birds. Methuen & Co., London. 409 p.

Lewin, V. 1963. Reproduction and development of young in a population of California Quail. Condor 65:249–278.

Ljunggren, L. 1969. Seasonal studies of Wood Pigeon populations. II. Gonads, crop glands, adrenals and the hypothalamo–hypophysial system. Viltrevy 6:41–126.

Lofts, B., and R. K. Murton. 1966. The role of weather, food and biological factors in timing the sexual cycle of Wood Pigeons. Brit. Birds 59:261–280.

MacIntyre, T. M., and J. R. Aitken. 1959. The performance of laying hens reared on restricted and full feeding programs. Can. J. Anim. Sci. 39:217–225.

Mackie, R. J., and H. K. Buechner. 1963. The reproductive cycle of the Chukar. J. Wildlife Manage. 27:246–260.

Mackowicz, R., J. Pinowski, and M. Wieloch. 1970. Biomass production by House Sparrow (*Passer d. domesticus* L.) and Tree Sparrow (*Passer m. montanus* L.) populations in Poland. Ekol. Polska 18:465–501.

Marrian, G. F., and A. S. Parkes. 1928. The effects of inanition and vitamin B deficiency on the testis of the pigeon. J. Roy. Microsc. Soc. 48:257–270.

Marshall, A. J. 1949. Weather and spermatogenesis in birds. Proc. Roy. Soc. London 119:711–716.

Marza, V. D., and E. V. Marza. 1935. The formation of the hen's egg. Q. J. Microsc. Sci. 78:133–249.

Meier, A. H., D. D. Martin, and R. MacGregor, III. 1971. Temporal synergism of corticosterone and prolactin controlling gonadal growth in sparrows. Science 173:1240–1242.

Milby, T. T., and D. H. Sherwood. 1956. The influence of feed intake during the growth period on subsequent performance of laying hens. Poultry Sci. 35:863–869.

Miller, G. R., A. Watson, and D. Jenkins. 1970. Responses of Red Grouse populations to experimental improvement of their food, p. 323–335. *In* A. Watson [ed.] Animal populations in relation to their food resources. Blackwell's, Oxford.

Morris, T. R., and A. V. Nalbandov. 1961. The induction of ovulation in starving pullets using mammalian and avian gonadotropins. Endocrinology 68:687–697.

Mosby, H. S., and C. O. Handley. 1943. The Wild Turkey in Virginia. Commission of Game and Inland Fisheries, Richmond, Virginia. 281 p.

Moustgaard, J. 1959. Nutrition and reproduction in domestic animals, p. 170–223. *In* H. H. Cole and P. T. Cupps [ed.] Reproduction in domestic animals. Vol. 2. Academic Press, New York.

Myrcha, A., and J. Pinowski. 1970. Weights, body composition, and caloric value of postjuvenal molting European Tree Sparrows (*Passer montanus*). Condor 72:175–181.

Myres, M. T. 1955. The breeding of the Blackbird, Song Thrush and Mistle Thrush in Great Britain. I. Breeding seasons. Bird Study 2:2–24.

Nalbandov, A. V., and M. F. James. 1949. Bood vascular supply of the chicken ovary. Am. J. Anat. 85:347–378.

Nice, M. M. 1937. Studies in the life history of the Song Sparrow. Part. 1. Trans. Linnaean Soc. N.Y. 4:1–247.

Paludan, K. 1951. Contributions to the breeding biology of *Larus argentatus* and *Larus fuscus*. Vidensk. Medd. Dansk Naturhist. Foren. 114:1–128.

Parker, J. E., and G. H. Arscott. 1964. energy intake and fertility in male chickens. J. Nutr. 82:183–187.

Parker, J. E., and B. J. McSpadden. 1943. Influence of feed restriction on fertility in male domestic fowls. Poultry Sci. 22:170–177.

Patterson, R. L. 1952. The Sage Grouse in Wyoming. Sage Books, Inc., Denver. 341 p.

Payne, R. B. 1969. Breeding seasons and reproductive physiology of Tricolored Blackbirds and Red-winged Blackbirds. Univ. Calif. Publ. Zool. 90:1–137.

Pinowski, J. 1968. Fecundity, mortality, numbers and biomass dynamics of a population of the Tree Sparrow. Ekol. Polska 16:1–58.

Raitt, R. J., and R. D. Ohmart. 1966. Annual cycle of reproduction and molt in Gambel Quail of the Rio Grande Valley, southern New Mexico. Condor 68:541–561.

Ricklefs, R. E. 1967. Relative growth, body constituents, and energy content of
nestling Barn Swallows and Red-winged Blackbirds. Auk 84:560–570.

Riddle, O. 1911. On the formation, significance and chemistry of the white and
yellow yolk of ova. J. Morphol. 22:455–491.

Riddle, O. 1916. Studies on the physiology of reproduction in birds. I. The occurrence and measurement of a sudden change in the rate of growth of the
avian ova. Am. J. Physiol. 41:387–396.

Riddle, O., and P. F. Braucher. 1933. Studies on the physiology of reproduction
in birds. XXXIII. Am. J. Physiol. 107:343–347.

Romanoff, A. L. 1943. The growth of the avian ovum. Anat. Rec. 85:261–267.

Romanoff, A. L., and A. J. Romanoff. 1949. The avian egg. John Wiley & Sons,
New York. 918 p.

Samuels, L. T. 1946. The relation of the anterior pituitary hormones to nutrition.
Recent Progr. Hormone Res. 1:147–176.

Schneider, A. J., B. B. Bohren, and V. L. Anderson. 1955. The effect of restricted
feeding on several genetically controlled characters of the fowl. Poultry Sci.
34:691–702.

Scott, H. M., and L. F. Payne. 1941. The influence of restricted food intake on
the reproductive performance of breeding turkeys. Poultry Sci. 20:395–401.

Singsen, E. P., L. D. Matterson, A. Kozeff, and L. D. Stinson. 1954. The effect
of feeding high and low efficiency rations to growing pullets on their subsequent laying performance. Poultry Sci. 33:1081.

Southern, N. H. 1970. The natural control of a population of Tawny Owls (*Strix
aluco*). J. Zool. 162:197–285.

Stieve, H. 1919. Die Entwicklung des Eierstockseises der Dohl (*Colaeus monedula*).
Ein Beitrag zur Frage nach den physiologischerweise im Ovar stattfinden
Rückbildungsvorgägen. Arch. Mikrosk. Anat. Abt. II. 92:137–288.

Stoddard, H. T. 1931. The Bobwhite Quail. Scribner, New York. 559 p.

Tangl, F. 1903. Beiträge zur Energetik der Ontogenese. 1. Mittheilung. Die Ent-
wicklungsarbeit im Vogelei. Pflügers Arch. Gesamte Physiol. 93:327–376.

Tanner, J. T. 1966. Control of the initiation of egg laying in temperate zone
birds, p. 111–112. *In* D. W. Snow [ed.] Abstracts of the XIV International
Ornithological Congress, Oxford.

Vaugien, L. 1959a. Cycle testiculaire naturel de la Linotte et cycles induits par
modification de l'illumination et de la période alimentaire journalière. C. R.
Séances Acad. Sci. (Paris) 249:1056–1058.

Vaugien, L. 1959b. Incidence de la durée journalière des repas sur la réponse
testiculaire du Moineau domestique illuminé vers la fin de l'hiver. C. R. Séances
Acad. Sci. (Paris) 248:3352–3354.

Vaugien, L. 1960a. Conséquences sur la spermatogenèse du moineau domestique
d'une brève période d'illumination et d'alimentation, au milieu de la nuit. C. R.
Séances Acad. Sci. (Paris) 250:3220–3222.

Vaugien, L. 1960b. L'appétit du moineau domestique s'abaisse graduellment
durant le cycle sexuel jusqu'à une valeur néfaste à la gonadostimulation
expérimentale par illumination. C. R. Séances Acad. Sci. (Paris) 251:1570–
1572.

von Haartman, L. 1971. Population dynamics, p. 391–459. *In* D. S. Farner and
J. R. King [ed.] Avian biology. Vol. 1. Academic Press, New York.

Walter, E. D., and J. R. Aitken. 1961. Performance of laying hens subjected to restricted feeding during rearing and laying periods. Poultry Sci. 40:345–354.

Warren, E. R., and R. M. Conrad. 1939. Growth of the hen's ovum. J. Agr. Res. 58:875–893.

West, G. C. 1960. Seasonal variation in the energy balance of the Tree Sparrow in relation to migration. Auk 77:306–329.

West, G. C. 1968. Bioenergetics of captive Willow Ptarmigan under natural conditions. Ecology 49:1035–1045.

DISCUSSION

Fred C. Zwickel

Fortunately for me, Professor King has allowed me to take an ecological approach by defining his presentation in terms of "ecologically or demographically significant reproduction." Since King, of necessity, has restricted his review to events leading up to and including incubation, I will do my best to stay within these limits. I, however, will make some more general comments about ecology and the energetics of reproduction, dividing my discussion into two main areas: specific comments raised by King's presentation and comments as to why energy may not limit reproduction in nature. These remarks will attempt to raise some questions for consideration and discussion.

COMMENTS RAISED BY KING'S PRESENTATION

It is very interesting that of the several energy areas that Professor King has been able to partition, he concludes that there is a significant energy cost only in the production of the ova. Even here, I wonder if birds may not be able to compensate for caloric restrictions within relatively wide limits through the use of endogenous reserves, perhaps within limits of food normally available in the field?

More specifically, it seems important that even after complete caloric restriction, restoration of adequate intake restores normal gonadal function, without permanent damage. Therefore, even though endogenous energy reserves may be utilized during laying, incubation, and other periods of energy stress, it is unlikely that caloric restrictions, short of starvation, will seriously threaten the survival of the individual. That survival of the breeding individual is important is also indicated by King's suggestion that gonadal function is stopped by neuroendocrine control if caloric restriction exceeds certain levels, i.e., prior to severe, and perhaps permanent, damage that might result if breeding continued. Such adaptation allows the individual to live and to breed at some future time. This suggests flexibility in the proximate timing of breeding and that caloric shortages during the reproductive period may not be demographically important beyond an annual, or even shorter-term, basis.

In addition to survival of the individual, the reproductive performance of calorically restricted birds is an important consideration. Breitenbach *et al.* (1963) report that alimentary efficiency is increased by caloric restriction and that the reproductive performance of such birds, when they are restored to an adequate diet, may exceed that of control birds. Hence, short-term shortage of energy may be of relatively little consequence to subsequent reproductive performance if it occurs prior to incubation, or perhaps even later. Since, in nature, the period of the breeding season is usually one of increasing food supply, caloric shortages are likely to be short term, except under very unusual circumstances.

Finally, although Professor King has not dealt in his presentation with the feeding and parental care of young, I cannot fully agree with his generalization that, "it is *obvious* that this phase is a second period of the breeding season . . . when caloric shortage *exerts* its major effects on reproductive potential" [italics mine]. I submit that this is arguable and will consider the issue at greater length below.

DOES ENERGY LIMIT REPRODUCTION IN NATURE?

Professor King has separated caloric costs of reproduction into several main areas. I will make a few comments about the energetics of reproduction in relation to its possible effects on both production and the size of the breeding population in the following year.

Sadleir (1969), discussing nutrition in relation to reproduction of mammals, has stated, ". . .the wild female mammal rarely meets with detrimental nutritional conditions, as selective pressures timing breeding have reduced the probability that these stages of reproduction will

occur at the time when little feed is available." Similarly, Lack (1968)
and Immelmann (1971), among others, have suggested that the timing
of breeding seasons in birds has evolved mainly in relation to the period
when hatched young can most profitably be raised and that food for
the young is the major selective agent. Thus, in birds, caloric shortage
may not be a critical factor during the posthatch period, at the proximate level. While there are some reports of starvation in nestlings of
altricial species (Cody, 1971), there is no evidence that this is determined calorically. In fact, Cody notes that the incidence of young
birds dying in the nest from starvation is "surprisingly low."

As for the prehatch period, Immelmann (1971) has noted that
some species begin breeding when environmental conditions are still
relatively severe. He also notes that some parts of the reproductive
cycle may be more or less independent of the environment. If selection has operated mainly at the level of hatched young, then perhaps
caloric shortage at the beginning of the breeding season is not normally
critical to reproduction. Other evidence that this may be so lies in
the fact that many birds, both migratory and nonmigratory, are at
their peak annual weight in spring, that many species are readily able
to renest if their first effort is destroyed, and that more northern
species, which presumably begin breeding under relatively severe
environmental conditions, tend to have larger clutches than do more
southern species (von Haartman, 1971; Cody, 1971).

Professor King has said that ". . .demographically significant reproduction culminates in self-sufficient offspring of adult size. . .
that are themselves capable of reproduction. . . ." This implies
a concern at the population level. There is still controversy as to
whether reproduction or mortality is the major determinant of breeding density (Lack, 1966; von Haartman, 1971; Christian and Davis,
1964; Wynne-Edwards, 1962). I must agree with Lack and von
Haartman, who argue that production, quantitatively, is almost
always more than sufficient to replace adult mortality. By this argument, mortality, rather than natality, is the important agent of
regulation. This view ignores caloric quality of the young, but I suspect that this may be beyond the scope of consideration here. Nevertheless, a failure to realize reproductive potential, numerically, even
if energy-determined, will not normally affect the breeding level in
the following year.

In closing, I would make a few general observations. With increasingly exploitive wildlife management and increasing pressures on land
use from soaring human populations, a further understanding of en-

ergetic requirements for reproduction is essential. And, since the breeding season is a period of relatively high energy costs, further study of the energetics of reproduction is essential for a full understanding of the functioning of ecosystems. At the same time, Professor King has indicated the capacity of birds to compensate proximately for periods of caloric stress during the breeding season. Furthermore, most investigators seem to agree that seasonal breeding is adaptive, especially in relation to food. These two points, plus empirical observations and experimental evidence from the field, suggest to me that reproduction is not normally significantly curtailed by caloric shortage under natural or seminatural conditions, when considered in relation to required recruitment to the subsequent breeding population. More experimental studies at the population level are needed to resolve this issue.

REFERENCES

Breitenbach, R. P., C. L. Nagra, and R. K. Meyer. 1963. Effect of limited food intake on cyclic annual changes in Ring-necked Pheasant hens. J. Wildlife Manage. 27(1):24–36.

Christian, J. J., and D. E. Davis. 1964. Endocrines, behavior, and populations. Science 146(3651):1550–1560.

Cody, M. L. 1971. Ecological aspects of reproduction, p. 461–512. *In* D. S. Farner and J. R. King [ed.] Avian biology. Vol. 1. Academic Press, New York.

Immelmann, K. 1971. Ecological aspects of periodic reproduction, p. 341–389. *In* D. S. Farner and J. R. King [ed.] Avian biology. Vol. 1. Academic Press, New York.

Lack, D. 1966. Population studies of birds. Clarenden Press, Oxford. 341 p.

Lack, D. 1968. Ecological adaptations for breeding in birds. Methuen, London. 409 p.

Sadleir, R. M. F. S. 1969. The role of nutrition in the reproduction of wild mammals. J. Reprod. Fertil. Suppl. 6:39–48.

von Haartman, L. 1971. Population dynamics, p. 391–459. *In* D. S. Farner and J. R. King [ed.] Avian biology. Vol. 1. Academic Press, New York.

Wynne-Edwards, V. C. 1962. Animal dispersion in relation to social behavior. Oliver and Boyd, Edinburgh. 653 p.

DISCUSSION

S. Charles Kendeigh

To measure and evaluate the energy resources available for reproduction, the total energy budget of a bird must be understood. Considerable work on this question has been done on the House Sparrow.
The energy balance varies with environmental temperature (Figure 1).
With fasting sparrows at complete rest, as at night, there is a zone of
thermal neutrality wherein basal metabolism remains constant and
body temperature is regulated by adjustments in rate of heat loss
(Hudson and Kimzey, 1966). Below the lower critical temperature of
21–22 °C, there is an increased heat production proportional to the
decline in temperature. Likewise, above the upper critical temperature of 37 °C, there is an increase in heat production due to uncon-
trollable rise in body temperature. During the daytime, no zone of
thermal neutrality is discernible and there is a linear increase in heat

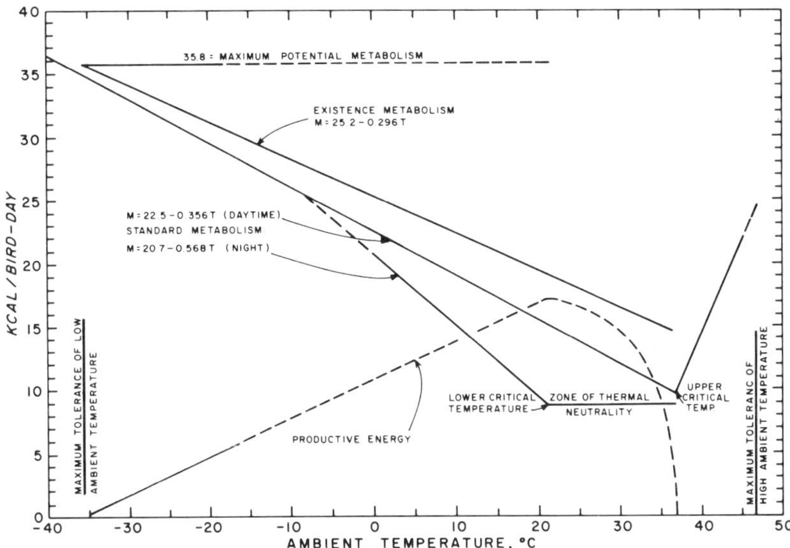

FIGURE 1 Energy responses of the House Sparrow to its thermal environment (from
Kendeigh, 1969).

production–i.e., standard metabolism–below the upper critical temperature.

Although basal and standard metabolism are of considerable physiological significance, they are seldom attained under natural conditions. More important ecologically is the measurement of existence metabolism, i.e., the energy balance of active, feeding, caged birds maintaining a constant weight (Kendeigh, 1969). Existence metabolism varies linearly from the upper critical temperature to the lower limit of temperature tolerance of −35 °C. At this extreme environmental temperature stress, they are mobilizing their maximum potential metabolism or heat production simply to maintain body temperature. It has been demonstrated that the maximum potential metabolism can be attained, at higher environmental temperatures, under stress of forced exercise (Kontogiannis, 1968) or egg-laying (El-Wailly, 1966).

The difference between the maximum potential energy that a species can attain and the amount that is required for existence at any environmental temperature is considered *productive* energy–this is what is available for reproduction. The availability of productive energy increases with temperature up to the zone of thermal neutrality. The capacity for sustained work becomes zero at the upper critical temperature because work produces extra heat and even in the case of resting birds exposed to environmental temperatures above this critical point, there is difficulty in dissipating the heat of basal metabolism. Hyperthermy develops at these high ambient temperatures that may lead to death of the bird. The environmental temperature at which the sparrow attains its maximum working capacity is approximately 22 °C. Below this temperature, extra energy is required for existence; above this temperature, the bird comes under heat stress (Kontogiannis, 1968). It may be more than coincidence that this point of greatest working capacity comes at the lower critical temperature. The decline in working capacity in the zone of thermal neutrality is probably inversely proportional to the increasing rate of evaporative cooling.

The initiation and continuance of reproductive activities depends on the availability of productive energy. The House Sparrow in Illinois begins egg-laying in April, but the laying of first clutches reaches a peak in early May and of second clutches in mid-June. There may even be third clutches, but in any event egg-laying terminates in early August. The normal ambient temperature for April in central Illinois is 10.6 °C; May, 16.5 °C; June, 21.8 °C, July 24.1 °C; and August, 23.0 °C. Temperatures during certain periods within each month are, of course, both higher and lower than these figures. But it is significant that egg-laying does not commence until productive energy is approaching its peak at 22 °C and stops soon after it goes beyond the peak.

The temperature, or the range of temperatures, at which a species has its greatest productive energy and working capacity is very significant, since it represents the optimum conditions for reproduction, affects the timing of events in the yearly cycle, determines the center of geographic distribution, regulates where populations may reach highest densities, and so on. If this temperature cannot be directly determined through experimentation, then the lower critical temperature might be taken as an approximation in those species where it is known (Kendeigh, 1969).

There is no question in my mind, contrary to the position taken by Professor King, that incubation carries an energy cost and that, if the energy is not available, reproduction will not take place. Heat is lost from the incubating bird to the eggs through the brood-patch. The brood patch develops only during the incubating period, becomes highly vascularized, and is clearly an adaptation for rapid and efficient transfer of heat from the body interior through the skin directly to the eggs. Although it has not been measured directly, there can be no doubt that the rate at which the heat is transferred is considerably greater than is the rate of heat loss from an equivalent area of ordinary skin on a nonincubating bird. Heat loss to the eggs must be compensated for if the bird is to maintain its body temperature; to me, this represents the energy cost of incubation. As I have pointed out (Kendeigh, 1963, p. 900), the amount of heat lost through the plumage for a bird at rest at summer ambient temperatures is relatively small and is entirely insufficient for maintaining the eggs at incubation temperature.

The energy cost of incubation can be measured by either direct or indirect methods; the latter will be considered in subsequent discussions. A direct method of measuring these costs is demonstrated in the work of El-Wailly (1966) with the Zebra Finch. The existence metabolism of caged pairs of this species was determined at five different constant temperatures (Figure 2). They were then supplied with nesting material, whereupon egg-laying and incubation began promptly. The difference between the energy metabolized during these activities and that metabolized before they began is considered the energy cost of the activities. The energy cost of egg-laying alone cannot be clearly defined, as it was combined with nest-building and

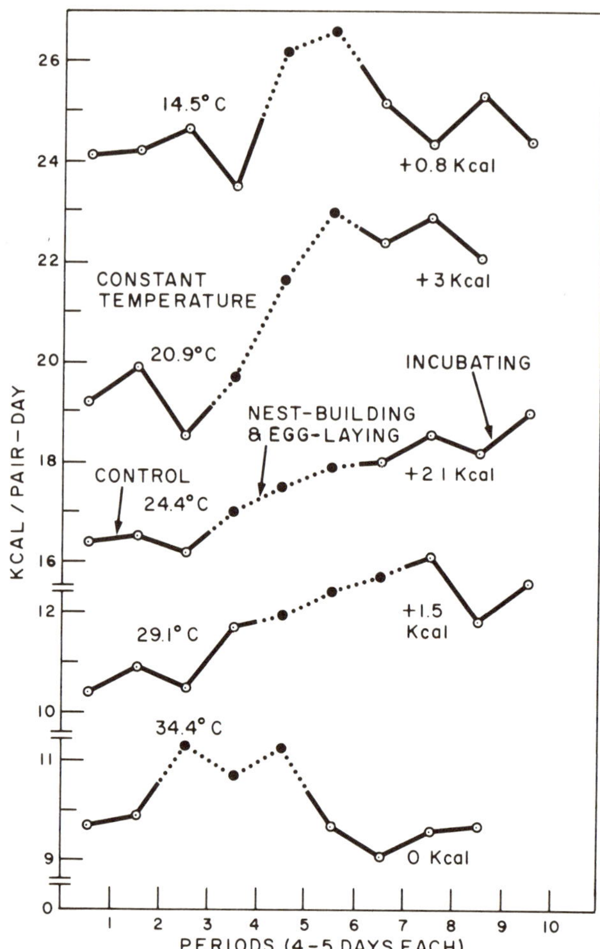

FIGURE 2   Metabolized energy of the Zebra Finch at various stages in the nesting cycle in relation to ambient temperature (El-Wailly, 1966).

beginning of incubating behavior, but at 34.4 °C it appears to be about 1.4 kcal/pair per day. The energy cost of incubation rises with decline of ambient temperature from zero at 34.4 °C—which approximates incubation temperature even without the bird on the eggs—to 3 kcal/pair per day at 20.9 °C. At 14.5 °C, existence metabolism before nesting began (24.0 kcal/pair per day) was already higher than during incubation at 20.9 °C (22.5 kcal/pair per day). The birds could not muster sufficient productive energy on the 12-hour photoperiod to maintain incubation temperatures; the eggs therefore did not hatch. The amount needed would have been about 4.8 kcal/pair per day or about 40 percent of the productive energy available—this was too high a cost and incubation was unsuccessful. Although locomotor activity was not monitored quantitatively, there was no increased activity observed at the lower temperatures. Although there is some heat production by the embryo, it is small and, in the case of the altrical House Wren, we were unable to find that it measurably affected the cooling rate of the eggs (Kendeigh, 1963) . Even in the much larger eggs of the semiprecocial Herring Gull, to be considered later, embryonic heat production does not become very significant in conserving energy expenditure by the adults during the first 20 days of the 30-day period (Drent, 1967).

A monthly energy budget throughout the year has been worked out for the House Sparrow that may serve as a sample of what may be done for other species (Figure 3). Monthly measurements of basal metabolism are from Miller (1939). Monthly values for standard and existence metabolism are calculated, using mean monthly temperatures in the equations presented by Kendeigh (1969). Excess locomotor activity of free existence was calculated from data given by Owen (1969) for the Blue-winged Teal (*Anas discors*). The energy cost of egg-laying was determined in a manner similar to that for the House Wren (Kendeigh *et al.*, 1956). The cost of incubation was calculated from an equation given by Kendeigh (1963) that will be discussed more fully later. The energy cost for raising the nestlings was assumed to be the same as for incubation, and the total reproductive costs were prorated over the entire breeding cycle. Blackmore (1969) has measured the energy cost of molting in the House Sparrow. The cost of adding weight in the species has been worked out by Kendeigh *et al.* (1969). The maximum potential metabolism throughout the year is based on limits of low temperature tolerance determined monthly by Barnett (1970). For more complete details on how this monthly energy budget was compiled see Kendeigh (in press). It may be noted that the species does not use all the energy that it has the capacity to mobilize. The difference between the maximum potential and the actual total metabolism used represents latent energy available for periods when temperatures drop below the mean and for emergencies.

A budget like this shows how the energy costs for incubation relate to the whole. During May, June, and July, the 3 months of most intensive reproductive activities, energy for reproduction makes up

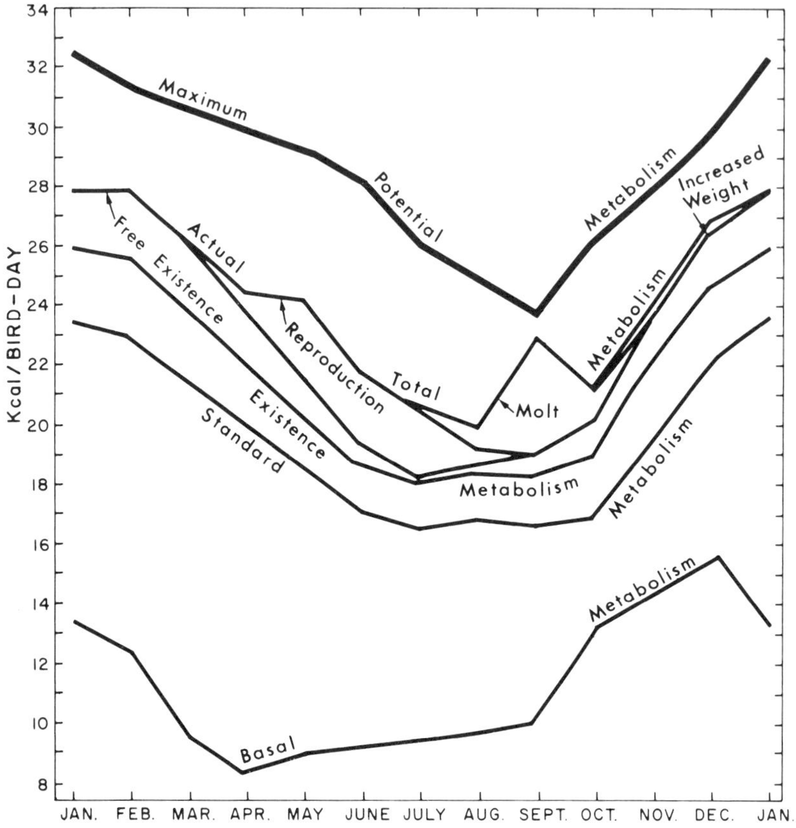

FIGURE 3   Monthly energy budgets of the House Sparrow throughout the year in central Illinois (from Kendeigh, in press).

only 11 percent of the total being used. In terms of the productive energy being used, however, it is 78 percent. Of the productive energy that could be mobilized, it constitutes 27 percent.

Since more measurements have been made of basal metabolism, of standard metabolism, and of existence metabolism than of energy requirements of free-living birds, attempts are often made to estimate the last from the first three. Ratios vary according to the time of the year, but for the House Sparrow during the breeding season (May, June, and July), the actual total metabolism used is 2.4 times basal metabolism, 1.5 times the prevailing standard metabolism, and 1.17 times the prevailing existence metabolism.

Finally, I would point out that, while the measurement of energy requirements and resources is a complex problem in physiological ecology, a knowledge of these factors is essential to understand how a species fits into its environmental niche, why it behaves the way it does, and how to maximize its productivity for man's benefit.

REFERENCES

Barnett, L. B. 1970. Seasonal changes in temperature acclimation of the House Sparrow, *Passer domesticus*. Comp. Biochem. Physiol. 33:559–579.

Blackmore, F. H., Jr. 1969. The effect of the temperature, photoperiod, and molt on the energy requirements of the House Sparrow, *Passer domesticus*. Comp. Biochem. Physiol. 30:433–444.

Drent, R. H. 1967. Functional aspects of incubation in the Herring Gull, *Larus argentatus* Pont. E. J. Brill, Leiden. 132 p.

El-Wailly, A. J. 1966. Energy requirements for egg-laying and incubation in the Zebra Finch, *Taeniopygia castonotis*.. Condor 68:582–594.

Hudson, J. W., and S. L. Kimzey. 1966. Temperature regulation and metabolic rhythms in populations of the House Sparrow *Passer domesticus*. Comp. Biochem. Physiol. 17:203–217.

Kendeigh, S. C. 1963. Thermodynamics of incubation in the House Wren, *Troglodytes aedon*, p. 884–904. *In* Proceedings, XIIIth International Ornithological Congress. American Ornithologists' Union, Louisiana State University, Baton Rouge.

Kendeigh, S. C. 1969. Energy responses of birds to their thermal environment. Wilson Bull. 81:441–449.

Kendeigh, S. C. Monthly variations in the energy budget of the House Sparrow throughout the year. *In* S. C. Kendeigh and J. Pinowski [ed.] Productivity, population dynamics and systematics of granivorous birds. Institute of Ecology, Polish Academy of Science, Warsaw. In press.

Kendeigh, S. C., T. C. Kramer, and F. Hamerstrom. 1956. Variations in egg characteristics of the House Wren. Auk 73:42–65.

Kendeigh, S. C., J. E. Kontogiannis, A. Mazac, and R. R. Roth. 1969. Environmental regulation of food intake by birds. Comp. Biochem. Physiol. 31:941–957.

Kontogiannis, J. E. 1968. Effect of temperature and exercise on energy intake and body weight of the White-throated Sparrow, *Zonotrichia albicollis*. Physiol. Zool. 41:54–64.

Miller, D. S. 1939. A study of the physiology of the sparrow thyroid. J. Exp. Zool. 80:259–285.

Owen, R. B., Jr. 1969. Heart rate as a measure of metabolism in Blue-winged Teal. Comp. Biochem. Physiol. 31:431–436.

INFORMAL DISCUSSION

W. O. WILSON: At 22 °C, *Coturnix* take about 20 percent more feed on egg-forming days than they do on non-egg-forming days, whereas they take 50 percent more feed at 10 °C.

With chickens, we know that the zone of thermal neutrality varies with acclimatization.

Finally, again with chickens, the best egg production occurs at 10 °C, but the best feeding efficiency for conversion of feed into eggs occurs at approximately 30 °C.

CONLEY: It seems to me that there is a distinction between absolute caloric requirements for adequate clutch development and for the individual parents and what we might refer to as general food supplies for population levels. The first is associated primarily with the physiology of reproduction and is related to laying high-weight eggs; the second has to do with the ecology and the demographics of recruitment. This distinction is dramatic in those species in which the parents feed on one type of food and the young feed on another type. Thus the physiology of getting the clutch out and getting it hatched would be one thing. Recruiting the young birds into the population and bringing them to breeding age would be quite a different question.

The need, then, is to distinguish between our ability to define energy requirements at an individual level, i.e., to get the clutch in the field, from the demographic consequences of ensuring recruitment.

HICKEY: Would any of our discussants or speakers care to comment on this?

SCOTT: I wish to comment on the first part of Dr. King's talk on caloric restriction and testicular activity. As I understood it, you restricted the total diet to 51 percent and 62 percent of *ad libitum*. I want to know if the protein, amino acids, vitamins, and minerals were all increased to double the amount that they were in the *ad libitum* diet, so that you knew you were restricting only energy or calories. It seems to me that protein restriction should have a very great effect because of the amino acid pattern that is needed. To call it caloric restriction alone seems to be nutritionally a little bit wrong.

IMMELMANN: I would like to refer to Figure 6 from Professor King's presentation. As one of the investigations that should be carried out in the near future, he mentioned clutch size in relation to season and environmental conditions. I would like to mention that, in recent years, several studies concerned with this question have been carried out with European passerines mainly with the Pied Fly-catcher *Ficedula hypoleuca* and several species of shrikes.* They

---

*Berndt, R., and W. Winkel. 1967. Die Gelegegrösse des Trauerschnäppers (*Ficedula hypoleuca*) in Beziehung zu Ort, Zeit, Biotop und Alter. Vogelwelt 88:97–136: Ullrich, B. 1971.

clearly show that, irrespective of the age of the bird, the clutch size decreases as the season proceeds. This is despite the fact that, as the laying period progresses, the available food supply for these mainly insectivorous birds greatly increases. This means that decreasing clutch size is probably not controlled by food supply, but by some other, possibly internal, factors.

The reason that has been discussed in these reports is that, for the offspring of later clutches, environmental conditions may start to deteriorate again during or shortly after their fledgling period. Therefore, the smaller size of clutches may enable the parents to provide the fewer young with sufficient nourishment. This might be an adaptation similar to those mentioned by Doctor Zwickel—i.e., that in many cases the breeding season is not primarily adapted to actual calorie intake by the female, but to an optimum chance of survival of the young.

HICKEY: This change in the size of the clutch according to the date laid is well known in galliforms like the Ring-necked Pheasant and the Bobwhite Quail. I might add that in April some of the females may be at their maximum state of lipid content.

I also wish to observe in connection with the energy required to produce a goose egg that modern American game management for geese is predicated upon the need to send the geese north with the maximum amount of fat. It is due to a winter feeding program that has been assiduously carried out that the goose population of North America is now at an all-time high.

NORTON: Like Doctor Conley, I was bothered by one conceptual fuzziness in presentations in this section. Both basic and applied studies in avian reproduction would be helped by making a clear distinction between bioenergetics of breeding and the energetics of reproduction in birds. The former is a concept encompassing physiology and autecology, whereas the latter concept introduces demography and synecology.

My studies on energetics of Arctic calidridine sandpipers illustrate the distinction. Arctic shorebirds undertake some bioenergetically extravagant processes during their breeding cycles, but such wastefulness may be vindicated in the long run by the reproductive performance of these birds. Sandpipers lay 4-egg clutches over a 4-day span; in smaller species, the output in egg mass actually exceeds the weight of the female. Incubation of these massive clutches confronts problems of heat loss, since the nests offer minimal insulation and the capricious arctic weather may make additional energy demands on incubating adults. Moreover, incubation takes place during a period preceding annual temperature maxima. Metabolic extravagance is further shown by species in this group that suffer the added expense of molting during or just after incubation and by extensive overlaps in feeding habitat and feeding styles among sympatric-breeding species.

Untersuchungen zur Ethologie und Ökologie des Rotkopfwürgers (*Lanius senator*) in Südwestdeutschland. Vogelwarte 26:1–77.

The overall reproductive strategy, on the other hand, may be energetically sound, since young hatch during a brief period when insect prey are abundant. Most mortality of birds on the breeding grounds is concentrated in the first few days of chick life, at which time the chicks must have access to small surface invertebrates in order to survive and grow. The necessary coincidence of shorebird hatching and insect emergence in the Arctic thus exacts a bioenergetic price from adults.

Studies of other species of shorebirds further emphasize the importance of looking at the whole reproductive strategy—as opposed to strictly breeding strategy—in evaluating the ecological commitments and energetic limitations upon bird populations. There are several species now known to be capable of laying successive 4-egg clutches, one each for incubation by male and female. Such a strategy raises the bioenergetic costs of both laying and incubation. But the return, in the form of reproductive success and recruitment to the population, must outweigh these costs. Double- or multiple-clutching would make particular strategic sense if found to represent a "hedging of bets" on optimum hatching time in relation to resources for the chicks. This strategy may appear in shorebirds breeding in an environment likely to fluctuate widely during the period between egg-laying and hatching. This system, in affording an early- and a late-hatching clutch, may make the difference between complete and only partial reproductive failure.

Role of the Environment in Reproduction as Source of "Predictive" Information

*KLAUS IMMELMANN**

Successful reproduction of birds and other species requires a number of ecological preconditions. On the one hand, the production of mature germ cells and the various behavioral activities connected with reproduction make special physiological claims on the adult organism; on the other hand, to ensure an optimal chance of survival, the young must be hatched at a time when the basic food situation is qualitatively and quantitatively favorable and when access to food is as easy as possible.

In the vast majority of climatic areas, optimum conditions are realized during only a part of the year. As a consequence, most species of birds have restricted breeding seasons and start nesting at about the same time each year.

The timing of restricted breeding touches two different groups of problems: the phylogenetic aspect, i.e., the development of species- and population-specific breeding seasons through natural selection and the current physiological aspect, i.e., the precise adjustment of the individual bird to the phylogenetically determined reproductive

*Dedicated to Professor Doctor Jürgen Aschoff on the occasion of his sixtieth birthday.

season. It is the second aspect that will receive most emphasis in this paper although a brief survey of the phenology of breeding seasons and their possible ecological determinants will contribute to a better understanding of actual timing processes.

The restriction of breeding to certain parts of the year is absent only in those climatic regions—like those on certain tropical islands and in certain tropical rain forests—where environmental conditions are equally, or almost equally, favorable during all months of the year. Even here, however, definite laying periods have been described (Ashmole and Ashmole, 1967; Ward, 1969) and similar timing requirements may pertain.

PHENOLOGY OF AVIAN REPRODUCTIVE SEASONS

From the literature, the following very broad outline of breeding seasons in birds can be compiled: In the temperate latitudes of Eurasia and America, the situation is highly uniform and almost all breeding occurs in spring and in early summer. In the tropics, on the other hand, breeding seasons of birds are very heterogeneous. In many areas, different ecological groups may breed at quite different times of the year and geographical variation tends to be far greater than in temperate zones. As a rule, the breeding season of tropical birds tends to be connected in some way with the seasonal distribution of rains; inhabitants of lowland forests with heavy annual precipitation tend to concentrate breeding in the drier months, whereas most birds in regions with a regular alternation between one long dry and one long wet season per year breed around the rainy season, with only a few specialists laying during the dry period. In areas with two wet and two dry seasons, some species breed twice a year, whereas others nest in the short rains and avoid the long rains. Double-broodiness similarly has appeared in some tropical and subtropical sea birds. Finally, certain species in different parts of the world do not possess regular breeding seasons but nest at irregular intervals. This last applies mainly to the inhabitants of the arid regions of central Australia, which are known as opportunistic breeders, taking advantage of favorable conditions created by the generally aperiodic rainfall and breeding irrespective of the time of the year [for reviews of avian breeding seasons, see Lack (1950), Moreau (1950), Lofts and Murton (1968), Immelmann (1971), Serventy (1971)].

ECOLOGICAL AND PHYLOGENETIC ASPECTS OF RESTRICTED BREEDING

As just noted, restricted breeding is a general phenomenon among birds, and very different breeding seasons occur in the different climatic regions of the world. These species- and population-specific breeding patterns have to be regarded as the result of specific environmental adaptations. Each species or population of birds is apparently adjusted to breed at the time(s) of the year at which it can raise young most efficiently. Young birds hatched at less advantageous times suffer a heavier mortality and rarely survive to reproductive age. The gene complexes of those pairs producing fledglings at the "wrong" time of the year therefore will be reduced or eliminated by natural selection. In general, such selection pressure is greatest in species with rather specialized feeding or nesting habits or that breed in areas where the favorable season is short. Consequently, those specialists always tend to have a brief and well-synchronized breeding season, whereas the nonspecialists may be less restricted and may display more individual variation.

For those environmental factors that control efficiency of breeding and, through natural selection, have led to the evolution of species- and population-specific breeding periodicities, Baker (1938) has introduced the term "ultimate causes." This has been slightly altered by Thomson (1950) into "ultimate factors," a term that has since been commonly used not only for birds but also for other animals, including invertebrates.

By far the most important ultimate factor for nearly all species of birds is the availability of an adequate food supply for parents and young. Many species, however, also are dependent on some additional environmental factors that may limit breeding and, hence, may lead to further restriction of the period of reproduction. These factors may include the necessary conditions for nesting, various climatic factors (temperature, rainfall, wind), or the number of competitive species in any one area.

NEED FOR PREDICTIVE INFORMATION

To summarize the evidence presented in the preceding sections, the temporal patterns of avian reproduction have been developed as an adaptation to the differences in ecological conditions at different times of the year. An adaptational system like this, of course, does require some sort of information that enables the individual bird to adjust itself to the respective environmental conditions and seasonal changes. Here, the following problem occurs: In almost every case, the annual cycle of a species of bird reaches its critical point at the time when the new generation leaves the nest and, after a while is released from parental care. At this time, on the one hand, the population density rises abruptly, so that a multiple of the previous number of individuals has to find sustenance in the same area. On the other hand, the more or less exposed and inexperienced young need particularly good facilities for retreat and shelter and need substantial and easily accessible nourishment. This leads to the conclusion that the breeding cycle of any species of bird in a seasonal climate must be synchronized with its changing environment in such a way that the optimal season coincides with the fledging of the young. This means that many ultimate factors operate toward the end of the reproductive period.

Fledging of the young, however, is always the end of a chain of events that includes, apart from the seasonal maturation of the gonads, nest-building, incubation, and care for the nestlings and— depending on the size of the species—may take several weeks or months. These processes, therefore, must occur in many cases before the period of optimal environmental conditions and therefore must be initiated a more or less long interval in advance. As a consequence, many species of birds start egg-laying under suboptimal living conditions, and some large birds of temperate and polar zones may even incubate while the snow is deep [for reviews, see Lack (1950), Prévost and Bourlière (1955)].

This again requires the organism to be informed well in advance of the optimal period concerning the imminent improvement of environmental conditions. It follows that, in many cases, the ultimate factors that have led to the phylogenetic development of favorable breeding seasons are not suitable as the sort of environmental information needed. With the possible exception of those climatic regions where optimal conditions last for many months and where some species of birds can thus afford to come into breeding condition only after favorable conditions have been reached, the ultimate factors usually will occur too late to serve as environmental information for the onset of reproductive activity. Consequently, the evolution of restricted breeding seasons necessarily also must incur the parallel development of a timing mechanism, causing the necessary physiologic and behavioral preparations for reproduction to occur in advance of optimum environmental conditions. In other words, the organism must possess the capacity to respond to "forewarning" stimuli that predict an improvement and thus set the aforementioned developmental cycle in motion at a point in time when the ultimate factors are not yet present.

SOURCES OF PREDICTIVE INFORMATION

Theoretically, there are two different sources for predictive information: rhythmic processes within the organism itself and external stimuli from its environment.

On principle, it is possible that the annual cycle of a bird, including the annual maturation and subsequent regeneration of the gonads, is governed by internal processes. They might well consist of regularly recurring endogenous rhythms with a period that corresponds to the natural sequence of environmental events, or of a nonrhythmic autonomous sequence of physiologic events induced by an external stimulus but running down without need of further stimuli until the next impulse is needed ("hourglass sytem"). Both processes could lead to the maturation of gonads and the initiation of reproductive behavior in advance of favorable breeding conditions and without immediate information from the environment.

The existence of regularly recurring endogenous periodicities in the control of reproductive cycles in birds has been presumed by several authors but has been seriously questioned by others on the basis that evidence in most cases is merely circumstantial (Farner, 1970b; Immelmann, 1971). In recent years, however, some long-term experimental studies clearly have revealed internal cycles for several functions (molt, body weight, gonad size, migratory restlessness) in several species of European warblers (Berthold *et al.*, 1971; Gwinner, 1972) (Figures 1 and 2) and further studies of endogenous timing mechanisms are urgently needed.

Endogenous periodicities of several months' duration have recently been termed "circennial" by Farner and Follett (1966) or "circannual" by Gwinner (1972). This corresponds to the term "circadian" for diurnal rhythms.

From the ecological standpoint, endogenous circennial periodicities can be expected mainly in two types of environment: In relatively

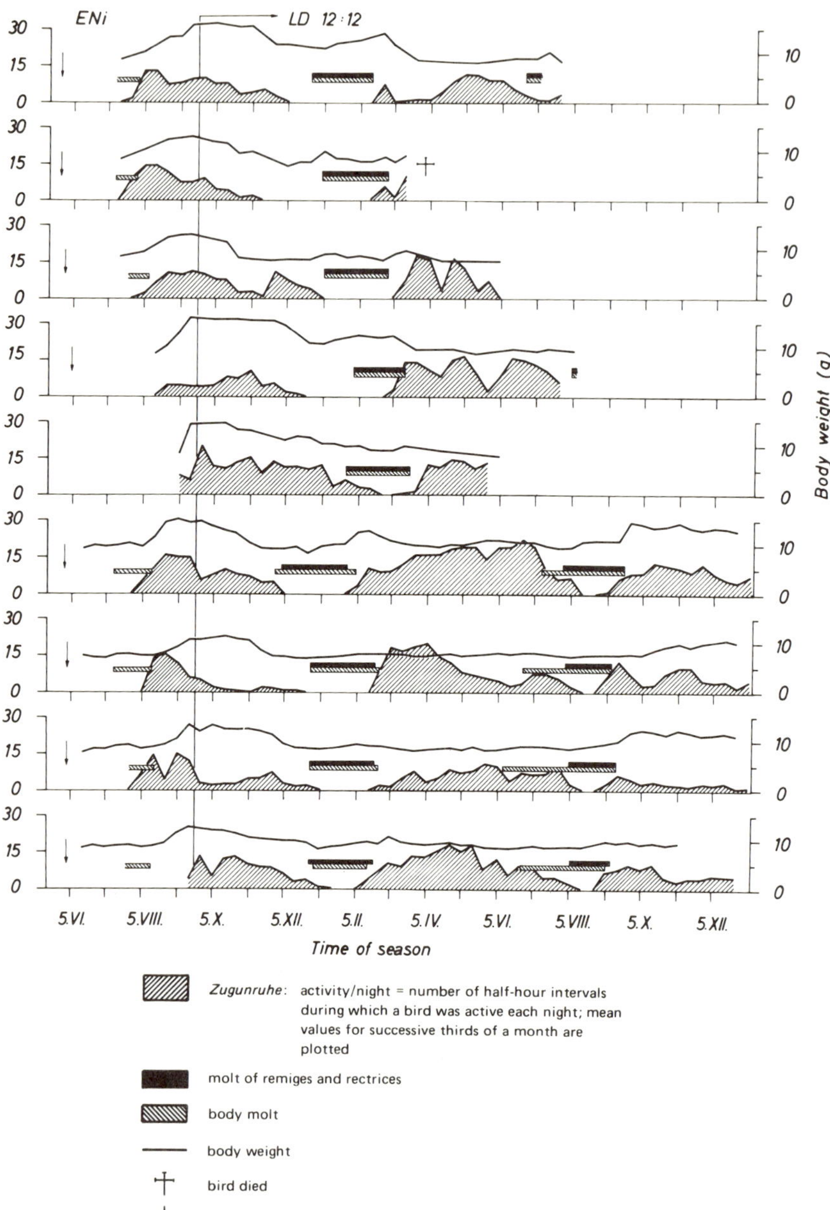

FIGURE 1   Variations in *Zugunruhe*, body weight, and molt in nine Willow Warblers (*Phylloscopus trochilus*) that were transferred in late September (vertical line) from natural light conditions (ENI) of their breeding grounds to a continuous LD 12:12 cycle. Bird No. 5 was the only subject not raised by hand; it was caught during the autumn migration. From Gwinner (1972).

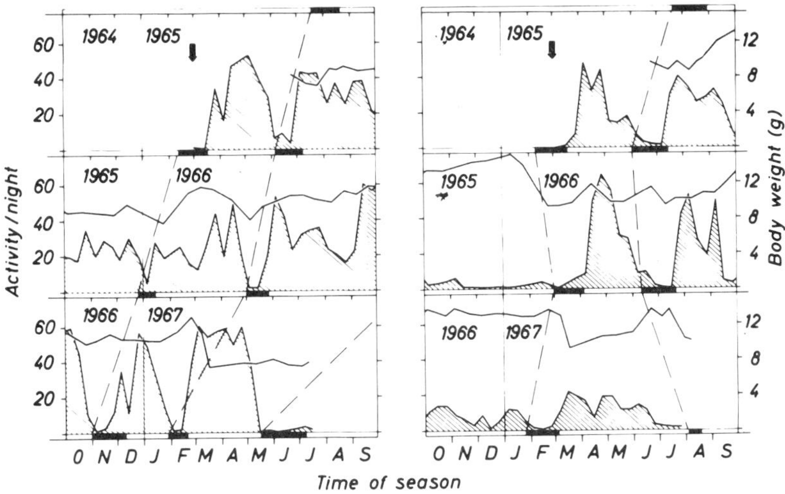

FIGURE 2 Variations in *Zugunruhe*, body weight, and molt in two Willow Warblers (*Phylloscopus trochilus*) that were kept from late February (1965) (arrow) for 27 months in a continuous L D 12:12 cycle. Successive years are displayed one beneath the other. The dashed lines connect the onsets of corresponding molts in successive years. Activity/night = number of 10-min intervals during which a bird was active each night; mean values for successive thirds of a month are plotted. Solid bar indicates molt. Other symbols as in Figure 1. From Gwinner (1972).

seasonless tropical areas, where conditions are favorable throughout the year and where—because synchronization is unnecessary—the organism may largely follow an internal sequence of alternate breeding and regeneration (Schreiber and Ashmole, 1970), and, second, in those seasonal climatic regions where optimal conditions for reproduction occur at almost exactly the same time every year. In the latter case, however, even if a species had evolved a 12-month cycle corresponding to the environmental changes, the cycle would gradually drift out of phase unless maintained by some kind of external regulation.

A second component of the timing mechanism that adjusts individual reproduction to the ultimately selected time of the year is the ability of the organism to react to environmental factors that predict this particular season in advance. Baker (1938) was the first to make a clear distinction between the ultimate factors responsible for the adaptation of breeding seasons and those factors that keep the adapted organism in conformity with the periodicity of its environment. The latter group was called "proximate causes," which was altered into "proximate factors" by Thomson (1950).

Proximate factors must possess two main attributes: They must be reliable and—due to the rather late invocation of the ultimate factors just mentioned—they must be predictive of forthcoming events. The longer the sequence of developments that culminates in the fledging of young, the more widely must ultimate and proximate factors diverge.

NATURE OF PROXIMATE FACTORS

In general, any factor that can be perceived by the organism and that in the course of the year changes in a way that is regularly connected with changing ecological conditions may function as a proximate factor. The evidence available, however, shows that only one or two environmental stimuli have gained general importance as proximate factors and that in this respect there is a profound difference between North Temperate and tropical birds.

PHOTOPERIOD

At middle and high latitudes, the most regular environmental event and, hence, the most reliable proximate factor for the initiation of reproductive activities is the annual change in the length of the day. As an environmental change that, furthermore, has a causal connection with the subsequent improvement of living conditions, it is ideal as a proximate factor. In the vast majority of temperate zone species, therefore, increasing photoperiod after the winter solstice is the most important stimulus for gonad development. Though many interspecific differences in photosensitivity have been described, there is no other environmental factor in any climatic region that is of comparable importance for the immediate control of annual cycles.

The significance of the increasing length of days was first proved by Rowan's (1926) classic experiments. In the meantime photoperiodic mechanisms have been demonstrated experimentally in about 60 species in different families (Farner and Follett, 1966; Farner, 1967). In some parts of the southern hemisphere where spring breeding is predominant, as in parts of southern Africa and eastern Australia (Winterbottom, 1963; Keast, 1968; Ridpath, 1972), photoperiodicity may also be involved, but experimental evidence is still lacking.

Possibly, for those few temperate zone species that regularly breed in autumn or start egg-laying before or around the winter solstice and must hence begin gonad development well ahead of the shortest day, decreasing rather than increasing day length may work as a proximate factor. Such a connection frequently has been suggested but, again, final proof is lacking.

In tropical areas, there must be much less use of day length as information in the control of reproductive cycles than in temperate regions. This is so not only because seasonal fluctuations in day length are small, but chiefly because there seems to be no immediate connection between changes in day length and subsequent changes in food supply and other environmental conditions. However, a number of tropical species have been found under experimental conditions to respond to photostimulation, but this does not necessarily imply that photoperiodism participates in the regulation of reproductive seasons under natural conditions (Farner, 1970a).

RAINFALL

As noted above, the breeding seasons of many tropical birds are associated with the seasonal distribution of rain. In these cases, rainfall may also be a valuable source of environmental information, and a stimulatory effect of precipitation on the reproductive activity of birds has indeed been reported for a number of mainly arid zone species in Australia, Africa, and tropical America (Keast and Marshall, 1954; Immelmann, 1963; Hoesch and Niethammer, 1940; Moreau, 1950; Brooke, 1966; Skutch, 1950; Snow, 1966).

Unfortunately, however, experimental evidence is still lacking and conclusions drawn thus far are based only on field observations. In contrast to the photoperiodic control of avian reproduction, therefore, there seems to be no good evidence as yet concerning the nature of the stimulus or combination of stimuli involved in rainfall as an informational source. It seems even possible that the mechanism is different in different ecological groups of tropical birds. For example, it has frequently been suggested that the relevant stimulus for the onset of reproductive activity is not provided by rainfall itself, but rather by the results of precipitation, such as green vegetation and improved food supply (Moreau, 1950; Marshall and Disney, 1957; Ridpath, 1972). This is quite possible, but may well apply mainly to those species that breed toward the end of the wet season. Early breeders must rely on an early stimulus to initiate gonadal development, well before the rains have caused any improvement in environmental conditions. On the contrary, for at least one group of birds, the granivorous species, the early parts of the wet season are a time of great food scarcity due to the germination of the seeds (Ward, 1965). If these birds do come into breeding condition during the early rains, it clearly must be the rainfall itself that provides the actual stimulus. Immediate reaction to the first, and often very slight, showers at the onset of the wet season has been observed in a number of species (Moreau, 1950; Keast and Marshall, 1954; Snow and Snow, 1964; Maclean, 1967). Even here, however, it is not known which stimulus exerts the most important influence. Factors such as thunder, lightning, sight of falling rain, and change in osmotic state due to the onset of the rains have been mentioned, but proof is lacking (Orr, 1945; Immelmann, 1963; Ward, 1965; Serventy and Whittell, 1967).

A stimulatory effect of rainfall also has been noted in the southern parts of North Temperate zones for a number of species that at other times may be under photoperiodic control, especially for late summer, autumn, and winter breeders in California and southern North America. This suggests that some species respond to both sets of factors known to function as a source of environmental information in avian reproduction.

As the improvement of food conditions in arid tropical regions is due almost exclusively to the influence of precipitation, rainfall itself is the earliest environmental factor functionally connected with the oncoming favorable season. This means that rainfall can have a significant predictive function; in this respect, it closely resembles photoperiodic control in North Temperate zones. As for reliability, however, rainfall lags far behind photoperiodic control. In many tropical areas, precipitation is rather erratic, and the first showers of the beginning wet season are frequently followed by extended dry spells, during which development of reproductive activities may be interrupted again. Consequently, rainfall, although probably initiating gonad development in many tropical bird species, does not necessarily lead to a continuous development of the processes connected with reproduction. In general, it can be said that in tropical climatic regions environmental information for avian reproduction is less reliable than in temperate zones.

OTHER FACTORS

Apart from photoperiod and rainfall, a variety of other environmental variables have also been described as possible proximate factors. In several species of Australian ducks and some other waterfowl, change in water level appears to control reproductive activities. In the nomadic ducks of interior Australia, an increase in water level has been found to be closely followed by sexual display. As these activities occur, even in areas where there had been no rain but where the rise in river flow had been caused by rainfall at the headwaters, it appears that increase in water level itself must be the relevant proximate factor (Frith, 1967).

Temperature is another environmental factor that, theoretically, could act as a source of information in those climatic regions where the improvement of living conditions is paralleled by increasing temperatures. This applies to temperate zones as well as to a variety of tropical areas. Especially, alterations of temperature seem to be effective and in many species of birds a positive correlation between temperature and the rate of gonad development has been shown (Burger, 1948; Farner and Wilson, 1957). These experimental investigations agree with a large amount of field data, indicating that many North Temperate birds breed later in a cold, late spring rather than in a warm and early one (Marshall, 1949; von Haartman, 1963; Lofts and Murton, 1968). A similar negative effect of low temperatures is found in the winter rain areas of Australia and southern Africa (Serventy and Marshall, 1957; Moreau, 1966) and, even in the tropics, a period of low minimum temperatures may lead to a low level in breeding (Snow and Snow, 1964).

At present, however, the evidence merely indicates the existence of a connection between ambient temperatures and reproductive activities. It does not prove that temperatures serve as a source of predictive information.

From the theoretical point of view, i.e., if the two required attributes of proximate factors are considered (see p. 128), temperature does not seem to be very suitable. In many climatic regions daily temperatures show frequent short-term fluctuations and, hence, do not possess a high degree of reliability. Only long-term averages could be expected to be somewhat more reliable, but no data about their possible informative role are at present available.

Under certain circumstances, food supply also has to be considered. Apart from being the most important ultimate cause for the evolution of breeding seasons, it also may have an immediate influence on the time of egg-laying by affecting the physiology of the adult bird directly. Consequently, it may provide some sort of environmental information. There is much circumstantial and some experimental evidence for the importance of nutritional factors in gonad develop-

ment (Marshall, 1949; Lofts and Murton, 1968; Schreiber and
Ashmole, 1970). A sudden improvement in the quality of food may
be especially effective, and the increase in its availability may be a
proximate trigger that stimulates gonad development (Kahl, 1966;
Ridpath, 1972).

Other factors that have been mentioned in the literature are increased sunshine (irrespective of day length); change in general appearance of the landscape; various stimuli from breeding habitat,
territory, nest site, and nest; and behavioral interactions (Moreau,
1950; Farner, 1959; Marshall, 1959; Lehrman, 1965; Hinde, 1967;
Lofts and Murton, 1968). In many cases, it also could be a combination of such factors that provided the necessary information.

SEXUAL DIFFERENCES IN THE RESPONSIVENESS TO PROXIMATE CONTROL; ROLE OF "SOCIAL ENVIRONMENT"

A large number of observations and experimental studies have clearly
shown that the role of the environment as a source of "predictive"
information may be different between sexes. This is particularly evident with regard to photoperiodic control. In many species under
experimental investigation, it has been found that, in the male, complete spermatogenesis can be induced merely by light stimulation.
The female, on the other hand, also responds to photostimulation
but the stimulus mostly causes only partial ovarian development; the
full reproductive state is achieved only under the influence of various
complementary factors (Farner, 1961; Farner *et al.*, 1966; Lofts and
Murton, 1968). This is in accord with the bulk of field data, indicating that in many North Temperate birds the gonads of the male
mature earlier in spring than do those of the female. In migratory
species, the males tend to arrive at the breeding grounds earlier and
with the gonads more advanced than in the female (Lehrman, 1959;
Lofts and Murton, 1968).

Recognition of sexual differences in the responsiveness to photoperiod leads to the question how the necessary environmental
information is acquired by the female. The relevant experimental data
suggest that this is probably done via the male. In many species of
birds, the female has been found to ovulate only after having been
together with a male for a certain amount of time. This suggests that
the male is more susceptible to environmental stimuli and that he, in
turn, stimulates the female by means of courtship and other behav-

ioral interactions. In other words, part of the environmental information will reach the female only indirectly.

A great advantage of a pattern like this—which may be why sexual differences in responsiveness have been developed and maintained through natural selection—lies in the fact that male courtship activities will lead not only to final ovarian development in the female but at the same time also to a precise synchronization between the members of a pair. This assumption is supported by geographic variations that seem to occur in the overall value of intrapair stimulation. This tends to be least in regions where a strong exogenous control of reproductive cycles induces at least a crude synchronization of the sexes, as in higher geographical latitudes. In the tropics, on the other hand, where powerful proximate factors are lacking, behavioral stimulation may be of considerable importance in keeping the pair together so that no time is lost in finding a sex partner and an early start of egg-laying can be achieved whenever favorable conditions occur. Hence, many tropical birds live in pairs throughout the year or at least engage in pair formation long before the actual breeding season. In most, special behavior patterns have been developed that serve to maintain the pair bond (Immelmann, 1963).

Even in temperate zones, however, females paired with their former mates tend to lay earlier than do those of pairs that had just been formed. Selection may thus favor retention of the same mate from one breeding season to the next and living in pairs throughout the year or at least from autumn to the next breeding season (Kalela, 1958; Coulson, 1966).

The stimulatory effect of male courtship behavior on the female has been known for a long time (Craig, 1911) and has since been confirmed in numerous experimental investigations. In general, visual stimuli seem to be most important in intrapair stimulation, and a male courting behind a glass plate has been found to induce ovulation in the female. Tactile stimulation during copulation and mutual caressing, as well as auditory signals, also have proved to be of great significance. Even without visual contact (as when played from a tape so as to be heard from a neighboring room), the voice of displaying males was capable of accelerating ovarian and oviduct development as well as nesting activities in the female and of initiating ovulation (Lehrman, 1959, 1965; Brockway, 1965; Hinde, 1967). Conversely, female courtship has also been found to have an influence on male spermatogenesis, but to a considerably smaller extent (Burger, 1953; Lehrman, 1959).

Altogether, the stimulatory function of intrapair behavior shows that, apart from other environmental factors, the "social environment" of the bird also may represent a source of information for the timing of reproductive activities. Social environment, however, is not necessarily restricted to the partners of a pair but, at least in colonial breeders, also involves other members of the species. In these cases, a similar exchange of information seems to take place between the breeding pairs. It has been found that beyond intrapair stimulation, courtship and other social activities also exert a stimulating and synchronizing effect within a flock or breeding colony. Again, evidence comes from field observations as well as from experimental investigations. In many sea birds and other colonial nesters, the onset of egg-laying is highly synchronous within the colonies or different parts of colonies, whereas it varies between different colonies or parts thereof (Lack and Emlen, 1939; Disney and Marshall, 1956; Orians, 1961; Medway, 1962; Hailman, 1964; Ward, 1965; Brown and Urban, 1969; Schreiber and Ashmole, 1970). In many species, the communal-breeding pairs also tend to start egg-laying earlier than do those nesting outside the colonies, and the same differences can be found between the center and the edge of a colony. Finally, it has been found that the performance of courtship or nesting behavior by one pair or group of pairs can immediately stimulate other pairs to do the same and can thus lead to closely synchronized activities (Hickling, 1959; Brown and Baird, 1965). Group stimulation has first been mentioned by Darling (1938) and, hence, is widely known as the "Fraser-Darling effect." Its occurrence has since been suggested for a great variety of species (Ashmole, 1963; Immelmann, 1963; Crook, 1964).

Group stimulation also has been demonstrated experimentally. In the Budgerigar (*Melopsittacus undulatus*), the opportunity to hear or see other pairs has been found to stimulate gonad development and various reproductive activities in both members of a pair as compared to neither being able to hear nor see others (Brockway, 1964). In the California Quail (*Lophortyx californicus*), males with sex hormone pellet implants not only came prematurely into sexual condition but also induced a similar acceleration in the untreated males of the covey (Emlen and Lorenz, 1942). In the Ring Dove (*Streptopelia risoria*), stimulation from the surrounding colony is "capable of producing acceleration of reproductive development over and above that induced by day length and by interaction with the mate" (Lott *et al.*, 1967).

It can be concluded that information from other pairs does contribute markedly to the timing of avian reproduction. As in intrapair relationships, such information—apart from supporting other environmental stimuli—may additionally lead to a certain amount of synchronization within the colony or parts thereof. The biological function of group stimulation and synchronous breeding seems to be manifold. It may contribute to reducing predation pressure and mutual interference between breeding pairs, and it may help to achieve quick response to favorable ecological conditions (p. 129) (Immelmann, 1963; Horn, 1968; Nelson, 1968).

COMPARATIVE EVALUATION OF PROXIMATE CONTROL

If the proximate factors known to control avian reproduction are compared, profound differences become apparent. Of environmental stimuli, photoperiod is by far the most reliable and most predictive source of information and is capable of exerting its influence even weeks and months ahead of optimal conditions. To a certain, though rather smaller, extent rainfall also can have a predictive function, but is much less reliable (see p. 130). Temperature is highly unreliable (see p. 131). Food supply can be reliable but does not provide much predictive information as it produces the relevant stimuli only when improvement of living conditions has already begun. Evidence as to the remaining environmental stimuli has not been substantiated to a degree that permits definite conclusions.

A further difference lies in the fact that photoperiod and rainfall, per se, are the only environmental stimuli acting as proximate factors only. In all other cases, the same event may be a proximate as well as an ultimate factor. Especially in those climatic regions that—due to extended periods of favorable ecological conditions—permit a rather long breeding season, the same factors that have caused the phylogenetic development of the relevant period of reproduction also may provide the necessary environmental information for its onset. Here, the limits between proximate and ultimate control are clearly imprecise.

DIFFERENCES IN DEMANDS ON ENVIRONMENTAL CONTROL

It must be pointed out here that environmental regulation of breeding seasons has two different aspects. First, the reproductive cycle as a whole has to be initiated, and this may have to occur a rather long time in advance of optimum conditions (p. 123). Second, the birds have to become adjusted to greater or smaller variations in season that may occur in almost any climatic region. From the different characters of proximate factors, it can be concluded that both functions can be accomplished by different stimuli. In North Temperate zones, for example, photoperiod has been found to be the primary timer initiating gonadal development and controlling its early stages. However, day length usually does not accommodate an immediate adjustment to slight annual variations in the onset of the favorable season. Photoperiodic stimulation, therefore, often leads only to a certain threshold of gametogenetic development at which additional factors (e.g., increasing temperatures, food supply, cover, and availability of favorable nesting sites and nesting materials) are required to bring the gonads into full functional maturity and to initiate actual nest-building and egg-laying. If the final environmental stimuli are lacking altogether (e.g., due to abnormal weather or food conditions), reproduction will not occur (Marshall, 1952). The same relations may also be expected for some tropical birds with regard to rainfall, and the subsequent results of precipitation (such as green vegetation and qualitative and quantitative improvement of food supply), but experimental evidence is lacking.

Those factors that modify the primary influence of day length and precipitation have been called "modifying factors" (Thomson, 1950), "subsidiary factors" (Burger, 1953), or "end stimuli" (Marshall, 1959). Their impact may be different among different species, and there may be several modifying factors acting successively on the same individual. As the subsidiary factors exert their influence considerably later than the primary timers and at times just prior to actual nest-building and egg-laying, they may be at least, in part, identical with the ultimate factors (see p. 123).

Stimuli from the social environment that have been cited as exerting some additional influence on the timing of avian reproduction (p. 133) always belong, of course, to the second group of stimuli, leading to a detailed final adjustment between the partners of a pair or, in some cases, between the members of a group. Under natural conditions, social stimuli probably never act per se, but rather serve to intensify and accelerate the effects of nonsocial proximate factors (Immelmann, 1963).

In principle, the second of the two functions of reproductive timing mechanisms, i.e., exact adjustment to stochastic variations in ecological conditions, can only be accomplished by environmental stimuli. The first task, however, i.e., the initiation and early control of the cycle, also could, from a theoretical point of view, be governed by internal factors, e.g., by a spontaneous onset of gonad development after an inactive period (see p. 125). Apart from inhabitants of relatively seasonless environments, which could rely largely on endogenous rhythms (p. 125–126), a cyclic resurgence of gonadal function may be of special importance for transequatorial migrants that in their winter quarters are exposed to "opposite" photoperiodic conditions. Here, internal rhythmical factors could serve—even under constant or decreasing day lengths—to start gonadal growth and to initiate a prenuptial migration that finally brings the bird under the photoperiodic control regime of North Temperate zones (Marshall, 1960; Curry-Lindahl, 1963).

At this point, it seems appropriate to comment on the term *Zeitgeber*, which is frequently used in connection with circennial events. The concept of the *Zeitgeber* was first introduced by Aschoff (1954) for factors that synchronize an endogenous periodicity with environmental changes of similar frequency. It has been applied mainly to circadian rhythms, which always seem to be based on recurrent autonomous cycles, but it has also been used for circennial periodicities (Aschoff, 1958; Immelmann, 1967). In the latter sense, it could be employed in two different ways: broadly using *Zeitgeber* as a synonym of circennial proximate factor and, narrowly applying it only to species for which the existence of endogenous cycles has been proven. The first approach seems to be justified if one assumes that some kind of intrinsic cyclic tendency as part of the regulation of reproduction is a rather wide-spread phenomenon among birds and if one prefers, therefore, to have a uniform term for all external factors that provide information for the timing of avian breeding seasons. In the sense of the original concept, however, narrow usage is appropriate. In this case, *Zeitgeber* is not a synonym of proximate factor but rather a special case of it that refers only to the external synchronization of endogenous rhythms, whereas the term proximate factor can also be employed if no internal rhythms are involved at all (Immelmann, 1972).

DEGREE OF DEPENDENCE ON PREDICTIVE CONTROL

The role of the environment as the source of predictive information has been found to be different in different climatic regions. There seem to be two situations in which the amount of proximate control is rather small. The first comprises tropical areas that have very extended periods of favorable ecological conditions. Here, predictive information does not always seem to be *necessary* but the birds could either follow an internally programmed sequence of events (p. 125–126) or they could be brought into breeding condition by the forthcoming favorable conditions themselves since even then there is time enough left to complete one or more breeding cycles before environmental conditions deteriorate again (see p. 136).

The second situation in which predictive environmental information plays a rather small role is encountered in those climatic regions where favorable ecological conditions occur at irregular and unpredictable intervals. Here, no anticipatory adjustment of the organism to the forthcoming optimum conditions for breeding is possible. This applies mainly to arid tropical regions with erratic distribution of rainfall, such as central and western Australia and certain parts of southern Africa (Maclean, 1970; Serventy, 1971). The inhabitants of these regions do not get any predictive information from the environment but have to respond directly to the favorable conditions as soon as they occur (p. 129).

The duration of favorable conditions in arid areas tends to be rather brief, so that a very rapid response to the occurrence of optimum conditions is required. In many arid-country birds special adaptations, therefore, have been developed that may contribute to shortening the time between the onset of favorable conditions and the start of nest-building and ovulation. For central and western Australia, for example, two main types of avian adaptations have been described, i.e., a certain amount of gonadal activity throughout the year and a high degree of sociability.

In the Australian Zebra Finch (*Poephila guttata*), Farner and Serventy (1960) noted a tonic gonadotropic activity of the hypothalamo–hypophysial system, except when inhibited by certain external factors (such as low temperatures and scarcity of water). Therefore, Zebra Finches with completely inactive gonads are hardly ever found (Serventy, 1971). As a result, the very first showers of a cyclone lead to an immediate outburst of courtship and nest-building. Some birds may even begin nest-building while rain is still falling, and ovulation has been observed to occur less than 2 weeks after the beginning of the rains. This situation applies even if it happens to be the very first rainfall after an extended period of drought (Immelmann, 1963). As similar behavioral reactions have also been observed in several other species in central Australia and, as similar and even shorter lag periods between the onset of rainfall and ovulation have been found both in Australia and southern Africa, e.g., lag periods as short as 6–10 days in several passerines in the Kalahari (Maclean, 1970), it can be assumed that similar gonadal adaptations as described for the Zebra Finch may be involved in quite a number of arid-country birds (Serventy, 1971). A hypothetical scheme for the control of reproductive activity in opportunistic breeders has been discussed by Farner (1967).

Another probable adaptation consists in the high degree of sociability that is to be seen in many tropical birds but is especially obvious in arid areas. Except for long-lasting and tight pair-bonds, the occurrence and biological significance of which has already been discussed (p. 133), many arid-country birds tend to breed in colonies. In this respect, even intraspecific differences have been observed. In the Australian Zebra Finch, for example, colonial breeding seems to be more pronounced in the arid interior of the continent than it is in coastal areas. As in pair formation, several behavior patterns are found that may serve to hold the group together (Immelmann, 1963). It seems possible, therefore, that the high degree of sociability has been developed to provide strong mutual stimulation within a pair or a group (p. 134) that may contribute to achieving an immediate response to favorable conditions and, thus, may enhance the effects of other environmental stimuli [see Serventy (1971) for further examples and discussion].

Altogether, it can be concluded that in climatic regions, where the environment does not provide much predictive information for avian reproduction, a number of physiological and behavioral adaptations have evolved to compensate for the lack of early environmental stimuli.

A nearly opposite timing mechanism for avian breeding seasons is encountered in another climatic region—the North Temperate zones. Due to a rather short period of optimum conditions and to the pronounced changes in day length that provide very precise and reliable environmental information, a very rigid form of proximate control has been developed. Here, ultimate and proximate factors, with the exception of some subsidiary factors (p. 136), are entirely different. As a consequence, most species of birds have become obligately dependent on the all-important proximate factor, day length, whereas the ultimate factors are no longer able to stimulate gonad growth and reproductive activities, except for an occasional influence at the final stages of gametogenesis (p. 135). An almost, if not completely, causal relationship has been found experimentally for a number of passerines of Europe, North America, and Asia, in which up to five gonadal cycles have been produced within 1 year by manipulation of day length (Miyazaki, 1934; Damsté, 1947; Wolfson, 1954). In other words, the annual maturation of gonads is caused by factors entirely different from those that played a part in the phylogenetic development of the particular breeding seasons. In these regions, the role of the environment as the source of predictive information has certainly reached a maximum and, due to its great reliability, photoperiodic control, which has evolved independently more than once within different groups of birds, has to be regarded as the most efficient timing mechanisms for avian reproduction.

The obligatory photoperiodic species of higher latitudes do represent only one possibility of photoperiodic control and in many residents of mid and low latitudes the significance of day length may be less [see Farner (1970a) for discussion]. It follows that there may be consistent differences according to the different needs of species or populations.

TERMINATION OF BREEDING

As mentioned before, the most critical time in the annual cycle of a species coincides approximately with the fledging of the young, which in turn represents the end of an extended sequence of physiological and behavioral events (p. 123). For this reason, reproductive activities, as a rule, must be initiated long before the period of optimum environmental conditions. For the same reason, however, breeding also must cease a more or less extended period ahead of deteriorating conditions to avoid the production of young at a time when their chance of survival is progressively reduced, as well as to render other energy-demanding processes (molt, migration) possible. The latter point is equally important for the young of the year and for the adult birds. It follows that, apart from the timing of the onset of reproductive activities, its timely termination is equally important. In contrast to the wealth of information about the initiation of breeding, however, little is known about the terminal timing mechanisms.

Farner (1967, 1970b) has discussed possible termination schemes. Discontinuation of gonadal function and reproductive activity may be due to a cessation of the stimulatory effects of proximate factors (decreasing instead of increasing day lengths, unavailability of water, etc.), to negative hormonal feedbacks on the hypothalamus from gonadal and hypophysial level, to the onset of photorefractoriness in photoperiodically controlled species, and, finally, to immediate negative influence from the environment (low temperatures, decreasing food supply, etc.). As for the initiation of the reproductive cycle, however, some sort of forewarning information is required in order to achieve a timely cessation of breeding activities and, for this reason, the last mentioned possibility will usually occur too late to serve as a terminal timing mechanism.

The term photorefractoriness refers to a period of unresponsiveness to stimuli of a quality and quantity that under other circumstances would be effective. It has been developed mainly in North Temperate birds, starting early in summer and terminating at a time when daily photoperiod has reached a nonstimulatory level. It leads to a discontinuation of gonad function while the days are still long and hence prevents the occurrence of new clutches in late summer and early fall. Photorefractoriness probably represents the most highly developed termination scheme. It is probably of multiple origin and has to be regarded as a special adaptation to the very rigid form of environmental control occurring in many photoperiodic species (see p. 139) (Farner, 1959, 1961, 1967; Farner and Follett, 1966; Hamner, 1968).

From the evidence available, no definite conclusions as to the distribution of the four mechanisms among different groups of birds are yet possible. As for the initiation of reproductive activities, interspecific differences in the kind of termination schemes as well as in the participation of environmental and internal control are likely to occur.

SUMMARY

Most species of birds have restricted reproductive periods, nesting at about the same time every year. Such breeding seasons have to be regarded as the result of specific environmental adaptations; every species and every population of birds is adjusted to breed at the time of the year at which it can raise young most efficiently. Consequently, great differences in breeding schemes occur among different climatic regions and different ecological groups of birds. Factors that have led, through natural selection, to the species- and population-specific breeding periodicities are called ultimate factors. The most important of these is the availability of an adequate food supply.

To raise young at the most favorable time of the year, however, most species require some sort of forewarning information to initiate the necessary physiological and behavioral preparations for reproduction well in advance of optimum living conditions. Ultimate factors, as a rule, operate too late to serve as a source of adequate predictive information. Therefore, most species have developed the capacity to respond to environmental factors that reliably predict the forthcoming optimum season. Such stimuli have been called proximate factors. In North Temperate zones, the most regular environmental event and, hence, the most reliable and most widely distributed proximate factor for the initiation of reproductive activities is the annual change in the length of the day. There is no other environmental factor in any climatic region that has gained a similar importance for the immediate control of avian annual cycles. In tropical regions, rainfall itself, or environmental changes connected with precipitation, have been found to be a widespread source of predictive information. Other proximate factors may be represented, for example, by changes in temperature and by various stimuli from habitat and nest site as well as from the social environment of the bird—i.e. by means of behavioral interactions of a pair or a breeding group. In many North Temperate species, the male has been found to be more susceptible to photoperiodic control and, in turn, to stimulate the female through courtship activities.

A second component of the actual timing mechanism for avian breeding seasons could be the regularly recurring endogenous rhythms, with a period that approximately corresponds to the natural sequence of environmental events. The existence of endogenous circennial rhythms has been suggested frequently, but has not been proved until very recently, when internal cycles for a number of functions have been proven experimentally in several species of European warblers. In cases in which endogenous rhythms participate in the annual regulation of breeding seasons, the proximate factors that keep the rhythm in phase with its changing environment are called *Zeitgebers*.

The degree of dependence upon proximate control is known to vary from one climatic region to another. In many tropical regions, it seems to be rather small, because predictive information may not be necessary or may not be available. In North Temperate zones, on the other hand, environmental control tends to be very rigid and many species of birds have become obligately dependent on the all-important proximate factor—day length.

In many cases, finally, proximate control of reproduction has been found not to be a uniform process but rather to consist of several steps. First, the reproductive cycle as a whole has to be initiated, and this may have to occur rather far in advance of optimum living conditions. Afterwards, the bird has also to become immediately adjusted to variations in season that may occur in almost every climatic region. Both tasks may be accomplished by different sets of proximate factors. In this case, the initiating stimuli are called primary timers. The adjusting stimuli are known as subsidiary factors.

REFERENCES

Aschoff, J. 1954. Zeitgeber der tierischen Tagesperiodik. Naturwissenschaften 41:49–56.
Aschoff, J. 1958. Tierische Periodik unter dem Einfluss von Zeitgebern. Z. Tierpsychol. 15:1–30.
Ashmole, N. P. 1963. The biology of the Wideawake or Sooty Tern (*Sterna fuscata*) on Ascension Island. Ibis 103b:297–364.
Ashmole, N. P., and M. J. Ashmole. 1967. Comparative feeding ecology of sea birds of a tropical oceanic island. Bulletin 24. Peabody Museum of Natural History, New Haven, Conn. 131 p.
Baker, J. R. 1938. The evolution of breeding seasons, p. 161–177. *In* G. R. de Beer [ed.] Evolution: Essays on aspects of evolutionary biology. University Press, Oxford.
Berthold, P., E. Gwinner, and H. Klein. 1971. Circannuale Periodik bei Grasmücken (*Sylvia*). Experientia 27:399.
Brockway, B. F. 1964. Social influences on reproductive physiology and ethology of Budgerigars (*Melopsittacus undulatus*). Anim. Behav. 12:493–501.
Brockway, B. F. 1965. Stimulation of ovarian development and egg laying by male courtship vocalization in Budgerigars (*Melopsittacus undulatus*). Anim. Behav. 13:575–578.
Brooke, R. K. 1966. Nuptial moult, breeding season and clutch size of Rhodesian Red Bishops (*Euplectes orix*) and congeners in relation to rainfall. Ostrich Suppl. 6:233–235.
Brown, L. H., and E. K. Urban. 1969. The breeding biology of the Great White Pelican (*Pelecanus onocrotalus roseus*) at Lake Shala, Ethiopia. Ibis 111:199–237.
Brown, R. G. B., and D. E. Baird. 1965. Social factors as possible regulators of *Puffinus gravis* numbers. Ibis 107:249–251.
Burger, J. W. 1948. The relation of external temperature to spermatogenesis in the male Starling. J. Exp. Zool. 109:259–266.
Burger, J. W. 1953. The effect of photic and psychic stimuli on the reproductive cycle of the male Starling, *Sturnus vulgaris*. J. Exp. Zool. 124:227–239.
Coulson, J. C. 1966. The influence of the pair-bond and age on the breeding biology of the Kittiwake Gull, *Rissa tridactyla*. J. Anim. Ecol. 35:269–279.

Craig, W. 1911. Oviposition induced by the male in pigeons. J. Morphol. 22:299–305.

Crook, J. H. 1964. The evolution of social organisation and visual communication in the weaver birds (Ploceinae). Behav. Suppl. 10:1–178.

Curry-Lindahl, K. 1963. Molt, body weights, gonadal development, and migration in *Motacilla flava*, p. 960–973. *In* Proceedings, XIII International Ornithological Congress, Ithaca, New York, 1962. American Ornithologists' Union, Baton Rouge.

Damsté, P. H. 1947. Experimental modification of the sexual cycle of the Greenfinch. J. Exp. Biol. 24:20–35.

Darling, F. F. 1938. Bird flocks and the breeding cycle. University Press, Cambridge. 124 p.

Disney, H. J. de S., and A. J. Marshall. 1956. A contribution to the breeding biology of the Weaver-Finch, *Quelea quelea* (Linnaeus) in East Africa. Proc. Zool. Soc. (London) 127:379–387.

Emlen, J. T., and F. W. Lorenz. 1942. Pairing response of free-living Valley Quail to sex-hormone pellet implants. Auk 59:369–378.

Farner, D. S. 1959. Photoperiodic and related control of annual gonadal cycles in birds, p. 716–750. *In* R. B. Withrow [ed.] Photoperiodism and related phenomena in plants and animals. American Association for the Advancement of Science, Washington, D.C.

Farner, D. S. 1961. Comparative physiology: Photoperiodicity. Ann. Rev. Physiol. 23:71–96.

Farner, D. S. 1967. The control of avian reproductive cycles, p. 107–133. *In* Proceedings, XIV International Ornithological Congress, Oxford, 1966. Blackwell's, Oxford.

Farner, D. S. 1970a. Day length as environmental information in the control of reproduction in birds, p. 71–91. *In* J. Benoit and J. Assenmacher [ed.] La photorégulation de la reproduction chez les oiseaux et les mammifères. Centre National de la Recherche Scientifique, Paris.

Farner, D. S. 1970b. Predictive functions in the control of annual cycles. Environ. Res. 3:119–131.

Farner, D. S., and B. K. Follett. 1966. Light and other environmental factors affecting avian reproduction. J. Anim. Sci. 25:90–118.

Farner, D. S., and D. L. Serventy. 1960. The timing of reproduction in birds in the arid regions of Australia. Anat. Rec. 137:354.

Farner, D. S., and A. C. Wilson. 1957. A quantitative examination of testicular growth in the White-crowned Sparrow. Biol. Bull. 113:254–267.

Farner, D. S., B. K. Follett, J. R. King, and M. L. Morton. 1966. A quantitative examination of ovarian growth in the White-crowned Sparrow. Biol. Bull. 130:67–75.

Frith, H. J. 1967. Waterfowl in Australia. Angus and Robertson, Sydney. 315 p.

Gwinner, E. 1972. A comparative study of circannual rhythms in warblers, p. 405–427. *In* M. Menaker [ed.] Biochronometry. National Academy of Sciences, Washington, D.C. 662 p.

Hailman, J. P. 1964. Breeding synchrony in the equatorial Swallow-tailed Gull. Am. Nat. 98:79–83.

Hamner, W. M. 1968. The photorefractory period of the House Finch. Ecology 49:211–227.

Hickling, R. A. O. 1959. The burrow excavation phase in the breeding cycle of the Sand Martin, *Riparia riparia*. Ibis 101:497–500.

Hinde, R. A. 1967. Aspects of the control of avian reproductive development within the breeding season, p. 135–153. *In* Proceedings, XIV International Ornithological Congress, Oxford, 1966. Blackwell's, Oxford.

Hoesch, W., and G. Niethammer. 1940. Die Vogelwelt Deutsch-Südwestafrikas. J. Ornithol. Sonderh. 88:1–404.

Horn, H. S. 1968. The adaptive significance of colonial nesting in the Brewer's Blackbird (*Euphagus cyanocephalus*). Ecology 49:682–694.

Immelmann, K. 1963. Tierische Jahresperiodik in ökologischer Sicht. Zool. Jahrb. Abt. Syst. 91:91–200.

Immelmann, K. 1967. Periodische Vorgänge in der Fortpflanzung tierischer Organismen. Stud. Gen. 20:15–33.

Immelmann, K. 1971. Ecologic aspects of avian periodicities, p. 341–389. *In* D. S. Farner and J. R. King [ed.] Avian biology. Academic Press, New York.

Immelmann, K. 1972. Erörterungen zur Definition und Anwendbarkeit der Begriffe Ultimate Factor, Proximate Factor und Zeitgeber. Oecologia 9:259–264.

Kahl, M. P. 1966. A contribution to the ecology and reproductive biology of the Marabou stork (*Leptoptilos cruminiferus*) in East Africa. J. Zool. 148:289–311.

Kalela, O. 1958. Über ausserbrutzeitliches Territorialverhalten bei Vögeln. Ann. Acad. Sci. Fenn. Ser. A. 42:1–42.

Keast, A. 1968. Moult in birds of the Australian dry country relative to rainfall and breeding. J. Zool. 155:185–200.

Keast, J. A., and A. J. Marshall. 1954. The influence of drought and rainfall on reproduction in Australian desert birds. Proc. Zool. Soc. (London) 124:493–499.

Lack, D. 1950. The breeding seasons of European birds. Ibis 92:288–316.

Lack, D., and J. T. Emlen. 1939. Observations on breeding behavior in Tricolored Red-Wings. Condor 41:225–230.

Lehrman, D. S. 1959. Hormonal responses to external stimuli in birds. Ibis 101:478–496.

Lehrman, D. S. 1965. Interaction between internal and external environments in the regulation of the reproductive cycle of the Ring Dove, p. 355–380. *In* F. Beach [ed.] Sex and behavior. Wiley, New York.

Lofts, B., and R. K. Murton. 1968. Photoperiodic and physiological adaptations regulating avian breeding cycles and their ecological significance. J. Zool. (London) 155:327–394.

Lott, D., S. D. Scholz, and D. S. Lehrman. 1967. Exteroceptive stimulation of the reproductive system of the female Ring Dove (*Streptopelia risoria*) by the mate and by the colony milieu. Anim. Behav. 15:433–438.

Maclean, G. L. 1967. A contribution to the biology of the Sociable Weaver, *Philetairus socius* (Latham). PhD thesis, Rhodes University, Grahamstown, South Africa.

Maclean, G. L. 1970. The breeding seasons of birds in the southwestern Kalahari. Ostrich Suppl. 8:179-192.

Marshall, A. J. 1949. Weather factors and spermatogenesis in birds. Proc. Zool. Soc. (London) 119:711-716.

Marshall, A. J. 1952. Non-breeding among arctic birds. Ibis 94:310-332.

Marshall, A. J. 1959. Internal and environmental control of breeding. Ibis 101: 456-478.

Marshall, A. J. 1960. The environment, cyclical reproductive activity, and behaviour in birds. Symp. Zool. Soc. (London) 2:53-67.

Marshall, A. J., and H. J. de S. Disney. 1957. Experimental induction of the breeding season in a xerophilous bird. Nature 180:647-649.

Medway, B. A. 1962. The swiftlets (*Collocalia*) of Niah Cave, Sarawak. Ibis 104:45-66.

Miyazaki, H. 1934. On the relation of the daily period to the sexual maturity and to the moulting of *Zosterops palpebrosa japonica*. Sci. Rep. Tôhoku Imperial Univ. 4th Ser. 9:183-203.

Moreau, R. E. 1950. The breeding seasons of African birds. Ibis 92:223-267, 419-433.

Moreau, R. E. 1966. The bird faunas of Africa and its islands. Academic Press, London. 424 p.

Nelson, J. B. 1968. Breeding behaviour of the Swallow-tailed Gull in the Galapagos. Behaviour 30:146-174.

Orians, G. M. 1961. Social stimulation within blackbird colonies. Condor 63:330-337.

Orr, R. T. 1945. A study of captive Galapagos finches of the genus *Geospiza*. Condor 47:177-201.

Prévost, J., and F. Bourlière. 1955. Sur le cycle reproducteur de quelques oiseaux antarctiques, p. 252-257. *In* Acta XI, Congressus Internationalis Ornithologicus, Basel, 1954.

Ridpath, M. G. 1972. The Tasmanian Native Hen *Tribonyx mortieri*. III. Ecology. CSIRO Wildlife Res. 17:91-118.

Rowan, W. 1926. On photoperiodism, reproductive periodicity and the annual migrations of birds and certain fishes. Proc. Boston Soc. Nat. Hist. 38:147-189.

Schreiber, R. W., and N. P. Ashmole. 1970. Sea-bird breeding seasons on Christmas Island, Pacific Ocean. Ibis 112:363-394.

Serventy, D. L. 1971. Biology of desert birds, p. 287-339. *In* D. S. Farner and J. R. King [ed.] Avian biology. Academic Press, New York.

Serventy, D. L., and A. J. Marshall. 1957. Breeding periodicity in western Australian birds: With an account of unseasonal nestings in 1953 and 1955. Emu 57:99-126.

Serventy, D. L., and H. M. Whittell. 1967. Birds of Western Australia. 4th ed. Lamb Publications, Perth. 440 p.

Skutch, A. F. 1950. The nesting season of Central American birds in relation to climate and food supply. Ibis 92:185-222.

Snow, D. W. 1966. Moult and the breeding cycle in Darwin's finches. J. Ornithol. 107:283-291.

Snow, D. W., and B. K. Snow. 1964. Breeding seasons and annual cycles of Trinidad land-birds. Zoologica 49:1–39.

Thomson, A. L. 1950. Factors determining the breeding seasons of birds: An introductory review. Ibis 92:173–184.

von Haartman, L. 1963. The nesting times of Finnish birds, p. 611–619. *In* Proceedings, XIII International Ornithological Congress, Ithaca, New York, 1962. American Ornithologists' Union, Baton Rouge.

Ward, P. 1965. The breeding biology of the Black-faced Dioch (*Quelea quelea*) in Nigeria. Ibis 107:326–349.

Ward, P. 1969. The annual cycle of the Yellow-vented Bulbul (*Pyenonotus goiavier*) in a humid equatorial environment. J. Zool. 157:25–45.

Winterbottom, J. M. 1963. Avian breeding seasons in southern Africa, p. 640–648. *In* Proceedings, XIII International Ornithological Congress, Ithaca, New York, 1962. American Ornithologists' Union, Baton Rouge.

Wolfson, A. 1954. Production of repeated gonadal, fat, and molt cycles within one year in the Junco and White-crowned Sparrow by manipulation of day length. J. Exp. Zool. 125:353–376.

DISCUSSION

H. J. Frith

I am in accord with the review by Professor Immelmann of the role of the environment in reproduction and with his conclusions. The most useful contribution I can make here is to expand on his brief remarks about the effect of changes in water level on the breeding of water birds in inland Australia. The passage in Immelmann's paper reads as follows:

In several species of Australian ducks and some other waterfowl, change in water level appears to control reproductive activities. In the nomadic ducks of interior Australia, an increase in water level has been found to be closely followed by sexual display. As these activities occur even in areas where there has been no rain but where the rise in river flow has been caused by rainfall at the headwaters, it appears that increase in water level itself must be the relevant proximate factor (Frith, 1967).

This passage is based on the results of work completed in 1962 and although nothing has been discovered to upset that premise, work done in the last decade has made it necessary to consider not only the water level itself but some of its effects as proximate factors. It also provided a more detailed understanding of the events in different species and individuals.

THE ENVIRONMENT

The inland rivers of southern Australia rise in regions of relatively high rainfall and flow westward across arid and semiarid plains. They are senile and form extensive systems of billabongs, meanders, and effluent streams. Their flow is not great, and many decrease in flow as they progress westward as water is drained off into effluent streams that disappear as a result of soakage and evaporation. Some rivers do not reach the sea and in dry weather most cease to flow; some dry up altogether.

In times of high rainfall on the catchments to the east, there is extensive flooding that can extend over many thousands of square miles of the inland plains. In times of average rainfall or snow melt on the catchments, the billabongs and effluent streams fill— sometimes very rapidly. These events create considerable areas of temporary waterfowl habitat. Some of this habitat survives for several years but most of it dries quickly and the billabongs remain empty until the next rise in river level, which in a particular area could be several years later. In the southern parts of the continent, the rainfall is mainly in winter, so there is a tendency for rises in river level to be most frequent in spring; this, however, can occur at any time.

MOVEMENTS

To cope with such an environment, many ducks and other water birds are nomadic (e.g., Frith, 1959; Carrick, 1962). They have the ability to move very widely in all directions, at any time, to find alternative habitat. The rainfall in the north of the continent is mainly in summer; in the south, it is mainly in winter. Thus, in some species there is a tendency to shift the center of gravity of the population north and south in summer and winter, respectively (Frith,

1963), but in others most movements are random (Frith, 1962). The breeding and movements of two species have been studied more extensively than others and these include an example of each type of movement. The Australian Black Duck (*Anas superciliosa*) has a tendency to regular movements; the Grey Teal (*A. gibberifrons*) is a complete nomad.

BREEDING

There is some flooding at some point in the continent in most years, but it varies very greatly in location, extent, and timing. If the breeding seasons of water birds and their movements were controlled by regular environmental factors, the birds would frequently miss favorable opportunities to breed and, at other times, would be prepared to breed at times when, locally, there was no suitable habitat. For these reasons, the breeding seasons of many water birds are geared to changes in water level and can occur at any time of the year.

A few species have highly regular breeding seasons irrespective of water level; some have regular breeding seasons but also can breed at any time should there be a significant increase in water level; some breed only when there is an increase in level.

When there is a major increase in water level due to rain or melting snow on the catchment hundreds of miles to the east, there is a regular sequence of breeding among the waterfowl on the inland plains. The Grey Teal begins increased sexual display as soon as the rise becomes evident and can lay eggs within 12 days at any season. The Black Duck, Hardhead (*Aythya australis*), Freckled Duck (*Stictonetta naevosa*), and Black Swan (*Cygnus atratus*), all of which need deeper water than the Grey Teal, delay breeding until the billabongs are full and some permanence of these conditions is ensured. The Pink-eared Duck (*Malacorhynchus membranaceus*), which feeds on plankton and accordingly breeds in shallow drying floodwater, delays ovulation until the floodwaters actually flow across the plains. A few species have no visible reaction to rising water level.

It is convenient to consider the birds in three groups.

SPECIES WITH REGULAR BREEDING SEASONS

The only two species with regular seasons are the Musk Duck (*Biziura lobata*) and the Blue-billed Duck (*Oxyura australis*). These two avoid the main hazards of the erratic environment by living only in the few permanent swamps that do exist. They are virtually sedentary, although the Blue-billed Duck does tend to move 100 miles or so south in winter.

The water level of these permanent swamps fluctuates considerably as they are either connected to a river or are fed by drainage water from irrigation systems. Nevertheless, examination of large series of gonads collected each month between 1962 and 1967 shows that the sexual cycle and the breeding season are regular. There is a regular pattern of increasing sexual activity from autumn through winter to spring, when breeding occurs. After breeding, there is collapse of the testes.

This cycle is followed during rising or declining water levels and at times of both high nutrition and low nutrition. Although there are no experimental data to settle the question, there is no reason to feel that the sexual cycle and breeding are not controlled by such fixed annual factors as changes in photoperiod.

SPECIES WITH BASICALLY REGULAR BREEDING SEASONS

Certain species have a regular breeding season as well as the ability to breed again at any time of the year, should favorable conditions occur. Examples are the Black Duck, the Australian Shoveler (*Anas rhynchotis*), the Hardhead, and the Freckled Duck; these species prefer the more permanent swamps and lagoons as habitat but are capable of living and breeding in floodwater when it occurs.

Each has been studied in the field by examining the gonads of samples collected monthly for periods of several years. In each there is a regular cycle of sexual activity similar to those reported for other birds in the North Temperate regions, i.e., some increase to full spermatogenesis and breeding in spring.

For example, in one study (Frith, 1959), the Murrumbidgee River Black Ducks were found to be sexually active and to have a breeding season in each year—even in 1957, when there was no increase in river level and, in fact, a serious drought. In the drought year 1957, the increase in testis size began at about the same time as in normal years and breeding began, as usual, in October. In 1956, a year of a record flood, testis size increased early, in April, and clutches were begun in May. There are similar data for other members of the group and similar results from another study in 1962 to 1967.

On the strength of this work, it was concluded that although the

Black Duck and the other species in this group retain the ability to breed in response to a change in water level at any time, fixed annual factors such as photoperiod cannot be ruled out as important agents in controlling the regular breeding.

Critical work needed to separate the effect of these and other factors is not possible under the variable conditions of the wild. Birds, therefore, are being held in captivity in light-controlled rooms, and experiments with proper controls are in progress. Certain results from this work by L. W. Braithwaite (unpublished data) can be cited here.

The birds have been subjected to various day lengths and levels of nutrition. The results indicate that the Black Duck has a regular sexual cycle that can be altered by changes in the photoperiod or level of nutrition. We have no reason to believe that it is not largely controlled by photoperiod in southern Australia where they are relatively large seasonal changes in daily photoperiod. However, all through the year, some individuals show gametogenic activity. Should there be a massive increase in water level, those that have completed the molt respond sexually and breed. If there is no increase in water level, they continue until spring, when the presumed photoperiodic effect leads to increased sexual activity and breeding. There is reason to believe that the change in water level operates through the nutrition of the males and females.

SPECIES WITH ERRATIC BREEDING SEASONS

The Black Swan, Grey Teal, and Pink-eared Duck have erratic breeding seasons. Samples have been taken of the Black Swan each month from 1963 to 1967. In each sample, there was a great variability of gonad condition; invariably, some males were producing sperm, some showed intermediate stages of spermatogenesis and some were inactive. Actual breeding occurred at any time, but there was a strong seasonal element, which appeared to be coincidental since rises in the water level of the swamp were commonest in spring when breeding occurred. There was strong evidence that each of the spring breeding periods and each of the out-of-season breeding periods was related to increases in the nutritional level of the swans. It has not yet been possible to perform critical experiments with Black Swans to test the thesis that breeding is ultimately controlled by changes in water level.

The Grey Teal is a highly nomadic species that is most common in semipermanent billabongs and floodwaters. Its sexual cycle and breeding season are erratic. In the study just noted the Grey Teal was sexually active on both the Lachlan and the Murrumbidgee rivers only when there was an increase in water level. Even in 1957, a drought year, a minor fresh in the river correlated to a brief outburst of sexual activity. In some of these outbursts of sexual activity, the response to changing water level was very rapid. Display intensified within a day or two of the time when the change was being detected, and eggs were laid in 12 days.

The change in water level that initiates breeding in Grey Teal also brings about a vast increase in the populations of the insect *Corixa*, the principal food of the ducklings. In this way, synchrony between the broods and abundant food is ensured.

Greater precision in the study of the breeding season of waterfowl was possible from 1963 to 1967 when regular samples were taken of nine species, including the Grey Teal, from the one large permanent swamp—the drainage area of an irrigation system, which was subject to drastic alterations in water level. In the Grey Teal, the relationship to water level was again clear, but there was one period of sexual activity and breeding that was not associated with a change in water level; instead, it occurred during a great increase in the abundance of food (Braithwaite and Frith, 1969). Braithwaite and I concluded that change in water level was the most powerful agent controlling the breeding season but that the nutrition of the birds also should be considered.

In recent work by L. W. Braithwaite on captive birds subjected to different light regimes and levels of nutrition, the results of the field work have been confirmed. In the Grey Teal, the situation is quite different from that in the Black Duck.

It is rare for an individual to be sexually inactive. Only under very poor conditions, that is, conditions corresponding to very low water level and poor nutrition, are all individuals in a population sexually inactive. In the wild at these times, nomadic movement would normally have occurred and the birds would have sought better habitat elsewhere in the continent.

At times when there is some water and rather more adequate nutrition, there is some low-level spermatogenesis in all individuals in the population and some low-level display. Should there be an increase in water level, with consequent improved nutrition, the whole population is ready to respond sexually very rapidly indeed; sexual display intensifies, which synchronizes all other individuals. Even in the final stages of wing molt, before the birds have regained efficient flight, there is intense display and active spermatogenesis.

The situation in the Grey Teal can be compared to that in the Pink-eared Duck, another very highly nomadic species. Here, the testis consists of constricted zones and expanded nodules that vary in size and number according to environmental conditions. In floods, the nodules are large and numerous; in drought, they are smaller. The constricted zones consist of sexually inactive cells. The nodules have enlarged tubules that are producing sperm at all times. Even in times unsuitable for breeding, there is continuous spermatogenesis (Braithwaite, 1969).

CONCLUSIONS

The waterfowl discussed seem to illustrate the various degrees of dependence on proximate factors discussed by Immelmann but in the one family of birds, in the same region, and even in the one swamp.

The regular breeders have little proximate control. As postulated, since they live in permanent rich swamps, there are very extended periods of favorable ecological conditions and little predictive information is necessary.

The second group represents those that need little predictive information for much the same reason; they prefer to live in the same permanent swamps. They, however, are able to use more temporary waters and to breed in it, provided it is not too temporary. The early rise in water level and increased nutrition provide the information that suitable breeding conditions are likely to occur and the sexual cycle is sufficiently long to ensure that these conditions are achieved before egg-laying occurs.

Some of the opportunistic breeders can secure little predictive information from the erratic changes in water level and need a very rapid response. These species keep themselves ready charged, as it were, by continuous spermatogenesis. Increased sexual activity is rapid and the associated display can lead to mutual stimulation and breeding of the whole population.

REFERENCES

Braithwaite, L. W. 1969. Testis cycle of a native duck. J. Reprod. Biol.
    19:390–391.
Braithwaite, L. W., and H. J. Frith. 1969. Waterfowl in an inland swamp in New
    South Wales. III. Breeding. CSIRO Wildlife Res. 14:65–109.
Carrick, R. 1962. Biology, movements and conservation of ibises
    (Threskiornithadae) in Australia. CSIRO Wildlife Res. 7:71–88.
Frith, H. J. 1959. Ecology of wild ducks in inland New South Wales. IV.
    Breeding. CSIRO Wildlife Res. 4:156–181.
Frith, H. J. 1962. Movements of the Grey Teal, *Anas gibberifrons* Müller
    (Anatidae). CSIRO Wildlife Res. 7:50–70.
Frith, H. J. 1963. Movements and mortality rates of Black Duck and Grey Teal
    in southeastern Australia. CSIRO Wildlife Res. 8:119–131.
Frith, H. J. 1967. Waterfowl in Australia. Angus and Robertson, Sydney. 315 p.

INFORMAL DISCUSSION

BEER: Do you know anything about the breeding seasons of the Black Swan in
New Zealand, where one does not have the unpredictable situation that Black
Swans encounter in Australia?

FRITH: I know little about the breeding of the Black Swan in New Zealand. But
in Australia, the Black Swan has a superficially regular breeding season. It
breeds in the early spring, but this is completely coincidental, because in
Australia it lives in the southern half of the continent where the rainfall tends
to be winter rainfall, and the water-level increases tend to be in the spring. If
you look at it closely, though, it is not a spring breeding; it is keyed to water
level.

I would think in New Zealand, where it does have a regular breeding season, if looked at closely it might show that something else is involved, too.

MURTON: I don't want to extend the discussion unnecessarily at this point for
I shall be dealing with this particular topic later. But I want to emphasize that
the Black Swan is a photoperiodic species—indeed, in a sense, all species are
photoperiodic in that they exhibit diurnal rhythms that are entrained by the
daily light regime. It will be shown that the Black Swan displays photore-
sponses at the Wildlife Trust, Slimbridge, England, that conform to the pattern of response shown by other swan species, including those from the
northern hemisphere. The responses noted at Slimbridge indicate a capacity

for an extended breeding season at its natural latitude. Does this appear in some years, perhaps in the form of repeat layings?

FRITH: Given suitable nutrition, which comes with high water level, the same individuals in the wild will breed continuously.

IMMELMANN: May I make a comment not pertinent to the Black Swan, but to other species? In the southern United States and in California, several species of Passeriformes have been found that, under normal conditions, are under photoperiodic control and breed in spring and early summer. But if there happen to be very extended rains during the usually dry summer months the birds also are stimulated by rainfall and a second reproductive season will occur in late summer or autumn. This means the same species may well be under photoperiodic control or under rainfall control, respectively, according to the external conditions. So these timing mechanisms are by no means mutually exclusive.

KENDEIGH: Professor Immelman is correct in that, in the center or optimal portion of the range of a nontropical species, photoperiod is most important in regulating the general time of initiating nesting activities. Temperature is of secondary importance, but its effect is superimposed on the photoperiod in determining the exact day at which eggs are laid. For instance, for the House Wren (*Troglodytes aedon*) in northern Ohio the photoperiod brings the gonads into condition for reproduction by mid-May, but whether egg-laying begins then, or 1–2 weeks later, is determined by a threshold of environmental temperatures averaging 15 °C over a 3-day period.

At the northern and southern boundaries of the range of a species, temperature may be equally as important as photoperiod in regulating time of nesting. Photoperiods in southern Canada are even more favorable in early May than in northern Ohio, but nesting by the House Wren is delayed or altogether curtailed because temperature does not reach the mean threshold for egg-laying until early to mid-June, which is very late. At the other extreme, in middle Georgia, near the southern boundary of the wren's breeding range, temperatures become favorable in April, before photoperiods have stimulated the gonads to full sexual maturity. The egg clutches laid there from late May to July are small, comparable to those laid farther north during the high temperatures of July. The breeding season may well be determined and regulated by the proper synchronization of photoperiod and temperature.*

GEE: I can cite another example of the effects of proximate and ultimate factors on reproduction. As physiologist with the Endangered Wildlife Research Program, I work with the Masked Bobwhite Quail (*Colinus virginianus ridgwayi*), formerly found in Southern Arizona but now restricted to an area in the State of Sonora, Mexico. Doctor W. O. Wilson and co-workers in a personal communication indicated that temperatures in excess of 27 °C may be

*Kendeigh, S. C. 1963. Regulation of nesting time and distribution in the House Wren. Wilson Bull. 75:418–427.

necessary for reproduction in the Masked Bobwhite. It now appears from our laboratory experiences that the Masked Bobwhite requires long daylight, warm temperatures, and high humidity to reproduce. We have been unable to obtain eggs under conditions of long day and warm temperatures when relative humidity is less than 25 percent, but have been successful in the same facilities, light, and temperature if relative humidity exceeds 50 percent.

An examination of the temperature and humidity records from the natural habitat and our out-of-doors enclosures in Laurel, Maryland, indicates a similar pattern. The residual wild population of Masked Bobwhite in Mexico reproduces in late summer and early fall. Adults captured from this wild population lay in May through September in Maryland. The most obvious difference between the two environments appears to be the lack of rain or less humidity in Mexico during April, May, and June. The average maximum temperature exceeds 27 °C from May through October in Mexico and from June through September in Maryland.

CAIN: Reproductive success in the case of the California Quail seems to be dependent on both photoperiod and soil moisture. In years that are particularly dry, the reproductive effectiveness is very low; the birds will generally have one hatch in the spring. In years of heavy rainfall, however, a hatch in the spring very often will be followed by a more successful hatch in the summer. According to a study by Doctor Leopold and associates at the University of California, Berkeley, the factor having highest correlation with reproductive success was the amount of groundwater that existed in April. It did not seem to be important whether heavy rains occurred over a short period or if moderate precipitation was spread over the entire winter. Somehow, the birds were measuring ground-level water, probably through some nutritional element that was available following favorable moisture conditions.

NORTON: I would like to bring out an intriguing example of proximate factors, apparently modifying the timing of reproduction. In the high Arctic, near Barrow, where I have worked on shorebirds, I think you have to recognize snowmelt as a highly important modifying factor that is normally closely tied to temperature. As soon as the snow cover begins to recede substantially, a "time bomb" starts clicking for a very well-demonstrated nexus between tundra arthropods and their emergence from a larval or pupal stage and the ability of the young, very small chicks to feed on the imagos of the arthropods.

Because of man's influence in certain places in the Alaskan Arctic, in association with the development of oil around Prudhoe Bay, we see the divorce of the snowmelt modifying factor on the timing of breeding from that of the temperature modifying factor as follows: Around Prudhoe there is an extensive, very highly built-up system of roads constructed of small- to medium-sized gravel. With strong wind, plus a good bit of traffic on the roads, as well as on gravel air strips, there is extensive dust fallout on the surrounding snow cover in the spring. As a result, the snow may actually begin to melt before the temperatures rise above freezing, through sublimation or actual melting. As the snow cover goes, it serves as a kind of releaser for the birds to begin breeding, particularly shorebirds in this case, and yet the temperatures are probably not suitable, and the arthropods are therefore unlikely to emerge on a normal schedule, tightly allied to the snowmelt.

Reproductive Endocrinology: The Hypothalamo-Hypophysial Axis

IVAN ASSENMACHER

As in other vertebrates, most of the various sources of information through which the environment controls avian reproduction act through specialized receptors annexed to the central nervous system; these convey information through nerve impulses to the diencephalic centers of the hypothalamus. At this critical level, neurosecretory cells transduce the nerve messages into chemical, hormonal information, which is transferred to hypophysial portal blood, within a complex neurohemal organ—the median eminence. From the hypophysial level on, the regulation of the reproductive organs and behavior occurs by a classical hormonal chain, involving the pituitary gonadotropic hormones and the gonadal steroid hormones, which act on their respective target organs through the general blood circulation (Figure 1).

PITUITARY CONTROL OF THE GONADAL FUNCTIONS

EFFECTS OF ADENOHYPOPHYSECTOMY

Since the pioneer work of Hill and Parkes (1934) in the domestic fowl, Benoit (1935) in the domestic Mallard (Pekin Duck) and Schooley *et al.* (1941) in the domestic pigeon, numerous data clearly

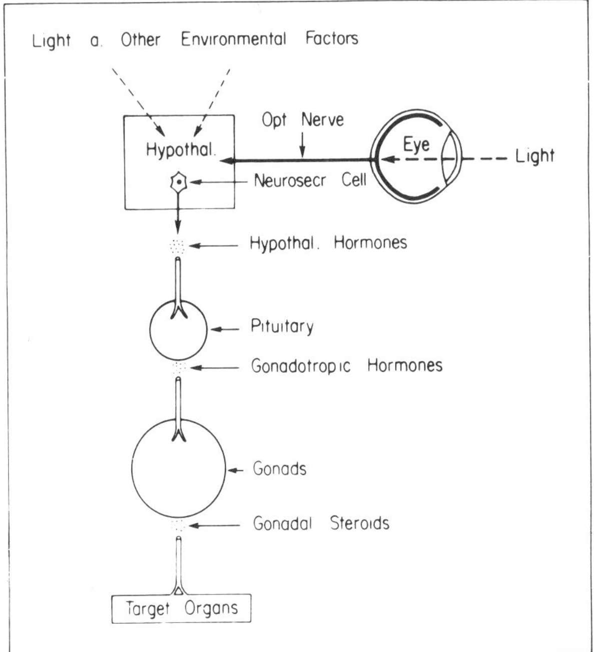

FIGURE 1   Diagrammatic representation of the control of sex organs by environmental factors.

have established that removal of the anterior pituitary leads to severe gonadal atrophy in both sexes. Hypophysectomized male birds show extensive decrease in testicular size due to shrinkage of the tubules, associated with a blockade of spermatogenesis and decreased Leydig-cell activity, which induces regression of the accessory sex organs. In the female, deprivation of the pituitary induces marked atresia of the ovarian follicles, together with involution of the medullary tissue.

Table 1 summarizes a typical experiment in three domesticated species. From the data on the male, the following conclusions may be drawn: (a) In the absence of the pituitary gland, gonadal development can be neither maintained (for birds that are hypophysectomized during the breeding season) nor obtained (for experimentally light-stimulated birds). It has been shown in sexually active ducks that gonadal regression after hypophysectomy is very drastic. Two weeks after operation, the atrophy is already complete. The effect of hypophysectomy is even more rapid in the laying hen; 6 days after the operation, ovary and oviduct will have undergone complete regression

TABLE 1 Effects of Adenohypophysectomy and Hypothalmic Pituitary Disconnection on Gonadal Weight in Three Bird Species[a]

Species and Treatment	Weight of Gonads (mg/100 g body weight[b])				
	Controls	Hypophysectomy[c]	Pituitary Autograft[d]	Transection of Portal Veins[e]	Destruction of Median Eminence[f]
Duck (male)					
Adult: Sexual activity (spring)	4878±516	58±13	60±18	88±34	73±17
	(8)	(7)	(11)	(5)	(6)
Adult: Artificial light-stimulated L L (fall)	4189±351	—	16±1	43±14	20±2
	(8)		(6)	(8)	(3)
Adult: Sexual quiescence (fall)	164±85	44±9	46±3	—	—
	(7)	(9)	(13)		
Prepuberal: Sexual quiescence (fall)	39±9	17±2	13±2	—	—
	(6)	(5)	(5)		
Pigeon					
Adult (male)	331±25	6.9±3.9	14.4±1.2		
	(125)	(40)	(56)		
Adult (female)	98.7±8.8	32.2±2.4	31.9±2.1		
	(106)	(33)	(39)		
Japanese Quail					
Prepuberal: Sexual quiescence (fall)	22.3±2.7	10.8±1.4	12.2±1.2		
	(52)	(37)	(39)		
Prepuberal: Artificial light-stimulated 18L–6D (fall)	1462±90	11.6±0.9	9.5±1.0		
	(38)	(19)	(14)		

[a] From Assenmacher (1958), Assenmacher and Baylé (1964), Baylé (1968).
[b] Values given as mean ±SD. Numbers within parentheses are number of animals.
[c] Postoperative delay for duck, 2-4 weeks; pigeon, 2-4 months; quail, 1 month.
[d] Postoperative delay for duck, 2-6 months; pigeon, 2-4 months; quail, 2-3 months.
[e] Postoperative delay for 1-7 months.
[f] Postoperative delay for 1-6 months.

(Nalbandov, 1966); (b) a definite gonadotropic control by the pituitary occurs before puberty, as sexually immature birds exhibit a 50 percent regression of their gonads after hypophysectomy; (c) similarly, a minimal gonadotropic activity persists during the seasonal quiescent phase of the sexual cycle, since hypophysectomy in adult ducks in the fall brings on further testicular regression to one-third the control values.

The hypophysectomy experiments therefore clearly demonstrate that, unlike the pituitary regulation of other peripheral endocrine glands, the gonadotropic component of the gland exerts, in birds as in mammals, a very early, permanent, and total control over the gonadal function. This control is exerted by the hypophysial gonadotropic hormones.

GONADOTROPIC HORMONES

Follicle-Stimulating Hormone and Luteinizing Hormone

Histophysiological Approach   During the past 10 years, various histophysiological procedures (e.g., photoperiodic stimulation, gonadectomy, or treatment with sexual steroids) have led to the identification of two types of gonadotropic cells within the anterior pituitary that may be reasonably assumed to be follicle-stimulating hormone (FSH) and luteinizing hormone (LH) cells. Using a large range of histochemical techniques, the so-called beta (FSH) and gamma (LH) cells have been demonstrated in the duck (Tixier-Vidal *et al.*, 1962; Tixier-Vidal, 1963), the domestic fowl, the guinea-cock, the House Sparrow (*Passer domesticus*) (Tixier-Vidal, 1963), the domestic pigeon (Tixier-Vidal, 1963; Tixier-Vidal and Assenmacher, 1966), the Orange Weaver (*Pyromelana franciscana*) (Tixier-Vidal, 1963; Gourdji, 1965), the Japanese Quail (*Coturnix coturnix japonica*) (Tixier-Vidal *et al.*, 1968), and the White-crowned Sparrow (*Zonotrichia leucophrys gambelii*) (Matsuo *et al.*, 1969).

Further evidence of the occurrence of two types of gonadotropic cells is derived from electron-microscope studies of the pituitary gland of the duck (Tixier-Vidal, 1965, 1970), the pigeon (Tixier-Vidal and Assenmacher, 1966), and the White-crowned Sparrow (Mikami *et al.*, 1969).

The basic importance of the identification of these two different types of pituitary cells, as probable sources of the two classical gonadotropic hormones, will be discussed later.

Another important point raised by these studies lies in the fact that in most species hitherto studied, the FSH and LH cells seem to be located in very specific areas of the anterior pituitary gland. As a matter of fact, earlier histologic studies based on differential staining properties of pituitary cells had already established that, within the avian adenohypophysis, a distinction can be made between a cephalic and a caudal lobe (Rahn and Painter, 1941; Wingstrand, 1951; Van Tienhoven, 1961). Recent observations, based on more sophisticated histochemical techniques combined with electron-microscope studies, have shown that both types of gonadotropic cells are usually located in different hypophysial lobes, the beta (FSH) cells being restricted to the cephalic lobe, while the gamma (LH) cells are present exclusively within the caudal lobe of the anterior pituitary. This peculiar distribution of the pituitary gonadotropic cells has been described in all species studied by Tixier-Vidal and co-workers, i.e., the domestic Mallard (*Anas platyrhynchos*), the female domestic fowl, the guinea-cock, the domestic pigeon, the Japanese Quail, the Orange Weaver, and the House Sparrow. However, the White-crowned Sparrow does not fit in with this model, as both types of gonadotrops are present in both lobes of the pars distalis (Matsuo *et al.*, 1969; Mikami *et al.*, 1969).

Gonadotropic Potency of Pituitary Extracts   After the first experiments of Riddle and Flemion (1928) with pigeons, and those of Domm (1931) with domestic fowl had shown the gonadotropic potency of pituitary implants, mammalian pituitary extracts were shown to elicit gonadal development in: (a) immature pigeons (Riddle, 1930; Evans and Simpson, 1934), ducks (Schockaert, 1931; Benoit and Aron, 1934), domestic fowl (Domm and Van Dyke, 1932; Schockaert, 1932; Jaap, 1934), and House Sparrow (Pfeiffer, 1947); (2) in seasonally quiescent grouse (Clark *et al.*, 1937) and European Robin (*Eithacus rubecula*) (Schildmacher, 1939); and (3) in Mallard drakes during the photorefractory period (Benoit *et al.*, 1950). On the other hand, avian pituitary extracts revealed their gonadotropic potency when tested either on immature mice (Benoit, 1935) or hypophysectomized rats (Meyer *et al.*, 1939) or, better, on chick testes (Evans *et al.*, 1940; Breneman, 1945), on hypophysectomized pigeons (Schooley *et al.*, 1941), or domestic fowl (Opel and Nalbandov, 1961; Mitchell, 1967).

Among modern bioassays for estimation of avian pituitary gonadotropins, the ^{32}P-uptake test by the chick testes (Breneman *et al.*, 1962) seems one of the most reliable, as far as total gonadotropic potency is concerned.

The bioassays of avian pituitary glands have shown many times that both gonadotropic hormones (or at least gonadotropic potencies), as found in mammals, i.e., FSH and LH, are also present in avian glands (Van Tienhoven, 1961; Nalbandov, 1966). On the other hand, from all of the experiments using either mammalian gonadotropins or, better, avian gonadotropic preparations, we may conclude that the general pattern of gonadal control by the pituitary gonadotropins, as established in mammals, also seems to hold for the avian gonad. LH, thus, is considered as the main stimulator of the endocrine function of the testis, while FSH appears as a prerequisite for at least the last steps of spermatogenesis. In female birds, FSH is thought to stimulate especially the development of the ovarian follicles and LH to induce ovulation (Figure 2).

Some attempts have been made to measure the respective gonadotropic potencies of the cephalic and caudal lobes of avian pituitary glands. In photoperiodically stimulated *Coturnix* quail, the gonadotropic activity (^{32}P-uptake test) was found to be about equal in the two lobes (Follett and Farner, 1966; Tixier-Vidal *et al.*, 1968). In the female domestic fowl, the cephalic lobe not only was found more potent (chick-testis assay) than the caudal lobe but this part of the pituitary gland also showed more significant fluctuations in its gonadotropic content during the breeding cycle; the cephalic lobe was 3 times as potent in nonlaying hens than in fowls sacrificed at the end of broodiness (Nakajo and Imai, 1961). More recently, Brasch and Betz (1971) have studied the regional distribution of FSH and LH activities in the cockerel pituitary by transplanting cephalic, middle, or caudal regions of the pars distalis of day-old chicks onto the chorioallantoic membrane of 10-day-old, partially decapitated ("hypophy-

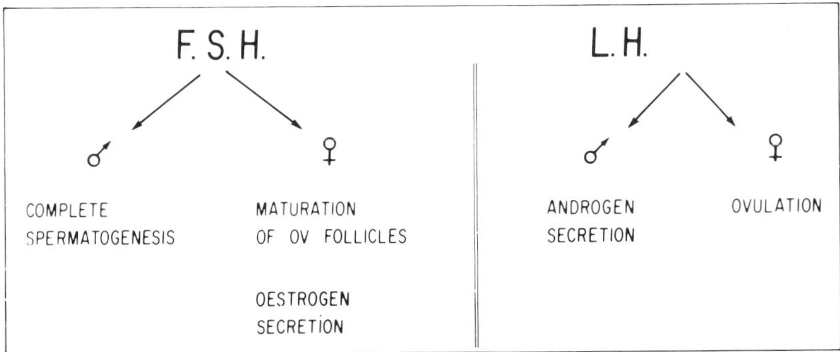

FIGURE 2   Effects of FSH and LH on avian reproductive organs.

sectomized") embryos. They observed that FSH activity is exclusively associated with the cephalic region of the pituitary, whereas a presumed LH activity originated from the caudal region of the gland. The latter result is in good agreement with the above-cited cytological data in the female domestic fowl (Tixier-Vidal, 1963).

Our present concepts of FSH and LH obviously must be re-evaluated in the light of the biological separation of highly purified avian FSH and LH preparations now in progress and the development of radioimmunoassays that will provide, as for other proteinaceous hormones, much more accurate data.

A first attempt to purify FSH and LH from domestic fowl was made by Fraps *et al.* (1947). But only recently has it become possible to obtain such purified hormones from the anterior pituitary of domestic fowl (Stockell-Hartree and Cunningham, 1969). This has opened the way to measurements of immunoreactive avian FSH and LH, and the first data on immunoreactive LH assays have been recently published for Japanese Quail by Follett *et al.* (1971).

The separation of different fractions of domestic fowl pituitaries—some of which are very potent in terms of FSH activity, with a very weak LH activity, and others of which bear only LH activity—affords a strong argument for the duality of gonadotropic hormones in birds, a problem that is still controversial even in mammals. This fits well with the morphological evidence of two different pituitary cell types involved in the synthesis of either FSH or LH. It, however, must be kept in mind that, in the most significant experiments (Furr and Cunningham, 1970), both hormonal activities have been measured on mammalian bioassays, i.e., the ovarian augmentation response in immature mice (Brown, 1955), as a specific FSH test, and the ovarian ascorbic acid depletion assay in pseudopregnant rats (Parlow, 1961) for LH activity. The accurate avian assay based on ^{32}P-uptake by the testes of 1-day-old chicks appears to be dependent on nonspecific effects of both FSH and LH.

We thus can conclude that while the cellular duality of avian FSH and LH seems to be reasonably assessed (although a decisive immunocytologic demonstration is not yet at hand), the well-advanced biochemical separation of both hormones from chick pituitaries should provide, in the near future, a better insight into the physiological specificity of the pituitary gonadotropins on their gonadal target organs.

Another unsolved problem related to the biochemical structure of the pituitary gonadotropins concerns the zoological specificity of the hormones. A number of investigators have compared the physiological potency of either mammalian or avian gonadotropic extracts, by either mammalian or avian bioassays. Their conclusions point to a certain degree of species specificity of gonadotropic hormones. Thus, avian bioassays are more sensitive to avian than to mammalian pituitary gonadotropic extracts (Evans *et al.*, 1940; Nalbandov *et al.*, 1951; Taber *et al.*, 1958; Opel and Nalbandov, 1961; Morris and Nalbandov, 1961; Mitchell, 1967). Conversely, mammalian bioassays react better to mammalian than to avian gonadotropins (Meyer *et al.*, 1939; Herrick *et al.*, 1962; Stockell-Hartree and Cunningham, 1969; Furr and Cunningham, 1970). Here, once more, data based on immunological techniques might yield a deeper insight into the problem. Recently, Follett *et al.* (1971), using labeled purified chick LH preparations that bound well to antiavian LH sera, observed no binding reactions of the avian hormone to antihuman or antibovine LH, and, conversely, human LH did not bind to antiavian LH sera. These findings indicate a very high immunological species specificity between human and chick LH that could, at least partially, account for the above-mentioned physiological species specificity.

Variations of Pituitary Gonadotropic Activity in Correlation with the Reproductive Cycle   From the foregoing, it may be seen how far we still are from definitely established assessments of the two gonadotropic hormones and their precise site of action, and from accurate measurements of their physiologic potency. In spite of this lack of information, we, however, can learn a great deal from a number of observations that were performed with more or less accurate techniques.

Mallard   Although no true quantitative statement of the variations in activity of the pituitary gonadotropic cells can be drawn, an extensive histological and ultrastructural study, often associated with an evaluation of the hormone content (measured on immature mice), provided a good estimate of the level of gonadotropic activity throughout the annual reproductive cycle (Tixier-Vidal *et al.*, 1962; Assenmacher *et al.*, 1962; Assenmacher and Tixier-Vidal, 1965). In the domestic Mallard reared under natural photoperiodic conditions, the testes start their seasonal development in December–January with mild enlargement of the tubules and a stimulation of the Leydig cells associated with an increased testosterone secretion (Garnier, 1971). They attain maximal spermatogenetic and androgenetic activity in

April–May, enter the regressive phase in June, and are almost quiescent from October to December. At the pituitary level, the gamma (LH) cells become active first, in December (Figure 3), attain a maximal activity in March–April, begin to regress in May, and are inactive from September to December. The beta (FSH) cells become active later on (February), i.e., at the time where the full spermatogenesis is attained. The maximum activity occurs in April–May, and, after a definite "storage" phase (June) that coincides with the testicular involution, the FSH cells regress from September to February. An evaluation of the gonadotropic content of the pituitary shows a high potency in April–May, an even higher content during the "storage" phase (June), followed by a marked decrease in fall. The apex of the breeding season seems thus correlated with a high FSH and LH activity, while the beginning of the early seasonal testosterone secretion is mainly associated with LH-cell activity, the peak in FSH-cell activity being concomitant with the terminal phase of spermatogenetic activation. Artificial photoperiodic stimulation during testicular quiescence stimulates simultaneous increases in testis weight and testosterone secretion (Jallageas and Assenmacher, 1970) together with the pituitary (Tixier-Vidal *et al.*, 1962), with changes analogous to those seen at the onset of the natural sexual activity but occurring much

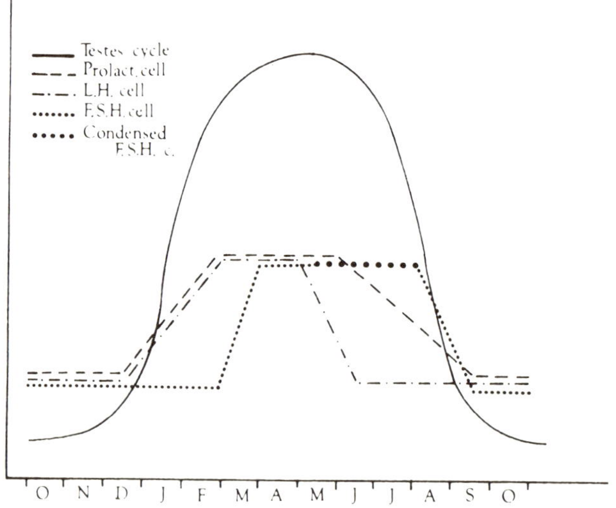

FIGURE 3 The annual cycle of testis and several anterior pituitary cell types in the male duck. From Tixier-Vidal *et al.* (1962) and Assenmacher and Tixier-Vidal (1965).

more explosively. After 2 weeks of constant light, the gonadotropic potency of the pituitary (^{32}P-uptake test) was found to have increased 36 times, but testosterone administration reduced the response 5 times (Kordon and Gogan, 1970), thus indicating a possible negative feedback of androgens on the gonadotropin secretion.

Ring-necked Pheasant (*Phasianus colchicus*)   Greeley and Meyer (1953) have investigated the annual cycle of gonadotropin activity in pituitaries of male pheasants. They have found an increase in gonadotropin content during vernal testicular growth.

Domestic Fowl   In their work on the fluctuations of pituitary gonadotropic potency in the domestic fowl, Nakajo and Imai (1961) have claimed that the total gonadotropic potency of the gland decreases from the nonlaying [3.2 chick units (c.u.)] to the laying phase (2.5 c.u.) and, later on, from the onset of brooding (1.9 c.u.) to the end of brooding time (1.0 c.u.). On the other hand, in a study on photoregulation in the ovulation cycle of the domestic fowl, very significant variations in LH levels (Parlow test) in the anterior pituitary and in the peripheral blood have been shown to occur during the preovulatory period and have been correlated to the elaboration and release of gonadotropin for follicular maturation (Fraps, 1970).

Japanese Quail   Several investigations have been devoted to the measurement of the gonadotropic content of the anterior pituitary in the Japanese Quail in relation to photoperiodically induced gonadal stimulation (Tanaka *et al.*, 1965; Follett and Farner, 1966; Tixier-Vidal *et al.*, 1968). These data deal with a correlative study on the pituitary gonadotropic content, together with a histochemical evaluation of the gonadotropic cell activity, in quail switched for 40 days from "short days" (6L–18D) to "long days" (20L–4D) and then back, for 30 days, to short days. Under these conditions, the testes went through a typical cycle, with a log-linear growth for the first 20 days of the experiment followed by a plateauing of their size once the maximum is attained. After the switch back to short days, the testes regressed and, after 30 days, testicular weight had nearly reached the initial values. The gonadotropin content (^{32}P-uptake test) of the anterior pituitary was increased by photostimulation to reach a maximum after 17 days. The suppression of photoperiodic stimulation did not affect enhanced gonadotropic content up to 30 days, at which time the concentration returned to its initial value. Castration only modified the gonadotropin content when combined with photoperiodic stimulation. The gonadotropin content of the pituitary paralleled the number and cytological activation of the beta (FSH) cells. Modifications of the gamma (LH) cells, although more difficult to interpret, were also closely related to the artificial gonadal cycle.

In more recent experiments, Follett *et al.* (1971) (Table 2), using a radioimmunoassay for plasma LH measurements, could show an eightfold increase in plasma LH in photostimulated quail, a further fourfold increase in castrated long-days birds, and the suppression of both stimulatory effects by testosterone injections. Furthermore, in still unpublished data, B. K. Follett measured the immunoreactive LH concentration in the plasma and pituitary of immature quail throughout an artificial photoperiodic controlled testicular cycle (Figure 4). The most salient observations follow: During the photostimulation sequences (27 days at 20L–4D), the testes started to grow at day 4 and reached their full development (>2 g) after 20 days stimulation; correlatively, the pituitary LH content increased steadily through these 3 weeks. On the other hand, the plasma-LH concentration remained low for the first 3 days; it then rose drastically at day 4 and stayed at a high and almost constant level throughout the stimulation period. During the photoperiodic inhibiting sequences (switch from 20L–4D to 30 days of 6L–18D), the testes remained developed for the first 8 days before undergoing regression. The pituitary-LH level increased within the first week of short days—suggesting "hormonal storage" reminiscent of the histological data in quail and ducks—and fell finally to low values after 2 weeks of photoperiodic inhibition. The plasma-LH level approximately paralleled

TABLE 2   Effects of Photoperiod, Castration, and Testosterone on Plasma-LH Concentration in Male Japanese Quail[a]

Treatment	LH (ng/ml)[b]
Intact 6L–18D	0.60 ± 0.06
Intact 18L–6D	5.04 ± 0.55
Castrated 18L–6D	19.51 ± 3.38
Intact 18L–6D + testosterone	0.35 ± 0.85
Castrated 18L–6D + testosterone	1.35 ± 1.53

[a] After Follett *et al.* (1971; unpublished data).
[b] Values given as mean ± standard deviation.

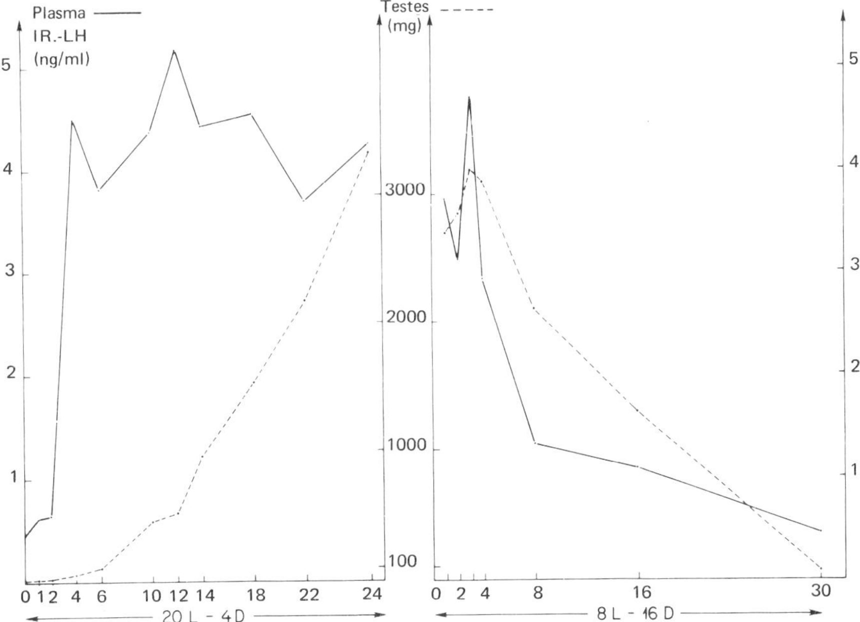

FIGURE 4   Photoperiodically induced testicular "cycle" and correlative fluctuations of plasma content in immunoreactive LH (IR-LH) in quail. From Follett *et al.* (1972).

the regression of the testes, i.e., no significant decrease was evident before 8 days.

We thus may conclude that the sexual cycle of birds living under the influence of fluctuating environmental factors is primarily directed by the seasonal variations in the amount of, and in the ratio between, both gonadotropins. It must be emphasized here that at no period of the sexual cycle is there any evidence of the secretion of one only of the two gonadotropic hormones.

Prolactin

Prolactin has many roles related to avian reproductive biology. Its role in stimulating the crop gland in columbid species, in inducing brood patch formation and incubation behavior, in timing migration and determining migratory restlessness—together with its various metabolic effects, related or not to migration—have been extensively reviewed (Riddle, 1963; Meites and Nicoll, 1966; Meier, 1969; King, 1970).

The only aspect of prolactin action that will be discussed here concerns its relationship to the gonadotropic hormones. Since Riddle's studies, in the 1930's, on endocrine regulation of incubation in the pigeon, in which a prolactin-dominated nursing phase follows the classical gonadotropin-controlled breeding phase, the concept of an antigonadotropic action of prolactin has evolved and has been considered for many years as a general rule among birds. In fact, since the first observation of Riddle and Bates (1933) on the drastic collapse of the male and female pigeon gonads, after mammalian prolactin injections, prolactin treatment has been shown to induce gonadal regression, or to inhibit the photoperiodic stimulation, in various bird species (domestic fowl: Breneman, 1942; Nalbandov, 1945; *Zonotrichia leucophrys pugetensis*: Bailey, 1950; *Fringilla coelebs, Chloris chloris,* and *Passer domesticus*: Lofts and Marshall, 1956; *Zonotrichia leucophrys nutallii* and *Melospiza melodia*: Meier and Dusseau, 1968; *Columba palumbus*: Lofts, 1970). However, once Laws and Farner (1960) had demonstrated that prolactin was unable to inhibit the photoperiodic gonadal stimulation in another passerine migratory species, *Zonotrichia leucophrys gambelii*, it became evident that the general concept of an antigonadotropic role of prolactin had to be revised. In the past 10 years, it has now been shown that, in a number of species, prolactin injections did not affect the photogonadostimulation [*Carpodacus mexicanus*: Hamner, 1968; *Zonotrichia albicollis* and *Junco hyemalis*: Meier and Dusseau, 1968; *Coturnix coturnix japonica*: Gourdji and Tixier-Vidal, 1968; even if the prolactin injections occur at various hours of the day (Gourdji, 1970) to avoid circadian variations of sensitivity to prolactin (Meier, 1969); *Pyromelana franciscana*: Gourdji, 1970; *Uroloncha punctulata*: Chandola, (unpublished data); *Columba livia*: Lofts, 1970].

On the other hand, according to Bates *et al.* (1937) (see Lofts, 1970), prolactin was considered to act on the gonads by the suppression of the pituitary gonadotropic hormones. But several recent data do not fit in with this view: In male ducks, the prolactin content of the pituitary is increased together with its gonadotropin content by artificial photostimulation (Assenmacher *et al.*, 1962; Gourdji, 1970); the same phenomenon occurs in castrated ducks (Gourdji, 1970); the pituitary prolactin content (Figure 5) (Gourdji and Tixier-Vidal, 1966a)—together with the epsilon (prolactin) cells, which are located in the cephalic lobe of the pituitary (Tixier-Vidal *et al.*, 1962)— displays an annual cycle in this species that strictly parallels the

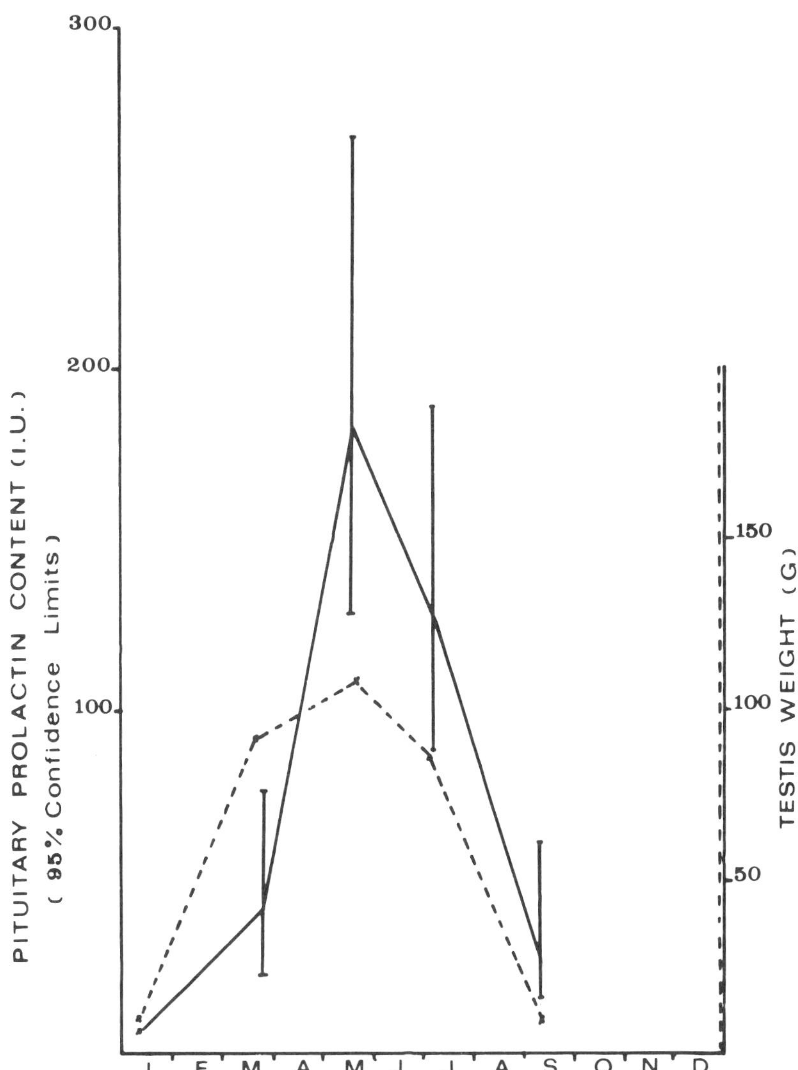

FIGURE 5   Annual testicular cycle and correlative fluctuations of anterior pituitary content in prolactin in the duck. From Gourdji and Tixier-Vidal (1966a) and Gourdji (1970).

testis cycle. The same parallel relationships between prolactin content and gonadal development also occur in the California Quail, *Lophortyx californicus* (Jones, 1969), and in the photostimulated Coturnix Quail (Gourdji, 1970).

It therefore seems certain that an antigonadal effect of prolactin in birds, whether migratory or not, can no longer be accepted as a general concept. Once purified avian prolactin preparations are at hand, radioimmunoassays of the plasma content of the hormone will permit measurement of the precise balance between prolactin, on the one hand, and FSH and LH, on the other hand, in those species in which a physiological antagonism between the two groups of pituitary hormones is assumed to occur. On the other hand, the role (other than a very probable metabolic one) of the vernal or photoperiodically induced prolactin activation, which occurs in several species together with gonadal stimulation, remains an open question.

THE HYPOTHALAMO–HYPOPHYSIAL LIAISON

THE MEDIAN EMINENCE

Anatomically, the median eminence constitutes the upper part of the neurohypophysial complex and, therefore, is located between the hypothalamus and the infundibular process (posterior hypophysis). Physiologically, this structure contains the proximal neurohemal organ of the neurophypophysial complex, where the hypothalamic neurohormones are stored and transferred into the portal system toward the adenohypophysis.

The microscopic structure of the avian median eminence has been given special attention in the past two decades. This is not only due to its primordial physiological importance in hypothalamo–hypophysial regulations but also because this relatively elementary structure may afford a simplified model for the comprehension of the complex and intermingle structure of its mammalian homolog (Wingstrand, 1951; Benoit and Assenmacher, 1953; Assenmacher, 1958; Kobayashi, 1970; Oksche, 1970).

In a cross section through the avian median eminence, three main layers may be schematically described (Figure 6): an inner layer, with the ependymal cells bordering the infundibular recess of the third ventricle, and, within the subependymal layer, some thin nerve tracts and scattered neurons belonging to the infundibular nucleus of

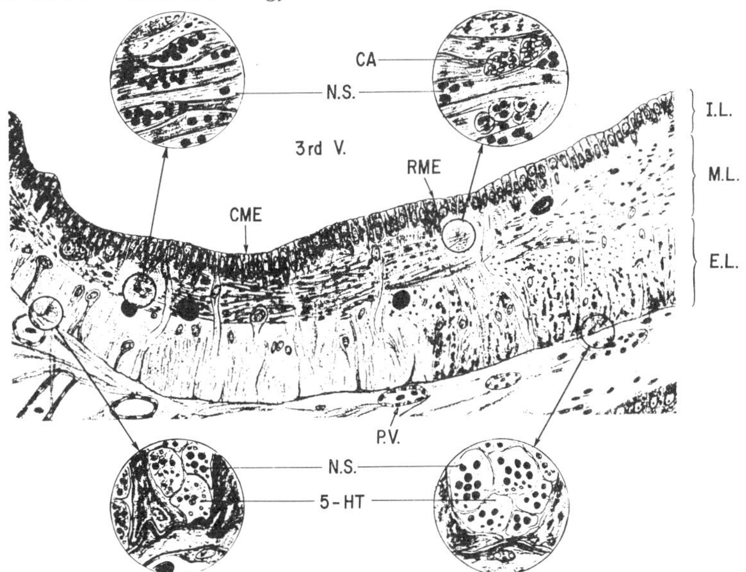

FIGURE 6 Diagrammatic representation of a longitudinal section through a duck median eminence. From A. Calas (unpublished data). RME, rostral median eminence; CME, caudal median eminence; I.L., internal layer; M.L., medial layer; E.L., external layer; 3rd V., third ventricle; N.S., neurosecretory granules (electron microscope); P.V., portal vessel (capillary); CA: catecholamines; 5-HT: 5-hydroxytryptammine (serotonin).

the hypothalamus; a medial layer, with coarse nerve fibers that run from the magnocellular nuclei of the anterior hypothalamus through the median eminence to the posterior lobe, carrying the posthypophysial hormones down to the distal neurohemal organ; an external layer that forms the proximal neurohemal organ and contains (a) a very dense frame of fine nerve fibers of various origins, which are mainly oriented perpendicularly to the outer border of the organ, (b) the coarse processes of the ependymal cells, which cross the three layers of the median eminence, giving the external layer a typical palisade-like aspect before ending on the basement membrane of the organ, and (c) the primary blood capillaries of the hypophysial portal system surrounded by their perivascular spaces.

The use of histochemical techniques (e.g., aldehyde-fuchsin), which reveal the macromolecular complex proteinaceous neurosecretory material originating mainly in the magnocellular divisions of the anterior hypothalamus, permits not only the tracing, on sagittal sections of the median eminence, of the hypothalamo–posthypophysial tract within the medial layer of the median eminence but also the identification for the external layer of two different areas: the rostral median eminence (RME), which extends grossly from the optic chiasma to the portal veins, and the caudal median eminence (CME). The rostral median eminence is peculiar in that aldehyde-fuchsin-positive neurosecretory material is present in large amounts in most fibers that radiate through the external layer of the median eminence up to the superficial blood capillaries.

Electron-microscope studies of the median eminence have been performed in several species: the White-crowned Sparrow (Bern *et al.*, 1966); the pigeon (Matsui, 1966; Peczely and Calas, 1970); the Mallard (Calas and Assenmacher, 1970), and the House Sparrow (Oehmke *et al.*, 1969). If one disregards minor species variations and keeps to the most salient aspects, three main types of nerve fibers may be seen within the median eminence: (a) axons containing large electron-dense granules, 1200–1900 Å in diameter. These axons are usually stained by the aldehyde-fuchsin technique, as seen in light microscopy; (b) axons including granules with detached membranes, or dense-core vesicles, 800–1000 Å in diameter, possibly associated with empty vesicles. Most of these axons are aldehyde-fuchsin-negative under light microscopy; (c) axons containing synaptic-like vesicles, 400–600 Å in diameter, without electron-dense inclusions.

Although all axon types may be present within any of the three layers of the median eminence, there is a marked predominance of b-type axons in the inner layer. Type-a axons are dominant within the medial layer and type-b and type-c axons prevail within the external layer. In the latter area, the large granules axons (type a), however, are frequent, particularly in the RME, and they may come, as others, especially in the Mallard, in very close contact to the basement membrane of the perivascular spaces, suggesting the possibility of a neurovascular transfer of their hormonal content.

In spite of very many investigations, all attempts to find precise correlations between the light-microscopically identified neurosecretory material and any of the multiple hypothalamo–adenohypophysial regulations have failed. However, new investigation techniques have evolved that have thrown new light on this crucial area in recent years. Using Falck's fluorescence microscope technique (Falck and Owman, 1965) for demonstrating monoaminergic nerve tracts [i.e., nerve tracts containing different kinds of substances, like noradrenaline (NA), dopamine (DA), or serotonin (5-HT), which are widespread within the central nervous system], it has been possible to trace aminergic tracts in the Japanese Quail (Follett and Sharp,

1968; Sharp and Follett, 1970), in *Carduelis chloris*, in the House Sparrow and in the Mallard (Oehmke, 1969; Oehmke *et al.*, 1969). In the median eminence, aminergic fibers are located mainly in the subependymal or inner layer; a more diffuse fluorescence appears in the external layer. In both of these areas, monoamineoxydase activity also has been demonstrated in the Tree Sparrow, *Passer montanus* (Matsui and Kobayashi, 1965), in the White-crowned Sparrow (Follett *et al.*, 1966), and in the Japanese Quail (Follett and Sharp, 1968).

In another line of investigation, radioautographic procedures at the ultrastructural level have been developed in the Mallard in order to study the specific content of the granules and vesicles of the axons of the median eminence (Calas and Droz, 1971; Calas, 1972). The technique is based on the ability of nerve endings to absorb and to concentrate, *in vitro* and *in vivo*, the monoamines they use to synthesize. From the work of Calas, it appears that NA, DA, and 5-HT are selectively fixed on type-b axons. However, the catecholamine (noradrenaline or dopamine) producing axons are electively located within the subependymal area, i.e., at the junction of the inner and the medial layer of the median eminence, while the indolamine (5-hydroxytryptamine) secreting axons are exclusively located near to the peripheral border of the external layer. Noradrenaline-containing axons often surround the scattered neurons of the infundibular nucleus in the inner layer of the median eminence, and both types (noradrenaline and 5-hydroxytryptamine) of aminergic fibers innervate the walls of the portal vessels.

In light of our present knowledge, type-a contain axons neurosecretory peptidergic material that is essentially associated with the posthypophysial hormones but seems also related to some kind of adenohypophysial regulations. Type-b axons represent a manifold class of neurons, including either peptidergic material (releasing factors?) or several kinds of monoamines. Radioautographic and other methods will have to provide further information on their content and topographical distribution within the median eminence, as they seem to be very much concerned with the hypothalamo–adenohypophysial control. Finally, insofar as type-c axons can be considered a separate entity (alternatively they could represent a cross section through a type-b axon at a level deprived of electron dense inclusions), they tentatively could be regarded as cholinergic nerve endings. Cholinesterase has already been demonstrated to occur mainly within the rostral median eminence in the White-

crowned Sparrow (Kobayashi and Farner, 1964) and in the Japanese
Quail (Follett and Sharp, 1968).

THE HYPOPHYSIAL PORTAL SYSTEM

The occurence of a portal system supplying the anterior pituitary
with blood, which first flows through a primary capillary network
located within the neurohemal organ of the median eminence, is
a general pattern among tetrapods. Even in most fish, various kinds
of neurovascular links between neuro- and adenohypophysis have
been described.

Compared to the mammalian model, the avian portal system shows
some minor pecularities that appear in Figure 7 (Wingstrand, 1961;
Benoit and Assenmacher, 1951, 1953; Assenmacher, 1952, 1958;
Assenmacher and Tixier-Vidal, 1965; Vitums *et al.*, 1964; Sharp and
Follett, 1969a; Dominic and Singh, 1969; Duvernoy *et al.*, 1969):
(1) The primary capillary network, from which the portal veins
originate, is, as a whole, restricted to the very superficial layer of
the median eminence, immediately under the thin tuberal part of the
adenohypophysis. Capillary "loops" that penetrate toward the inner
layers of the median eminence are exceptional. (2) The portal veins
run directly, usually within a compact tract, from the median emi-

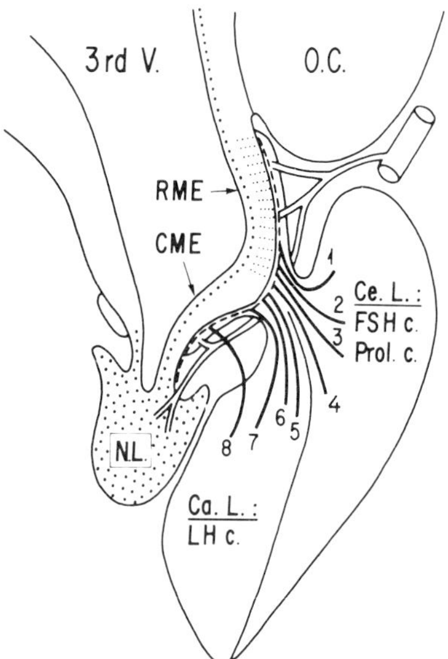

FIGURE 7 Diagrammatic representation
of a mid-sagital section through an avian
hypophysial complex. 3rd V., third ventricle; O.C., optic chiasma; RME, rostral
median eminence; CME, caudal median
eminence; N.L., neural lobe (posterior
pituitary); Ce.L., cephalic lobe; Ca.L.,
caudal lobe of the anterior pituitary;
FSH c., FSH cells; Prol.c., prolactin cells;
LH c., LH cells; 1–4, anterior group of
portal veins; 5–8, posterior group of portal
veins.

nence to the anterior pituitary, instead of following the infundibular stalk as in mammals. (3) The portal veins consist of distinct anterior and posterior groups, which drain the primary capillary network of the rostral and of the caudal median eminence, respectively, toward the cephalic and the caudal lobes of the anterior pituitary. Such peculiar anatomical relationships of the portal vessels appear to be of special interest since there is evidence of some diversity in the neurons that approach (end on ?) the portal vessels, respectively, within the rostral and the caudal parts of the median eminence and an equal diversity in the distribution of the specific cell types between the two lobes of the anterior pituitary, as described above for the gonadotrops. Similar, though more complex, topographical arrangements have been described in mammals. In lower vertebrates, only the gonadotrops that are the most dependent on the central nervous system—and eventually the metamorphosis-promoting thyrotropic cells—are located at the very "strategic" point of entry of the portal vessels within the pituitary [see review in Assenmacher and Tixier-Vidal (1965)]. From an evolutionary standpoint, it is interesting to note that a special topographical distribution of *all* cell types of the anterior pituitary, in relation to portal vessels from different origins, seems to have evolved once the modulating role of the central nervous system on *all* pituitary functions became more extensive and delicately tuned.

EFFECTS OF HYPOTHALAMO–ADENOHYPOPHYSIAL DISCONNECTION

FSH and LH

Complete surgical separation of the anterior pituitary from the hypothalamus has been obtained in several species, by transection of the portal veins (Mallard: Assenmacher and Benoit, 1953; Benoit and Assenmacher, 1953; Assenmacher, 1958, 1970; domestic fowl: Shirley and Nalbandov, 1956); ectopic hypophysial autografting (Mallard: Assenmacher, 1958, 1970; Assenmacher and Baylé, 1964; pigeon and Japanese Quail: Baylé and Assenmacher, 1966, 1967; Baylé, 1968; domestic fowl: Ma and Nalbandov, 1963), or total destruction of the median eminence (duck: Benoit and Assenmacher, 1953; White-crowned Sparrow: Wilson and Farner, 1965; Stetson, 1969; Japanese Quail: Sharp and Follett, 1969b). Whatever the surgical procedure is, hypothalamo–adenohypophysial disconnection results in a rapid, complete, and irreversible gonadal atrophy, which evokes adenohypophysectomy (see Table 1). The gonadotropic cells have been studied in pituitary autografts (Tixier-Vidal, 1970). In the Mallard and Japanese Quail, the FSH cells were regressed and the LH cells were either undetectable (Quail) or they showed no evidence of secretory function (Mallard). In the pigeon, both types of gonadotrops could be identified, although they were present in reduced number.

Some attempts have been made at more selective disconnections. In Mallards, the transsection of only the anterior portal vessels was as effective in inducing gonadal atrophy and blockade of gonadal photostimulation as was complete hypothalamo–hypophysial disconnection (Assenmacher and Benoit, 1953; Assenmacher, 1958, 1970). Within the pituitary, some FSH cells still could be demonstrated. They had a condensed quiescent aspect; by contrast, the LH cells were not reduced in number, but they appeared similarly condensed (Tixier-Vidal, 1970).

In other experiments, selective lesions in the median eminence were designed. In the drake, stereotaxic lesions, selectively located within the rostral median eminence and involving the hypothalamo–posthypophysial tracts, have been claimed to cause gonadal atrophy (Gogan *et al.*, 1963). Lesions in the rostral median eminence, however, were without gonadal effect in the White-crowned Sparrow (Wilson and Farner, 1965) or in the Japanese Quail (Sharp and Follett, 1969b), while in the latter experiment the posterior median eminence seemed more involved in the gonadotropic regulation.

To destroy selectively localized areas in such an intermingled structure as in the median eminence is obviously a difficult task. Further detailed and diversified investigations will be needed before any general conclusions, e.g., in terms of species specificity, will be possible.

Prolactin

The question of prolactin has been extensively reviewed in recent years (Nicoll *et al.*, 1970; Tixier-Vidal and Gourdji, 1972). Only the most salient results will be discussed here.

At first sight, hypothalamo–hypophysial disconnection leads to opposite results in mammals and in birds: The prolactin secretion is drastically enhanced in the former, while it remains low in the latter.

In vivo experiments are scarce, as, apart from the pigeon, birds lack any specific target organ for prolactin. In fact, ectopic pituitary autografts in pigeons left the crop sac in the same resting state as in the hypophysectomized controls (Table 3) (Baylé and Assenmacher, 1965). The graft itself showed no activation of the prolactin cells (Tixier-Vidal, 1970). Pituitary autografts also have been investigated for prolactin cells in Japanese Quail and Mallard (see Tixier-Vidal, 1970). In both species, the stimulatory effect of "long days" photoperiods was inhibited. In the absence of photostimulation, however, the prolactin cells of the duck pituitary autotransplants display indisputable histologic and ultrastructural evidence of stimulation.

This peculiar behavior of the Mallard prolactin cells after hypothalamo–hypophysial disconnection was also found in *in vitro* experiments. From a number of pituitary culture experiments that have been performed along with prolactin bioassays, all agree that no consistent autonomous synthesis and release of prolactin may be maintained in the pigeon (Meites *et al.*, 1963; Nicoll, 1965; Gala and Reece, 1965; Gourdji, 1970), the Tricolored Blackbird, *Agelaius tricolor* (Nicoll, 1965), or the Japanese Quail (Gourdji, 1970). The cultured Mallard pituitary behaves in a different way. Although a marked and steady secretion of prolactin within the medium can be measured after 2 weeks, the prolactin content of the explant remains at its initial level. The renewal time of the hormone was estimated at 24–48 hours. Moreover, such explants were shown to consist almost exclusively of actively functioning prolactin cells (Tixier-Vidal and Gourdji, 1965; Gourdji, 1970).

Even though these results have not been confirmed by Nicoll *et al.* (1970) on another strain of Mallards and using a poorer culture medium, Tixier-Vidal's observations, which have been repeated,

TABLE 3   Effects of Adenohypophysectomy and of Pituitary Autograft on the Pigeon Crop Sac[a]

Treatment	Number	Weight of Crop Sac (g)	Height of Crop Sac Epithelium (μm)
Adult controls	33	1.531 ± 0.081	33.12 ± 1.73
Nursing controls	25	8.877 ± 0.949	1026.00 ± 98
Hypophysectomy (after 1–4 months)	9	0.860 ± 0.052	27.21 ± 2.64
Pituitary autograft (after 4–5 months)	15	0.776 ± 0.087	33.85 ± 0.29

[a]After Baylé and Assenmacher (1965).

appear acceptable. The peculiar, partially autonomous behavior of
the prolactin function with respect to hypothalamic control in
Mallards thus tends to temper the general statement on the lack of
autonomy of avian prolactin secretion.

RELEASING FACTORS IN AVIAN MEDIAN EMINENCE EXTRACTS

Gonadotropin-Releasing Factors

The hypothalamic disconnection experiments indicate that the
pituitary gonadotropic function is as dependent in birds as in
mammals on the passage of neurohormones into the portal vessels
within the median eminence. In fact, *in vivo* infusion of domestic
fowl with median eminence extracts from the same species induced
ovulation (Clark and Fraps, 1967; Opel and Lepore, 1967; Fraps,
1970). Similarly, median eminence extracts from Japanese Quail
produced an *in vitro* release of gonadotropin by domestic fowl or
Quail pituitaries, as measured by the ^{32}P-uptake test (Follett, 1970).
Due to the lack of specificity of the ^{32}P-uptake test toward FSH
and LH, there is no information as to whether there is an avian
LH-RF and an FSH-RF. In mammals it has been shown recently that
the highly purified, or even synthesized LH-RF has both activities
(Schally *et al.*, 1971). On the other hand, the specificity of the
gonadotropin-releasing factors seems rather low, for ovulation in
domestic fowl has also been obtained with beef median eminence
extract by Fraps's group, whereas *in vitro* rat pituitaries released
gonadotropin (Parlow test) with domestic median eminence extract
(Jackson and Nalbandov, 1969) or with pigeon median eminence
(Ishi *et al.*, 1970).

Prolactin-Releasing Factor

According to the results of the hypothalamo–hypophysial separation
studies, the main hypothalamic hormone regulating avian prolactin
secretion should be a releasing hormone. As a matter of fact, such
a prolactin-releasing factor (P-RF) can be extracted from the median
eminence of pigeon (Kragt and Meites, 1965), of brooding Tricolored
Blackbirds (Nicoll, 1965), Mallards (Gourdji and Tixier-Vidal,
1966b), Japanese Quail and chicks (Kragt, 1966), and turkeys (Chen
et al., 1968). All of these preparations have been tested *in vitro* on
corresponding pituitaries.

The peculiar reaction of the Mallard prolactin function to the deprivation of the hypothalamic link, however, can lead one to postulate the occurrence of a second hypothalamic hormone that would exert an inhibiting action on prolactin secretion as in mammals. This hypothetical avian prolactin-inhibiting factor (P-IF) would have in most species, except the Mallard, a very secondary role toward the predominant P-RF. It must be kept in mind here that in mammals, where a P-IF has been demonstrated for years as the only hypothalamic hormone regulating prolactin secretion, recent evidence has been obtained in the rat of the occurrence of a P-RF, together with the predominant P-IF (Nicoll *et al.*, 1970).

POSSIBLE INVOLVEMENT OF AN AMINERGIC CONTROL OF GONADOTROPIN RELEASE

After nearly 20 years, during which peptidergic control mechanisms of pituitary functions were almost exclusively considered, the probable involvement of aminergic mechanisms in the regulation of adenohypophysial activities has suddenly returned to fashion. The reason lies not only in the demonstration of aminergic fiber systems in close relationship to the peptidergic neuroendocrine tracts within hypothalamus and median eminence, but more generally, in a marked shift of interest, in neurophysiological studies from electrophysiological toward biochemical investigations.

As a matter of fact, aminergic and also cholinergic mechanisms had been presumed for years to participate in the control of ovulation. For ovulation in the domestic fowl, as in mammals, could be blocked by either antiadrenergics such as dibenamine and dibenzyline (Van Tienhoven *et al.*, 1954), or their congener SKF-501 (Zarrow and Bastian, 1953), or by such anticholinergics as atropine (Zarrow and Bastian, 1953).

More recently, two further pharmacological agents have been shown to interfere with gonadotropin regulation: reserpine, a depletor of brain amine stores, and 6-hydroxydopamine (6-OH-DA), a noradrenalin congener that has been shown to cause emptying and degeneration of catecholamine-containing nerve terminals within the avian median eminence (Calas, 1971).

Depending on the dosage, reserpine induces a more or less complete gonadal atrophy in the pigeon (Khazan *et al.*, 1960; Assenmacher and Baylé, 1964), the domestic fowl (Hagen and Wallace, 1961), and the White-crowned Sparrow (Brown and

Mewaldt, 1967). In the Mallard, reserpine inhibits the photogonadal
response (Assenmacher *et al.*, 1961) (Figure 8) and depresses the
FSH cells and, to a less extent, the LH cells, while stimulating the
prolactin cells (Tixier-Vidal and Assenmacher, 1962). It also lowers
the gonadotropic content of the pituitary and increases its prolactin
content (Assenmacher *et al.*, 1962).

On the other hand, 6-hydroxydopamine, when injected intra-
ventricularly in long-day photostimulated male Japanese Quail,
caused a marked (75 percent) reduction of the testicular response,
compared to solvent-injected controls (Figure 9) (Assenmacher and
Boissin, 1972).

The few data available give but a poor insight as to where and how
central aminergic control mechanisms may be related with the
hormonal peptidergic hypothalamic regulation of gonadotropic
function. Nonetheless, they constitute arguments in favor of such
a dual regulatory mechanism, which fits in with the neuroanatomical
data. It will be the task of the years ahead to explore whether the

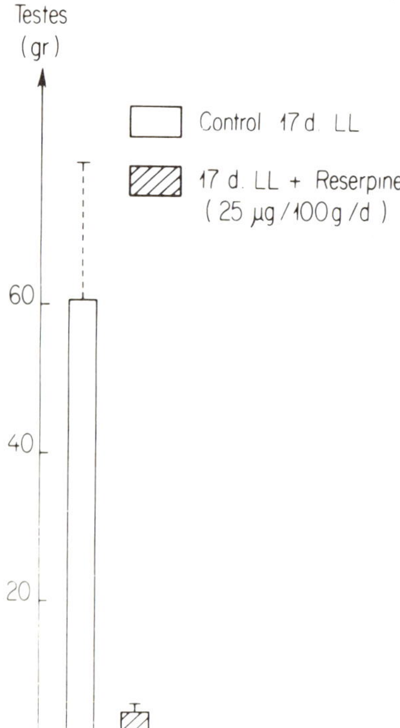

FIGURE 8 Inhibiting effects of reserpine
administration on light-induced testicular
growth in the duck. From Assenmacher
et al. (1961).

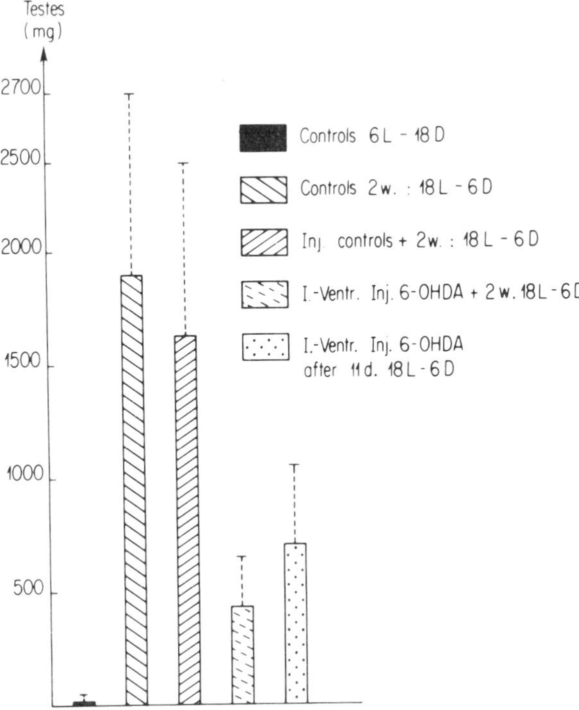

FIGURE 9   Inhibiting effect of intraventricular injection of 6-hydroxydopamine (6-OH-DA) on light-induced (2 weeks at 18L–6D) testicular growth in the Japanese Quail. From Assenmacher and Boissin (1972).

aminergic system regulates the synthesis, or more probably the release, of the peptide hormones.

CONCLUSION

We may summarize the role of the hypothalamo–hypophysial axis on the avian reproduction as follows:

1.   The seasonal production of sexual cells and of steroid hormones by the gonads is controlled by two main gonadotropic hormones from the anterior pituitary: FSH and LH. Except for some migratory species, the precise role of prolactin in the reproductive cycle is still obscure.

2.   An intact hypothalamo–hypophysial liaison, including the portal system and several kinds of specific nerve terminals in the median eminence, is an absolute prerequisite for any gonadotropic function. All the nervous stimuli by which the environment controls the gonadotropic function lead finally to a modulation of the synthesis and transfer of specific hypothalamic hormones within the portal system.

3.   The physiology of the various components of the very complex transfer organ that is the median eminence is still very unclear. At least two of these components seem to play a major role here—the peptidergic nerve endings and the aminergic nerve terminals.

REFERENCES

Assenmacher, I. 1952. La vascularisation du complexe hypophysaire chez le canard domestique. I. La vascularisation du complexe hypophysaire adulte. II. Le développement morphologique de l'appareil vasculaire hypophysaire. Arch. Anat. Microsc. Morphol. Exp. (Paris) 41:69–152.

Assenmacher, I. 1958. Recherches sur le contrôle hypothalamique de la fonction gonadotrope préhypophysaire chez le canard. Arch. Anat. Microsc. Morphol. Exp. (Paris) 47:447–472.

Assenmacher, I. 1970. Importance de la liaison hypothalamo–préhypophysaire dans le contrôle nerveux de la reproduction chez les oiseaux, p. 167–191. *In* J. Benoit and I. Assenmacher [ed.] La photorégulation de la reproduction chez les oiseaux et les mammifères. Editions du C.N.R.S., Paris.

Assenmacher, I., and J. D. Baylé. 1964. Répercussions endocriniennes de la greffe hypophysaire ectopique chez le canard mâle. C.R. Acad. Sci. (Paris) 259:3848–3850.

Assenmacher, I., and J. Benoit. 1953. Répercussions de la section du tractus porto-tubéral hypophysaire sur la gonadostimulation par la lumière chez le canard domestique. C.R. Acad. Sci. (Paris) 236:2002–2004.

Assenmacher, I., and J. Boissin. 1972. Circadian endocrine and related rhythms in birds. Gen. Comp. Endocrinol. Suppl. 3:489–498.

Assenmacher, I., and A. Tixier-Vidal. 1965. Hypothalamic–pituitary relations. Excerpta Med. Int. Congr. Ser. 83:131–145.

Assenmacher, I., A. Tixier-Vidal, and J. D. Baylé. 1961. Inhibition du réflexe photo-sexuel par la réserpine chez le canard mâle. C.R. Soc. Biol. (Paris) 155:2235–2240.

Assenmacher, I., A. Tixier-Vidal, and J. Boissin. 1962. Contenu en hormones gonadotropes et en prolactine de l'hypophyse du canard soumis à un traitement lumineux et réserpinique. C.R. Soc. Biol. (Paris) 156:1555–1559.

Bailey, R. E. 1950. Inhibition with prolactin of light induced gonad increase in White-crowned Sparrows. Condor 52:247–251.

Bates, R. W., I. Riddle, and E. L. Lahr. 1937. The mechanisms of antigonad action of prolactin in adult pigeons. Am. J. Physiol. 119:610–614.

Baylé, J. D. 1968. Importance relative des différents niveaux de commande des régulations neuroendocriniennes chez les oiseaux. Thèse Sciences Naturelles, Montpellier No.–CNRS A02191. 213 p.

Baylé, J. D., and I. Assenmacher. 1965. Absence de stimulation du jabot de pigeon après autogreffe hypophysaire. C. R. Acad. Sci. (Paris) 261:5667–5670.

Baylé, J. D., and I. Assenmacher. 1966. Répercussions endocriniennes de l'autogreffe hypophysaire chez le pigeon. J. Physiol. (Paris) 58:203.

Baylé, J. D., and I. Assenmacher. 1967. Contrôle hypothalamo–hypophysaire du fonctionnement thyroidien chez la caille. C. R. Acad. Sci. (Paris) 264:125–128.

Benoit, J. 1935. Rôle de l'hypophyse dans l'action stimulante de la lumière sur le développpment testiculaire chez le canard. C. R. Soc. Biol. (Paris) 118:672

Benoit, J., and M. Aron. 1934. Sur le conditionnement hormonique du développement testiculaire chez les oiseaux. Injection d'extrait hypophysaire chez le canard. Remarques sur divers éléments d'interprétation des expériences. Influence de l'âge. C. R. Soc. Biol. 116:215.

Benoit, J., and I. Assenmacher. 1951. Etude préliminaire de la vascularisation de l'appareil hypophysaire du canard domestique. Arch. Anat. Microsc. Morphol. Exp. 40:27–45.

Benoit, J., and I. Assenmacher. 1953. Rapports entre la stimulation préhypophysaire et la neurosécrétion chez l'oiseau. Arch. Anat. Microsc. Morphol. Exp. 42:334–386.

Benoit, J., P. Mandel, F. X. Walter, and I. Assenmacher. 1950. Sensibilité testiculaire aux hormones gonadotropes hypophysaires chez le canard domestique au cours de la période de régression testiculaire saisonnière. C. R. Soc. Biol. (Paris) 144:1400–1403.

Bern, H. A., R. S. Nishioka, L. R. Mewaldt, and D. S. Farner. 1966. Photoperiodic and osmotic influences in the ultrastructure of the hypothalamic neurosecretory system of the White-crowned Sparrow, *Zonotrichia leucophrys gambelii*. Z. Zellforsch. 69:198–227.

Brasch, M., and T. W. Betz. 1971. The hormonal activities associated with the cephalic and caudal regions of the cockerel pars distalis. Gen. Comp. Endocrinol. 16:241–256.

Breneman, W. R. 1942. Action of prolactin and estrone on weights of reproductive organs and viscera of the cockerel. Endocrinology 30:609–615.

Breneman, W. R. 1945. The gonadotropic activity of the anterior pituitary of cockerels. Endocrinology 36:190–199.

Breneman, W. R., F. J. Zeller, and R. O. Creek. 1962. Radioactive phosphorus uptake by chick testes as an end-point for gonadotropin assay. Endocrinology 71:791–798.

Brown, I. L., and L. R. Mewaldt. 1967. Effects of reserpine on the White-crowned Sparrow (*Zonotrichia leucophrys gambelii*). Brit. J. Pharmacol. Chemother. 30:251–257.

Brown, P. S. 1955. The assay of gonadotrophin from urine of nonpregnant human subjects. J. Endocrinol. 13:59–64.

Calas, A. 1971. Identification de diverses catégories axonales dans la neuro-

hypophyse des oiseaux. Abstract of the 6th European Congress of Comparative
Endocrinology, Montpellier (France), 2–7 August 1971. No. 35.

Calas, A. 1972. Capture et rétention de monoamines dans des fibres nerveuses de
l'éminence médiane. Etude *in vivo* chez le canard par radioautographie
à haute résolution. C. R. Acad. Sci. 274:925–927.

Calas, A., and I. Assenmacher. 1970. Ultrastructure de l'éminence médiane de
canards (*Anas platyrynchos*). Z. Zellforsch. 109:64–82.

Calas, A., and B. Droz. 1971. Synthèse et capture de monoamines dans des fibres
nerveuses de l'éminence médiane étude *in-vitro* chez le canard par autoradi-
ographie à haute résolution. C. R. Acad. Sci. 272:1890–1892.

Chen, C. L., E. J. Bixler, A. I. Weber, and J. Meites. 1968. Hypothalamic stimulation of prolactin release from the pituitary of turkey hens and poult. Gen.
Comp. Endocrinol. 11:489.

Clark, C. E., and R. M. Fraps. 1967. Induction of ovulation in chickens with
median eminence extracts. Poultry Sci. 46:1245–1246.

Clark, L. B., S. L. Leonard, and G. Bump. 1937. Light and the sexual cycle of
grouse birds. Science 85:339–340.

Dominic, C. J., and R. M. Singh. 1969. Anterior and posterior groups of portal
vessels in the avian pituitary. Gen. Comp. Endocrinol. 13:22–26.

Domm, L. V. 1931. Precocious development of sexual characters in the fowl by
homeoplastic hypophyseal implants. I. The male. Proc. Soc. Exp. Biol. Med.
29:308–309.

Domm, L. V., and H. B. Van Dyke. 1932. Precocious development of sexual
characters in the fowl by daily injections of hebin. I. The male. II. The female. Proc. Soc. Exp. Biol. Med. 30:349–353.

Duvernoy, H., F. Gainet, and J. G. Koritké. 1969. Sur la vascularisation de
l'hypophyse des oiseaux. J. Neuro-Visc. Rel. 31:109–127.

Evans, H. M., and M. E. Simpson. 1934. The response of the gonads of immature
pigeons to various gonadotropic hormones. Anat. Rec. 60:405–421.

Evans, J. S., L. Hines, R. Varney, and F. C. Koch. 1940. The comparative assay
of gonadotropic substances on rats, mice, and chicks. Endocrinology 26:1005.

Falck, B., and C. Owman. 1965. A detailed methodological description of the
fluorescence method for the cellular demonstration of monoamines. Acta
Univ. Lund 7:1–23.

Follett, B. K. 1970. Gonadotropin-releasing activity in the quail hypothalamus.
Gen. Comp. Endocrinol. 15:165–179.

Follett, B. K., and D. S. Farner. 1966. Pituitary gonadotropins in the Japanese
Quail (*Coturnix coturnix japonica*) during photoperiodically induced gonadal
growth. Gen. Comp. Endocrinol. 7:125–131.

Follett, B. K., and P. J. Sharp. 1968. Adrenergic and cholinergic systems in the
hypothalamus of the Japanese Quail, *Coturnix coturnix japonica*. Arch. Anat.
Histol. Embryol. Norm. Exp. 51:213–222.

Follett, B. K., H. Kobayashi, and D. S. Farner. 1966. The distribution of monoamine oxidase and acetylcholinesterase in the hypothalamus and its relation
to the hypothalamo–hypophysial neurosecretory system in the White-crowned
Sparrow, *Zonotrichia leucophrys gambelii*. Z. Zellforsch. 75:57–76.

Follett, B. K., C. G. Scanes, and F. J. Cunningham, 1971. A radioimmunoassay
for avian luteinizing hormone. J. Endocrinol. 51:v.vI.

Fraps, R. M. 1970. Photoregulation in the ovulation cycle of the domestic fowl, p. 281–306. *In* J. Benoit and I. Assenmacher [ed.] La photorégulation de la reproduction chez les oiseaux et les mammifères. Editions du C.N.R.S., Paris.

Fraps, R. M., H. L. Fevold, and B. H. Neher. 1947. Ovulatory response of the hen to presumptive luteinizing and other fractions from fowl anterior pituitary tissue. Anat. Rec. 99:571–572.

Furr, B. J. A., and F. J. Cunningham. 1970. The biological assay of chicken pituitary gonadotrophins. Brit. Poultry Sci. 11:7–13.

Gala, R. R., and R. P. Reece. 1965. "In vitro" lactogen production by anterior pituitaries from various species. Proc. Soc. Exp. Biol. Med. 120:263–264.

Garnier, D. H. 1971. Variations de la testostérone du plasma périphérique chez le canard Pékin au cours du cycle annuel. C. R. Acad. Sci. (Paris) 272:1665–1668.

Gogan, F., C. Kordon, and J. Benoit. 1963. Retentissement de lésions de l'éminence médiane sur la gonadostimulation du canard. C. R. Soc. Biol. (Paris) 157:2133–2136.

Gourdji, D. 1965. Modifications des types cellulaires hypophysaires impliqués dans le cycle sexuel annuel chez l'ignicolore mâle. Gen. Comp. Endocrinol. 5:682.

Gourdji, D. 1970. Contribution à l'étude de la prolactine hypophysaire chez les oiseaux. Thèse Sciences Naturelles, Paris, No. CNRS AO 4716. 165 p.

Gourdji, D., and A. Tixier-Vidal. 1966a. Variations du contenu hypophysaire en prolactine chex le canard Pékin mâle au cours du cycle sexuel et de la photo-stimulation. C. R. Acad. Sci. (Paris) 262:1746–1749.

Gourdji, D., and A. Tixier-Vidal. 1966b. Mise en évidence d'un contrôle hypothalamique stimulant de la prolactine hypophysaire chez le canard. C. R. Acad. Sci. (Paris) 263:162–165.

Gourdji, D., and A. Tixier-Vidal. 1968. Quelques aspects nouveaux du rôle de la prolactine chez les oiseaux, p. 231–241. *In* M. Fontaine [ed.] La spécificité zoologique des hormones hypophysaires. Editions du C.N.R.S., Paris.

Greeley, F., and R. K. Meyer. 1953. Seasonal variations in testis-stimulating activity of male pheasant pituitary glands. Aukland 70:350–358.

Hagen, P. and A. C. Wallace. 1961. Effect of reserpine on growth and sexual development of chickens. Brit. J. Pharmacol. Chemother. 17:267–275.

Hamner, W. M. 1968. The photorefractory period of the House Finch. Ecology 49:211–227.

Herrick, R. B., W. H. McGibbon, and W. H. McShan. 1962. Gonadotropic activity of chicken pituitary glands. Endocrinology 71:487–491.

Hill, R. I., and A. S. Parkes. 1934. Hypophysectomy of birds. III. Effects on gonads, accessory organs and head furnishings. Proc. Roy. Soc. London, Ser. B 116:221–236.

Ishii, S., A. K. Sarkar, and H. Kobayashi. 1970. Ovarian ascorbic acid-depleting factor in pigeon median eminence extracts. Gen. Comp. Endocrinol. 14:461–466.

Jaap, R. G. 1934. Gonad-stimulating potency of individual pituitaries. Poultry Sci. 14:237–246.

Jackson, G. L., and A. V. Nalbandov. 1969. Luteinizing hormone releasing activity in the chicken hypothalamus. Endocrinology 84:1262–1265.

Jallageas, M., and I. Assenmacher. 1970. Testostéronémie du canard photostimulé ou soumis à des injections répétées de testostérone. C. R. Soc. Biol. (Paris) 164:2338.

Jones, R. E. 1969. Epidermal hyperplasia in the incubation patch of the California Quail, *Lophortyx californicus*, in relation to pituitary prolactin content. Gen. Comp. Endocrinol. 12:498–501.

Khazan, N., F. G. Sulman, and H. Z. Winnik. 1960. Effect of reserpine on pituitary–gonadal axis. Proc. Soc. Exp. Biol. Med. 105:201–204.

King, J. R. 1970. Photoregulation of food: Intake and fat metabolism in relation to avian sexual cycles, p. 365–385. *In* J. Benoit and I. Assenmacher [ed.] La photorégulation de la reproduction chez les oiseaux et les mammifères. Editions du C.N.R.S. Paris.

Kobayashi, H. 1970. Fine structure and adrenergic mechanism of the median eminence in relation to gonadotropic activity of adenohypophysis, p. 193–210. *In* J. Benoit and I. Assenmacher [ed.] La photorégulation de la reproduction chez les oiseaux et les mammifères. Editions du C.N.R.S. Paris.

Kobayashi, H., and D. S. Farner. 1964. Cholinesterase in the hypothalamo–hypophysial neurosecretory system of the White-crowned Sparrow, *Zonotrichia leucophrys gambelii*. Z. Zellforsch. 63:965–973.

Kordon, C., and F. Gogan. 1970. Interaction du feed-back et de la photostimulation dan les régulations gonadotropes chez les mammifères et les oiseaux, p. 325–364. *In* J. Benoit and I. Assenmacher [ed.] La photorégulation de la reproduction chez les oiseaux et les mammifères. Editions du C.N.R.S., Paris.

Kragt, C. 1966. Study on the neuroendocrine control of prolactin release in mammals and birds. PhD thesis, Michigan State University, East Lansing.

Kragt, C., and J. Meites. 1965. Stimulation of pigeon pituitary prolactin release by pigeon hypothalamic extracts *in vitro*. Endocrinology 76:1169–1176.

Laws, D. F., and D. S. Farner. 1960. Prolactin and the photoperiodic testicular response in White-crowned Sparrows. Endocrinology 67:279–281.

Lofts, B. 1970. Cytology of the gonads and feed-back mechanisms with respect to photosexual relationships in male birds, p. 307–324. *In* J. Benoit and I. Assenmacher [ed.] La photorégulation de la reproduction chez les oiseaux et les mammifères. Editions du C.N.R.S. Paris.

Lofts, B., and A. J. Marshall. 1956. The effects of prolactin administration on the internal rhythm of reproduction in male birds. J. Endocrinol. (London) 13:101–106.

Ma, R. C. S., and A. V. Nalbandov. 1963. Discussion, p. 306–311. *In* A. V. Nalbandov [ed.] Advances in neuroendocrinology. University of Illinois Press, Urbana.

Matsui, T. 1966. Fine structure of the posterior median eminence of the pigeon, *Columba livia domestica*. J. Fac. Sci. Univ. Tokyo IV. 2:49–70.

Matsui, T., and H. Kobayashi. 1965. Histochemical demonstration of monoamine oxidase in the hypothalamo–hypophysial system of the Tree Sparrow and the rat. Z. Zellforsch. 68:172–182.

Matsuo, S., A. Vitums, J. R. King, and D. S. Farner. 1969. Light microscope studies of the cytology of the adenohypophysis of the White-crowned Sparrow, *Zonotrichia leucophyrs gambelii*. Z. Zellforsch. 95:143–176.

Meier, A. H. 1969. Diurnal variations of metabolic response to prolactin in lower vertebrates. Gen. Comp. Endocrinol. Suppl. 2:55–62 (5th Int. Cong. Comp. Endocrinol. Delhi 1967).

Meier, A. H., and J. W. Dusseau. 1968. Prolactin and the photoperiodic gonadal response in several avian species. Physiol. Zool. 41:95–103.

Meites, J., and C. H. S. Nicoll. 1966. Adenohypophysis: Prolactin. Ann. Rev. Physiol. 28:57–88.

Meites, J., C. H. S. Nicoll, and P. K. Talwaker. 1963. The central nervous system and the secretion and release of prolactin, p. 238–277. *In* A. V. Nalbandov [ed.] Advances in neuroendocrinology. University of Illinois Press, Urbana.

Meyer, R. K., C. H. Mellish, and H. S. Kupperman. 1939. The gonadotropic and adrenotropic hormones of the chicken hypophysis. J. Pharmacol. Exp. Ther. 65:104–114.

Mikami, S. I., A. Vitums, and D. S. Farner. 1969. Electron microscopic studies on the adenohypophysis of the White-crowned Sparrow, *Zonotrichia leucophrys gambelii*. Z. Zellforsch. 97:1–29.

Mitchell, M. E. 1967. Stimulation of the ovary in hypophysectomized hens by an avian pituitary preparation. J. Reprod. Fert. 14:249–256.

Morris, T. R., and A. V. Nalbandov. 1961. The induction of ovulation in starving pullets using mammalian and avian gonadotropins. Endocrinology 68:687–697.

Nakajo, S., and K. Imai. 1961. Gonadotropin content in the cephalic and the caudal lobe of the anterior pituitary in laying, nonlaying and broody hens. Poultry Sci. 40:739–744.

Nalbandov, A. V. 1945. A study of the effects of prolactin on broodiness and on cock testes. Endocrinology 36:251–258.

Nalbandov, A. V. 1966. Hormonal activity of the pars distalis in reptiles and birds, p. 295–316. *In* G. W. Harris and B. T. Donovan [ed.] The pituitary gland. Vol. I. Butterworths, London.

Nalbandov, A. V., R. K. Meyer, and W. H. McShan. 1951. The role of a third gonadotropic hormone in the mechanism of androgen secretion in chicken testes. Anat. Rec. 110:475–494.

Nicoll, C. S. 1965. Neural regulations of adenohypophysial prolactin secretion in tetrapods. J. Exp. Zool. 158:203–210.

Nicoll, C. S., J. A. Parson, R. P. Fiorindo, and C. T. McKenee. 1970. Assay of hypothalamic factors which regulate prolactin secretion, p. 115–144. *In* J. Meites [ed.] Hypophysiotropic hormones of the hypothalamus. Assay and chemistry. Williams & Wilkins Co., Baltimore.

Oehmke, H.-J. 1969. Topographische Verteilung der Monoamin-fluoreszenz im Zwischenhirn-Hypophysensystem von *Carduelis chloris* und *Anas platyrhynchos*. Z. Zellforsch. 101:266–284.

Oehmke, H.-J., J. Priedkalns, M. Vaupel von Harnack, and A. Oksche. 1969. Fluoreszenz und electronmikroskopische Untersuchungen am Zwischenhirn-Hypophysensystem von *Passer domesticus*. Z. Zellforsch. 95:109–133.

Oksche, A. 1970. Discussion, p. 206–207. *In* J. Benoit and I. Assenmacher [ed.] La photorégulation de la reproduction chez les oiseaux et les mammifères. Editions du C.N.R.S., Paris.

Opel, H., and P. D. Lepore. 1967. Ovulating hormone-releasing factor in chicken hypothalamus. Poultry Sci. 46:1302.

Opel, H., and A. V. Nalbandov. 1961. Follicular growth and ovulation in hypophysectomized hens. Endocrinology 69:1016–1028.

Parlow, A. F. 1961. Bioassay of pituitary luteinizing hormone by depletion of ovarian ascorbic acid, p. 300–310. *In* A. Albert [ed.] Human pituitary gonadotrophins. Charles C Thomas, Springfield, Ill.

Péczely, P., and A. Calas. 1970. Ultrastructure de l'éminence médiane du pigeon (*Columba livia domestica*) dans diverses conditions expérimentales. Z. Zellforsch. 111:216–345.

Pfeiffer, C. A. 1947. Gonadotrophic effects of exogenous sex hormones on the testis of sparrows. Endocrinology 41:92–104.

Rahn, H., and B. T. Painter. 1941. A comparative histology of the bird pituitary. Anat. Rec. 79:297–311.

Riddle, O. 1930. Studies on pituitary functions. Endocrinology 15:307–314.

Riddle, O. 1963. Prolactin in vertebrates. Function and organization. J. Nat. Cancer Inst. (U.S.A.) 5:1039–1110.

Riddle, O., and R. W. Bates, 1933. Concerning anterior pituitary hormone. Endocrinology 17:689, 698.

Riddle, O., and F. Flemion. 1928. Studies on the physiology of reproduction in birds. XXVI. The rôle of the anterior pituitary in hastening sexual maturity in Ring Doves. Am. J. Physiol. 87:110–123.

Schally, A. V., A. Arimura, Y. Baba, M. G. Nair, H. Matsuo, T. W. Reeding, L. Debeljuk, and W. F. White. 1971. Purification and properties of the LH and FSH-releasing hormone from porcine hypothalami. Abstr. of the Fifty-third Meeting of the Endocrine Society, June 24–26, 1971, San Francisco.

Schildmacher, H. 1939. Über die künstliche Aktivierung der Hoden einiger Vogelarten im Herbst durch Belichtung und Vorderlappen Hormone. Biol. Zentralbl. 59:653–657.

Schockaert, J. 1931. Response of the male genital system of the immature domestic duck to injections of anterior pituitary substances. Anat. Rec. 50:381–397.

Schockaert, J. 1932. Le coq impubère comme test de l'hormone gonadotrope préhypophysaire. C. R. Soc. Biol. (Paris) 111:1095.

Schooley, J. P., O. Riddle, and R. W. Bates. 1941. Replacement therapy in hypophysectomized juvenile pigeons. Am. J. Anat. 69:123.

Sharp, P. J., and B. K. Follett. 1969a. The blood supply to the pituitary and basal hypothalamus in the Japanese Quail *Coturnix coturnix japonica*. J. Anat. (London) 104:227–232.

Sharp, P. J., and B. K. Follett. 1969b. The effects of hypothalamic lesions on gonadotrophin release in Japanese Quail (*Coturnix coturnix japonica*). Neuroendocrinology 5:205–218.

Sharp, P. J., and B. K. Follett. 1970. The adrenergic supply within the avian hypothalamus. *In* Proceedings of the 5th International Symposium on Neurosecretion, Kiel, 1969. Springer-Verlag, Berlin.

Shirley, H. V., and A. V. Nalbandov. 1956. Effects of transecting hypophysial stalks in laying hens. Endocrinology 58:694–700.

Stetson, M. H. 1969. The role of the median eminence in control of the photoperiodically induced testicular growth in the White-crowned Sparrow, *Zonotrichia leucophrys gambelii*. Z. Zellforsch. 93:369–394.

Stockell-Hartree, A. S., and F. J. Cunningham. 1969. Purification of chicken pituitary follicle stimulating hormone and luteinizing hormone. J. Endocrinol. 43:609–616.

Taber, E., M. Clayton, J. Knight, D. Gambrell, J. Flowers, and C. Ayers. 1958. Ovarian stimulation in the immature fowl by desiccated avian pituitaries. Endocrinology 62:84.

Tanaka, K. M., F. B. Mather, W. O. Wilson, and L. Z. McFarland. 1965. Effects of photoperiods on early growth of gonads and on potency of gonadotropins of the anterior pituitary in *Coturnix*. Poultry Sci. 44:662–665.

Tixier-Vidal, A. 1963. Histophysiologie de l'adénohypophyse des oiseaux, p. 255–274. *In* J. Benoit and C. da Lage [ed.] Cytologie de l'adénohypophyse. Editions du C.N.R.S., Paris.

Tixier-Vidal, A. 1965. Caractères ultrastructuraux des types cellulaires de l'adénohypophyse du canard mâle. Arch. Anat. Morphol. Exp. 54:719–780.

Tixier-Vidal, A. 1970. Cytologie hypophysaire et relations photosexuelles chez les oiseaux, p. 211–232. *In* J. Benoit and I. Assenmacher [ed.] La photo-régulation de la reproduction chez les oiseaux et les mammifères. Editions du C.N.R.S., Paris.

Tixier-Vidal, A., and I. Assenmacher. 1962. Effets d'un traitement isolé ou combiné à la réserpine et à la lumière permanente, sur la préhypophyse du canard mâle. C. R. Soc. Biol. (Paris) 156:37–43.

Tixier-Vidal, A., and I. Assenmacher. 1966. Etude cytologique de la préhypophyse du pigeon pendant la couvaison et la lactation. Z. Zellforsch. 69:489–519.

Tixier-Vidal, A. and D. Gourdji. 1965. Evolution cytologique ultrastructurale de l'hypophyse du canard en culture organotypique. Elaboration autonome de prolactine par les explants. C. R. Acad. Sci. (Paris) 261:805–808.

Tixier-Vidal, A., and D. Gourdji. 1972. Cellular aspects of the control of prolactin secretion in birds. Gen. Comp. Endocrinol. Suppl. 3:51–64.

Tixier-Vidal, A., M. Herlant, and J. Benoit. 1962. La préhypophyse du canard Pékin au cours du cycle annuel. Arch. Biol. (Liège) 73:317–368.

Tixier-Vidal, A., B. K. Follett, and D. S. Farner. 1968. The anterior pituitary of the Japanese Quail *Coturnix coturnix japonica*. The cytological effects of photoperiodic stimulation. Z. Zellforsch. 92:610–635.

Van Tienhoven, A. 1961. Endocrinology of reproduction in birds, p. 1088–1169. *In* W. C. Young [ed.] Sex and internal secretions. Vol. III. Williams & Wilkins Co., Baltimore.

Van Tienhoven, A., A. V. Nalbandov, and H. W. Norton. 1954. Effect of dibenamine on progesterone induced and "spontaneous" ovulation in the hen. Endocrinology 54:605–611.

Vitums, A., S. Mikami, A. Oksche, and D. S. Farner. 1964. Vascularization of the hypothalamo–hypophysial complex in the White-crowned Sparrow, *Zonotrichia leucophrys gambelii*. Z. Zellforsch. 64:541–569.

Wilson, F. E., and D. S. Farner. 1965. Effects of hypothalamic lesions on testicular growth. Fed. Proc. 24:129.

Wingstrand, K. G. [ed.] 1951. The structure and development of the avian pituitary. C.W.K. Gleerup, Lund. 316 p.

Zarrow, M. X., and J. W. Bastian. 1953. Blockade of ovulation in the hen with adrenolytic and parasympatholytic drugs. Proc. Soc. Exp. Biol. Med. 84:457–459.

DISCUSSION

Charles W. Dane

Professor Assenmacher has certainly presented a thorough discussion of the hypothalamo–hypophysial axis. To provide a meaningful response from the viewpoint of a field biologist, I must extend the framework of my remarks beyond the central portion of the axis to include the gonads. But first, I have some supplemental remarks on Doctor Assenmacher's presentation. Then, I will discuss specific phenomena that relate to the hypothalamo–hypophysial axis and are important to wildlife management.

Doctor Assenmacher has noted that the hypothalamus is the principal neural structure governing the rate of gonadal development and the length of the refractory period. This center is not an independent regulatory structure, but is affected by neural transmissions from higher centers and by endogenous and exogenous chemicals. Both the hypothalamus and pituitary are subject to negative and positive feedback, and numerous studies indicate that sex steroids and thyroid hormones can inhibit photoperiodically induced gonadal development. Perhaps greater knowledge of these influencing chemicals will aid in selecting drugs to stimulate laying in difficult-to-propagate birds.

As for releasing factors produced in the hypothalamus, it is important to note the presumed anatomical integrity of the neurohumoral tracts from specific areas in the hypothalamus through the median eminence and portal vessels to the anterior pituitary. Thus, even if the FSH- and LH-releasing factors were chemically similar or identical, the specific gonadotropin released from the anterior pituitary would be governed by the area of the hypothalamus stimulated.

The disagreements that still exist on the exact structures involved in a specific neurohumoral tract are understandable in view of the complexity of this axis, particularly the median eminence, and Doctor Assenmacher has given us a better insight into these complexities. In assessing the meaning of numerous lesioning studies that have been made in this area, differences between species and in duration of the study must also be considered.

Assenmacher has mentioned that transection of the anterior portal vessels, or lesions in the anterior median eminence, in domestic Mallards results in gonadal atrophy. Transections made in the anterior median eminence of the Mallard by Stetson (1971) did not prevent photoperiodically induced testicular growth. The work of Graber *et al.* (1967) with domestic cockerels and of Stetson (1969) and Wilson (1967) with White-crowned Sparrows indicates involvement of the posterior median eminence with testicular growth. In Japanese Quail, apparently both the anterior and posterior median eminence may be involved with testis development (Sharp and Follett, 1969).

Assenmacher has already noted the general pattern of separation of gonadotrops within the pituitary, with FSH-secreting cells located in the cephalic lobe and LH-secreting cells in the caudal lobe. The exception to this anatomical division is acknowledged for the White-crowned Sparrow, but there appears to be some disagreement as to whether this separation exists in the domestic fowl. I understand that this lack of agreement between Mikami (1958 as cited by Stetson, 1971) and Tixier-Vidal (Tixier-Vidal *et al.*, 1967, 1968) may be due to a difference in distinguishing between FSH- and TSH-secreting cells.

While discussing the pituitary, I might mention two other hormones directly involved with reproduction. Prolactin and its involvement with incubation will be discussed in other papers, but I certainly agree that more studies of levels and rhythms of the prolactin-inhibiting factor are needed. It is also interesting to note that testosterone may stimulate production of prolactin in ducks (Tixier-Vidal, 1969) and that in the Ring Dove (Stern and Lehrman, 1969) testosterone sensitizes neural centers to progesterone-induced incubation. Another pituitary hormone, vasotocin, is involved in oviposition and is thought to be the antidiuretic hormone, but Van Tienhoven (1968, p. 174) indicates that work by Ralph (1959, 1960) suggests that the antidiuretic function is controlled by the supraoptic nuclei and that the oviposition-inducing function is produced in the preoptic nuclei. Unlike the gonadotropins, vasotocin is stored in the neural lobe.

I agree with Assenmacher in emphasizing the need for specific plasma gonadotropin assay techniques. Reported work indicates a lower gonadotropin level in the pituitary of laying and brooding birds—a conclusion that is supported by work with the female Mallard by R. S. Donham and C. W. Dane (unpublished data), who noted at least a 50 percent reduction between prebreeding and laying levels. Caution, however, is warranted in this interpretation because of several daily LH fluctuations during the laying period.

On the subject of annual histological and hormonal changes, one might expect the number of gonadotrops, pituitary levels of gonadotropins, testis weights, and plasma testosterone levels to fluctuate with some synchrony, reaching a peak at time of breeding. However, the combination of findings by Tixier-Vidal and Benoit (1962), Garnier and Attal (1970), and Garnier (1971) suggests otherwise. The number of pituitary FSH cells and the weight of the testis reach their peak at the same time during the laying period, but gonadotropin levels are higher during the "storage phase" when testis weight is beginning to decline. The number of pituitary LH cells, which begin to increase before pituitary FSH cells, reaches its peak at the height of testis weight, but testosterone levels are high in September and October, when the number of pituitary LH cells and pituitary gonadotropin activity are low. The timing of this testosterone peak recalls the earlier work of Höhn (1947), who presented data in support of the correlation between an increase in testicular interstitial cells (presumably including Leydig cells) and the beginning of fall courtship behavior in the Mallard. Lack of synchrony between gonadotropin indices and plasma testosterone suggests a seasonal difference in tissue sensitivity.

Change in tissue sensitivity is but one example of the more complex interactions that affect every phase of the breeding cycle. Diurnal changes of other hormones, synergistic effect of two or more hormones, latent responses, seasonal changes in the phase of diurnal rhythms of hormones, and change in sensitivity to a given stimulus further complicate our understanding.

How can studies of the hypothalamo–hypophysio–gonadal axis provide information useful to wildlife management? Many laboratory and field studies have been directed to initiation of laying or the nesting chronology of numerous avian species. Certainly, optimum habitat and environment are important in terms of egg production and survival of young, and much has been learned of the primary external information that is responsible for the primary timing of breeding (Farner, 1967) and of such essential environmental requirements as temperature, nest site, and social interactions that govern the final reproductive development. These subjects are discussed by others at this symposium.

Fewer definitive field or laboratory studies have been conducted, however, on termination factors, that is, on terminal timing mechanisms that preclude further laying, on factors preventing renesting or redevelopment of testes after some stage of regression, and on factors that shorten the refractory period. It is hoped that studies of the hypothalamo–hypophysio–gonadal axis, in conjunction with studies of breeding will provide a better understanding of when to expect poor production and suggest ways to increase production in captive populations, including rare and endangered species. As one example, can clutch size be increased or the breeding season extended?

Under certain environmental stresses, follicular atresia occurs. In instances where ovaries are fully developed, the initial changes may be similar to those induced by removal of the pituitary. Such rapid follicular atresia of enlarged follicles has been observed following a severe late spring storm (Dane, 1971), after which redevelopment of the ovary apparently occurs. A different type of atresia may occur in waterfowl that, after prolonged subthreshold stimulation, fail to lay. Presumably, redevelopment of the ovary after such atresia does not occur during that nesting season.

Studies of the exact timing of the gonadotropic changes associated with termination of laying also might be enlightening. Among indeterminate layers, some such stimulus as number of eggs or surface area of eggs leads to a specific clutch size. In these species, properly timed removal of eggs increases the number of eggs that are laid in succession. Possibly, manipulation of hormone levels would increase clutch size of determinate, as well as indeterminate, layers.

In game-bird species that characteristically raise one brood per year (i.e., grouse and waterfowl), the percentage of females that are able to renest after loss of the first nest is critical to total production. Waterfowl biologists need to investigate under what conditions females will move to new areas and nest successfully after having lost their first nests in an area where the habitat subsequently deteriorates and no longer provides the environmental conditions essential for breeding. It would appear that much information about physiological changes associated with renesting could be gained through extensive study of the phenomenon reported by Fredrickson (1969), who observed that removal of eggs during incubation induced the American Coot (*Fulica americana*) to leave incubation responsibilities to her mate for a period during which her ovaries underwent redevelopment and she laid additional eggs.

More information is needed on factors responsible for the timing of testicular regression. Such information, for example, could be important in determining the possible value of excess males—a subject to be discussed later. Farner (1967) has suggested that negative feedback resulting from elevated levels of androgen or gonadotropin may affect the hypothalamus or pituitary and thus be involved in gonadal regression and part of the terminal timing mechanism. Thyroidectomy prevented testicular regression in the Starling, *Sturnus vulgaris* (Woitkewitch, 1940, as cited by Van Tienhoven, 1968), and the Spotted Munia, *Uroloncha punctulata* (Thapliyal and Pandha, 1965), and Van Tienhoven (1968) has speculated on a possible role of the thyroid in the regulation of the refractory period. Research at our laboratory indicates that such physiological changes as body weight, thyroid weight, and testicular regression in the male are correlated with the nesting stage of his female. Doctor McKinney's opening remarks on comparative length and strength of pair bonds in the Shoveler (*Spatula clypeata*) and Pintail (*Anas acuta*) suggest a possible correlation between pair bond duration and the period of functional testes.

Regression of testis appears to be due to a lack of circulating gonadotropin. The first stage of testicular regression in amphibians was described by Brian Lofts (1961), who noted the appearance of lipids in Sertoli cells, a condition that I and others have noted in waterfowl and other avian species. This raises the question of whether at this stage or at some previous or subsequent stage the regression of the testes can be reversed by the behavior of the renesting female.

A postjuvenal refractory period exists in some species (e.g., *Zonotrichia leucophrys*) (Farner and Follett, 1966) but not in others (e.g., *Z. capensis*) (Miller, 1959); and, if it occurs in female Mallards, it is terminated before 6 months of age (Dane, unpublished data). A postbreeding refractory period is not present in some species, e.g., Bobwhite (*Colinus virginianus*) (Kirkpartrick, 1959). Some species require a short photoperiod before termination of the refractory period, but the Mallard does not need to be exposed to short days before the hypothalamus can again be stimulated by photoperiod (Van Tienhoven, 1968). In an earlier discussion, Frith and Immelmann noted that in aperiodic breeders a substantial fraction of the birds in a population have partially developed gonads during the nonbreeding season, which enables them to respond more rapidly to improved habitat conditions. More experimental studies of the refractory period in game birds are needed.

REFERENCES

Dane, C. W. 1971. Effect of spring storm on waterfowl mortality and breeding activity, p. 258–267. *In* A. O. Haugen [ed.] Proceedings of symposium on snow and ice in relation to wildlife and recreation, Ames, Iowa, February 11–12, 1971. Iowa Cooperative Wildlife Research Unit, Iowa State University, Ames.

Farner, D. S. 1967. The control of avian reproductive cycles. Proc. of 14th Int. Ornithol. Congr. (1966) 14:107–133.

Farner, D. S., and B. K. Follett. 1966. Light and other environmental factors affecting avian reproduction. J. Anim. Sci. 25 (Suppl.):90–115.

Fredrickson, L. H. 1969. An experimental study of clutch size of the American Coot. Auk 86:541–550.

Garnier, D. H. 1971. Variations de la testosterone du plasma périphérique chez le canard Pékin au cours du cycle annuel. C. R. Séances Acad. Sci. (Paris) Ser. D 272:1665–1668.

Garnier, D. H., and J. Attal. 1970. Variations de la testostérone du plasma testiculaire et des cellules interstitielles chez le canard Pékin au cours du cycle annuel. C.R. Séances Acad. Sci. (Paris) Ser. D 270:2472–2475.

Graber, J. W., A. I. Frankel, and A. V. Nalbandov. 1967. Hypothalamic center influencing the release of L H in the cockerel. Gen. Comp. Endocrinol. 9:187–192.

Höhn, E. O. 1947. Sexual behavior and seasonal changes in the gonads and adrenals of the mallard. Proc. Zool. Soc. London 117:281–304.

Kirkpatrick, C. M. 1959. Interrupted dark period: Tests for refractoriness in Bobwhite Quail hens. *In* R. B. Withrow [ed.] Photoperiodism and related phenomena in plants and animals (proceedings of a conference, October 27–November 2, 1957). Publication 55. American Association for the Advancement of Science, Washington, D.C. 903 p.

Lofts, B. 1961. The effects of follicle-stimulating hormone and luteinizing hormone on the testis of hypophysectomized frogs (*Rana temporaria*). Gen. Comp. Endocrinol. 1:179–189.

Mikami, S. I. 1958. The cytological significance of regional patterns in the adenohypophysis of the fowl. J. Fac. Agr. (Iwate Univ.) 3:473–545.

Miller, H. 1959. Response to experimental light increments by Andean Sparrows from an equatorial area. Condor 61:344–347.

Ralph, C. L. 1959. Some effects of hypothalamic lesions on gonadotrophin release in the hen. Anat. Rec. 134:411–431.

Ralph, C. L. 1960. Polydipsia in the hen following lesions in the supraoptic hypothalamus. Am. J. Physiol. 198:528–530.

Sharp, P. J., and B. K. Follett. 1969. The effect of hypothalamic lesions on gonadotropin release in Japanese Quail (*Coturnix coturnix japonica*). Neuroendocrinology 5:205–218.

Stern, J. M., and D. S. Lehrman. 1969. Role of testosterone in progesterone-induced incubation behavior in male Ring Doves (*Streptopelia risoria*). J. Endocrinol. 44:13–22.

Stetson, M. A. 1969. The role of the median eminence in control of photoperiodically induced testicular growth in the White-crowned Sparrow, *Zonotrichia leucophrys gambelii*. Z. Zellforsch. Mikrosk. Anat. 93:369–394.

Stetson, M. H. 1971. Control mechanisms in the avian hypothalamo–hypophysial–gonadal axis. PhD thesis, University of Washington, Seattle. 196 p.

Thapliyal, J. P., and S. K. Pandha. 1965. Thyroid–gonad relationship in Spotted
    Munia, *Uroloncha punctulata*. J. Exp. Zool. 158:253–261.
Tixier-Vidal, A. 1969. New results on prolactin secretion in birds, p. 45–46. *In*
    Proceedings of seminar on hypothalamic and endocrine function in birds.
    International House of Japan, Tokyo.
Tixier-Vidal, A., and J. Benoit. 1962. Influence de la castration sur la cytologie
    préhypophysaire du canard mâle. Arch. Anat. Microscopique Morphol. Exp.
    51:265–286.
Tixier-Vidal, A., B. K. Follett, and D. S. Farner. 1967. Identification cytologique
    et fonctionelle des types cellulaires de l'adénohypophyse chez le caille mâle,
    Coturnix coturnix japonica, sourmise à différentes conditions expérimentales.
    C.R. Séances Acad. Sci. Ser. D (Paris) 264:1739–1742.
Tixier-Vidal, A., B. K. Follett, and D. S. Farner. 1968. The anterior pituitary of
    the Japanese Quail, *Coturnix coturnix japonica*. The cytological effects of
    photoperiodic stimulation. Z. Zellforsch. 92:610–635.
Van Tienhoven, A. 1968. Reproductive physiology of vertebrates. W. B.
    Saunders Company, Philadelphia–London–Toronto. 498 p.
Wilson, F. E. 1967. The tubero-infundibular neuron system: A component of
    the photoperiodic control mechanism of the White-crowned Sparrow,
    Zonotrichia leucophrys gambelii. Z. Zellforsch. Microsk. Anat. 82:1–24.
Woitkewitsch, A. A. 1940. Dependence of seasonal periodicity gonadal changes
    on the thyroid gland in *Sturnus vulgaris*. C.R. Acad. Sci. (URSS) 27:741–745.

DISCUSSION

Brian Lofts

Professor Assenmacher has provided us with a succinct dissertation
on the role of the hypothalamo–hypophysial system and its regulation of the functional activity of the gonads. This mechanism, together with the neurohypophysis discussed by Doctor Follett,
provides a chain of command linking the environment with the
reproductive potentialities of the animal. The first part of Professor
Assenmacher's lecture dealt with the pituitary–gonadal axis and the
release of gonadotropic hormones; it is on this particular aspect that
I first wish to expand.

Much of my own work has been concerned with gonadal cycles.
With the advent of new histochemical techniques and the develop-

ment of electron microscopy in recent years, we have been able to gain much more information about the functional activity of the avian testis. Testicular activity, of course, is linked with pituitary activity, and changes observed in the former also reflect events occurring further along the chain. In the sectioned testis, two major components are easily observable—i.e. the seminiferous tubules within which the propagation of the germ cells takes place and, dispersed in the interstices between them, the glandular interstitial tissue that is the endocrine component producing the steroid sex hormones. It is generally believed, as Professor Assenmacher has indicated, that testicular regulation is controlled by the luteinizing hormone (LH), which stimulates the secretory activity of the inter-stitial tissue and the follicle-stimulating hormone (FSH), which regulates the activity of spermatogenesis in the germinal tissue. In short, LH controls steroid production and FSH controls production of spermatozoa. In recent years, we have obtained additional infor-mation about testicular activity and we now realize that the basic mechanisms of the pituitary—gonad axis described by Professor Assenmacher seem no longer as clear cut as they once did.

It has long been known that, when a seasonally breeding bird enters its so-called postnuptial refractory phase, the testes rapidly regress and, in many species, there is a massive buildup of cholesterol-rich lipoidal material within the seminiferous tubules (Figure 1). Cholesterol is one of the precursors used in the synthesis

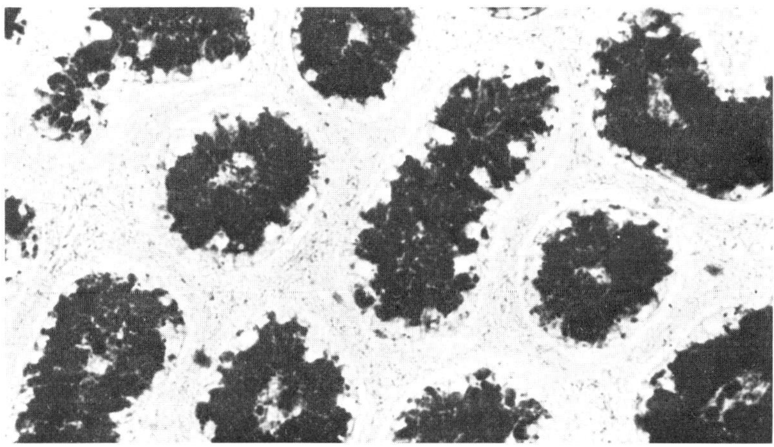

FIGURE 1 Section of the postnuptial testis of the Weaver Finch, *Quelea quelea*. The seminif-erous tubules are regressed and filled with a mass of cholesterol-rich lipoidal material. × 270.

of androgenic and other steroids; several years ago, in a series of collaborative experiments with the late Professor A. J. Marshall, we speculated that this tubular material might be indicative of an endocrine function, i.e., that the interstitial tissue may not be the only locus of steroid production within the testis (Lofts and Marshall, 1959). This phenomenon has since been shown to occur also in a number of other seasonally breeding vertebrates (Lofts, 1968). One feature that appears to be common to all is that this heavily lipoidal stage undergoes a seasonal depletion at a time when spermatogenetic recrudescence occurs in the germinal epithelium. Thus, there is a clear-cut seasonal cycle, involving a waxing and waning of cholesterol-rich lipoidal material in the seminiferous tubules, as there is also in the adjacent interstitial tissue.

Electron microscopy has demonstrated that steroid-producing tissues in mammals have characteristic ultrastructural features—such as smooth endoplasmic reticular membranes and mitochondria that have tubular rather than lamellate cristae—that provide additional parameters for their detection. When the testicular tissue of a bird is examined by electron microscopy, the tubule Sertoli cells also seem to have the fine structure normally associated with a steroid-producing cell. Thus, it can be seen from Figure 2 that the Sertoli cell cytoplasm of a light-stimulated quail contains much smooth

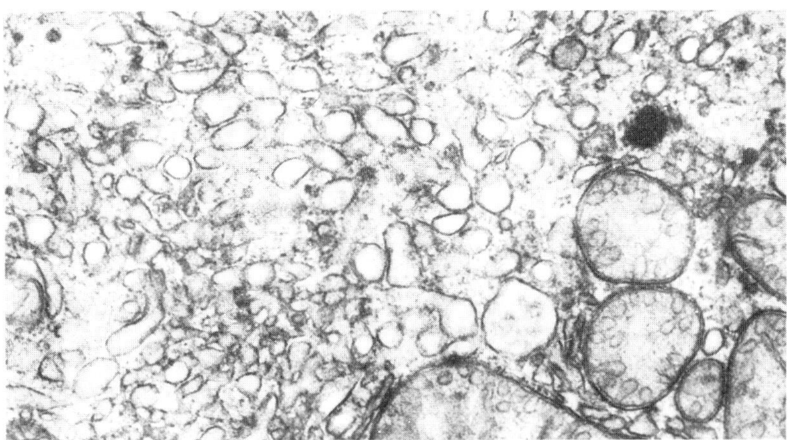

FIGURE 2   Electron micrograph of a testicular Sertoli cell of a light-stimulated Japanese Quail, *Coturnix c. japonica*. The cell shows the ultrastructural characteristics of a steroid-producing tissue, with smooth vesiculate endoplasmic reticulum and mitochondria with tubular cristae. × 40,500.

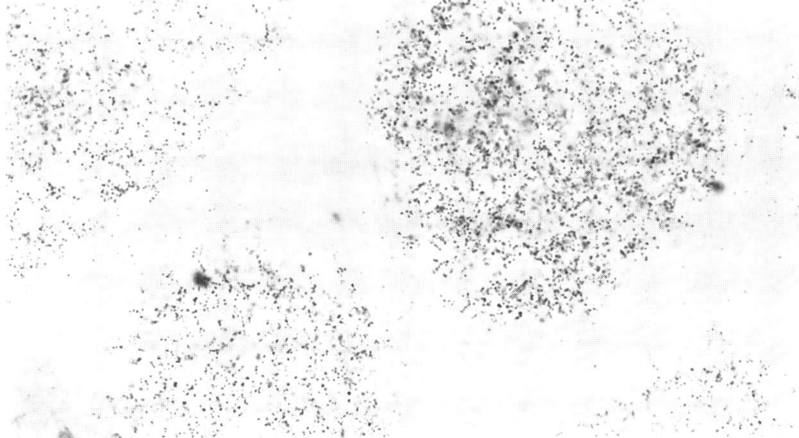

FIGURE 3   Testis of a Tree Sparrow, *Passer montanus*, showing marked formizan deposits in the seminiferous tubules indicating strong Δ^5-3β-hydroxysteroid dehydrogenase activity, and hence steroid production, within these structures. × 340.

vesiculate endoplasmic reticulum and numerous mitochondria with tubular cristae. This lends support to the suggestion of an endocrine role in this tissue. In addition, histochemical tests have revealed that some of the enzyme systems known to be essential to the biosynthesis of steroid hormones also are present within the tubules, as well as in the adjacent interstitium. In the European Tree Sparrow, *Passer montanus*, for example, we recently have shown that a very dense granulation indicating the presence of 3β-hydroxysteroid dehydrogenase (Figure 3) can be demonstrated within the seminiferous tubules. Furthermore, 17β-hydroxysteroid dehydrogenase also has been shown to occur. Thus, there now appears to be unequivocal evidence that the seminiferous tubules, as well as the interstitial tissue, are a locus for steroid biosynthesis within the gonad. Therefore, when one is speculating on possible regulatory mechanisms for such phenomena as the annual reproductive cycles or the development of the photorefractory phase, one must think not only about possible feedback from an interstitial component but also from another component of the testis. In other words, if these are asynchronous, there may be feedback from different sources at different stages of the cycle.

For many years, it has been tacitly assumed that the gonads have no direct involvement in the evolution of the photorefractory state, and contemporary thought suggests that the seat of photorefractoriness lies entirely within the hypothalamic region. However, it may be premature to exclude the pituitary–gonad axis from involvement in this phenomenon since the testicular structures of the postnuptial photorefractory bird, and the similarly sexually regressed but light-sensitive bird differ. This has been demonstrated in wild Mallards collected during the postnuptial refractory stage and divided into two groups: One is placed under long summer photoperiods to maintain the refractory condition and the other on short photoperiods to break photorefractoriness. After 10 weeks of such treatment (Lofts and Coombs, 1965), the gonads of both groups remained spermatogenetically inactive, but differed physiologically. In the latter aspect, exposure to a stimulatory photoperiod indicated that the photorefractoriness had been terminated in the short-day birds, who responded by gonadal expansion, but not in the birds maintained on long photoperiods. Although at the end of 10 weeks the testes of both groups were fully regressed and morphologically identical, they were quite distinct histologically. The seminiferous tubules in all cases were regressed, with only a peripheral lining of spermatogonia, but the condition of the interstitial tissue differed significantly between the two experimental groups. In the birds kept under a long 16-hour photoperiod, the interstitium appeared more extensive but not as lipoidal as did the interstitial tissue of the group that had regained its photosensitivity. In the latter case, the cells were much more lipoidal and the nuclei more rounded. I am now carrying out a similar investigation on the Teal, *Anas crecca*, in collaboration with T. Nicholls. We find that at the ultrastructural level, the Leydig cells of previously refractory birds that have been induced to regain their photosensitivity by exposure to an 8L – 16D light regime are well defined, with large amounts of smooth membrane and numerous lipid droplets. In the 18L–6D group, on the other hand, the Leydig cells are more spindle-shaped, lack in such smooth endoplasmic reticulum, and are less lipoidal. This difference in interstitial cell structure suggests that during the refractory period the pituitary–gonad axis may not be as inactive as had been thought heretofore and may be secreting at some level.

It has become clear from the work of Hamner (1963, 1966) on the House Finch, *Carpodacus mexicanus*, and the work of others (outlined by Follett, page 230, this volume) (see also Lofts *et al.*, 1970) that the photoperiodic response of birds may depend only on light received during a restricted sensitive phase of a circadian-based rhythm. Only when the natural photoperiod is sufficiently long to engage with this sensitive period will the hypothalamo–hypophysial system be stimulated and light experienced outside this period has no gonadal stimulating effect. Thus in winter, when short daylengths prevail, the light-sensitive phase is coincident with night and the birds remain sexually regressed. But with the lengthening days of spring and early summer, the light eventually coincides with the light-sensitive phase and stimulates the neuroendocrinological system, with consequent release of gonadotropic hormones that brings the bird up into breeding condition. Some 3 years ago, Doctor Murton and I carried out some light-flashing experiments on Greenfinches, *Chloris chloris*, and measured the plasma levels of LH as well as the development of the gonads. The birds were kept under a daily photoperiod of 6 hours, with an additional 1-hour light flash as an interruption of the night. We found that a peak of LH activity was induced by light flashes early in the scotophase and that this peak occurred in birds in which the testes were still regressed and spermatogenetically inactive. When the light flash was experienced later in the night, the LH levels declined but spermatogenesis was stimulated. The interstitial histology closely agreed with the plasma LH levels. In short, we appeared to be getting a separation of LH release from that of FSH release. The fact that the LH release occurred in response to light experienced early in the night and that FSH release (as judged by spermatogenetic development) took place in those groups where light was experienced later in the night suggests that the lengthening daily photoperiods early in the year engages the circadian rhythm responsible for the release of LH prior to that responsible for the release of FSH. This interpretation is in agreement with the data discussed by Professor Assenmacher, where there is cytological evidence that the pituitary of the Mallard releases LH before FSH.

REFERENCES

Hamner, W. M. 1963. Diurnal rhythm and photoperiodism in testicular recrudescence of the House Finch. Science 142:1294–1295.

Hamner, W. M. 1966. Photoperiodic control of the annual testicular cycle in the House Finch, *Carpodacus mexicanus*. Gen. Comp. Endocrinol. 7:224–233.

Lofts, B. 1968. Patterns of testicular activity, p. 239–304. *In* E. J. W. Barrington and C. B. Jørgensen [ed.] Perspectives in endocrinology: Hormones in the lives of lower vertebrates. Academic Press, London and New York.

Lofts, B., and C. F. J. Coombs. 1965. Photoperiodism and the testicular refractory period in the Mallard. J. Zool. London 146:44–54.

Lofts, B., and A. J. Marshall. 1959. The post-nuptial occurrence of progestins in the seminiferous tubules of birds. J. Endocrinol. 19:16–21.

Lofts, B., B. K. Follett, and R. K. Murton. 1970. Temporal changes in the pituitary–gonadal axis. Mem. Soc. Endocrinol. 18:545–575.

INFORMAL DISCUSSION

MURTON: I believe Professor Assenmacher talked in terms of follicle-stimulating hormone (FSH) affecting ovarian function and causing the release of estrogen, whereas luteinizing hormone (LH) was supposedly responsible for the release of androgen in the male. Some behavior-endocrine studies of feral pigeons (*Columba livia*) that we made showed that exogenous mammalian FSH caused androgen-type displays and, by inference, could have achieved this by releasing androgen from the Leydig cells.*

We have independent evidence to suggest that distinct FSH- and LH-type gonadotropins are secreted by the avian pituitary.† More recently my colleagues and I have found that when a male and female pigeon are initially paired and begin courtship, plasma IR-LH titers are low (as determined by a radioimmunoassay). At this stage, FSH- and androgen-type displays are most prevalent, and there are various grounds for believing endogenous androgen concentrations to be high.‡ Plasma LH titers in the male reach a peak at time of egg-laying by the female and stay high throughout the incubation phase. That is, maximum LH activity does not coincide with the peak androgenic phase. There are many other circumstantial indications that, although LH may be involved in Leydig cell development and the early biosynthesis of steroids, it alone is not responsible for androgen release and, perhaps, not specifically in androgen synthesis per se.

LOFTS: The evidence indicates that the mechanisms that regulate testicular

---

*Murton, R. K., R. J. P. Thearle, and B. Lofts. 1969. The endocrine basis of breeding behaviour in the feral pigeon (*Columba livia*). 1. Effects of exogenous hormones on the pre-incubation behaviour of intact males. Anim. Behav. 17:286–306.

†Murton, R. K., B. Lofts, and N. J. Westwood. 1970. The circadian basis of photoperiodically controlled spermatogenesis in the Greenfinch *Chloris chloris*. J. Zool. London 161:125–136.

‡Hutchinson, J. B. 1970. Influence of gonadal hormones on the hypothalamic integration of courtship behaviour in the Barbary Dove. J. Reprod. Fert. Suppl. 11:15–41.

function are similar throughout the vertebrates. The indication for birds is that they are very similar to those of mammals. I have not worked to any great extent on the ovarian tissue. There is just one point on which I could comment here. Murton has already touched on it. Our understanding of the control of ovarian function is that FSH stimulates the follicles and, when the follicles develop, they release their estrogenic hormone—i.e., FSH is stimulating a steroid-producing tissue. Whereas on the male side, the LH stimulates the steroidogenic interstitial tissue. This might suggest a difference in control between the two sexes, but one must bear in mind that FSH also is apparently stimulating a steroid secreting tissue in the testes, and this makes it more understandable. There is not, therefore, a great difference in the control of the two sexual components.

IMMELMANN: At the end of his discussion, Doctor Dane stated that further investigations about the refractory period of birds are urgently needed. I would like to mention that one of my co-workers, Roland Sossinka, has just finished a study of the refractory period in the Australian Zebra Finch. Through histological and biochemical investigations, he was able to prove that in this arid-country species there is no juvenile refractory period at all, but gametogenesis already starts while the birds are still in the nest and continues gradually until full maturity is reached at less than 3 months of age.

This, of course, is understandable from the biology of the Australian Zebra Finch, which begins breeding very early if conditions are favorable. I just wanted to mention that there is now one more species for which a definite lack of juvenile refractory period has been proven.

FARNER: I think that I would add that it is questionable whether the adults of this species have a true refractory period, because we have carried breeding pairs through 15 consecutive clutches without any interruption. So at least if refractoriness does occur, it must be some sort of condition that is exposed environmentally rather than being built into the system.

STETSON: I was very interested in the description of the studies on the Mallard performed by Doctor Lofts, especially the testicular situation in the refractory mallards, in which it was shown what appeared to be a hypertrophy of interstitial tissue. Yet when the testes were subjected to the histochemical techniques used, no lipoidal material of any type could be demonstrated. Specifically, I am inquiring if any enzymic activity was demonstrated in this tissue? I found similar-appearing testes in lesioned Japanese Quail yet could not demonstrate any type of enzymic activity. I just assumed that we had for some reason a massive hypertrophy of fibrocytes or similar tissue.

LOFTS: The enzymic studies have not yet been completed, so I can't answer that question. We recently have sent off some material to Doctor Follett's laboratory, where it is being examined by electron microscopy (EM). So far as I can understand, the electron microscopy seems to be in fairly close agreement with the evidence from light microscopy. The EM observations show that, after emergence, lipid accumulates and the testes are probably ready to begin the next cycle. But on the histochemical side, I do not yet have the information.

OKSCHE: I have a question for Professor Lofts, actually two questions. First, what type of enzymic reactions did you mean—some kind of specific reaction in lipid metabolism or some reactions indicating general cellular activity? Second, what happens to the connective tissue? This is an additional component of the testes besides the tubules and the Leydig cells. Does this interstitial tissue increase in volume? How is it reduced? Are cells destroyed? How are these reactions in connective tissue controlled? Does it depend on the hormones, too, or is it a different mechanism?

LOFTS: Doctor Trevor Nichols has been doing some electron microscopy studies on the interstitial tissue in the Japanese Quail under various light-stimulatory conditions, and I have seen the rough draft of his manuscript. His observations show the evolution of an interstitial secretory cell from fibroblast-like precursors. Under the EM what appears to be an interstitial fibroblast-like cell with a spindle-shaped nucleus develops into a full fledged secretory Leydig cell under heavy light stimulation, and it is quite conceivable that at the end of the secretory cycle such a cell may revert back into a fibroblast-like form. So it would be very difficult to distinguish what might be a Leydig precursor from an actual true connective tissue cell. It is very difficult to answer that particular question.

With regard to the enzyme studies, the enzymes I referred to are specific steroid dehydrogenase enzymes, which can be visualized by the Wattenberg technique in which tetrazolium becomes reduced to insoluble formizan; the technique is quite well established in steroid endocrinology now. The two that I showed you, specifically the Δ^5-3β-hydroxysteroid dehydrogenase and the 17β-hydroxysteroid dehydrogenase, catalyze certain chemical reactions along the metabolic pathways in the production of androgenic and estrogenic steroids.

There is one point here, of course. As in the ovarian cycle, it has been established by incubation studies that the testes produce estrogenic steroids as well as androgenic ones. This is of significance in some of the behavioral patterns that Murton mentioned. It is quite conceivable, I suppose, that the seminiferous-tubule component may be one of the sites that produce the estrogenic steroids.

FOLLETT: One point worth stressing is that we are still unclear as to the exact roles of the two avian gonadtropins in the breeding cycle. This is largely because the purified avian hormones have been unavailable until recently and therefore all deductions as to the biological function of LH and FSH in birds has rested on work with mammalian hormones.

One fact to emerge has been that mammalian LH or FSH often demonstrates both an LH-like and an FSH-like action in the bird.

I cannot help much, but injections of purified avian LH do lead to differentiation and hypertrophy of the Leydig cells within the testis of the Japanese Quail. In addition there are effects on tubule growth.* Also, it is interesting that Leydig-cell differentiation from fibroblasts in young quail (30 days of age) does not take place until they have been exposed to long days for 3 days.[†]

A time-lag of 2 days also is found before there is a rise in plasma L H* (see Figure 4 in Assenmacher's contribution). When mature quail are returned to short days, the first ultrastructural evidence for testicular regression again coincides with the time when plasma L H falls (Figure 4). All of these data point to a role for L H in regulating the interstitium.

F. WILSON: In my laboratory, we have been investigating the mechanism that initiates spontaneous testicular regression in the Tree Sparrow (*Spizella arborea*). We have found a region in the hypothalamus that is very sensitive to testosterone propionate; this region is immediately dorsal to the anterior median eminence. If we implant testosterone propionate into that region while Tree Sparrows are on short days and then, on the day after operation, transfer them to long days, we find that photoperiodically induced testicular growth is prevented or retarded drastically. Moreover, if we implant testosterone propionate 3 weeks after transfer to long days, during which time the testes have grown considerably, we find virtually complete testicular regression 4 weeks later in a sizeable proportion of birds. If, on the other hand, the antiandrogen cyproterone is implanted on the day before birds are transferred to long days (that is, while days are still short) and cyproterone is then maintained in place in the testosterone-sensitive region of the hypothalamus during 17 weeks of long-day stimulation, spontaneous testicular regression can be retarded, or indeed prevented, in which case both testes and seminal sacs are in full breeding condition.

FARNER: I think that in closing this section, I will take just a couple of moments to make a few comments, because there is a very significant development that may have escaped some of us. Also I wish to make subsequently a comment on one part of Professor Assenmacher's presentation.

There now is ample evidence, I think, to show that circulating sex steroids in males, produced by the testes, have a negative feedback effect on the hypothalamic mechanism that controls the output of luteinizing hormone; it probably also affects the production and release of F S H.

I have always been willing to concede that this may be part of the mechanism that causes photorefractoriness; however, a simple model that involves only negative feedback by sex steroids, no matter what kind of linear or nonlinear functions one assumes, says that a long-day bird is going to have an

---

*Follett, B. K., C. G. Scanes, and T. J. Nicholls. 1972. The chemistry and physiology of the avian gonadotrophins, p. 193–212. Colloqué sur les hormones glycoprotéiques hypophysaires. INSERM, Paris.

[†]Nicholls, T. J., and G. P. Graham. 1972. The ultrastructure and differentiation of testicular Leydig cells in the Japanese Quail. Biol. Reprod. 6:179–192.

oscillating system that runs indefinitely through testicular cycles. There is no way of escaping this without a more complex model.

So this brings us to what I think is a very significant report presented here, namely, that of Brian Lofts, that identifies a possible—I am not completely convinced yet—but a possible second source of sex steroid, namely, the testicular tubule. This could have a time course and characteristics sufficiently different than that of the Leydig cells so that one, perhaps, using these two serially related sources of testosterone, could make a relatively simple model for the development of photorefractoriness.

The final comment that I would like to make concerns something that Professor Assenmacher did not develop as much as he could have because of lack of time. This concerns the so-called antigonadal effect of prolactin, which has been the source of a great deal of argument in the literature. At present, as Professor Assenmacher rightly pointed out, we sit in a situation in which we say that prolactin inhibits gonadal function by inhibiting output of gonadotropic hormones in some species. We can make a very logical argument as to why this is useful. But it turns out that, in other species, this does not happen.

What I wish to confess for all who work in this field is that we may be dealing with a hopeless series of artifacts. The reason I say this is that Albert Meyer, initially in my laboratory, and, subsequently, more extensively in his own at Louisiana State University, has shown clearly that the effect of prolactin is fantastically dependent upon the time of the day that it is released by the pituitary or injected. Injections made at 12 hours difference in time during the day, in fact, can produce completely opposite effects. So it will be necessary, before we generalize too much about the antigonadal effect of prolactin, to re-examine and re-do most of the experiments in the literature. This is not particularly uncharacteristic of avian neuroendocrinology, or even of neuroendocrinology in general.

The Neuroendocrine Regulation of Gonadotropin Secretion in Avian Reproduction

BRIAN K. FOLLETT

The environmental factors most strongly influencing reproductive success in birds are the availability of food and the risk of predation (Lack, 1968). Each species, therefore, has undergone appropriate ecological, behavioral, and physiological adaptations to improve its breeding potential. One of the most marked physiological adaptations has been to restrict breeding to the most favorable season of the year, invariably the one when there is most food. The annual cycle of breeding is most obvious at mid- and high latitudes but, contrary to a once widely held view, species of tropical and subtropical regions are also seasonal breeders (Lack, 1968; Marshall, 1970).

In most species, reproductive preparations must begin long before the breeding season, and birds have evolved the ability to use some environmental cue (proximate factor) to trigger off these processes. The nature of this cue (or cues) is immaterial, providing it ensures that breeding occurs at the best time of the year. Many environmental factors seem to be used by tropical species, including rainfall, temperature, the availability of nest material and of nest sites, and the abundance of food (Marshall, 1970). Localized factors tend to predominate in the tropics as they do in relation to birds living in arid regions, where rainfall seems to act most often as the proximate inducer

(Farner, 1967; Marshall, 1970). At mid- and high latitudes a single
environmental factor, the day length, is by far the most consistent indicator of the season and seems to have been adopted by virtually all
species [for reviews see Farner (1967, 1970), Wolfson (1966, 1970),
Farner and Follett (1966), Lofts and Murton (1968), Lofts *et al.*
(1970)]. This primary source of predictive information can be modified by such secondary factors as temperature, which can accelerate
or retard gonadal growth. There is no evidence in wild species, however, that temperature can act in the absence of a stimulatory photoperiod. Some strains of Japanese Quail (*Coturnix coturnix japonica*)
exhibit gonadal growth under short daily photoperiods (Follett,
1969b) and this may be temperature-dependent, since the transfer of
such sexually mature males to a low temperature room (5–9 °C) results in testicular atrophy (Kato and Konishi, 1968). The overriding
importance of day length may be judged from the fact that increasing
the light period restored the quail to full sexual maturity.

Physiological and behavioral changes initiated by the action of the
proximate factors are many and varied. The most important of these
is to stimulate growth of the reproductive organs, their associated
tracts, and such structures as the brood patch. Other changes may include, at a relatively early stage in the reproductive cycle, the induction of migratory activity and its associated hyperphagia. Later, new
patterns of behavior appear, including the acquisition and defense of
territory, pairing, courtship, nest-building, and incubation behavior.
The interaction between the sexes at these points in the cycle is important, particularly for the female, and the presence of a male is a
crucial external factor (Lehrman, 1961; Hinde and Steel, 1966). It t
is also clear from studies with the canary that other factors such as
nest-building modify the final stages of ovarian development (Hinde
and Steel, 1966).

Some sense may be made of these varied physiological changes if
it is kept in mind that most, if not all, result from hormone action.
The growth of the gonads and the synthesis of the sex steroids depend upon the gonadotropic hormones (FSH, LH) of the pituitary
complex (see Assenmacher, this volume). The sex steroids, in turn,
regulate other aspects of gonadal function, such as the later stages of
spermatogenesis and the synthesis of yolk. In addition, they cause the
reproductive tract to grow and, in conjunction with prolactin, cause
brood-patch development. A number of hormones (prolactin, adrenal
steroids, gonadotropins, androgens) are involved in migratory fattening and behavior (Meier and Farner, 1964; Meier *et al.*, 1965; King,

1970; Meier and Martin, 1971). It also is evident that much reproductive behavior, both before and after egg-laying, is dependent upon hormones. The endocrines most likely to be involved are the gonadotropins, prolactin, and the sex steroids (e.g., Lehrman, 1961, 1965; Hinde and Steel, 1966; Komisaruk, 1967; Hutchison, 1967, 1970, 1971; Murton *et al.*, 1969a). To a major extent, therefore, the reproductive biology of birds can be viewed as an exquisite example of control by a neuroendocrine system. External information is perceived by the sense organs, is passed to the hypothalamus, and there acts via the hypothalamo–hypophysial axis to initiate pituitary hormone release. These hormones act either alone or with other pituitary and steroid hormones to produce the various changes necessary to bring a bird into breeding. Many target organs are involved, including the brain, where the hormones act to trigger specific patterns of behavior. According to this view, then, the control of bird reproduction does not occur through a number of independent causal channels but through a limited series of interrelating chains. Figure 1 attempts to summarize the many interrelationships and underlines the central position of the hypothalamus and of the neuroendocrine step in the overall process.

Modern theories of pituitary control in vertebrates (Harris and Donovan, 1966; Martini and Ganong, 1966; Halász, 1969; Dodd *et al.*, 1971) envisage the transfer of information from the environment to the pituitary as occurring in several stages (Figure 1). At the highest level comes integration and modulation of the incoming information. This is the most difficult stage to understand because it involves a host of inputs ranging from external factors—such as day length, temperature, food abundance, presence of male, and the state of the nest—to such internal environmental factors as the sex steroid level. Virtually nothing is known as to where these integration processes occur in the brain. Parts of the process, such as the monitoring of plasma steroid levels, may take place within the hypothalamus, but one imagines that most occur in a number of extrahypothalamic centers. Ultimately, however, information in the form of both stimulatory and inhibitory neuronal pathways enters the hypothalamus. It seems probable that these pathways pass through a nervous relay center in the hypothalamus where the information is perhaps collated and condensed. Finally, the information is passed to specific neurosecretory neurones whose cell bodies lie in the basal hypothalamus and whose axons terminate in the median eminence on the primary capillary plexus. These neurons play the key role in transforming in-

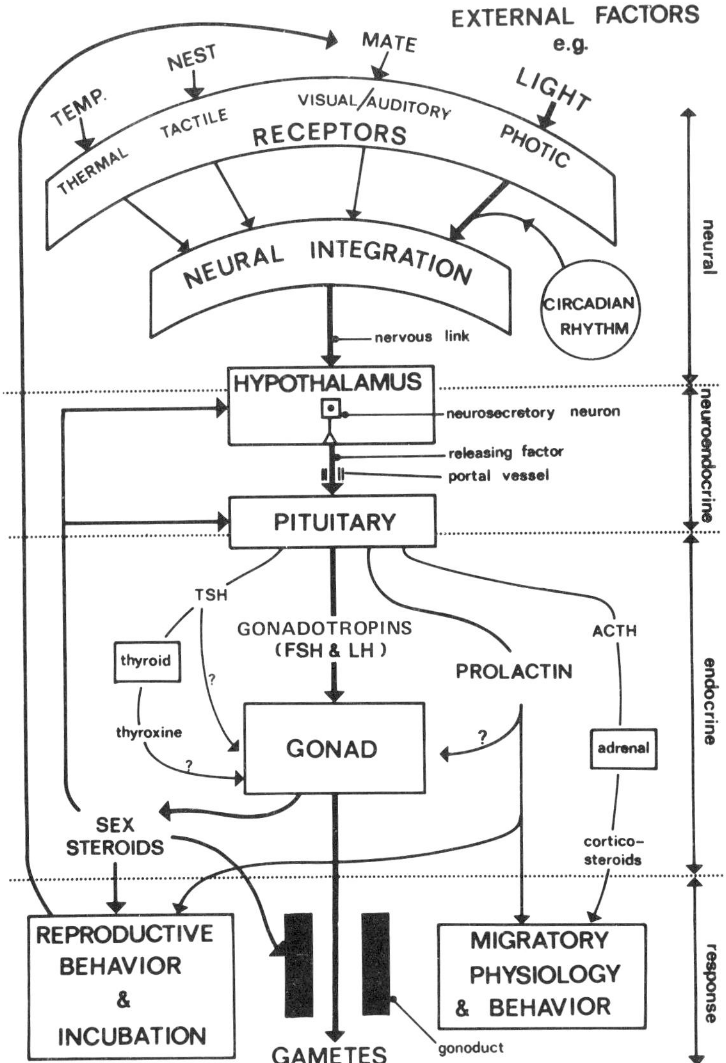

FIGURE 1   Generalized scheme showing the interrelationships between external and internal stimuli and the neuroendocrine control of avian reproduction.

coming neural information into the secretion of neurohormones.
They thus represent the primary level of control. The neurohormones
then act on the cells of the adenohypophysis to stimulate the secretion of the gonadotropins (or of prolactin, TSH, ACTH, etc.). The
system, of course, is infinitely more complex than as here outlined
and one wonders to what extent "model-building" is a useful exercise. With all its limitations, however, it does focus attention on those
parts of the system where our ignorance is particularly great.

The intention of this communication is to consider this neuroendocrine chain in the avian breeding pattern that involves the photoperiodic control of reproduction in temperate-zone species.

THE HYPOTHALAMUS: ITS STRUCTURE AND ITS FUNCTION IN AVIAN PHOTOPERIODISM

The central role of the hypothalamus was established by the studies
of Benoit and Assenmacher in the Pekin duck (*Anas platyrhynchos*).
They showed that transplantation of the pituitary gland to a distal
site caused as complete a gonadal atrophy as did hypophysectomy.
Surgical sectioning of the hypophysial portal vessels, or destruction
of the median eminence, led to testicular collapse and blocked the
photoperiodic response (Benoit and Assenmacher, 1955; Benoit,
1961; Assenmacher, 1958, 1970). Essentially similar results have
been obtained in the domestic fowl (Shirley and Nalbandov, 1956),
pigeon (Baylé and Assenmacher, 1966), and the Japanese Quail
(Baylé and Assenmacher, 1967).

In physical terms, the hypothalamus forms only a small portion of
the bird's brain (Figure 2). It is bounded dorsally by the hypothalamic sulcus, while its rostral and caudal limits lie just anterior to the
optic chiasma and immediately posterior to the mammillary bodies,
respectively. Ventrally it is bounded by the median eminence and the
pars nervosa (Figure 3). Its particular significance in regulating the
pituitary is that it acts as an endocrine organ and yet remains an integral part of the brain. This is reflected in its structure. As in mammals, it is possible to distinguish between a medial region that ad-
joins the third ventricle and contains many cell bodies and a lateral
region characterized by the fiber tracts (e.g., medial forebrain bundle) that run through it. These tracts are important because they appear to be the primary pathways whereby many of the extrinsic connections are made between the hypothalamus and other brain centers

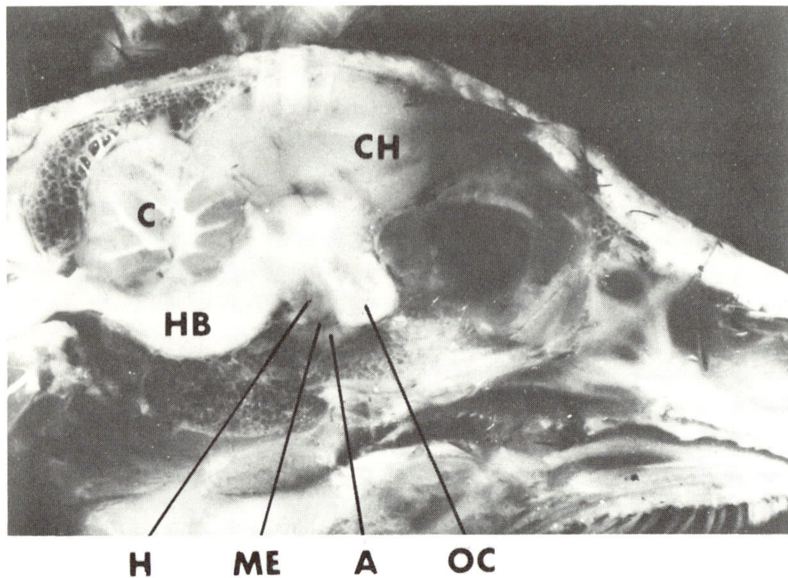

FIGURE 2 A bisected quail head showing the relative size and disposition of the different parts of the brain. A, adenohypophysis; C, cerebellum; CH, cerebral hemisphere; H, hypothalamus; HB, hind brain; ME, median eminence; OC, optic chiasma.

(Raisman, 1966). The medial regions, however, are more directly involved in the endocrine functions of the hypothalamus and the neuroanatomy of this area has been thoroughly studied (Huber and Crosby, 1929; Kuhlenbeck, 1937; Wingstrand, 1951; Oksche, 1962, 1965; Farner *et al.*, 1966; Oehmke, 1968, 1969, 1971; Oksche *et al.*, 1970; Dodd *et al.*, 1971). The cell bodies of the medial region can be divided into a number of groups; names, often drawn from the mammalian literature, have been assigned to these various nuclei (Figure 3). The key feature of the hypothalamus is that the axons of many of the cell bodies terminate in the median eminence and form a network of neurosecretory fiber systems. The best characterized of these is the Gomori-positive or AF-positive neurosecretory system, so named because the neurons are stainable with Gomori's chrome-alum hematoxylin or with aldehyde fuchsin (AF). The cell bodies of this system lie in two pairs of nuclei (paraventricular and supraoptic) located in the anterior hypothalamus. The axons from these cells form a tract that passes caudally into the median eminence (Figure 3). Here it divides to terminate in two neurosecretory depots, a large one

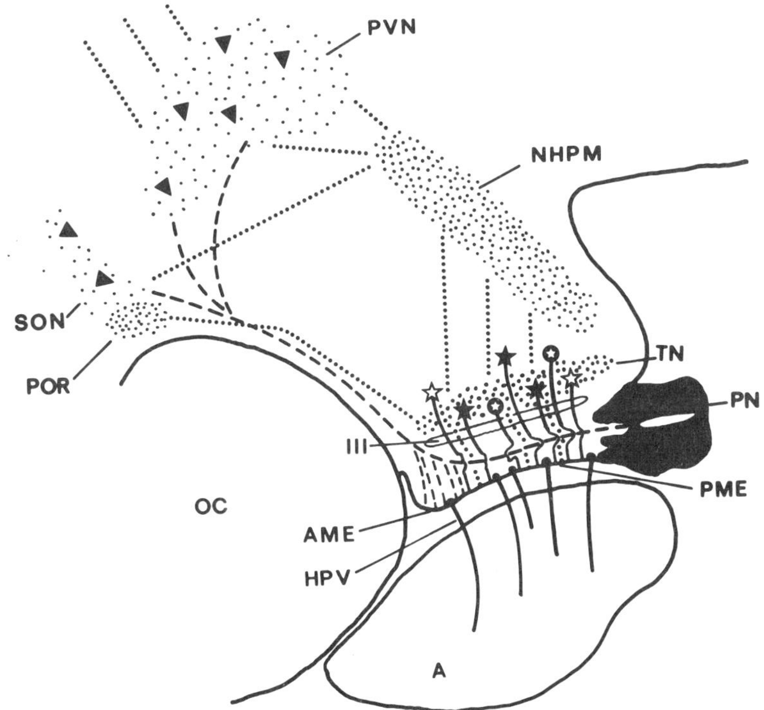

FIGURE 3 A simplified diagram of the hypothalamo–hypophysial system of the Japanese Quail to show the major neural and neurosecretory pathways. The relative size of the median eminence has been increased for the sake of clarity. The AF-positive neurosecretory system, the cell bodies of which lie in the paraventricular (PVN) and supraoptic (SON) nuclei, is shown as ▲ — — — — —. It terminates in the anterior median eminence (AME) and the pars nervosa (PN) of the pituitary. From the posterior hypothalamus arises a second neurosecretory system with axons passing in the tuberoinfundibular tract to terminate in both the anterior and the posterior (PME) divisions of the median eminence. This system contains a number of different neurosecretory neurons (⊕ ☆ ★), each perhaps responsible for the secretion of a particular releasing factor (neurohormone) from the median eminence into the hypophysial portal vessels (HPV). The cell bodies lie within the various divisions of the infundibular nucleus of which only the basal region is shown in the figure—the tuberal nucleus (TN). In addition, the hypothalamus contains many catecholamine-containing fibers and terminals. The regions with a high concentration of terminals are indicated by the shaded areas. They include two nuclei in the posterior hypothalamus—the tuberal nucleus (TN) and the nucleus hypothalamicus posterior medialis (NHPM). Many terminals also are visible in the nuclei (PVN, SON) of the AF-positive neurosecretory system. Fiber tracts are visible linking these regions together (. . . . .), as well as an adrenergic component of the tuberoinfundibular tract.

A, anterior pituitary (adenohypophysis); OC, optic chiasma; POR, preoptic recess; III, third ventricle.

in the *pars nervosa* and a secondary store in the anterior median eminence. The primary function of this system is to produce and release the neurohypophysial hormones (vasotocin, oxytocin) from the pars nervosa. These hormones are involved in osmoregulation and possibly also in oviposition (Follett, 1969a). Recent electron microscopic evidence suggests that the AF-positive neurosecretion stored in the anterior median eminence may be different from that in the pars nervosa (Figure 3; Bern *et al.*, 1966; Oehmke *et al.*, 1969).

The second neurosecretory system is less homogeneous and as a result more complex (Figure 3). It derives from a number of nuclei lying in the posterior basal hypothalamus. The cell bodies of these neurons do not stain with aldehyde fuchsin and they are often referred to collectively as belonging to the parvocellular nuclei (Farner *et al.*, 1966; Dodd *et al.*, 1971). The axons of these cell bodies form an extensive tuberoinfundibular tract that passes into the median eminence (Figure 4). An attempt is made in Figure 5 to compare some of the terminologies used for the various nuclei. In passeriforms the primary basal nucleus is called the infundibular nucleus. Detailed analyses in-

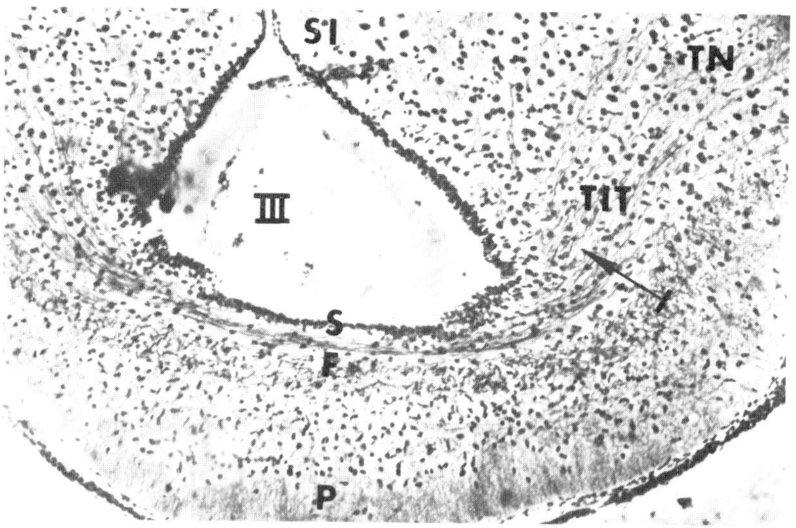

FIGURE 4 Transverse section through the quail median eminence to show part of the tuberoinfundibular neurosecretory tract (TIT). Many fibers (←) can be seen coming from the tuberal nucleus (TN) to enter the subependymal layer of the median eminence (S); these fibers terminate in the palisade layer of the median eminence (P). No fibers are visible in the stratum cellulare internum (SI) although fluorescent tracts can be demonstrated here (see Figures 6 and 9). III, third ventricle. From Sharp and Follett (1968).

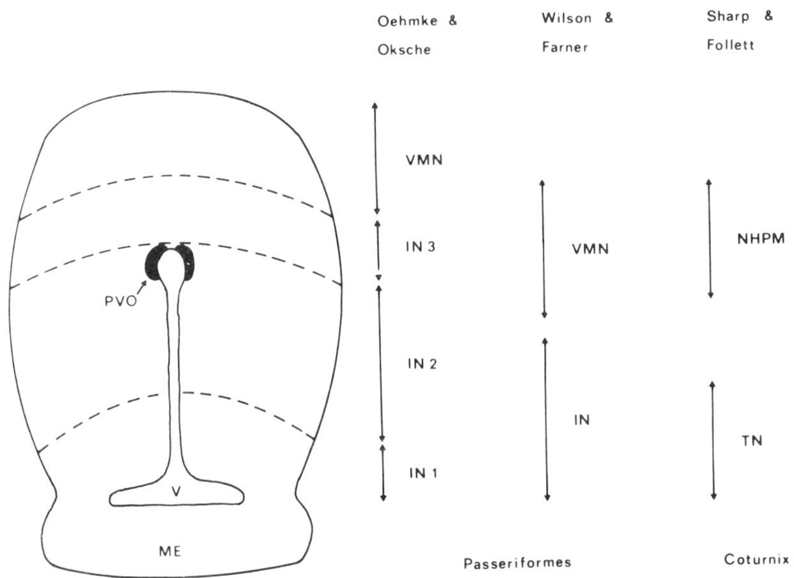

FIGURE 5   An attempt to compare the terminology employed by various research groups in defining the parvocellular nuclei of the posterior hypothalamus. Drawn from Oehmke (1968), Wilson (1967), and Sharp and Follett (1968, 1970).

IN, infundibular nucleus (1, 2, and 3 refer to subdivisions of this nucleus); NHPM, nucleus hypothalamicus posterior medialis; TN, tuberal nucleus; VMN, ventromedial nucleus.

dicate that it is divisible into three horizontal zones and possibly also into further subdivisions (Oehmke, 1968, 1971; Oksche *et al.*, 1970, 1972). Dorsally, the infundibular nucleus becomes practically continuous with other nuclear formations such as the ventromedial nucleus. A slightly different nomenclature has been used for the galliform species such as the domestic fowl and the Japanese Quail (Huber and Crosby, 1929; Van Tienhoven and Juhasz, 1962; Graber *et al.*, 1967; Sharp and Follett, 1968, 1969b). Under this arrangement, the tuberal nucleus corresponds to the basal zone of the infundibular nucleus (Figure 5), while the posterior medial hypothalamic nucleus (NHPM) (Figure 3) overlaps with the dorsal parts of the infundibular nucleus and with the ventromedial nucleus.

It is apparent from these comments that the neurosecretory neurons in the tuberoinfundibular tract can no longer be regarded as belonging to a homogeneous population. This becomes virtually certain if physiological results are considered. One now envisages that the tract contains a number of separate neurosecretory fiber systems, each perhaps secreting a different neurohormone (Figure 3). It might be hoped that individual fiber systems will be shown to derive from particular cell groups within the basal hypothalamus but this is as yet unproven. The great majority of fibers derive from the infundibular nucleus, but some probably also originate from the ventromedial nucleus (Oksche *et al.*, 1970). Some recent studies (Oksche *et al.*, 1972) suggest that photoperiodic stimulation of White-crowned Sparrows leads to a significant increase in the nuclear volume density of cell bodies within the basal infundibular nucleus.

The avian hypothalamus also abounds in catecholamine-containing nerve fibers and terminals that can be visualized histochemically (Fuxe and Ljunggren, 1965; Sharp and Follett, 1968, 1970; Warren, 1968; Oehmke, 1969; Oehmke *et al.*, 1969; Oksche *et al.*, 1970; Dodd *et al.*, 1971). The distribution seems similar in all bird species so far examined and is shown for *Coturnix* in Figures 3 and 6. A concentration of adrenergic nerve terminals was found in two regions of the posterior hypothalamus, the tuberal nucleus (NT) and the posterior medial nucleus (NHPM). The two regions are linked by adrenergic tracts and the NHPM also is connected by lateral connections to the forebrain bundle and anteriorly with the paraventricular nucleus. In addition, the tuberoinfundibular tract has an adrenergic component and a few amine fibers terminate in the median eminence. The functions of these adrenergic innervations are still unknown. Apparent adrenergic synapses have been noted on cell bodies in the infundibular nuclei (Priedkalns and Oksche, 1969), suggesting adrenergic regulation of at least a part of the neurosecretory cell population in this region, but the origin of these fibers is unknown. Similarly, the cell bodies providing the adrenergic terminals in the NHPM and in the tuberoinfundibular tract are also unknown. It seems likely that much of the adrenergic innervation functions in a purely nervous manner but the component of the tuberoinfundibular tract may be neurosecretory. What also needs to be known is the distribution of cholinergic systems within the hypothalamus. There is no *prima facie* reason why the neuronal circuits regulating reproduction should be adrenergic rather than cholinergic!

The anatomical studies on the hypothalamus, while providing a general understanding of how the adrenohypophysis is controlled, have been unable to identify the specific neurosecretory system(s) involved in gonadotropin release. To achieve this, a physiological approach has been used in which discrete electrolytic lesions are placed

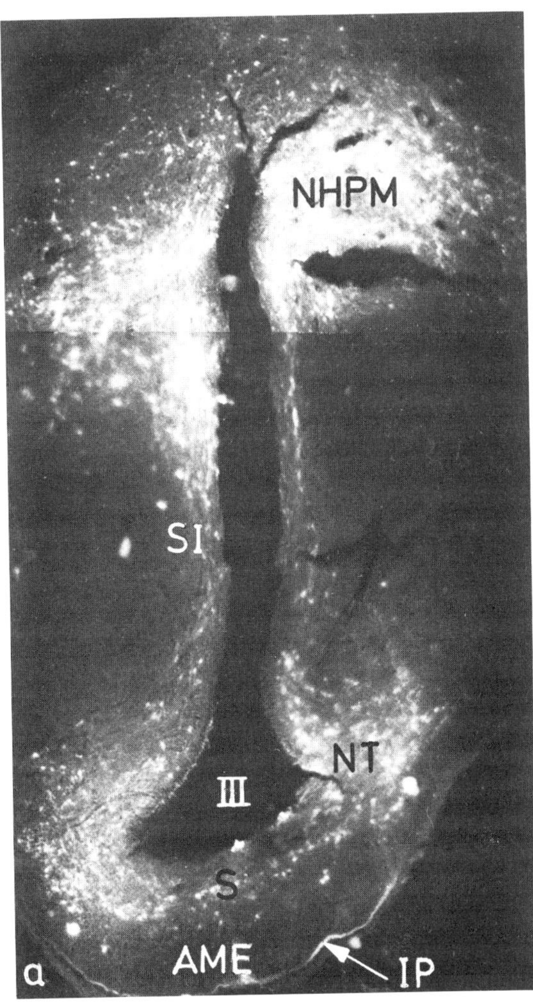

FIGURE 6 Transverse section through the anterior median eminence (AME) to show the main catecholamine-containing areas and tracts in *Coturnix*. Note the tracts running in the stratum cellulare internum (SI) from the region of the nucleus hypothalamicus posterior medialis (NHPM) into the tuberal nucleus (TN). From Sharp and Follett (1968).

IP, intima pia; S, subependymal layer; III, third ventricle.

throughout the hypothalamus. The effects on gonadotropin release are then assessed by weighing and sectioning the gonads.

The median eminence constitutes the final common pathway through which the neurohormones are released and its destruction blocks gonadotropin secretion (Assenmacher, 1958; Shirley and Nalbandov, 1956; Wilson, 1967; Sharp and Follett, 1969b; Stetson, 1969a). It would be of rather more interest to know which areas within the hypothalamus are involved and this has been studied in five species: the Pekin Duck (Assenmacher, 1958, 1970; Benoit and Assenmacher, 1955; Gogan *et al.*, 1963); the domestic fowl (Graber *et al.*, 1967; Kanematsu *et al.*, 1966); *Coturnix* (Sharp and Follett, 1969b; Stetson, 1969a); the White-crowned Sparrow, *Zonotrichia leucophrys gambelii* (Wilson, 1967; Stetson, 1969b); and the Tree Sparrow, *Spizella arborea* (Wilson and Hands, 1968). The results are not wholly consistent, but certain facts have become clear. Electrolytic destruction of the tuberal nucleus (ventral region of the infundibular nucleus) blocks testicular growth in the Japanese Quail (Figures 7 and 8), White-crowned Sparrow, and the domestic fowl. Such lesions invariably destroy the origins of the tuberoinfundibular tract but have no effect on the integrity of the AF-positive neurosecretory system, or of the median eminence. Lesions that destroy the AF-

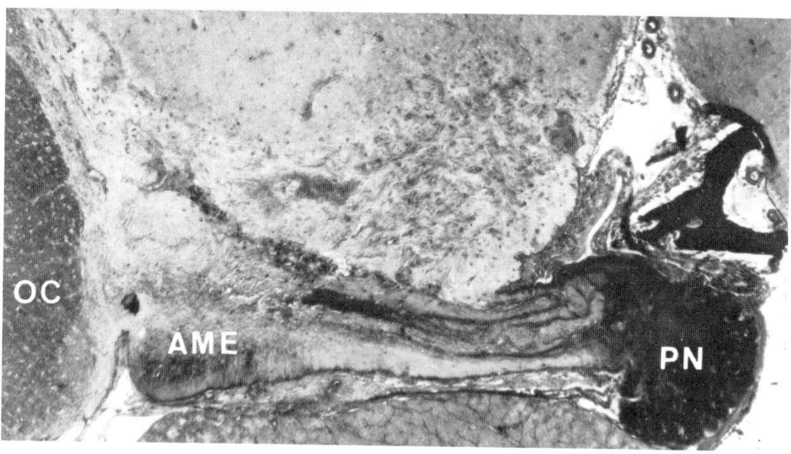

FIGURE 7   Median sagittal section through the hypothalamus of a male quail with an electrolytic lesion in the tuberal nucleus. The AF-positive neurosecretory system is intact with material visible in the anterior median eminence (AME) and the pars nervosa (PN). After photostimulation for 10 days, the combined testicular weight was 28 mg (c.f. sham-operated controls of about 1,000 mg). From Sharp and Follett (1969a). OC, optic chiasma.

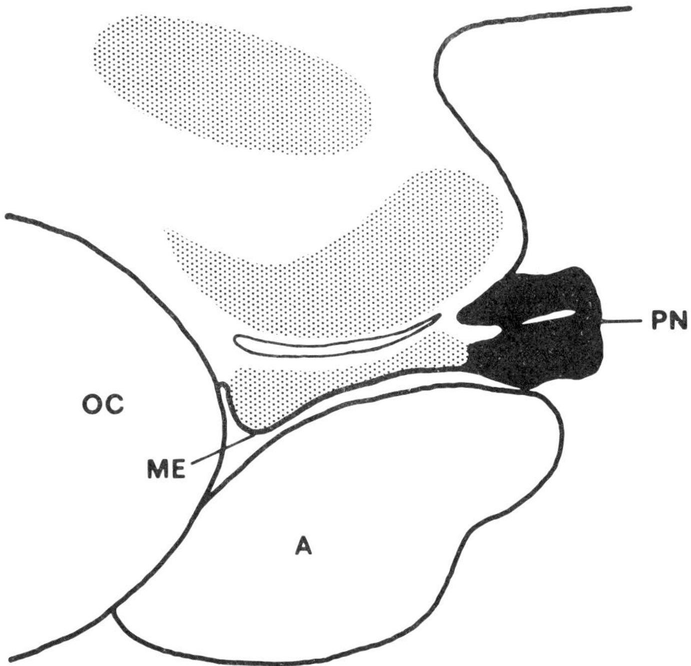

FIGURE 8   Simplified diagram showing the localization of the hypothalamic lesions that block photo-induced testicular growth in the Japanese Quail.

A, anterior pituitary (adenohypophysis); ME, median eminence; OC, optic chiasma; PN, pars nervosa of the pituitary.

positive tract before it enters the median eminence have no apparent effect on testicular growth and maintenance in these same species as well as in the Tree Sparrow. These results clearly indicate that the neurosecretory system derived from the parvocellular nuclei regulates gonadotropin release and are consistent with comparable studies in mammals (Harris and Donovan, 1966; Halász, 1969), amphibians, and fish (for review see Dodd *et al.*, 1971). However, in the Japanese Quail at least, the situation is more complex. Lesions dorsal to the tuberal nucleus are also effective in blocking gonadotropin release (Figure 8). The region destroyed always includes part of the NHPM nucleus as well as the paraventricular organ. In all cases the tuberoinfundibular system appeared intact, with catecholamines visible in the tuberal regions and median eminence (Figure 9).

The relative roles of the two separate regions are as yet unclear. It is possible, of course, that the tuberal nucleus is not involved in the

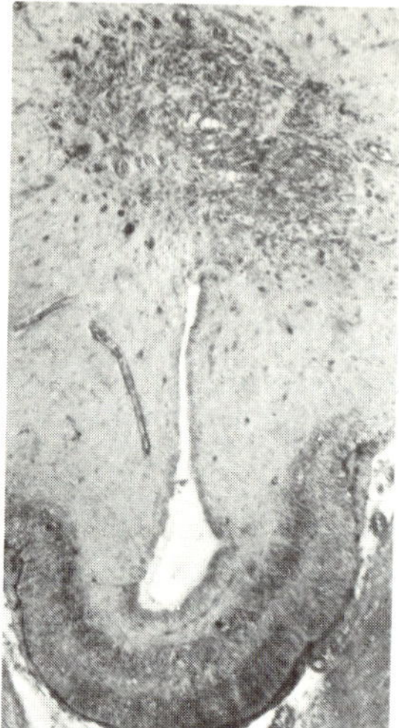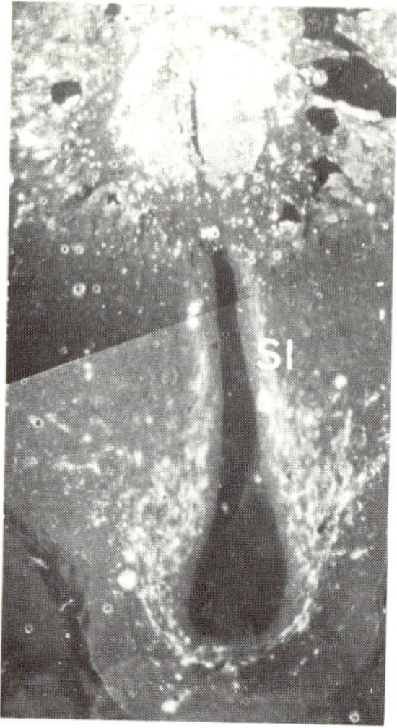

FIGURE 9   Transverse sections through the basal hypothalamus of
two Japanese Quail with lesions in the posterior medial hypothalamic
nucleus. Left: stained with aldehyde fuchsin. Right: fluorescence
technique to visualize catecholamines. Note that the lesion does not
affect the tuberal nucleus or the catecholamine-containing component of the tuberoinfundibular tract. The AF-positive neurosecretory tracts are also intact. Combined testicular weights after
photostimulation were 17 and 9 mg, respectively. SI, stratus
cellulare internum.

photoperiodic response and that lesions here are effective only because they sever direct neurosecretory links between the dorsomedial
region and the median eminence (Figure 10a). There is some anatomical evidence for such a direct tract (Wingstrand, 1951). An alternative hypothesis would be that the tuberal nucleus and the tuberoinfundibular tract contained the neurosecretory elements of the control
system. These neurosecretory fibers would only be activated if neural
stimulation arrived from the dorsomedial area (Figure 10b, c). Such
stimulation could act at a number of sites. Nerves might synapse with
cell bodies in the tuberal nucleus; electron microscopic evidence exists for adrenergic synapses in this region (Priedkalns and Oksche,
1969). Under this arrangement (Figure 10b), depolarization of the
appropriate neurosecretory neuron could lead directly to the secretion of the releasing factor into the portal vascular system. This
would be analogous to the way in which the AF-positive neurosecretory system is thought to be stimulated (Douglas, 1968; Heller,

1971). Another possibility is that neuronal control is exercised at the level of the median eminence via axo–axonal synapses between nerves and the terminals of the neurosecretory axons (Figure 10c, d). Such synapses, involving adrenergic axons, have been seen in the vertebrate median eminence (Kobayashi and Matsui, 1967; Fuxe and Hökfelt, 1969; Kobayashi *et al.*, 1970), and regulation of LH-RF secretion in mammals by this method has been suggested (Schneider and McCann, 1970; Fuxe and Hökfelt, 1970). In the present context such a nervous mechanism might be mediated by nerves coming directly from the dorsomedial region to the median eminence (Figure 10c), or via

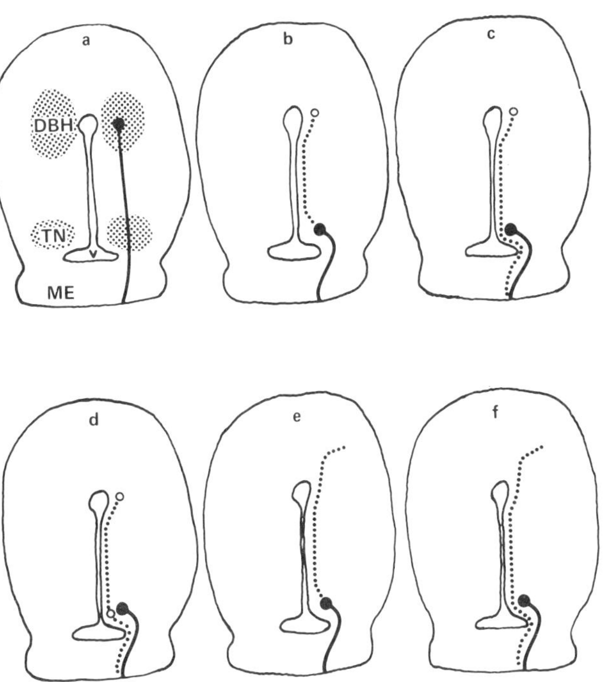

FIGURE 10 A series of diagrams to show some of the possible ways in which the neurosecretory system could be stimulated to secrete gonadotropin-releasing factors into the portal vascular system. All drawings are of transverse sections through the basal hypothalamus of the quail. A thick black line is used to indicate the neurosecretory neuron (●, cell body; ———, axon); dotted lines indicate purely nervous links (○, cell body; · · · · · ·, axon). Other alternatives, of course, are possible. The two regions where lesions block gonadotropin release in quail (see Figure 8) are stippled.

DBH, dorsobasal hypothalamic region; ME, median eminence; TN, region of the tuberal nucleus; V, third ventricle.

synapses with other nerves in the tuberal nucleus (Figure 10d).

The range of possible control mechanisms must also extend to include the dorsomedial region. It is attractive to imagine this region as a complex relay center, as one of the integrative components in the neuroendocrine chain, possibly fulfilling a role analogous with that of the anterior hypothalamus in the control of mammalian estrus cycles (see Halász, 1969). The NHPM has connections (Figure 3) both within the hypothalamus and extrinsically to other brain centers and, if lesioned, gonadotropin secretion ceases. However, this is still speculation and the arguments applied to the tuberal nuclear lesions also hold for the dorsomedial region. Perhaps lesions stop gonadotropin release only because they interrupt neural circuits running from other brain centers to the tuberal nucleus. If this applies—and there is strong evidence in mammals that part of the adrenergic system, including the noradrenaline-containing nerves of the tuberoinfundibular system, is of extrahypothalamic origin (see Fuxe and Hökfelt, 1969; Bjorklund *et al.*, 1969)—then additional alternatives can be suggested (Figure 10e, f).

The question as to which, if any, of these schemes is involved in regulating gonadotropin secretion is important, and a series of papers has been published on the situation in mammals (see Schneider and McCann, 1970; Fuxe *et al.*, 1969; Fuxe and Hökfelt, 1970; Kamberi *et al.*, 1971; Ahren *et al.*, 1971). These indicate that the adrenergic systems play a crucial role in regulating the secretion of the gonadotropin-releasing factor(s) and prolactin-inhibiting factor (P-IF). A major difficulty is that one school believes the adrenergic system to be stimulatory, while another considers it inhibitory. Unfortunately, avian neuroendocrinology has not advanced sufficiently to resolve this point and, while the anatomical studies are valuable, they seem to indicate that all schemes are possible. One physiological approach would be to seek out where the gonadotropin-releasing factors are made in the hypothalamus; this could help to distinguish between alternative (a) in Figure 10 and the others. This point can be investigated by implanting pituitary fragments in the hypothalamus and, subsequently, studying their cytology. Using this method, my colleague Dr. Sharp has found that pituitary basophils (the cell type responsible for the production of FSH, LH, and TSH) regress if implants are placed in the region of the posterior medial nucleus (Figure 11) but remain stimulated if located in the tuberal nucleus (Sharp, 1972). In addition, estimations of hypothalamic LH-RF content in Japanese Quail by an *in vitro* technique show a greater

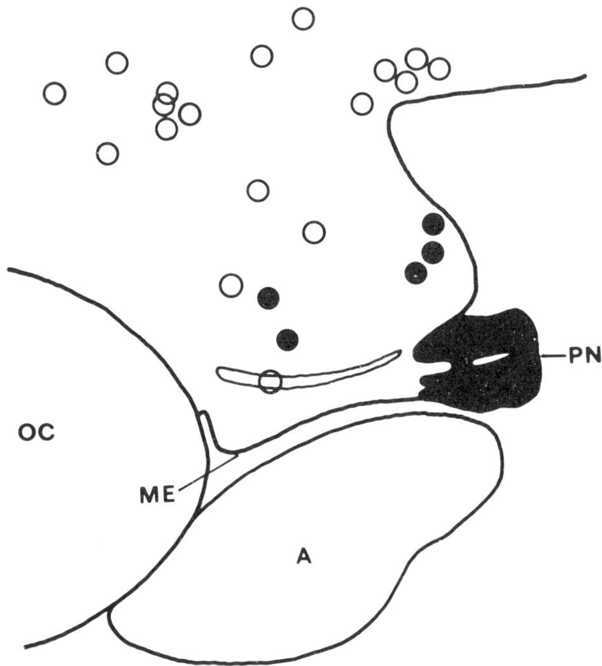

FIGURE 11   Diagram to show where in the hypothalamus pituitary
implants remain cytologically active. Fragments of pituitary tissue
were taken from sexually mature Japanese Quail and implanted into
the hypothalamus of male quail maintained on long day lengths.
Birds were killed 14 days later, and the activity of the basophils in
the implant was assessed cytologically. From Sharp (1972). ○,
pituitary inactive; ●, pituitary active; A, anterior pituitary; ME,
median eminence; OC, optic chiasma; PN, pars nervosa.

concentration in the ventral basal hypothalamus than in more dorsal
regions (P. M. Smith and P. J. Sharp, unpublished data). Although
additional experiments are required, these data suggest that the
NHPM does not elaborate LH-RF and that one of the other schemes
may be more likely. As yet there is no evidence to distinguish
between these but the problem may yield to more sophisticated
analyses of experiments with lesions.

Whether adrenergic mechanisms are involved remains unanswered.
The similarity between the areas containing catecholamines and those
where lesions block gonadotropin release (Figures 3 and 7) makes it
tempting to implicate them, but attempts to correlate the intensity
of monoamine fluorescence with photoinduced gonadal development

yielded equivocal results (Sharp, 1970). Implants of reserpine in the tuberal nuclear regions did not alter the rate of testicular growth (Sharp and Follett, 1969a). Recent data, however, show that intraventricular injections of 6-hydroxydopamine into sexually mature quail caused a marked reduction in testicular weight (see Assenmacher, this volume, Figure 9, p. 182).

Occasionally, lesioned birds are obtained in which the gonad shows signs that the secretion of one rather than both gonadotropins has been interrupted (Graber *et al.*, 1967; Stetson, 1969a; Sharp, 1970). This could be very important, since there is ecological and physiological evidence for a temporal separation of LH and FSH secretion in species such as the Greenfinch (Lofts and Murton, 1968; Murton *et al.*, 1969c). Unfortunately, it is difficult to reproduce such lesions and the evidence is still equivocal. The advent of specific radioimmunoassays for the avian gonadotropins (Follett *et al.*, 1972) should facilitate renewed attack on this problem.

The extensive studies in the Pekin Duck are at variance with those discussed above, for lesions that destroy the AF-positive neurosecretory system block testicular growth. The data are very convincing (see Assenmacher, 1958; Gogan *et al.*, 1963), and it is indeed possible that species differences do exist, particularly if photoperiodic controls have evolved separately on a number of occasions. On the whole, however, one inclines to the view that the hypothalamic mechanisms are relatively conservative and that another explanation for the differences can be found. This is mere supposition and the problem remains unresolved.

The final stage in the neuroendocrine chain is the secretion of the gonadotropin-releasing neurohormone(s) from the median eminence into the portal vascular system. The mechanisms involved in these secretory processes and the fine structure of the median eminence have been reviewed recently and will not be considered here (Kobayashi and Matsui, 1969; Kobayashi *et al.*, 1970). There is substantial physiological evidence for a gonadotropin-releasing factor in hypothalamic extracts of domestic fowl (Clark and Fraps, 1967; Opel and Lepore, 1967; Jackson and Nalbandov, 1969; Jackson, 1971) and of Japanese Quail (Follett, 1970, 1971; Smith and Follett, 1972). Most studies have concentrated on LH-RF and it is not known whether separate releasing factors exist for the two gonadotropins [recent evidence from mammals suggests that LH-RF and FSH-RF may be the same peptide—a GTH-RF (Schally *et al.*, 1971)]. Neither the catecholamines (Figure 12) nor the neurohypophysial hormones can stimulate

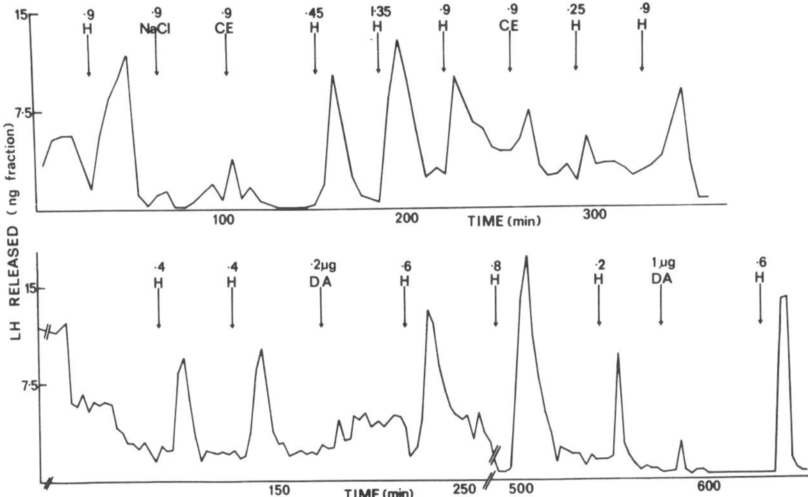

FIGURE 12 Two experiments to show the effect of quail hypothalamic extracts on the release of pituitary LH. Four quail pituitaries were mounted in a small chamber and Ringer's solution pumped over them at a constant rate. LH was measured in the effluent by radio-immunoassay (Follett *et al.*, 1972). Different doses of hypothalamic (H) and cerebral cortical extracts (CE), used as control, were introduced into the chamber at the times indicated by the arrows. All doses are expressed in terms of equivalents of one quail basal hypothalamus (1–1.5 mg).

Note the rapid increase in LH secretion occurring soon after the introduction of any hypothalamic extract. The responses are dose-dependent. Secretion falls to basal levels once the hypothalamic extract has left the chamber. Cortical extracts and dopamine (DA) elicit a minor release only of LH, as does isotonic saline (NaCl). From Smith and Follett (1972).

gonadotropin release directly, and domestic fowl LH-RF appears to be a small peptide (Jackson, 1971). These data support the various conclusions derived from lesioning experiments. When LH-RF acts on the pituitary it causes a rapid release of LH (Figure 12) that is dose-dependent. Once the neurohormone is removed from the vicinity of the pituitary, LH secretion returns to a low level. These effects are not necessarily surprising but they do underline the dependence of gonadotropin secretion upon output from the hypothalamus.

INTEGRATIVE FACTORS IN AVIAN PHOTOPERIODISM

The main features of the gonadotropin control system are probably common to most species, and the basis for the range of breeding patterns depends primarily on the types of external and internal information used to activate the hypothalamus. It is at this level, rather than in the final neuroendocrine pathway, that adaptive evolution has occurred. The complexity of the processes is such that in no species is it known precisely how the information is integrated and passed to the neurosecretory system. This is especially true during the final stages of breeding, when stimuli of many types interact to complete reproductive development (Hinde and Steel, 1966; Lehrman, 1961; Lack, 1968). Relatively more is known about the photoperiodic induction of gonadal growth but, as will be apparent, knowledge is still scant.

THE PERCEPTION OF LIGHT

Firm evidence exists for an extraretinal photoreceptor in birds, dating from Benoit's observations in 1935 that long day lengths continue to induce testicular development in ducks even after total severance of the optic nerves or removal of both eyes. Subsequent investigations led him to conclude that the photoreceptor lies in the hypothalamus, is responsive to light over a wide spectral range, and is highly sensitive (Benoit, 1961, 1964, 1970; Benoit and Assenmacher, 1966). By implanting "optic fibers" (diameter, 0.3 mm) in the hypothalamus, Kordon [quoted in Benoit (1970)] has been able to stimulate testicular development with white light of an intensity of 10^{-5} phot. The most sensitive areas lie in the ventromedial, supraoptic and paraventricular nuclei; implants outside this region are ineffective.

In the House Sparrow, *Passer domesticus*, blinding does not alter the rate of photo-induced testicular growth, suggesting that the retina is not involved in photoreception (Menaker and Keatts, 1968; Underwood and Menaker, 1970; Menaker, 1971). Confirmation derives from other experiments (Menaker *et al.*, 1970) in which intact sparrows were exposed to long days of low-intensity white light (7–14 lux). Whenever the feathers were plucked from the surface of the head, the testes grew rapidly, but this did not occur if the skull's opacity was increased by injecting India ink beneath the skin. Since the eyes were intact it was concluded that retinal light reception played no role in stimulating gonadotropin release. The same appears true for the Japanese Quail, another species in which removal of the eyes has little effect on gonadotropin release (Oishi *et al.*, 1966; Sayler and Wolfson, 1968b), and possibly also for the canary (Munns,

1971). In *Coturnix*, Homma and Sakakibara (1972) could induce testicular growth on short days by implanting small disks, coated with a radioluminous paint, at various points in the brain. Implants in areas such as the hypothalamus, thalamus, and behind the eye were effective, but no growth occurred if the disks lay close to the pineal, in front of the retina, or on the upper surface of the brain. Coating the skull with radioluminous paint prevented testicular regression when Japanese Quail were transferred to short day lengths (Kato *et al.*, 1967).

The nature and exact localization of the photoreceptor remains an enigma. It is tempting to speculate that some types of hypothalamic lesions may block gonadotropin release by interfering with the photoreception mechanism, but this remains speculation. The pineal gland, although important in other vertebrate groups (Wolstenholme and Knight, 1971), appears not to act as a photoreceptor in birds. Its removal has no effect on testicular growth or maintenance in the Japanese Quail (Homma *et al.*, 1967; Sayler and Wolfson, 1967, 1968a; Renzoni, 1967), White-crowned Sparrow (Kobayashi, 1969), or the House Sparrow (Menaker *et al.*, 1970). Injections of melatonin are largely without effect (Homma *et al.*, 1967; Sayler and Wolfson, 1968a). In female Japanese Quail, pinealectomy, or removal of the superior cervical ganglion, temporarily upsets vitellogenesis for a few days (Sayler and Wolfson, 1968a, b; McFarland *et al.*, 1968). Recently, the issue has been reopened with the discovery that pinealectomy of ducks in the spring inhibits the normal growth of the testes at this time (Cardinali *et al.*, 1971). The possibility of an indirect role also exists, since pinealectomy in the House Sparrow leads to the disappearance of the circadian rhythm of locomotor activity in constant darkness (Gaston and Menaker, 1968; Binkley *et al.*, 1971). If this applies generally to the bird's circadian biology, it could be important (see below).

The results with sparrows and quail differ from those with the Mallard, for which the latter Benoit believes that the retina is also involved in the photoreception mechanism (see Benoit, 1970; Menaker *et al.*, 1970). Resolution of this point is needed, particularly as it might affect a bird exposed to the high light intensities of the field. Also pertinent here is the question of direct nervous links between the retina and the hypothalamus. Evidence for (Blümcke, 1961; Bons and Assenmacher, 1969) and against (Cowan *et al.*, 1961; Oksche, 1970) a retino-hypothalamic tract is available. No firm con-

clusions can be drawn, but as Oksche (1970) has stressed, such tracts
need not be necessary, as the retina could influence the hypothalamus
indirectly via the visual centers in the diencephalon.

MEASUREMENT OF DAY LENGTH

However the photic information is received, a mechanism must exist
in the brain to measure its duration and to activate the neuroendocrine pathways when day lengths exceed a species-specific level. This
system does not operate on an all-or-none basis, since the rate of
gonadotropic secretion and of gonadal growth increases as day length
becomes greater. Current theories as to how this is achieved emphasize a circadian rhythm of "photoresponsivity" and are derived from
a more generalized model of day length measurement [for reviews see
Farner and Follett (1966), Wolfson (1966), Farner (1967, 1970),
Lofts *et al.* (1970); Follett (1973)]. In its simplest form, the model is
essentially a coincidence device between light and some direct or indirect rhythm of "photoresponsivity," the phase of which is controlled by the driving light cycle [but also see Wolfson (1970)]. Such
rhythms, first demonstrated in the House Finch, *Carpodacus mexicanus*, by Hamner (1963, 1964, 1966), appear widespread and have
been described in other passeriforms (Farner, 1965; Menaker, 1965;
Wolfson, 1965; Menaker and Eskin, 1967; Murton *et al.*, 1970b;
Murton *et al.*, 1970c) and in the Japanese Quail (Follet and Sharp,
1969; Follett, 1973). Their general form is shown in Figure 13. It is
not the duration of the daily photoperiod that is important in inducing gonadal growth but its relationship to the underlying cycle of
"photoresponsivity." Thus, light pulses in *Coturnix* are effective only
if they are given between 11.5 and 16 hours after dawn (Figure 13);
pulses at other times fail to induce significant growth. Such rhythms
of photosensitivity serve to explain how the breeding cycle of the animal is timed in nature, for in winter the short day lengths do not
coincide with the photoresponsive phase and the gonads remain regressed. As day length increases it comes to coincide with the rhythm
and causes gonadotropin release. Further increases in day length maximize the degree of coincidence and lead to a faster rate of maturation.

The characteristics of the photoresponsive rhythm may form the
physiological basis for many of the intra- and interspecific differences
that exist in the photoperiodic response (Lofts and Murton, 1968;

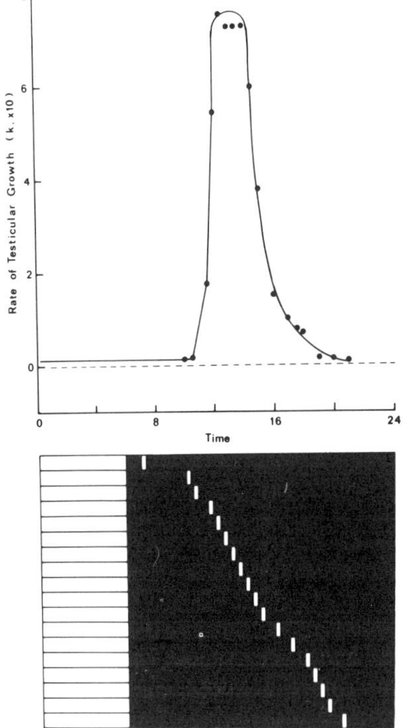

FIGURE 13 The results of an experiment designed to show the circadian rhythm of "photoresponsivity" in the Japanese Quail. Each group of birds was exposed to a total of 6.25 hours light per day given as a main period of 6 hours followed by a 15-minute pulse at different times during the 18-hour dark period. By itself, a 6.25-hour daylength given as one complete photoperiod constitutes a short day and is nonstimulatory in this species. Each point shows the rate of growth for a group receiving the 15-minute light period at the times indicated on the bars below the curve. The results emphasize that it is not the duration of the daily photoperiod that is important in causing gonadotropin release but where the light falls relative to the bird's "photoresponsive" phase. From Follett and Sharp (1969).

Farner, 1970; Lofts *et al.*, 1970). For example, its exact position within the 24-hour cycle may determine whether a species responds to the relatively short day lengths of early spring or requires longer photoperiods before gonadotropic secretion is initiated. Such differences might explain the results Dol'nik (1963) obtained in Chaffinches, *Fringella coelebs*, and by Lofts *et al.* (1967) in pigeons, in which marked differences have been found in the critical day lengths needed to stimulate gonadal growth. Naturally, many variants are possible, including not only phase but also amplitude of the rhythm, slope characteristics, and duration. It would be fascinating to know if the subtle differences in the patterns of gonadal growth seen, for example, in the rosy finches (*Leucosticte*) (King and Wales, 1965), have their basis in circadian functions; the topic requires active research.

Nothing is known of the nature of the rhythm, its location, and its relationship with any "master clock" (Hamner and Enright, 1967). There is some evidence that, once temporal coincidence occurs between light and the rhythm, the secretion of gonadotropin-releasing factors soon occurs (Follett and Sharp, 1969) as well as L H (Nicholls and Follett, 1973).

A further complication has been added by the suggestion that LH and FSH may be controlled by separate timing mechanisms. By exposing Greenfinches (*Chloris chloris*) to "interrupted-night" treatments, Murton (Murton *et al.*, 1969b, 1970b) obtained evidence that LH is secreted independently from FSH, depending on whether the photoschedule mimicked a barely stimulatory day length or one of 16 hours duration. Such a separation could not be found in the House Sparrow (Murton *et al.*, 1970c). Apart from their possible ecological importance, these results imply separate hypothalamic control mechanisms for the two gonadotropins.

In most species, breeding is terminated abruptly by a spontaneous regression of the gonads at a time when the bird is experiencing normally stimulatory photoperiods. Since such birds cannot be photostimulated experimentally, they are said to be photorefractory. This is but one form of a general postnuptial refractoriness in birds characterized by a temporary inability of the proximate inducers to stimulate gonadotropin release. This blockage most probably occurs at the level of the hypothalamus, but the pituitary and gonad may also be involved. A number of theories have been advanced to account for the development of refractoriness, including alterations in the underlying circadian biology and the involvement of steroid feedback. No one theory has been proven, however, and the matter remains open (see Farner and Follett, 1966; Hamner, 1968; Lofts *et al.*, 1970; Murton *et al.*, 1970b; Wilson, 1970; Meier *et al.*, 1971; Cusick and Wilson, 1973).

THE ROLE OF SEX STEROIDS

Many observations suggest that, as in mammals, sex steroids can modify gonadotropin secretion. Castration, if combined with photostimulation, leads to hypertrophy of the pituitary gonadotropin-producing cells (for review see Tixier-Vidal and Follett, 1973) and to increases of twentyfold or more in the concentration of plasma LH (Follett *et al.*, 1972). The effects on pituitary gonadotropin content are more variable (cf Benoit *et al.*, 1950; Tixier-Vidal *et al.*, 1968; Kordon and Gogan, 1970; Gogan, 1970; Stetson and Erickson, 1971) but this probably reflects the time scale of the experiments rather than any fundamental species differences. Treatment with exogenous androgens or estrogens suppresses the pituitary and leads to a decrease in pituitary and plasma gonadotropin concentrations (Follett *et al.*, 1972).

The focus of this negative feedback appears to be neural, since implants of testosterone in the hypothalamus block testicular growth in ducks (Kordon and Gogan, 1964, 1970; Gogan, 1968), Tree Sparrows (Wilson, 1970; Cusick and Wilson, 1973), and Japanese Quail (C. Boddy and B. K. Follett, unpublished data). In all three species, the most effective site is in the basal infundibular nucleus (Figure 14), the same region where electrolytic lesions block gonadotropic release. Implants in the median eminence seem less effective. Partial suppression also was found in the duck when testosterone was implanted in more anterior hypothalamic regions. The consistent involvement of this area in the Mallard makes it even more important that detailed comparisons are made of the neuroanatomy in the various species used experimentally. No further information is available on steroid-sensitive elements in the hypothalamus and the manner in which the blockade is effected is unknown. In mammals, the pituitary itself is often an additional site of feedback; this also may be true for birds. Implants of testosterone in the pituitary invariably altered gonadotropin secretion (Figure 14) but rarely blocked it completely (Wilson, 1970; Cusick and Wilson, 1973; C. Boddy and B. K. Follett, unpublished data). Also, testosterone seems to affect gonadotropin release in an *in vitro* incubation of duck pituitaries (Gogan, 1970).

While these data show the potential for negative feedback, they do not prove its existence in the normal breeding cycle. A basic difficulty is that, when gonadotropin secretion is maximal in the spring, there is a rising level of plasma androgens (Garnier and Attal, 1970; Jallageas and Assenmacher, 1970; Garnier, 1971). During photo-induced testicular growth in the Japanese Quail, the plasma LH concentration remains high and constant, while that of testosterone increases fivefold. Such results suggest that feedback is minimal during the growth of the gonads. This may be achieved physiologically by lowering the sensitivity of the steroid-responsive neurons and there is some evidence for this in the Mallard (Gogan, 1970) and the Tree Sparrow (Wilson, 1970; Cusick and Wilson, 1973). At a later stage in the cycle, an increase in steroid-sensitivity could lead to testicular regression. This seems possible since implanting cyproterone in the hypothalamus of the Tree Sparrow prevented the spontaneous collapse of the testes

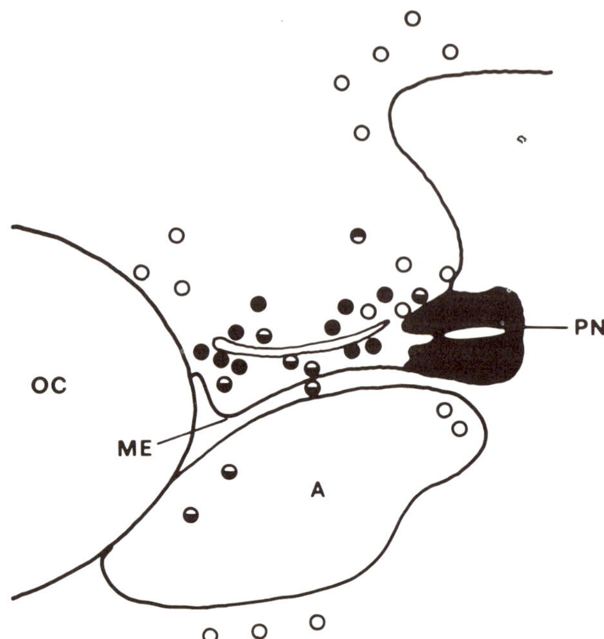

FIGURE 14   A diagram of the hypothalamus of the Japanese Quail
showing the location of testosterone implants and their effect on
testicular growth. Implants of cholesterol anywhere had no effect on
the rate of photoinduced testicular growth. From C. Boddy and
B. K. Follett (unpublished data).
   ○, no effect (normal growth); ◒, partial suppression of gonadotropin
secretion; ●, considerable or total suppression; A, adenohypophysis;
ME, median eminence; OC, optic chiasma; PN, pars nervosa.

that occurs normally (Wilson, 1970; Cusick and Wilson, 1973). Cyproterone is a potent antiandrogen and could be expected to block
any natural feedback.

The sex steroids also act on hypothalamic structures involved with
behavior. More particularly, local implants of androgens or progesto-
gens in the preoptic and anterior hypothalamic areas of the dove can
induce a range of courtship behavior (Hutchison, 1967, 1971;
Komisaruk, 1967). A final area of steroid feedback is that involved
with the ovulation cycle in the female bird. Anterior hypothalamic
implants of progesterone can induce ovulation in the domestic hen
(Ralph and Fraps, 1960), and most theories of ovulation control in-

volve progestins and estrogens in causing a burst of LH secretion (Fraps, 1970).

REFERENCES

Ahrén, K., K. Fuxe, L. Hamberger, and T. Hökfelt. 1971. Turnover changes in the tubero-infundibular dopamine neurons during the ovarian cycle of the rat. Endocrinology 80:1415–1424.

Assenmacher, I. 1958. Recherches sur le contrôle hypothalamique de la fonction gonadotrope préhypophysaire chez le canard. Arch. Anat. Microsc. Morphol. Exp. 47:448–572.

Assenmacher, I. 1970. Importance de la liaison hypothalamo–hypophysaire dans le contrôle nerveux de la reproduction chez les oiseaux, p. 167–192. *In* La photorégulation de la reproduction chez les oiseaux et les mammifères. C.N.R.S., Paris.

Baylé, J-D., and I. Assenmacher. 1966. Repercussions endocriniennes de l'auto-greffe hypophysaire chez le pigeon. J. Physiol. (Paris) 58:203–204.

Baylé, J-D., and I. Assenmacher. 1967. Contrôle hypothalamo–hypophysaire du fonctionnement thyroidien chez la caille. C.R. Hebd. Séances Acad. Sci. (Paris) 264:125–128.

Benoit, J. 1961. Opto-sexual reflex in the duck: Physiological and histological aspects. Yale J. Biol. Med. 34:97–116.

Benoit, J. 1964. The role of the eye and the hypothalamus in the photostimulation of the gonads in the duck. Ann. N.Y. Acad. Sci. 117:204–216.

Benoit, J. 1970. Étude de l'action des radiations visibles sur la gonadostimulation, et de leur pénétration intra-crânienne chez les oiseaux et les mammifères, p. 121–146. *In* La photorégulation de la reproduction chez les oiseaux et les mammifères. C.N.R.S., Paris.

Benoit, J., and I. Assenmacher. 1955. Le contrôle hypothalamique de l'activité préhypophysaire gonadotrope. J. Physiol. (Paris) 47:427–567.

Benoit, J., and I. Assenmacher. 1966. Recherches sur la photosensibilité des récepteurs nerveux superficiel et profond dans la gonadostimulation par les radiations visibles chez le canard Pékin impubère. C.R. Hebd. Séances Acad. Sci. (Paris) 262:2750–2752.

Benoit, J., I. Assenmacher, and F. X. Walter. 1950. Résponses du mécanisme gonado-stimulant a l'eclairement artificiel et de la préhypophyse aux castrations bilaterale et unilatérale, chez le canard domestique mâle, au cours de la période de regression testiculaire saisonnière. C.R. Séances Soc. Biol. 144:573–577.

Bern, H. A., R. S. Nishioka, L. R. Mewaldt, and D. S. Farner. 1966. Photoperiodic and osmotic influences on the ultrastructure of the hypothalamic neurosecretory system of the White-crowned Sparrow, *Zonotrichia leucophrys gambelii*. Z. Zellforsch. Mikrosk. Anat. 69:198–227.

Binkley, S., E. Kluth, and M. Menaker. 1971. Pineal function in sparrows: Circadian rhythms and body temperature. Science 174:311–314.

Björklund, A., B. Falck, F. Hromek, Ch. Owman, and K. West. 1969. Distribution and identification of the monoamine fibre systems in the rat neurohypophysis by means of stereotaxic and microspectrofluorimetric techniques. Brain Res. 17:1–23.

Blümcke, S. 1961. Vergleichend experimentell-morphologische Untersuchungen zur Frage einer retino-hypothalamischen Bahn bei Huhn, Meerschweinchen und Katze. Z. Zellforsch. Mikrosk. Anat. 67:469–513.

Bons, N., and I. Assenmacher. 1969. Présence de fibres rétiniennes dégenérées dans la région hypothalamique supra-optique du canard aprés section d'un nerf optique. C.R. Hebd. Séances Acad. Sci. 269:1535–1538.

Cardinali, D. P., A. E. Cuello, J. H. Tramezzani, and J. M. Rosner. 1971. Effects of pinealectomy on the testicular function of the adult male duck. Endocrinology 89:1082–1093.

Clark, C. E., and R. M. Fraps. 1967. Induction of ovulation in the chicken with median eminence extracts. Poultry Sci. 46:1245–1246.

Cowan, W. M., L. Adamson, and T. P. S. Powell. 1961. An experimental study of the avian visual system. J. Anat. (London) 95:545–563.

Cusick, E. K., and F. E. Wilson. 1973. On control of spontaneous testicular regression in Tree Sparrow (*Spizella arborea*). Gen. Comp. Endocrinol. In press.

Dodd, J. M., B. K. Follett, and P. J. Sharp. 1971. Hypothalamic control of pituitary function in submammalian vertebrates. Adv. Comp. Physiol. Biochem. 4:114–223.

Dol'nik, V. R. 1965. A quantitative study of vernal testicular growth in several species of finches (*Fringillidae*). Dokl. Akad. Nauk. SSSR 149:370–372.

Douglas, W. W. 1968. Stimulus-secretion coupling: The concept and clues from chromaffin and other cells. The First Gaddum Memorial Lecture. Brit. J. Pharmacol. 34:451–474.

Farner, D. S. 1965. Circadian systems in the photoperiodic responses of vertebrates, p. 357–369. *In* J. Aschoff [ed.] Circadian clocks. North Holland, Amsterdam.

Farner, D. S. 1967. The control of avian reproductive cycles, p. 107–133. *In* D. W. Snow [ed.] Proceedings of the XIV International Ornithological Congress. Blackwell's, Oxford.

Farner, D. S. 1970. Daylength as environmental information in the control of reproduction of birds, p. 71–91. *In* La photorégulation de la reproduction chez les oiseaux et les mammifères. C.N.R.S. Paris.

Farner, D. S., and B. K. Follett. 1966. Light and other environmental factors affecting avian reproduction. J. Anim. Sci. 25(Suppl.):90–118.

Farner, D. S., F. E. Wilson, and A. Oksche. 1966. Neuroendocrine mechanisms in birds, p. 529–582. *In* L. Martini and W. F. Ganong [ed.] Neuroendocrinology. Vol. 2. Academic Press, New York.

Follett, B. K. 1969a. Effects of neurohypophysial hormones and their synthetic analogues on lower vertebrates, p. 321–350. *In* H. Heller and B. T. Pickering [ed.] International encyclopaedia of pharmacology and therapeutics. Section 41. Pergamon Press, Oxford.

Follett, B. K. 1969b. Strain differences in photoperiodic responses in *Coturnix*,

p. 68. *In* Seminar on hypothalamic and endocrine functions in birds. International House, Japan.

Follett, B. K. 1970. Gonadotropin-releasing activity in the quail hypothalamus. Gen. Comp. Endocrinol. 15:165–179.

Follett, B. K. 1971. The neurohormones regulating the adenohypophysis, p. 86–103. *In* The Neurohypophysis. Ciba Foundation Study Group No. 39. Churchill Livingstone, London.

Follet, B. K. 1973. Circadian rhythmicity and time measurement in avian photoperiodicity. *In* J. S. Perry and I. W. Rowlands [ed.] The environment and reproduction in mammals and birds. J. Reprod. Fertil. Suppl. No. 15. In press.

Follett, B. K., and P. J. Sharp 1969. Circadian rhythmicity in photoperiodically induced gonadotropin release and gonadal growth in the quail. Nature 223:968–971.

Follett, B. K., C. G. Scanes, and F. J. Cunningham. 1972. A radioimmunoassay for avian luteinizing hormone. J. Endocrinol. 52:359–378.

Fraps, R. M. 1970. Photoregulation in the ovulation cycle of the domestic fowl, p. 281–306. *In* La photorégulation de la reproduction chez les oiseaux et les mammifères. C.N.R.S., Paris.

Fuxe, K., and T. Hökfelt. 1969. Catecholamines in the hypothalamus and pituitary gland, p. 47–96. *In* W. F. Ganong and L. Martini [ed.] Frontiers in neuroendocrinology, 1969. Oxford, New York.

Fuxe, K., and T. Hökfelt. 1970. Participation of central monoamine neurons in the regulation of anterior pituitary function with special regard to the neuroendocrine role of the tubero-infundibular dopamine neurons, p. 192–205. *In* W. Bargmann and B. Scharrer [ed.] Aspects of neuroendocrinology. Vol. V. International Symposium on Neurosecretion, August 1969, Kiel, Germany. Springer-Verlag, Berlin.

Fuxe, K., and L. Ljunggren. 1965. Cellular localization of monoamines in the upper brain stem of the pigeon. J. Comp. Neurol. 125:355–382.

Fuxe, K., T. Hökfelt, and O. Nilssvon. 1969. Castration, sex hormones and tubero-infundibular dopamine neurons. Neuroendocrinology 5:107–120.

Garnier, D. H. 1971. Variations de la testostérone du plasma périphérique chez le canard Pékin au cours du cycle annuel. C.R. Hebd. Séances Acad. Sci. 272:1665–1668.

Garnier, D. H., and J. Attal. 1970. Variations de la téstosterone du plasma testiculaire et des cellules interstitielles chez le canard Pékin au cours du cycle annuel. C.R. Hebd. Séances Acad. Sci. 270:2472–2475.

Gaston, S., and M. Menaker. 1968. Pineal function: The biological clock in the sparrow? Science 160:1125–1127.

Gogan, F. 1968. Sensibilité hypothalamique à la testostérone chez le canard. Gen. Comp. Endocrinol. 11:316–327.

Gogan, F. 1970. Rétroaction des stéroides sexuels sur les fonctions gonadotropes de l'oiseau, p. 351–362. *In* Neuroendocrinologie. C.N.R.S., Paris.

Gogan, F., C. Kordon, and J. Benoit. 1963. Retentissement de lésions de l'eminence médiane sur la gonadostimulation du canard. C.R. Séances Soc. Biol. 157:2133–2136.

Graber, J. W., A. I. Frankel, and A. V. Nalbandov. 1967. Hypothalamic center influencing the release of LH in the cockerel. Gen. Comp. Endocrinol. 9:187–192.

Halász, B. 1969. The endocrine effects of isolation of the hypothalamus from the rest of the brain, p. 307–342. *In* W. F. Ganong and L. Martini [ed.] Frontiers in neuroendocrinology, 1969. Oxford, New York.

Hamner, W. H. 1963. Diurnal rhythm and photoperiodism in testicular recrudescence of the House Finch. Science 142:1294–1295.

Hamner, W. H. 1964. Circadian control of photoperiodism in the House Finch demonstrated by interrupted-night experiments. Nature 203:1400–1401.

Hamner, W. H. 1966. Photoperiodic control of the annual testicular cycle in the House Finch, *Carpodacus mexicanus*. Gen. Comp. Endocrinol. 7:224–233.

Hamner, W. H. 1968. The photorefractory period of the House Finch. Ecology 49:211–227.

Hamner, W. H., and J. T. Enright. 1967. Relationship between photoperiodism and circadian rhythms of activity in the House Finch. J. Exp. Biol. 46:43–61.

Harris, G. W., and B. T. Donovan [ed.]. 1966. The pituitary gland. 3 Vols. Butterworths, London.

Heller, H. 1971. The neuro-endocrine control of water metabolism. Mem. Soc. Endocrinol. 18:447–464.

Hinde, R. A., and E. Steel, 1966. Integration of reproductive behaviour of female canaries. Symp. Soc. Exp. Biol. 20:401–426.

Homma, K. 1969. Gonadal development in the Japanese Quail after local photic stimulation to the brain p. 56–57. *In* Seminar on hypothalamic and endocrine functions in birds. International House of Japan, Tokyo.

Homma, K., and Y. Sakakibara. 1972. Encephalic photoreceptors and their significance in photoperiodic control of sexual activity in the Japanese Quail, p. 333–341. *In* M. Menaker [ed.] Biochronometry. National Academy of Sciences, Washington, D.C.

Homma, K., L. Z. McFarland, and W. O. Wilson. 1967. Response of the reproductive organs of the Japanese Quail to pinealectomy and melatonin injections. Poultry Sci. 46:314–319.

Huber, G. C. and E. C. Crosby. 1929. The nuclei and fiber paths of the avian diencephalon, with consideration of the telencephalic and certain mesencephalic centers and connections. J. Comp. Neurol. 48:1–225.

Hutchison, J. B. 1967. Initiation of courtship by hypothalamic implants of testosterone propionate in castrated doves (*Streptopelia risoria*). Nature 216:591–592.

Hutchison, J. B. 1970. Differential effects of testosterone and oestradiol on male courtship in Barbary Doves (*Streptopelia risoria*). Anim. Behav. 18:41–51.

Hutchison, J. B. 1971. Effects of hypothalamic implants of gonadal steroids on courtship behaviour on Barbary Doves (*Streptopelia risoria*). J. Endocrinol. 50:97–113.

Jackson, G. L. 1971. Purification of luteinizing hormone-releasing factor, Biol. Reprod. 5:107.

Jackson, G. L., and A. V. Nalbandov. 1969. Luteinizing hormone releasing activity in the chicken hypothalamus. Endocrinology 84:1262–1266.

Jallageas, M., and I. Assenmacher. 1970. Testostéronémie du canard photostimulé ou soumis á des injections répétées de testostérone. C.R. Séances Soc. Biol. 164:2338–2341.

Kamberi, I. A., R. S. Mical, and J. C. Porter. 1971. Effects of anterior pituitary perfusion and intraventricular injection of catecholamines on FSH release. Endocrinology 88:1003–1011.

Kanematsu, S., T. Sonoda, M. Kii, and Y. Kato. 1966. Effects of hypothalamic lesions on the gonad of the hen. Jap. J. Vet. Sci. 28(Suppl.):451.

Kato, M., and T. Konishi. 1968. The effect of light and temperature on the testicular growth of the Japanese Quail. Poultry Sci. 47:1052–1056.

Kato, M., Y. Kato, and T. Oishi. 1967. Radioluminous paints as activator of photoreceptor systems studied with swallow-tail butterfly and quail. Proc. Jap. Acad. 43:220–223.

King, J. R. 1970. Photoregulation of food intake and fat metabolism in relation to avian sexual cycles, p. 365–379. *In* La photorégulation de la reproduction chez les oiseaux et les mammifères. C.N.R.S., Paris.

King, J. R., and E. E. Wales. 1965. Photoperiodic regulation of testicular metamorphosis and fat deposition in three taxa of Rosy Finches. Physiol. Zool. 38:49–68.

Kobayashi, H. 1969. Pineal and gonadal activity in birds, p. 72. *In* Seminar on hypothalamic and endocrine functions in birds. International House of Japan, Tokyo.

Kobayashi, H., and T. Matsui. 1967. Synapses in the rat and pigeon median eminences. Endocrinol. Jap. 14:279–283.

Kobayashi, H., and T. Matsui. 1969. Fine structure of the median eminence and its functional significance, p. 3–46. *In* W. F. Ganong and L. Martini [ed.] Frontiers in neuroendocrinology, 1969. Oxford, New York.

Kobayashi, H., T. Matsui, and S. Ishii. 1970. Functional electron microscopy of the hypothalamic median eminence. Int. Rev. Cytol. 29:281–381.

Komisaruk, B. R. 1967. Effects of local brain implants of progesterone on reproductive behaviour in Ring Doves. J. Comp. Physiol. Psychol. 64:219–224.

Kordon, C., and F. Gogan. 1964. Localisation par une technique de microimplantation de structures hypothalamique résponsables du feed-back par la testostérone chez le canard. C.R. Séances Soc. Biol. 158:1795–1798.

Kordon, C., and F. Gogan. 1970. Interaction du feed-back et de la photostimulation dans les régulations gonadotropes chez les mammifères et les oiseaux, p. 325–346. *In* La photorégulation de la reproduction chez les oiseaux et les mammifères. C.N.R.S., Paris.

Kuhlenbeck, H. 1937. The ontogenetic development of the diencephalic centers in a bird's brain (chicken) and comparison with the reptilian and mammalian diencephalon. J. Comp. Neurol. 66:23–75.

Lack, D. 1968. Ecological adaptations for breeding in birds. Methuen, London. 409 p.

Lehrman, D. S. 1961. Hormonal regulation of parental behaviour in birds and infrahuman mammals, p. 1268–1382. *In* W. C. Young [ed.] Sex and internal secretions. Vol. 2. Williams & Wilkins, Baltimore.

Lehrman, D. S. 1965. Interaction between internal and external environments in the regulation of the reproductive cycle of the Ring Dove, p. 355–381. *In* F. A. Beach [ed.] Sex and behavior. Wiley, New York.

Lofts, B., and R. K. Murton. 1968. Photoperiodic and physiological adaptations regulating avian breeding cycles and their ecological significance. J. Zool. (London) 155:327–394.

Lofts, B., R. K. Murton, and N. J. Westwood. 1967. Interspecific differences in photosensitivity between three closely related species of pigeons. J. Zool. (London) 151:17–25.

Lofts, B., B. K. Follett, and R. K. Murton. 1970. Temporal changes in the pituitary–gonadal axis. Mem. Soc. Endocrinol. 18:545–577.

Marshall, A. J. 1970. Environmental factors other than light cycles involved in the control of sexual cycles in birds and mammals, p. 53–70. *In* La photorégulation de la reproduction chez les oiseaux et les mammifères. C.N.R.S., Paris.

Martini, L., and W. F. Ganong [ed.]. 1966. Neuroendocrinology. 2 Vols. Academic Press, New York.

McFarland, L. Z., K. Homma, and W. O. Wilson. 1968. Superior cervical ganglionectomy in the Japanese Quail. Experientia 24:245–247.

Meier, A. H., and D. S. Farner. 1964. A possible endocrine basis for premigratory fattening in the White-crowned Sparrow, *Zonotrichia leucophrys gambelii*. Gen. Comp. Endocrinol. 4:584–595.

Meier, A. H., and D. D. Martin. 1971. Temporal synergism of corticosterone and prolactin controlling fat storage in the White-throated Sparrow, *Zonotrichia albicollis*. Gen. Comp. Endocrinol. 17:311–318.

Meier, A. H., D. S. Farner, and J. R. King. 1965. A possible endocrine basis for migratory behaviour in the White-crowned Sparrow. Anim. Behav. 13:453–465.

Meier, A. H., D. D. Martin, and R. MacGregor. 1971. Temporal synergism of corticosterone and prolactin controlling gonadal growth in sparrows. Science 173:1240–1242.

Menaker, M. 1965. Circadian rhythms and photoperiodism in *Passer domesticus*, p. 385–395. *In* J. Aschoff [ed.] Circadian clocks. North Holland, Amsterdam.

Menaker, M. 1971. Rhythms, reproduction and photoreception. Biol. Reprod. 4:295–308.

Menaker, M., and A. Eskin. 1967. Circadian clock in photoperiodic time measurement: A test of the Bünning hypothesis. Science 157:1182–1185.

Menaker, M., and H. Keatts. 1968. Extraretinal light perception in the sparrow. II. Photoperiodic stimulation of testis growth. Proc. Nat. Acad. Sci. U.S.A. 60:146–151.

Menaker, M., R. Roberts, J. Elliott, and H. Underwood. 1970. Extraretinal light perception in the sparrow. III. The eyes do not participate in photoperiodic photoreception. Proc. Nat. Acad. Sci. U.S.A. 67:320–325.

Munns, T. J. 1971. Effect of different photoperiods on melatonin synthesis in the pineal gland of the canary (*Serinus canarius*). Diss. Abstr. 31:1288-B.

Murton, R. K., R. J. P. Thearle, and B. Lofts. 1969a. Endocrine basis for breeding behaviour in the feral pigeon (*Columba livia*). 1. Effects of exogenous hormones on the pre-incubation behaviour of intact males. Anim. Behav. 17:286–306.

Murton, R. K., K. D. Bagshawe, and B. Lofts. 1969b. The circadian basis of specific gonadotropin release in relation to avian spermatogenesis. J. Endocrinol. 45:311–312.

Murton, R. K., B. Lofts, and N. J. Westwood. 1970a. Manipulation of photorefractoriness in the House Sparrow (*Passer domesticus*) by circadian light regimes. Gen. Comp. Endocrinol. 14:107–113.
Murton, R. K., B. Lofts, and N. J. Westwood. 1970b. The circadian basis of
photoperiodically controlled spermatogeneis in the Greenfinch, *Chloris
chloris.* J. Zool. (London) 161:125–136.
Murton, R. K., B. Lofts, and A. H. Orr. 1970c. The significance of circadian
based photosensitivity in the House Sparrow, *Passer domesticus.* Ibis 112:
448–456.
Nicholls, T. J., and B. K. Follett. 1973. Daily rhythms of LH release in quail
when gonadal development is initiated by long daylengths. J. Reprod. Fert.
In press.
Oehmke, H.-J. 1968. Regionale Strukturunterschiede im Nucleus infundibularis
der Vögel (Passeriformes). Z. Zellforsch. Mikrosk. Anat. 92:406–421.
Oehmke, H.-J. 1969. Topographische Verteilung der Monoaminfluoreszenz im
Zwischenhirn-Hypophysensystem von *Carduelis chloris* und *Anas platyrhyn-
chos.* Z. Zellforsch. Mikrosk. Anat. 101:266–284.
Oehmke, H.-J. 1971. Vergleichende neurohistologische Studien am Nucleus infun-
dibularis einiger australischer Vögel. Z. Zellforsch. Mikrosk. Anat. 122:122–
138.
Oehmke, H.-J., J. Priedkalns, M. Vaupel-von Harnack, and A. Oksche. 1969.
Fluoreszenz- und elektronenmikroskopische Untersuchungen am
Zwischenhirn-Hypophysensystem von *Passer domesticus.* Z. Zellforsch.
Mikrosk. Anat. 95:109–133.
Oishi, T., T. Konishi, and M. Kato. 1966. Investigations on photorecepting mech-
anism to control gonadal development in Japanese Quail. Environ. Control
Biol. 3:87–90.
Oksche, A. 1962. The fine nervous, neurosecretory, and glial structure of the me-
dian eminence in the White-crowned Sparrow. Mem. Soc. Endocrinol.
12:199–208.
Oksche, A. 1965. The fine structure of the neurosecretory system of birds in relation to its functional aspects, p. 167–171. *In* Proceedings of Second International Congress of Endocrinology, London, 1964. North Holland, Amsterdam.
Oksche, A. 1970. Retino-hypothalamic pathways in birds and mammals, p. 151–
161. *In* La photorégulation de la reproduction chez les oiseaux et les mammifères. C.N.R.S., Paris.
Oksche, A., H.-J. Oehmke, and D. S. Farner. 1970. Weitere Befunde zur Struktur
und Funktion des Zwischenhirn-Hypophysensystems der Vögel, p. 261–273.
In W. Bargmann and B. Scharrer [ed.] Aspects of Neuroendocrinology. Vth
International Symposium on Neurosecretion, August 1969, Kiel, Germany.
Springer-Verlag, Berlin.
Oksche, A., P. Zimmermann, and H.-J. Oehmke. 1972. Morphometric studies of
tubero-eminential systems controlling reproductive functions, p. 142–153. *In*
K. M. Knigge, D. E. Scott, and A. Weindl [ed.] Brain-endocrine interaction.
Median eminence: Structure and function. Karger, Basel.
Opel, H., and P. D. Lepore. 1967. Ovulating hormone-releasing factor in the
chicken hypothalamus. Poultry Sci. 46:1302.

Priedkalns, J., and A. Oksche. 1969. Ultrastructure of synaptic terminals in the nucleus infundibularis and nucleus supraopticus of *Passer domesticus*. Z Zellforsch. Mikrosk. Anat. 98:135-147.

Raisman, G. 1966. Neural connexions of hypothalamus. Brit. Med. Bull. 22:197-201.

Ralph, C. L., and R. M. Fraps. 1960. Induction of ovulation in the hen by injection of progesterone in the brain. Endocrinology 66:269-272.

Renzoni, A. 1967. La fisiologia dell'epifisi negli uccelli. 1. Pinealectomia in *Coturnix coturnix japonica*. Soc. Ital. Biol. Sper. 43:585-588.

Sayler, A., and A. Wolfson. 1967. Avian pineal gland: Progonadotrophic response in the Japanese Quail. Science 158:1478-1479.

Sayler, A., and A. Wolfson. 1968a. Influence of the pineal gland on gonadal maturation in the Japanese Quail. Endocrinology 83:1237-1246.

Sayler, A., and A. Wolfson. 1968b. Role of the eyes and superior cervical ganglia on the effects of light on the pineal and gonads of the Japanese Quail. Arch. Anat. Histol. Embryol. 51:213-222.

Schally, A. V., A. Arimura, A. J. Kastin, H. Matsuo, Y. Baba, T. W. Redding, R. M. G. Nair, and L. Debeljunk. 1971. Gonadotrophin-releasing hormone: One polypeptide regulates secretion of luteinizing and follicle-stimulating hormones. Science 173:1036-1038.

Schneider, H. P. G., and S. M. McCann. 1970. Dopaminergic pathways and gonadotropin releasing factors, p. 177-191. *In* W. Bargmann and B. Scharrer [ed.] Aspects of neuroendocrinology. Vth International Symposium on Neurosecretion, August 1969, Kiel, Germany. Springer-Verlag, Berlin.

Sharp, P. J. 1970. The hypothalamic control of gonadotrophin release in the Japanese Quail (*Coturnix coturnix japonica*). PhD thesis, University of Leeds.

Sharp, P. J. 1972. Pituitary implants in the hypothalamus of *Coturnix* Quail. J. Endocrinol. 53:329-330.

Sharp, P. J., and B. K. Follett. 1968. The distribution of monoamines in the hypothalamus of the Japanese Quail, *Coturnix coturnix japonica*. Z. Zellforsch. Mikrosk. Anat. 90:245-262.

Sharp, P. J., and B. K. Follett. 1969a. The effect of reserpine on the pituitary-gonadal axis in quail, p. 43. *In* Seminar on hypothalamic and endocrine function in birds. International House of Japan, Tokyo.

Sharp, P. J., and B. K. Follett. 1969b. The effect of hypothalamic lesions on gonadotrophin release in Japanese Quail (*Coturnix coturnix japonica*). Neuroendocrinology 5:205-218.

Sharp, P. J., and B. K. Follett. 1970. The adrenergic supply within the avian hypothalamus, p. 95-103. *In* W. Bargmann and B. Scharrer [ed.] Aspects of neuroendocrinology. Vth International Symposium on Neurosecretion, August 1969, Kiel, Germany. Springer-Verlag, Berlin.

Shirley, H. V., and A. V. Nalbandov. 1956. Effects of transecting hypophyseal stalks in laying hens. Endocrinology 58:694-700.

Smith, P. M., and B. K. Follett. 1972. A luteinizing hormone-releasing factor in the quail hypothalamus. J. Endocrinol. 53:131-138.

Stetson, M. H. 1969a. Hypothalamic regulation of FSH and LH secretion in male and female Japanese Quail. Am. Zool. 9:104.

Stetson, M. H. 1969b. The role of the median eminence in control of photoperi-
    odically induced testicular growth in the White-crowned Sparrow (*Zonotrichia
    leucophrys gambelii*). Z. Zellforsch. Mikrosk. Anat. 93:369–394.
Stetson, M. H., and J. E. Erickson. 1971. Endocrine effects of castration in
    White-crowned Sparrows. Gen. Comp. Endocrinol. 17:105–114.
Tixier-Vidal, A., and B. K. Follett. 1973. The adenohypophysis. *In* D. S. Farner
    and J. R. King [ed.] Avian biology. Vol. 3. Academic Press, New York.
Tixier-Vidal, A., B. K. Follett, and D. S. Farner. 1968. The anterior pituitary of
    the Japanese Quail (*Coturnix coturnix japonica*). The cytological effects of
    photoperiodic stimulation. Z. Zellforsch. Mikrosk. Anat. 92:610–635.
Underwood, H., and M. Menaker. 1970. Photoperiodically significant photore-
    ception in sparrows: Is the retina involved? Science 167:298–301.
Van Tienhoven, A., and L. P. Juhasz. 1962. The chicken telencephalon, dien-
    cephalon and mesencephalon in stereotaxic coordinates. J. Comp. Neurol.
    118:185–197.
Warren, S. P. 1968. Primary catecholamine fibers in the ventral hypothalamus of
    the White-crowned Sparrow, *Zonotrichia leucophrys gambelii*. Master's thesis,
    University of Washington.
Wilson, F. E. 1967. The tubero-infundibular neuron system: A component of the
    photoperiodic control mechanism of the White-crowned Sparrow, *Zonotrichia
    leucophrys gambelii*. Z. Zellforsch. Mikrosk. Anat. 82:1–24.
Wilson, F. E. 1970. The tubero-infundibular region of the hypothalamus: A focus
    of testosterone sensitivity in male Tree Sparrows (*Spizella arborea*), p. 274–
    286. *In* W. Bargmann and B. Scharrer [ed.] Aspects of neuroendocrinology.
    Vth International Symposium on Neurosecretion, August 1969, Kiel, Ger-
    many. Springer-Verlag, Berlin.
Wilson, F. E., and G. R. Hands. 1968. Hypothalamic neurosecretion and photo-
    induced testicular growth in the Tree Sparrow, *Spizella arborea*. Z. Zellforsch.
    Mikrosk. Anat. 89:303–319.
Wingstrand, K. G. 1951. The structure and development of the avian pituitary.
    C. W. K., Gleerup, Lund.
Wolfson, A. 1965. Circadian rhythm and the photoperiodic regulation of the an-
    nual reproductive cycle in birds, p. 370–378. *In* J. Aschoff [ed.] Circadian
    clocks. North-Holland, Amsterdam.
Wolfson, A. 1966. Environmental and neuroendocrine regulation of annual
    gonadal cycles and migratory behavior in birds. Rec. Prog. Hormone Res.
    22:177–239.
Wolfson, A. 1970. Light and darkness and circadian rhythms in the regulation of
    annual reproductive cycles in birds, p. 93–120. *In* La photorégulation de la re-
    production chez les oiseaux et les mammifères. C.N.R.S., Paris.
Wolstenholme, G. E. W., and J. Knight [ed.]. 1971. The pineal gland. Ciba Foun-
    dation Symposium. Churchill, London. 416 p.

DISCUSSION

R. K. Murton

Doctor Follett has provided an excellent, comprehensive, and precise exposition on the role of the hypothalamus in regulating reproductive activity. I propose now to illustrate how the physiological mechanisms he described become translated to suit the ecological requirements of the subject. In so doing, I shall pose some new problems, the solutions to which must in turn help illuminate the kind of hypothalamic processes that occur. This work has been done in collaboration with Doctor Janet Kear of the Wildfowl Trust, Slimbridge, and will be published in full detail elsewhere (Murton and Kear, in press).

A large captive collection of waterfowl (Anatidae) is maintained by the Trust and most of the species find conditions conducive to breeding. In paired birds, the production of a normal clutch in a nest can be regarded as indicating that the male has attained full spermatogenetic development and is able to produce those hormones and displays necessary to stimulate the female to accept copulation and to perform appropriate behavior leading to nest-building, oviduct development, and ovulation. This is an important point, since it is usually the case that females, unlike males, are unable to achieve full reproductive development in isolation. The date of production of first eggs is known for many, if not all, years back to about 1950, and there is surprisingly little variability in laying date of the swans and true geese (mean coefficient of variation, 11.9 percent; range, 2.7–53.9). Each species produces eggs at a characteristic time. Those whose natural breeding range is in the far north lay sequentially later than those from lower latitudes; this circumstance pertains despite the fact that many of the subjects have been reared in captivity. The mean date of first eggs for all years available for each species can be ascribed a number by designating the days of the calendar year as 1–365. The midlatitude of the natural breeding range is the mean of the most northern and most southern extremes at which the species is known to nest. In Figure 1, the results of plotting the mean date of laying for eight species of swans (*Cygnus* and *Coscoroba*) and species of true geese (*Anser* and *Branta* combined) against the midlatitude of the natural breeding range is shown as calculated regression lines.

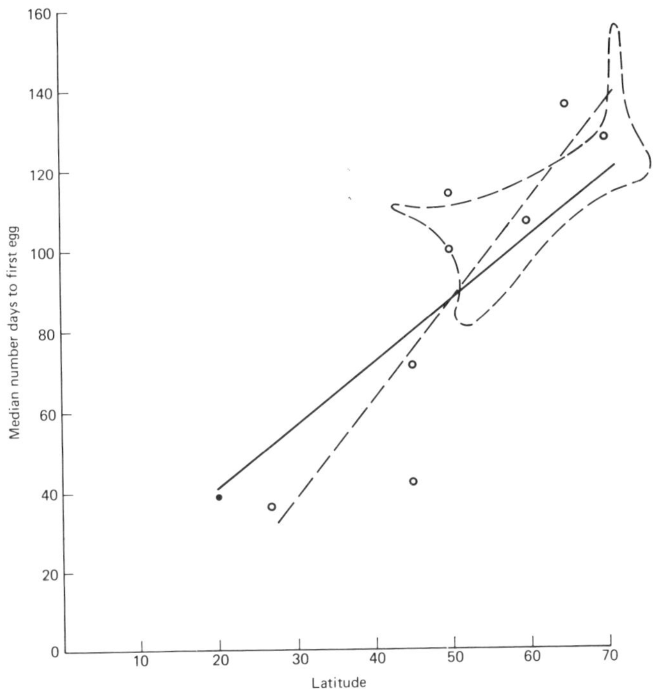

FIGURE 1 Number of days to egg-laying for various swan and goose species in the collection of the Wildfowl Trust, Slimbridge. Day number 1 = January 1, etc. Open circles depict the laying date of eight species of swans plotted according to the midlatitude of the natural breeding range. The regression line is plotted as a dashed line (r_6= 0.882, P>0.01; y = – 35.44 + 2.47x). The solid line is the regression for 17 species of *Anser* and *Branta* not plotted with the exception of the solid circle, indicating the Hawaiian Goose. The dotted line encloses all the plots for the remaining 16 species (r_{15} = 0.930, P>0.001; y = 8.35 + 1.59x).

It can be shown that the breeding dates of wildfowl at Slimbridge are not related to temperature and, since the food supply and most other variables are held constant, it seems reasonable to postulate that Figure 1 represents a photoperiodic effect. Indeed, there are other grounds for believing this to be so. The regressions account for 77 percent and 86 percent of the variability in the breeding date of the swans and geese, respectively, and it is remarkable that for any group of closely related species there should be so little variability. Ecologically, conditions become suitable for breeding progressively later with increase in latitude, but the strict relationships shown in Figure 1 suggest that some relatively invariable underlying mechanism is also involved. Swans feed primarily on aquatic weeds but their young are

reared on water-dwelling invertebrates. There is good evidence that
swans are ancestral to the geese (Johnsgard, 1965, 1967; Kear, 1970),
and Figure 1 shows that this long-evolved group occupies a wide range
of latitudes.

Ancestral swan species nesting at high latitudes must have encountered rich supplies of grass following the spring melting of snow; the
protein content of grass in spring at high latitudes is so great
(Fridriksson, 1960) that it is completely adequate for the growth of
goslings. Thus, a niche existed for a species able to exploit green
vegetation, but with the additional capacity of breeding sufficiently
early to enable the young to utilize the same food source. Kear and I
believe that a fundamental change in the photoperiodic response
mechanism of ancestral swans was necessary to take advantage of this
opportunity but that, once this major change occurred, the emergent
geese could radiate to occupy the new niche. As Figure 1 shows, for
any particular photoperiod (latitude), geese achieve full reproductive
capacity sooner than swans, but the nature of the response mechanism
is such that this is true only above 50°N lat. Below latitude 50°N,
swans must be potentially more effective grazers than are geese, and
Figure 1 shows that, with one exception, geese do not occur as breeding species further south. The exceptional Hawaiian Goose, *Branta
sandvicensis*, must certainly have evolved from northern stock and
has been able to find a habitat on the grass-covered laval slopes of an
oceanic island that is also relatively unsaturated in terms of its faunal
composition.

As Doctor Follett has intimated, the photoresponse mechanism of
birds is now known to depend on relatively limited periods of light
sensitivity (photoinducible phase) that become activated according
to an underlying and endogenous circadian timing mechanism
(Bünning, 1960; Hamner, 1963, 1966; Murton *et al.*, 1969, 1970;
Lofts *et al.*, 1970). The observations shown in Figure 1 should be
reconcilable in terms of such a mechanism. We have suggested that the
fundamental "circadian clock," because it must control so many
physiological functions, is not readily amenable to genetic change. An
alternative hypothesis would be that the body contains many separate
"clocks," each controlling separate functions; but such complexity
seems unlikely. Figure 2 depicts a circadian (cycle about 24 hours)
oscillator of characteristic amplitude (Murton and Kear, in press).

Different species could have a particular threshold relative to the
oscillator such that, if light impinges on them during the period when
the oscillator is above the threshold, a response can occur. The three

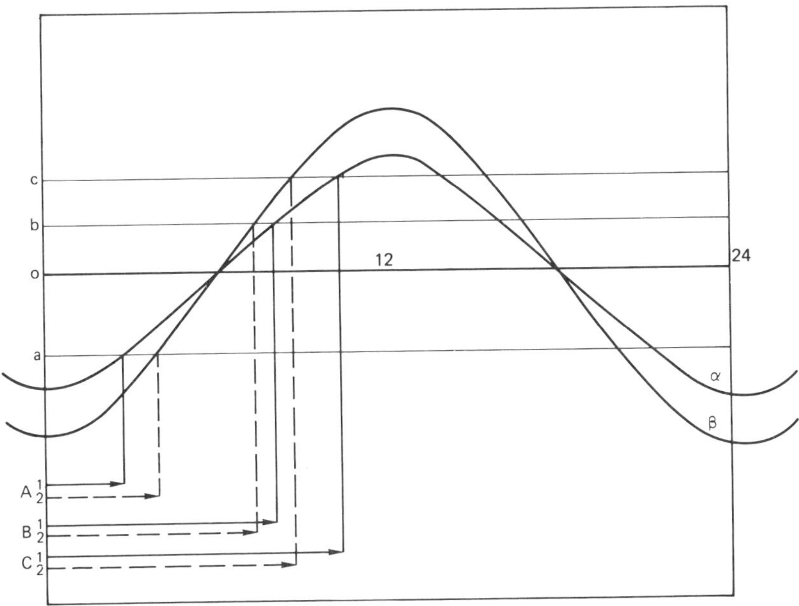

FIGURE 2　Two different hypothetical circadian oscillators α and β. Species A_1, B_1, and C_1 have thresholds of sensitivity a, b, and c to the oscillator α so that, if light impinges on the subject when the oscillator is above the threshold, a response can occur. The level of the oscillator depicted by the central horizontal line is marked off at 0, 12, and 24 hours to indicate the daily time scale. Clearly, species A_1 will become activated by shorter days than species B_1 assuming that time 0 represents dawn (the *Zeitgeber*). Species A_2, B_2, and C_2 are closely related members of a distinct group with thresholds a, b, and c to the oscillator β. Species B_2 and C_2 would be able to respond to light earlier than B_1 or C_1 and so could breed earlier in the year (geese compared to swans), but A_2 must breed later than A_1. From Murton and Kear (in press).

species A_1, B_1, C_1 represented in Figure 2 vary in photosensitivity. They can respond to short, medium, or long days and, thus, have the capacity to breed at different times in the year, depending on their threshold of response designated a, b, or c. Evidently, these three species would exhibit responses that, when plotted, would show a straight-line relationship (cf Figure 1). A completely new relationship would become apparent if an evolutionary change occurred in the amplitude of the circadian oscillator such that a new series of species A_2, B_2, and C_2 could emerge, each having thresholds a, b, and c. We feel that this kind of mechanism explains the observed differences between swans and geese.

Although birds, like other animals, exhibit a considerable adaptive radiation in occupying a wide range of ecological niches, there are nevertheless physiological limitations that influence the ability of species or groups to take advantage of all the possibilities open to them. Knowlege of these processes is not only of academic value in understanding the nature of the photoresponse mechanism, but it is of considerable interest in terms of conservation. The ability of a species to adapt to a changed environment resulting from the activities of man may well be restricted consequent on some deep-seated physiological factor that is not readily amenable to adaptive change.

REFERENCES

Bünning, E. 1960. Circadian rhythms and time—measurement in photoperiodism. Cold Spring Harbor Symp. Q. Biol. 25:249–256.

Fridriksson, S. 1960. Eggjahvitumagn og lostaetni Túngrasa (Protein content and palatability of cultivated grasses in Iceland) Rit Landbúnadardeildar, B—Flokkur, No. 12, p. 1–27.

Hamner, W. H. 1963. Diurnal rhythm and photoperiodism in testicular recrudescence of the House Finch. Science 142:1294–1295.

Hamner, W. H. 1966. Photoperiodic control of annual testicular cycle in the House Finch, *Carpodacus mexicanus*. Gen. Comp. Endocrinol. 7:224–233.

Johnsgard, P. A. 1965. Handbook of waterfowl behavior. Cornell University Press, Ithaca, N.Y.

Johnsgard, P. A. 1967. Animal behavior. Brown, Dubuque, Iowa.

Kear, J. 1970. The adaptive radiation of parental care in waterfowl, p. 357–392. *In* Social behaviour in birds and mammals. Academic Press, London.

Lofts, B., B. K. Follett, and R. K. Murton. 1970. Temporal changes in the pituitary–gonadal axis. *In* Hormones and the environment. Mem. Soc. Endocrinol. 18:545–575.

Murton, R. K., and J. Kear. The nature and evolution of the photoperiodic control of reproduction in certain wildfowl (Anatidae). J. Reprod. Fertil. Suppl. (In press)

Murton, R. K., K. D. Bagshawe, and B. Lofts. 1969. The circadian basis of specific gonadotropin release in relation to avian spermatogenesis. J. Endocrinol. 45:311–312.

Murton, R. K., B. Lofts, and N. J. Westwood. 1970. The circadian basis of photoperiodically controlled spermatogenesis in the Greenfinch *Chloris chloris*. J. Zool. (London) 161:125–136.

DISCUSSION

A. Oksche

I can add only a few neuroanatomical data to the precise review by Follett and comment mainly on the hypothalamo–hypophysial systems of the passerine species—e.g., White-crowned Sparrow, House Sparrow, and Greenfinch. My remarks will be based on neurohistological and fluorescence-microscopic studies, conducted partly also by Dr. H.-J. Oehmke and other members of our group.

Figure 1 summarizes the present state of our knowledge of the hypothalamo–hypophysial axis in passerine birds. In contrast to the Japanese Quail (see B. K. Follett, this volume), the infundibulum of the passeriforms displays two very distinct neurohemal areas, the anterior and posterior divisions of the median eminence. Widely separate bundles of vessels of the portal vascular system lead from these two areas into the cephalic and caudal lobe of the adenohypophysis (pars distalis). In the avian hypothalamus, anterior (preoptic, supraoptic, suprachiasmatic, and pretuberal) and posterior (tuberal) areas must be distinguished. For a detailed description of the pathways connecting the above-mentioned hypothalamic divisions with the median eminence, see Figure 1 (cf Oksche *et al.*, 1970, 1971, 1972). I would like to emphasize that the specifically stainable (Gomori-positive) connections to the anterior median eminence contain smaller elementary granules than does the neurosecretory pathway to the neural lobe (see legend, Figure 1; cf Oehmke *et al.*, 1969; Calas and Assenmacher, 1970). This smaller type of granule may be associated with some kind of releasing factors or similar hypothalamic agents.

Dr. Follett's presentation was very much focused on the posterior division of the avian hypothalamus, which is called tuber cinereum. Dr. Follett referred to a part of this complex as the "tuberal nucleus." There is no doubt that this part of the tuber is the neuroendocrine effector of light-dependent testicular growth reactions also in passerine birds. This interpretation is supported by lesion work carried out by F. E. Wilson (1967) and Stetson (1969) on different passerine species.

To avoid semantic problems with respect to the terminology of avian tuberal nuclei, I am going to use the term "tuberal nucleus" in the sense of the presentation of Dr. Follett. Follett's tuberal nucleus

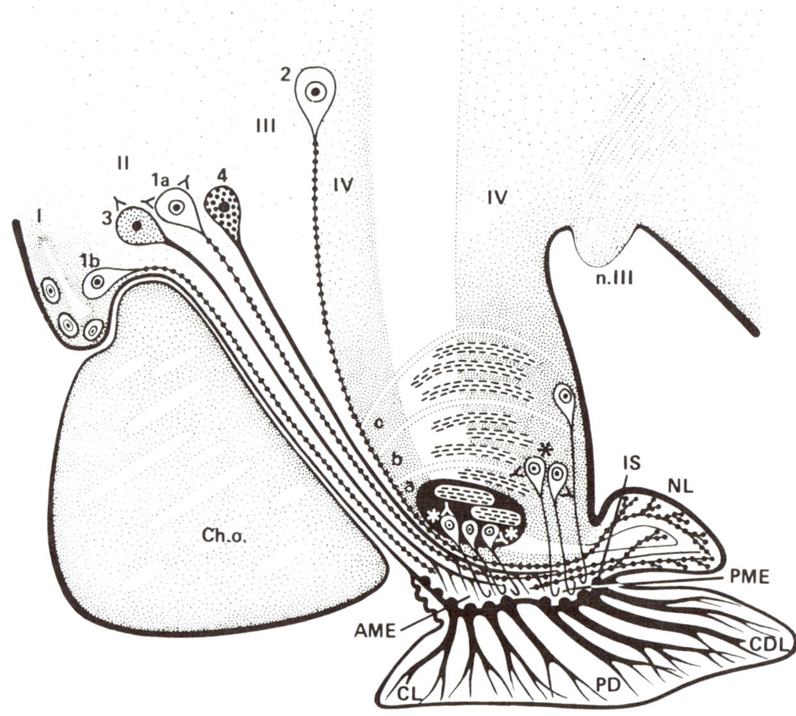

FIGURE 1   Diagrammatic view of the hypothalamo–hypophysial system in passerine birds.
(Drawing by Miss D. Vaihinger, Giessen.) I, preoptic; II, supraoptic and suprachiasmatic;
III, pretuberal; and IV, tuberal regions of the hypothalamus. The neurosecretory pathway
arising in 1 a, b (supraoptic) and 2 (paraventricular) nuclei terminates in the neural lobe (NL).
The selectively stainable (Gomori-positive) material of the anterior median eminence (AME)
is produced by neurons that are located in the preoptic or/and suprachiasmatic region of the
hypothalamus (see 3 and 4). A part of this system (3) enters the rostral portion of the
anterior median eminence. The elementary granules of these axons belong to the 1300–1500
Å class (the granules of the posterior lobe system are approximately 2000 Å in diameter).
Another component (4) of the Gomori-positive eminential system runs along with the
common (posterior lobe) neurosecretory pathway and supplies in a cascade-like manner the
central and caudal portion of the anterior median eminence. At the border of the anterior
and posterior divisions of the median eminence, most of these axons have left the common
Gomori-positive pathway; only a few palisade fibers are found within the posterior median
eminence (PME) and the infunibular stem (1S). The infundibular nucleus (a,b,c,) is only
a part of the tuberal complex; the area occupied by the tuberal complex is marked by
monoamine fluorescence in Falck-Hillarp preparations. a is the basal portion of the
"infundibular nucleus" [called "tuberal nucleus" by Wingstrand (1951) and Sharp and
Follet (1970).] A posterior medial nucleus could not be clearly delineated in Falck-Hillarp
preparations of the passerine hypothalamus although Huber and Crosby have described this
nucleus in their Nissl preparations (*Passer domesticus*). Note the mosaiclike arrangements of
the cell clusters in the infundibular (tuberal) nucleus. They consist of neurons of different
types; most of them are secretory. The basal nuclear area (a) has very abundant axonal

is apparently identical with the n. tuberis of Wingstrand (1951). Further, this nucleus is consistent with the basal infundibular nucleus in Oehmke's work (see Oksche *et al.*, 1972). The nomenclature of the avian hypothalamus derives much from the excellent pioneer work of Huber and Crosby (1929). However, Huber and Crosby described the hypothalamic nuclei of birds topographically, relating them to their anteroposterior or mediolateral position in the brain and, in so doing, introduced terms specific for cell aggregates of the avian hypothalamus. Nevertheless, the tuberal nuclei of birds are anatomical and functional homologs of some mammalian, amphibian, and teleostian formations. For this reason I prefer to use the term "infundibular nucleus" for the apparent avian homolog of the mammalian infundibular (= arcuate) nucleus. This nucleus shows the above-mentioned basal portion and two additional dorsal divisions. The upper (dorsal) border of the infundibular nucleus is approximately the plane connecting the vertex of the optic chiasma with the point where the oculomotor nerve leaves the brain. The tuberal complex of birds covers a relatively larger area and exceeds dorsally the level of the infundibular nucleus; in Falck-Hillarp preparations the entire area of the tuberal complex shows monoamine fluorescence. In mammals, the ventromedial (= principal) and the dorsomedial nucleus are juxtaposed with the paraventricular nucleus. The homologs of the mammalian ventromedial and dorsomedial nuclei can hardly be delineated in avian tuberal complex. The posterior medial nucleus mentioned by Follett protrudes partly into an area dorsad to the infundibular region. I am not quite sure whether his posterior medial nucleus (Japanese Quail) is completely identical with the n. hypothalamicus posterior medialis of Huber and Crosby (House Sparrow). Although taxonomic polynomialism should be avoided by more precise comparative studies, no difficulties of communication will arise if everybody presents a clear topographic diagram of the region analyzed in his experimental work.

The anatomical projection of the "tuberal nucleus" (according to Follett's definition) on the median eminence is represented by the

connections with the anterior and the posterior median eminence. In the posterior part of the infundibulum, axons could be traced to the median eminence also from the more dorsal subdivisions of the infundibular nucleus (b and c). The perikarya of the tuberal neurons are embedded in a neuropile very rich in axo–somatic and axo–dendritic synapses. Point-to-point connections seem to exist between topographically related areas of the infundibular (tuberal) nucleus and the independent portal vascular bundles penetrating from the AME into the cephalic (CL) and from the PME into the caudal lobe (CDL) of the *pars distalis* of the adenohypophysis (PD). N.III, oculomotor nerve; Ch.o., optic chiasm.

tuberoinfundibular tract. The nerve cells forming this pathway extend from the rostral border of the anterior median eminence to the posterior wall of the infundibulum. It should be emphasized that the tuberoinfundibular pathway is visible only in silver impregnations of highest technical quality (cf Oksche *et al.*, 1970). There is no doubt that ependymal and glial cells play an important role in transport phenomena within the median eminence. This capacity has been proven in experiments with tracer substances (cf Kobayashi, 1972). However, I feel that the predominance of neuroglial elements has been overestimated by all authors who failed to demonstrate the very discrete tuberoinfundibular fibers in their silver impregnations. There is no reason to assume that the nerve terminals are not significant primarily for the release of active agents into the portal circulation. A considerable number of tuberoinfundibular fibers are also received by the anterior median eminence. The projection of the rostral portion of the tuberoinfundibular tract is restricted to the anterior median eminence.

An interesting aspect of Follett's paper was his remark that some kind of integration and modulation of incoming external and internal information takes place in the tuberal nucleus, where neural circuits important for neuroendocrine regulations may exist. There is some functional evidence for stimulating and inhibitory pathways. One of our most exciting findings on the avian tuber was the observation that secretory tuberal neurons show numerous axo–somatic and axo–dendritic synapses (Priedkalns and Oksche, 1969). These nerve cells are embedded in a neuropile that is extremely rich in different synaptic structures, some having characteristics of an inhibitory connection. The complete reconstruction of even a small tuberal area from electron micrographs would be of great value for physiological considerations. Some kind of circuit model may show up and lead to a type of analysis as used by Kandel (1970) in his work with invertebrate nervous systems (e.g., *Aplysia*). The organization of the vertebrate brain, however, is more complex and neuroendocrine systems differ from the circuits investigated by Kandel and his associates. Nevertheless, some basic thoughts of Kandel should be adapted for work in avian neuroendocrinology. The stimulation of circumscribed areas of the avian hypothalamus selected on the basis of their anatomical structure might lead to certain characteristic functional (neuroendocrine) effects.

As pointed out earlier, the avian tuber is formed by different nuclear divisions. The recent cytoarchitectonic investigations of the avian tuberal complex by Oehmke (1971) have shown that there are distinct interspecific differences in the subnuclear topography of this region, e.g., the pattern of distribution of tuberal neurons is very different in the wader, *Calidris acuminata*, and in the estrildid finch, *Poephila guttata*. However, there is a common basic principle of cytoarchitectonical organization. The "tuberal nuclei" (and other nuclei of the tuberal complex) consist of mosaically arranged cell clusters that are composed of different types of neurons. There is evidence from fluorescence microscopy that most of these perikarya do not produce biogenic amines. Some of the interposed nerve cells may produce releasing factors. The ultrastructural characteristics of releasing-factor-producing cells and the precise diameter of their granules are not as yet known. Morphometric studies with point-counting methods have shown that clusterlike subunits of tuberal nuclei may increase their activity independently of other adjacent cell groups (cf Oksche *et al.*, 1972).

A similar mosaiclike pattern of different neurons—i.e., classical peptidergic (Gomori-positive) neurosecretory cells, aminergic endings, and neurons probably capable of forming releasing factors—seems to occur in the preoptic portion of the avian hypothalamus. In neuroanatomical terms this hypothalamic region is still a *terra incognita*, although different functional effects have been reported after lesioning the anterior hypothalamus in birds. Priedkalns and Oksche (1969) noted that the neuropile of the pre-and supraoptic region resembles the tuberal neuropile ultrastructurally; it is extremely rich in synaptic connections of different types. We feel that the pre-optic region should be investigated neuroanatomically with great care to explain some differences between the response of the gonadotropic hypothalamo–hypophysio axis in the Mallard (cf Benoit and Assenmacher, 1970) and in the White-crowned Sparrow (Wilson, 1967; Stetson, 1969). From our neuronanatomical charts, one can conclude that the anterior portion of the tuberal complex, which is connected to the median eminence through fine tuberoinfundibular fibers, can be traced very close to the lateral supraoptic and the para-ventricular nuclei. In the preoptic and the suprachiasmatic region of the passerine hypothalamus, the Gomori-positive fiber systems of the anterior median eminence seem to originate. Recently, Clattenburg *et al.* (1972) have reported postcoital changes in the secretory activity of the suprachiasmatic nucleus of the rabbit; the granules of these perikarya fall in the vicinity of 1500 Å. One can assume that a ho-molog of this nucleus also occurs in birds.

A great challenge for the neuroanatomists is the question of the precise position and ultrastructure of the avian deep hypothalamic photoreceptor (cf Benoit and Assenmacher, 1970; Underwood and Menaker, 1970). It is completely open to debate whether the cells responsible for this type of reaction contain some kind of (intra-cytoplasmic ?) array of lamellae characteristic of different types of photosensitive cells. In the light-sensitive sixth abdominal ganglion of the crayfish, *Cambarus virilus*, Hama (1961) discovered neurons with intracytoplasmic stacks of lamellae. If physiologists could pinpoint the light-sensitive region of the avian hypothalamus, efforts should be made to reconstruct this area from electron micrographs.

Are neuroanatomical studies of any value for experimental work in avian neuroendocrinology? In the past they have helped to improve lesion work through brain charts and stereotaxic coordinates (cf Van Tienhoven and Juhász, 1962). Some of the more recent methods of quantitative cytology—e.g., microspectrography for identification of biogenic amines and morphometry for analysis of hypothalamic cell aggregates of different composition and properties—could prove to be of fundamental functional value.

REFERENCES

Benoit, J., and I. Assenmacher [ed.]. 1970. La photorégulation de la reproduction chez les oiseaux et les mammifères. (Colloques internationaux du C.N.R.S., No. 172.) Edition du C.N.R.S., Paris. 588 p.

Calas, A., and I. Assenmacher. 1970. Ultrastructure de l'éminence médiane du canard (*Anas platyrhynchos*). Z. Zellforsch. Mikrosk. Anat. 109:64–82.

Clattenburg, R. E., R. P. Singh, and D. G. Montemurro. 1972. Post-coital ultrastructural changes in neurons of the suprachiasmatic nucleus of the rabbit. Z. Zellforsch. Mikrosk. Anat. Cell Tissue Res. 125:448–459.

Hama, K. 1961. A photoreceptor-like structure in the ventral nerve cord of the crayfish, *Cambarus virilus*. Anat. Rec. 140:329–336.

Huber, G. C., and E. C. Crosby. 1929. The nuclei and fiber paths of the avian diencephalon with consideration of telencephalic and certain mesencephalic centres and connections. J. Comp. Neurol. 48:1–225.

Kandel, E. R. 1970. Nerve cells and behavior. Sci. Am. 223:57–70.

Kobayashi, H. 1972. Median eminence of the hagfish and ependymal absorption in higher vertebrates, p. 67–78. *In* K. M. Knigge, D. E. Scott, and A. Weindl [ed.]. Brain—endocrine interaction. Median eminence: Structure and function (International Symposium, Munich, 1971). Karger, Basel.

Oehmke, H. -J. 1971. Vergleichende neurohistologische Studien am Nucleus infundibularis einiger australischer Vögel. Z. Zellforsch. Mikrosk. Anat. 122:122–138.

Oehmke, H. -J., J. Priedkalns, M. Vaupel-von Harnack, and A. Oksche. 1969.
Fluoreszenz- und elektronenmikroskopische Untersuchungen am
Zwischenhirn-Hypophysensystem von *Passer domesticus.* Z. Zellforsch.
Mikrosk. Anat. 95:109–133.

Oksche, A., H. -J. Oehmke, and D. S. Farner. 1970. Weitere Befunde zur
Struktur und Funktion des Zwischenhirn-Hypophysensystems der Vögel,
p. 261–273. *In* W. Bargmann and B. Scharrer [ed.] Aspects of neuroendocrinology (V International Symposium on Neurosecretion). Springer, Berlin,
Heidelberg, New York.

Oksche, A., H. -J. Oehmke, and D. S. Farner. 1971. Neuro-anatomical problems
of detection and localization of neurones producing neurohormones and
releasers, with special reference to the avian hypothalamo–hypophysial
system. *In* H. Heller and K. Lederis [ed.] Subcellular organization and function
in endocrine tissues. Mem. Soc. Endocrinol. 19:903–908.

Oksche, A., P. Zimmermann, and H. -J. Oehmke. 1972. Morphometric studies
of tubero-eminential systems controlling reproductive functions, p. 142–153.
In K. M. Knigge, D. E. Scott, and A. Weindl [ed.] Brain–endocrine interaction.
Median eminence: Structure and function (International Symposium, Munich,
1971). Karger, Basel.

Priedkalns, J., and A. Oksche. 1969. Ultrastructure of synaptic terminals in
nucleus infundibularis and nucleus supraopticus of *Passer domesticus.* Z.
Zellforsch. Mikrosk. Anat. 98:135–147.

Sharp, P. J., and B. K. Follett. 1970. The adrenergic supply within the avian
hypothalmus, p. 95–103. *In* W. Bargmann and B. Scharrer [ed.] Aspects of
neuroendocrinology (V Int. Symp. Neurosecretion). Springer, Berlin, and
Heidelberg, New York.

Stetson, M. H. 1969. The role of the median eminence in control of photoperiodically induced testicular growth in the White-crowned Sparrow,
Zonotrichia leucophrys gambelii. Z. Zellforsch. Mikrosk. Anat. 93:369–394.

Underwood, H., and M. Menaker. 1970. Photoperiodically significant photoreception in sparrows: Is the retina involved? Science 167:298–301.

Van Tienhoven, A., and L. P. Juhász. 1962. The chicken telencephalon,
diencephalon and mesencephalon in stereotaxic coordinates. J. Comp. Neurol.
118:185–198.

Wilson, F. E. 1967. The tubero-infundibular neuron system: A component of
the photoperiodic control mechanism of the White-crowned Sparrow,
Zonotrichia leucophrys gambelii. Z. Zellforsch. Mikrosk. Anat. 82:1–24.

Wingstrand, K. G. 1951. The structure and development of the avian pituitary.
Lund, Gleerup. p. 3–316.

INFORMAL DISCUSSION

FARNER: Professor Oksche's presentation demonstrates the extent of efforts that are being made to try to close the gap between classical and contemporary neuroanatomy and neuroendocrinology and physiology. You can see that there are still problems, but I think that truly significant progress is being made. I hope that similar progress can be made in bridging the gap between neuroendocrinology and ethology. It is in fact now in progress, but the problems here are much greater than those experienced between neuroendocrinology and classical and contemporary neuroanatomy.

ABEL: I would like to comment on Follett's attempts to localize the steroid feedback centers in the brain. We have been doing freeze-dry autoradiography of tissues in an attempt to localize the sites of estrogen, progesterone, and corticosterone feedback in the brains of the Pekin Duck. With these techniques, we have been unable to demonstrate any significant localization of radiochemical in the median eminence or in any other portion of the hypothalamus. The pituitary gland, however, does incorporate a small but significant amount of each compound, while the limbic system incorporates large amounts of each. Estrogen was localized predominantly in large steroid-binding neurons in the amygdala, whereas corticosterone and progesterone were taken up by neurons in the septum, hippocampus, and hyperstriatum dorsale and ventrale. I think it would be profitable, in light of these data, to implant these steroids stereotaxically into various parts of the limbic system and to determine what effect such implants have on gonadotrophic function.

FOLLETT: We have followed up the steroid implant work by injecting tritiated testosterone into Japanese Quail, and the results tend to agree with your data, showing a greater localization of label in the pituitary gland than in the hypothalamus.

Another point should be made. Specifically, steroid implants directly in the pituitary gland also have a partial degree of feedback.

F. WILSON: I should like to support Follett's comments. It seems clear from available information on both birds and mammals that the anterior pituitary accumulates steroids much more effectively than the hypothalamus. This is, as Doctor Follett indicated, an enigma, and I am not equipped to resolve it.

There are two points I should like to make by way of comment on the very excellent presentations of Doctor Follett and Professor Oksche. First, both have intimated, and I should like to emphasize, that there are multiple opportunities for input of information–both neural and hormonal–to the basal tuberal region. Second, there is the possibility that neurons in the basal tuberal region may be variably sensitive to such inputs. Our data on the Tree Sparrow lead us to speculate that androgen sensitivity of the hypothalamus varies during the annual testicular cycle–which prompts me to suggest that

the negative feedback mechanism may operate about different set points at
different times of the year.

FARNER: I dislike the necessity of confronting this additional complexity, but
I think that all, or almost all, experience with hypothalamic control mechanisms, including thermoregulation, leads us to understand that hypothalamic
set points are variable. Just how they are changed according to physiologic
condition apparently differs with the mechanism with which one is dealing.

HOMMA: Doctor Follett showed in Figure 1 how avian reproduction is controlled by interplay of numerous neural and hormonal factors. In contrast to
the complexity seen at the lower half of the figure, the upper half contained
only four arrows that indicate merely the route of inflow of external signals.
This tells us that there still is a large gap in our knowledge about the upper
half—the control mechanism operated by the central nervous system above
the hypothalamic level—a field that had been monopolized by behaviorists.

In the study of photoperiodic responses, the stimulatory aspects have been
extensively studied, but the mechanism of the inhibitory aspects, such as that
of photorefractoriness, has not been well investigated. In our previous experiments with direct illumination of the brain of *Coturnix*, the results as a whole
agreed well with a hypothesis that retinal participation is not essential for induction of gonadal growth.* A question arose, however, concerning certain
results with female *Coturnix* that had been stimulated by implanted light
source. Removal of the internal light source from the brain did not terminate
egg production, even though the birds were maintained under short-day light
regimen. This suggested, at least in this species, that the inhibitory mechanism invoked by short day is not simply a reversal of the stimulatory
process.

To investigate this problem further, we began a series of experiments in
which blinded and normal birds were exposed to long- and short-day photoperiods under otherwise identical environmental conditions. Similar experiments have been carried out concomitantly at the University of California,
Davis, by Professor W. O. Wilson. The results of these two experiments, done
quite independently, were in good agreement on the point that the retina
must play a significant role in the photoperiodic gonadal responses, probably
as the receptor of inhibitory information when birds were exposed to different photoperiods.† Details of these experiments will be published elsewhere,
but it should be noted that once egg-laying had been initiated by long-day
treatment, blinded hens continued egg production irrespective of the environmental photoperiods, but normal hens readily responded to the environmental
photoperiods.

The next question concerns the character of such retinal inhibition—

---

*Homma, K., and Y. Sakakibara. 1971. Encephalic photoreceptors and their significance in
photoperiodic control of sexual activity in Japanese Quail, p. 333-341. *In* M. Menaker [ed.]
Biochronometry. National Academy of Sciences, Washington, D.C.

†Homma, K., W. O. Wilson, and T. D. Siopes. 1972. Eyes have a role in photoperiod control of sexual activity of *Coturnix*. Science 178:421-423.

whether the signal is neural or chemical or both. The results of an experiment done in our laboratory with rats suggested that the synthesis of certain neuro-hormones, for instance serotonin, is facilitated in the retina under dark conditions. They might mediate the inhibitory signal to the brain.* However, for *Coturnix* we have no evidence that adequately explains the nature of inhibition through the retina.

W. O. WILSON: We have found that, depending on their previous photoperiod, blinded birds may respond differently to light stimuli. With apologies to Konrad Lorenz, we refer to this response as "photoprinting." It may be similar to "imprinting."

FARNER: I would like to make a brief comment on one of Professor Homma's experiments that I had not known about earlier. First, you may recall that he has removed both eyes from Japanese Quail and has found that they respond to photostimulation. This is consistent with what we know about some other species. Now, the interesting new development here is that when he removes blinded birds from long days they continue to perform just as though they were on long days. This raises the whole question as to whether we are dealing with a single component system or a double component system that has both a "turn-on" mechanism that involves long daily photoperiods and a separate "turn-off" mechanism.

There is another set of experiments in the literature that has been largely overlooked, because it was tucked away in a review paper by Professor Albert Wolfson.† He discovered that, if he photostimulated intact birds with long day lengths, then placed them into continuous darkness, and if they maintained long circadian periods of activity, they would continue to maintain gonadal function just as though they were still on long days. This also suggests that there may be also a component that turns off the activity in response to short daily photoperiods.

HAILMAN: I am very intrigued by this extraretinal light perception, because in ethology we face similar problems. My question, I suppose, is directed to anyone who could possibly answer it. Have you used the tool that those who study vision use to solve such problems? Their approach is mainly to look at the action spectrum of the response (i.e., they use a series of stimuli of different wavelengths of equal quantum intensity in order to find out the spectral response). This would tell us what kind of a structure we might look for in the brain. If it has a spectral response, say, like a density distribution of a visual pigment, then we would expect a classical photoreceptor structure. However, it may not have that kind of spectrum at all. We know, for instance, that melanin has a fast photoresponse, that can be recorded electrically. We

---

*Takahashi, M., C. Kyw-sik, and Y. Suzuki. 1971. Light activation of gonadal function restrained by continuous dark treatment in male rats. Endocrinol. Jap. 18(2):195–203.

†Wolfson, A. 1966. Environmental and neuroendocrine regulation of annual gonadal cycles and migratory behavior in birds. Recent Prog. Hormone Res. 12:177–244.

do not know how it is used in various visual systems. Have the relevant studies been done on the action spectrum of this response and, second, are there any indications of melanin or other light-sensitive pigments somewhere in the central nervous system that could be a clue to the site of this receptor?

FARNER: First, I should say that this matter of nonretinal reception was demonstrated almost 35 years ago by Professor Benoit with ducks. These results, although quoted occasionally in the literature, have been largely ignored. Most of the experiments on the action spectra of photoperiodic responses in birds have been unsatisfactory for one or two reasons. First, the incident energy has not been controlled carefully. But, even in the cases in which the incident energy is controlled carefully, one still faces the matter of tissue absorption, since the site of the actual receptor or primary photochemical event is unknown. Actually, in an intact bird, with fairly carefully controlled energy incidence for the wavelengths used, one finds about the same response all the way from green to extreme red, but not far red.

I think that such data are relatively meaningless. More recent experiments have involved the use of optic fibers, and now we have Professor Homma's very ingenious experiments in which he, with a stereotactic apparatus, inserts fluorescent disks into the brain hoping in this way to localize the site of reception, thereby giving the electron microscopist a clue as to the areas that he should examine.

HOMMA: I would like to answer the question by Doctor Hailman as to what pigment is involved in this special brain photoreceptor. This has been a question of general interest for many years. If the pigment is similar to those found in the retina, its extraction should be achieved by the use of those procedures generally accepted in the investigation of retinal photosensitive pigments. However, so far as I am aware, attempts to isolate such pigment from avian brain, especially from the hypothalamus, have failed and have, understandably, therefore not been described in the literature.

Next, a possible role of melanin and closely related compounds may be excluded on the basis of the indirect, but strong, evidence that albino mutants of *Coturnix*, either completely or incompletely devoid of melanin, do respond to environmental photoperiod in almost the same pattern as that of normally pigmented birds.

Porphyrin is one of the compounds that might possibly be involved in encephalic photoreception, since the concentration is relatively high in the central nervous system and since absorption maxima of coproporphyrins range from 620 to 630 nm.* In this connection, high concentration of porphyrin in the Haderian gland of the rat is interesting, but homologous glands in the chicken and quail do not contain a detectable amount of porphyrin when examined with ultraviolet light. In the rat, the implication of porphyrin

---

*Klüver, H. 1944. On naturally occurring porphyrins in the central nervous system. Science 99:482–484.

in extraretinal photoreception in neonatal pups has been questioned in relation to the photodependent changes in the pineal,* but this does not seem to
negate possible participation of porphyrin in encephalic photoreception in
avian species. The weak points of this hypothesis are that radiation sources
that can induce porphyrin to emit visible light are not effective in induction
of gonadal growth and that sites rich in prophyrin are not always correlated
to gonadal activity.

FARNER: What do your experiments with implanted luminescent disks tell us
about the action spectrum of photoperiodically induced gonadal function?

HOMMA: This is a very important question for which I do not have a satisfactory answer. With the implantation technique I have tested, the sources of
available visible light from radioluminescent paints—orange-yellow and blue.
Both were effectively implanted at essentially the same sites in the brain.
However, the orange-yellow disks were effective over greater distances than
the blue disks. Therefore, the generally known action spectrum for environmental light on avian gonadal growth seems to be more a matter of tissue absorption as a function of wave length than a function of the wavelength requirement of the encephalic photoreceptor.

A weak point in the use of radioluminous paints in the study of action
spectra is the broad spectrum of the light emitted from orange-yellow paint,
which is most effective in gonad stimulation.† If we could have a series of
paints, each with a much narrower spectrum, we might obtain more precise
information. At present, it may be safe to say that we can expect encephalic
photoreception throughout the visual spectrum.

ASSENMACHER: I would like to comment briefly on the duck experiments,
that Professor Benoit performed in the late 1930's and with which I was later
associated. Using white light, low-light intensities are more effective, in terms
of gonadotropic response, in intact than in blinded (transection of the optic
nerve) ducks. However, if receiving the same light energy, the hypothalamic
area of the brain is more sensitive as a photoreceptor in this respect than is
the retina. If led directly to the hypothalamus by quartz rods, any wavelength
in the visible spectrum, i.e., between 400 and 800 nm, is active as a gonadotropic stimulus, while monochromatic light acting on the head of intact animals exhibits a stimulating effect only in the orange-red region of the spectrum or, with lower energies, in the bright red (625–650 nm). On the other
hand, when used at very low energies, red light is more active in intact than in
blinded ducks. For the gonadotropic response to light, the retina thus seems
less sensitive than the hypothalamic photoreceptors, and the retinal sensitivity

*Wetterberg, L., A. Yuwiler, R. Ulrich, E. Geller, and R. Wallace. 1970. Harderian gland:
Influence on pineal hydroxyindole-*O*-methyltransferase activity in neonatal rats. Science
170:194-196.

†Homma, K., and Y. Sakakibara. 1971. Encephalic photoreceptors and their significance in
photoperiodic control of sexual activity in Japanese Quail, p. 333-341. *In* M. Menaker [ed.]
Biochronometry. National Academy of Sciences, Washington, D.C.

could be restricted to the red wavelengths; for the visual function of the eye, however, the retinal sensitivity is maximal in the yellow range of the spectrum as it is in mammals.

DYER: I would like to have more information on the pineal, especially in view of the fact that the University of Texas studies—as far as I know—have apparently conclusively shown that the pineal is important in certain circadian rhythms, notably temperature regulation. I would like to know from Doctor Follett what evidence exists that the pineal exerts no control over the brainstem in terms of a photoreceptive response.

FOLLETT: It remains a difficult problem. Many workers have pinealectomized birds at just about every point in the breeding cycle and in no case has anyone produced more than a transitory effect. However, as you stress, the pineal may play some role in regulating circadian periodicities and, since it is believed that one such rhythm is involved in photoperiodic time-measurement, the pineal could have a subtle effect. In an attempt to answer this problem, Doctor Sharp and I pinealectomized a number of sexually immature quail and placed them under photoperiodic schedules designed to locate the phase of the underlying photoinducibility rhythm. We were hoping that pinealectomy might alter the phase of this rhythm but the peak was located in a similar position to that found for sham-operated controls. In brief, nobody has yet come up with anything really involving the pineal as a *major* component in the photoperiodic testicular response.

OKSCHE: I agree, in principle, with Doctor Follett's functional comments on the avian pineal organ. I would like to add only a brief outline of the anatomical situation. In cyclostomes, fishes, amphibia, and lizards the pineal organ is a functioning photoreceptor supplied with sensory cells that resemble retinal cones.* In birds, the data are quite contradictory. The avian pineal organ fails to show electrical responses to illumination. This seems to be in good agreement with the vestigial character of the pinealocyte outer segments. On the other hand, the passerine pineal is rich in nerve cells, synaptoid contacts, and axon bundles that form a tract directed toward the brain. Furthermore, the light-induced synthesis of melatonin in the avian pineal organ does not depend on intact lateral eyes and superior cervical sympathetic ganglia. So the problem of the avian pineal organ is still very enigmatic.† There is no evidence that the avian pineal organ is even partly identical with the deep hypothalamic receptor. Even the neuroendocrine role of the avian pineal organ, which is very rich in 5-HT, is still quite unclear with respect to the biological effects.

---

*Dodt, E., M. Ueck, and A. Oksche. 1971. Relation of structure and function: The pineal organ of lower vertebrates, p. 253-278. *In* V. Kruta [ed.] F. E. Purkyne Centenary Symposium, Prague 1969. Universita Jana Evangelisty Purkyne, Brno.

†Oksche, A., H. Kirschstein, H. Kobayashi, and D. S. Farner. 1972. Electron microscopic and experimental studies of the pineal organ in the White-crowned Sparrow, *Zonotrichia leucophrys gambelii.* Z. Zellforsch. 124:247–274.

The Natural History
of Incubation

RUDOLF DRENT

The subject of this paper is to review the adaptiveness of incubation behavior. Clearly, we are concerned with the totality of parent and egg; each of these components deserves attention. The study of adaptiveness can logically be considered as two questions: (1) What is the survival value of a specific aspect of incubation behavior? (2) How is the behavior organized to contribute to a solution; what external clues elicit or modify it; what internal factors are of influence; and how are these to be understood as a control system?

It must be recognized that little direct evidence bearing on the survival value of particular aspects of parental activity has been derived from field studies, but in many cases insights as to possible functions can be had by referring to the many experimental studies that have been made on artificial incubation. Extensive research aimed at improving the hatching success of eggs incubated artificially [reviewed by Landauer (1967) and Lundy (1969)] has identified four physical factors of overriding importance: turning of the eggs, temperature, humidity, and ventilation (i.e., O_2 and CO_2 levels). Maximal hatch is obtained only when each of these variables is carefully controlled.

THE MECHANICAL ENVIRONMENT

Anyone who has watched an incubating bird for any length of time will have noted periodic interruptions during which the parent rises, peers down at the eggs, and makes a sweeping movement with the bill, arching the neck and drawing the bill back among the eggs toward the belly (a movement reminiscent of the way many birds retrieve eggs that have rolled out of the nest). One result of this egg-poking behavior, particularly noticeable in large clutches, is that the eggs are shifted relative to one another in the nest bowl. This feature impressed de Reaumur, who carried out observations with marked eggs in the nest of a domestic fowl. Over 200 years ago he suggested (de Reaumur, 1751) that the function of this behavior was to promote an even distribution of warmth among the eggs of the clutch. This is a perfectly valid suggestion when we bear in mind the large differences in internal temperature that have been measured between central and peripheral eggs; Huggins (1941) reported an average difference of 5.6 °C in a clutch of the Mallard, *Anas platyrhynchos*, and Mertens (1970) found a similar gradient in a clutch of the Blue Tit, *Parus caeruleus.*

Redistribution of eggs to counteract the temperature gradients existing in the nest cannot be the only function of this behavior, however, since even in modern incubators with virtually identical temperatures throughout, a certain amount of tilting or turning is beneficial to hatching rate (Robertson, 1961b; Kaltofen, 1961). It is generally agreed (Lundy, 1969) that a certain amount of egg movement is especially critical in the first half of the incubation period, and various investigators have emphasized the significance of this behavior in reducing the incidence of premature adhesions involving the extraembryonic membranes. These early adhesions can lead to various difficulties later in development, such as disruption in the uptake of the egg white, or aberrant positioning within the egg that reduces the probability of successful hatching [see reviews by Robertson (1961) and Kaltofen (1961)]. Egg-turning in the wild is doubtless of significance in this way also, but an additional and unexpected effect of egg-turning was discovered by Lind (1961) in the first large-scale attempt to measure variations in egg position in the field.

By marking each of 44 eggs with a series of small spots spaced at 45-deg. intervals around the blunt pole and by recording which of the spots was uppermost during subsequent nest inspections, Lind was able to show that in the Black-tailed Godwit, *Limosa limosa*, there

was a tendency for each egg to assume a stable position as incubation
progressed, i.e., the same mark tended to be uppermost. Lind correlated this finding with the progressive development of weight asymmetry of the egg as the embryo develops, such that the egg, when
placed in a dish of water, will settle in a definite position. Once the
extraembryonic membranes have fused with the inner-eggshell membranes, and egg content thus becomes fixed with respect to the shell,
this asymmetry will be reflected in a restriction in the degree of movement of the egg about its long axis. By imitating egg-turning behavior
by using a thin stick, Lind was able to show that the eggs did not necessarily revolve about their long axes as might intuitively be supposed. On the contrary, egg-turning behavior released the individual
egg momentarily from the constraints posed by friction from the surrounding eggs, so that it was free to adopt the position of equilibrium
dictated by its weight asymmetry. It, of course, is essential in studying this phenomenon to consider each egg individually. It is because
the data were pooled indiscriminantly that the basic asymmetry of
the frequency distribution of the individual eggs escaped Tinbergen
(1948, 1953) in his earlier investigation of this problem. When new
measurements were collected for the Herring Gull (Drent, 1970), the
situation was found to be altogether comparable to that in the
Godwit, as will be clear from examination of the figures shown here.

The method of measurement is illustrated in Figure 1: Only movements around the long axis of the egg are considered, and one obtains
a frequency distribution showing where a randomly chosen part of
the shell (marked by a thin ink line) is positioned with respect to the
horizon in the course of incubation. End-on views (i.e., viewed from
the blunt pole) are presented for three individual eggs in Figure 2;
Compared to the position of these same eggs when placed in a water
dish, the restriction of movement of the egg in the nest can be considered a consequence of its weight asymmetry. This hypothesis was
checked by observing how the position of the egg was altered by the
actions of the parent Herring Gull. A tunnel was excavated beneath
the nest and, by replacing the nest floor with a transparent perspex
bowl of appropriate shape, observations could be done with the aid
of a flashlight from beneath without disturbing incubation. Movements carried out when the bird was settling on the eggs to incubate
were found to influence egg position as well as the egg-poking behavior already described. In all cases, the net effect of these actions was to leave the position unchanged if the egg was already in
the equilibrium position and to return it to that position if it hap-

pened to deviate. The bird's behavior therefore could be considered corrective.

Holcomb (1969), in his study of egg-turning in 12 passerine species, used such huge and conspicuous markings to identify the eggs that his work, as he realized, in fact constituted an experiment on the influence of foreign color patches (red and orange fingernail polish) on turning behavior. In the 9 species for which enough observations were collected to allow statistical analysis, it appeared that exposure to the foreign color blotches more often led the bird to make adjustments of egg position than when these blotches were not visible (i.e., turned down in the nest). Visual stimulation when the bird rises at the nest, or returns from an absence, thus appears to be one of the factors eliciting egg-turning behavior, but we cannot conclude anything about the influence of the parent on the position of the eggs in the course of undisturbed incubation from these observations.

What significance might the progressive restriction in egg position

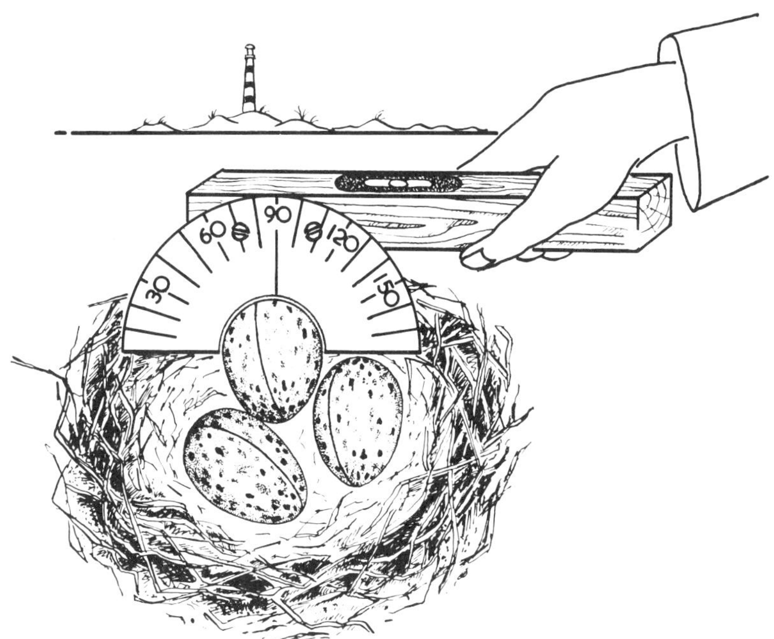

FIGURE 1 Method of measuring egg position in the nest of the Herring Gull, *Larus argentatus*. During nest inspections, a protractor notched to receive the egg was placed over each of the eggs in turn, and the position of an ink reference mark on the shell noted to the nearest 10 degree in relation to the horizon. (Note that the protractor is attached to a level.)

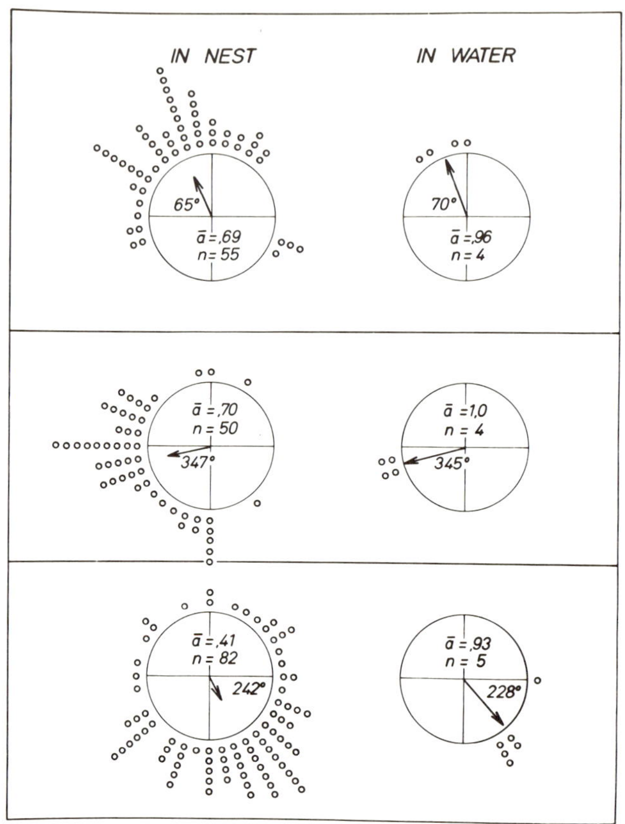

FIGURE 2. Position measurements on three individual eggs of the
Herring Gull. The eggs are viewed from the blunt pole, and each dot
represents one observation on the whereabouts of an ink line passing
through the poles: on the left, in the nest during incubation; on the
right, the same egg when placed in a water dish. The arrow shows
mean position (arrow length is proportional to reliability of the
mean). From Drent (1970).

that we have seen in both the Godwit and the Herring Gull (Figure 3)
have for the developing embryo? If the shell is peeled away from that
part of the egg generally lying uppermost in the nest (or floating uppermost in a water dish), the pattern shown in Figure 4 is revealed.
At the earliest stage, the yolk mass is still free to revolve within the
shell, so the embryo is always uppermost. As pointed out by
Ragosina (1963), the lighter portion of the yolk bears the embryo,
which is oriented as shown in the illustration in the majority of cases.

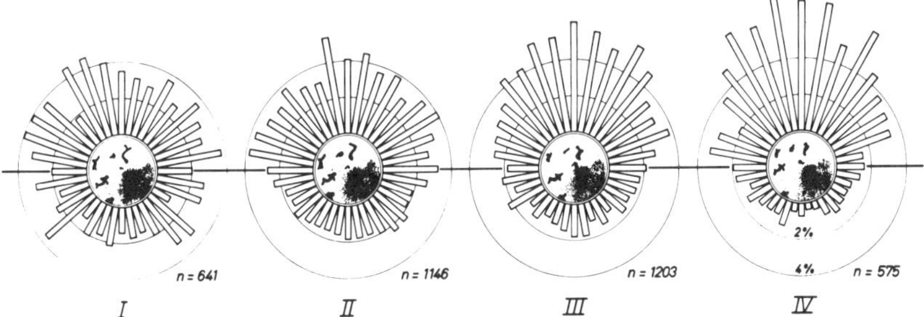

FIGURE 3 Position of the Herring Gull egg in the nest during incubation, measured as indicated in Figure 1. The egg is viewed from the blunt pole; the heavy line represents the horizon. Shown in the form of a percentage frequency diagram is the position of that 10-degree sector of each egg that lies uppermost when the egg is in its mean position. Incubation has been subdivided into four periods. The figure is based on 3,565 observations on 51 eggs. From Drent (1970).

The head faces away from the observer—described as von Baer's rule—if the blunt pole is held to the left; Fargeix (1963) has recently reviewed the data for five species. Eventually the extraembryonic membranes and the shell membranes fuse; by about age 14–15 days. embryo position is fixed in relation to the shell in the Herring Gull. Generalizing from the few species investigated so far, fixation takes place approximately midway through incubation. Due to subsequent

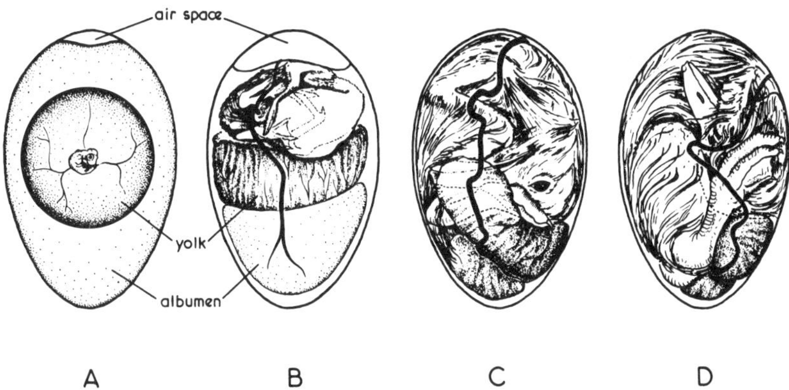

FIGURE 4 Position of the Herring Gull embryo within the egg, as seen when that part of the shell lying uppermost in the nest is chipped away. Approximate ages: A, 5 days; B, 16–17 days; C, 21–22 days; D, 26 days (the egg hatches on day 27). In B, the head of the embryo is visualized as being transparent, in order to show the position of the limbs. From Drent (1970).

changes in the position of the embryo, the center of gravity may change in some eggs after this time. The extent to which a repositioning of the egg can lead to active or passive adjustments in embryo position is a point currently under investigation (Drent, 1970; Oppenheim, 1970, 1972; Kovach, 1970).

During the whole of incubation, then, there is a predictable relation between the orientation of the embryo and the horizontal plane. At first, the contents revolve to bring the embryo uppermost; later, the frequent egg-turning behavior of the parent allows the egg to assume its equilibrium position. This fixed orientation has immediate relevance to the problems faced in escaping from the shell during hatching and, perhaps, also for the attainment of the prehatching position.

Hatching itself is a relatively quick operation, but successful emergence depends in large measure on attainment of the correct starting position [see the discussion of malpositions in Landauer (1967, p. 205), even minor aberrations can greatly impair the probability of successful hatch]; this is the end result of a train of events commencing many days earlier. The prehatching position illustrated for the Herring Gull in Figure 4D—where the tarsal joints are located in the pointed pole of the egg, with the toes resting against the head, and the head is withdrawn and tucked beneath the right wing, with the bill directed toward the air space in the blunt pole—is in fact typical of birds in general [the essential features are identical in *Fulica atra, Limosa limosa, Larus ridibundus, Ardea cinerea, Somateria mollissima, Aptenodytes forsteri*, (Drent, 1970), *Gallus domesticus* (Hamburger and Oppenheim, 1967) *Anas platyrhynchos* (Oppenheim, 1970) (Figure 5), and *Coturnix coturnix* (Vince, 1969)]. The single exception to date, the megapode *Alectura lathami*, in keeping with the aberrant positioning of the embryo, also shows a peculiar mode of hatching (Baltin, 1969) that will be commented on later.

The standard prehatching position is attained from the preceding head-between-legs stance (Figure 4C) by a gradual withdrawal of the head from the yolk mass followed by a pivoting movement of the head that draws the bill tip up between the body and the right wing. Hamburger and Oppenheim (1967) term this latter process "tucking," and have studied it in great detail in the domestic fowl, using sophisticated observational techniques. [This approach has since been extended to nine other species (Oppenheim, 1972).] It is during tucking that a new type of coordinated muscular activity commences, a pattern that occurs again during actual hatching. In this sense, tuck-

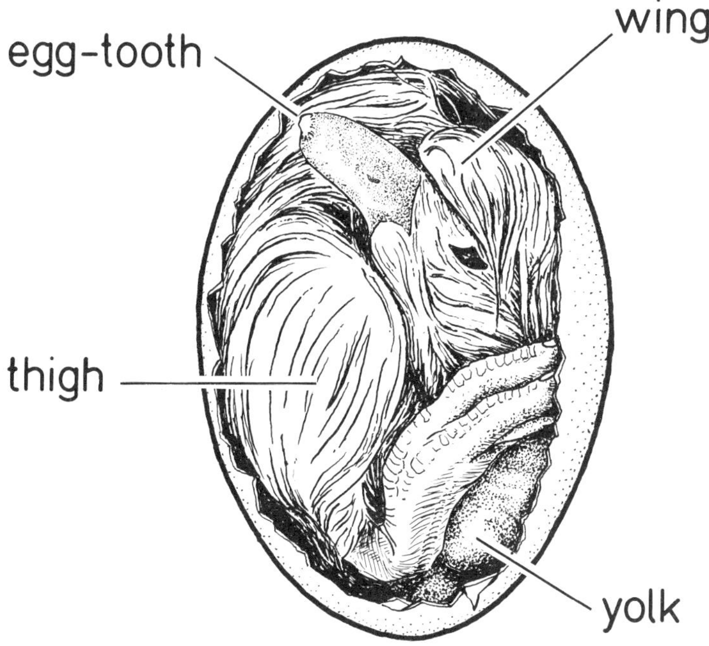

FIGURE 5 Prehatching position of embryo of the Mallard, *Anas platyrhynchos*, as seen when that part of the shell lying uppermost in the nest is chipped away. Note the close similarity with Figure 4 D. Drawn from a photograph in Oppenheim (1970), courtesy of Ballière, Tindall, and Cassell.

ing ushers in a new phase of embryonic activity, and its implications for correct attainment of the position that forms the starting point for escape from the shell have been stressed by Hamburger and Oppenheim. Penetration of the membranes separating the bill from the air chamber is the next significant event and, in general, this coincides with the commencement of lung ventilation, which, in the remaining interval, gradually will usurp the respiratory role of the chorioallantois. The available data, showing when lung ventilation starts in relation to the total incubation period, are summarized in Table 1. In some cases, the first respiratory movements may precede membrane penetration. Pipping, the formation of the first cracks in the shell, generally follows, but in a few species—the Herring Gull is one of these—the first breaks on the shell surface may occur before the bill has worn its way through the membranes. Vince (1969), has designated this as "false pip." Observations on the external surface of the egg, made during periodic inspection visits to the nest, showed

TABLE 1 Chronology of Hatching in Selected Species

Species	Incubation Period (days)	Initiation, in hours, before Emergence			Reference
		Breathing	Pipping	Rotation	
Domestic fowl	20–21	28–35	12.5	0.8	Vince, 1970; Oppenheim, 1972
Colinus virginianus	23	55	44	0.6	Vince, 1970; Oppenheim, 1972
Coturnix coturnix	16–17	40	10	—	Vince, 1970
Uria aalge	32	35	22	4.8	Tschanz, 1968

(Figure 6) that a series of bumps started to appear about 64 hours
before hatch, preceding the formation of the first hole by some 35
hours. The observations of Oppenheim (1972) on specially prepared
eggs of the Laughing Gull, *Larus atricilla*, placed in a viewing incubator, make clear that this track of bumps in fact reveals the path of the
bill-tip as the head is withdrawn during tucking (compare Figure 4C
and Figure 6A). Oppenheim relates this to the relatively long bill and
fragile eggshell and has noted similar scoring of the shell during tucking in the pigeon; from Lind's (1961) description, this preliminary
breakage also occurs in the Black-tailed Godwit. In the majority of
species, tucking leaves no trace on the shell and the bill tip punctures
the membranes and gains access to the air cell with the egg still intact.
[See for example, the very complete description given for the domestic fowl by Robertson (1961a), who traced head movements on the
basis of x-ray photography.]

Hatching itself is usually inaugurated quite suddenly (the stimuli

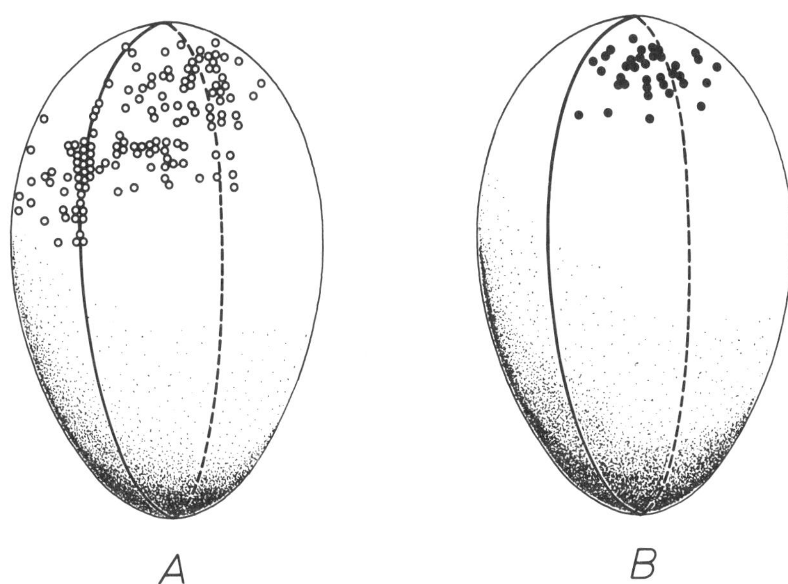

FIGURE 6   Surface views of the egg of the Herring Gull during hatching. A shows the progress of cracking in relation to the first crack made (64 hours before hatch), which is plotted
along the solid line. Each dot represents an observation on the position of the bill-tip. The
broken line shows that portion of the egg that, on the average, lies uppermost in the nest.
B shows the site of the first hole (28 hours before hatch); each dot represents the measurement for a single egg (36 in all). From Drent (1970).

that bring this about are still far from clear) (see Vince, 1970) when vigorous thrusts of the beak against the shell are accompanied by alternating treading movements of the legs that impart a rotary component to the movements of the embryo. As a result, the hole broken in the shell is gradually extended in a circular fashion, the embryo rotating counterclockwise as viewed from the blunt pole (Figure 7). The degree of rotation varies greatly from species to species, depending in large measure on the strength of the shell and shell membranes: Tschanz (1968) noted that, if the embryo *Uria aalge* had not managed to sever the cap neatly in its first bout, it would proceed to rotate through a second revolution and even a third; Oppenheim (1972) experimentally induced additional rotation in *Larus atricilla* by taping that portion of the shell already cut. Eventually, the cap is lifted off, lid-like, by vigorous heaving movements of the shoulders coupled with extension of the legs, and the embryo emerges. In the Herring Gull, more than 24 hours elapse between the making of the first hole and emergence, but the time devoted to the climax movements (to employ Oppenheim's term) described here is only a fraction of that time. How short the actual hatching process is, in relation to the long period elapsing while the embryo is in the prehatching position, is shown in Table 1.

Although Wetherbee and Bartlett (1962) indicate that hatching in

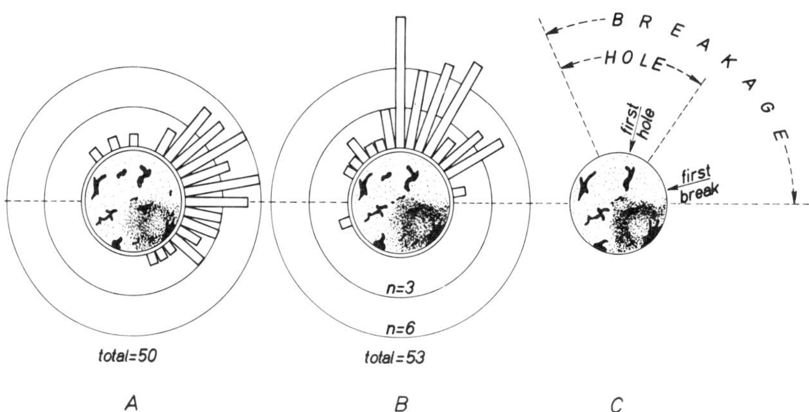

FIGURE 7   The egg of the Herring Gull viewed from the blunt pole during hatching (the horizon is indicated by the dotted line). A, first crack in shell; B, first hole; C, summary for 53 eggs, showing means from A and B as well as the mean arc of breakage (115 deg) and span of the hole (60 deg) attained before the cap is broken off and hatching accomplished. From Drent (1970).

the waders (*Philohela minor* and *Catoptrophorus semipalmatus*) that they investigated was rather unusual in that the blunt pole of the egg appeared to be torn off by convulsing movements of the embryo after only limited slit formation by the working of the bill, it seems from the brief description given that the general pattern shows great similarity to other species that tend to rotate little before emergence, namely, various gulls and *Limosa limosa.* Undoubtedly highly aberrant, however, is the mode of hatching in the megapodes—a remarkable group of galliforms of the Australasian region known as moundbuilders on account of the incubation practices followed by certain species. In all species, the egg is buried and is incubated by heat supplied by either the sun, fermentation, volcanic heat, or a combination of these (Frith, 1962). Hatching has been described in *Leipoa ocellata* (Frith, 1962) and *Alectura lathami* (Baltin, 1969). Prior to hatching, the embryo remains in the head-between-legs position, and the shell is burst by the combined action of legs and wings, without the help of the bill, which lacks an egg-tooth at this stage (in earlier embryonic life one is present). It is possible that this unique method of shattering the egg is related to the circumstances, wherein the egg stands vertically in a relatively densely packed mound.

Returning now to the question of relevance of egg-turning behavior for the hatching process, it is noteworthy that in all cases examined thus far, the position that the embryo assumes in the egg, together with the weight asymmetry of the egg, result in the first hole formed being uppermost. It is tempting to speculate that this has functional significance: Tschanz (1968), for example, who described this relation in *Uria aalge*, suggests that this might be important in assuring the fetus an adequate supply of oxygen, since the ledges on which the Guillemots nest are often extremely slimy at hatching time and upside-down eggs might be in danger of suffocating. It is also possible that there is less danger of the eggshell being crushed when the hole is uppermost and obvious to the parent: Birds incubating hatching eggs settle in an entirely different stance and appear to apply less pressure to the eggs. In any case, a certain amount of mortality occurs at this stage by virtue of eggs being flattened by the parent (personal observations on Herring Gulls). Again, it might be important for the embryo to maintain a certain position with respect to gravity, on account of requirements for adequate lung breathing, for example. Clearly, it would be rewarding to carry out research on the ways in which the position of the egg influences the assumption of the correct prehatching position of the embryo and how it influences the process of hatching itself. It is suggestive that, when air breathing has begun and the embryo is capable of vocalization, one of the stimuli eliciting a type of cheeping known as the distress call, is to turn the egg from the resting position. Würdinger (1970 and in preparation) experimented with *Anser indicus*: When the pipped egg was turned upside down, the fetus started to call; the effect on the parent was to return to the nest immediately followed by egg-turning behavior, after which the parent settled to incubate. Recordings of this distress call, when played back to parents sitting quietly, caused them to rise at the nest and engage in egg-poking.

So far, the hatching process has been discussed as an interaction between the behavior of the parent and the needs of the embryo. Recently, however, it has become clear that there is in addition, a communication among the embryos of a clutch, effecting a synchronization of hatch in certain species.

Many years ago, Heinroth (1938) pointed out that in natural circumstances the entire clutch (11–13 eggs) of the Mallard tends to hatch within 2 hours [more recently, Bjärvall (1967) has indicated a range from 3 to 8 hours], an astonishing degree of synchronization when one considers the many factors contributing to variations in length of incubation. Of course, this synchronization would have adaptive significance in species such as the Mallard in which the mother leads the young from the nest together, not to return, for late-hatchers would thereby be left behind (personal observations on Mallard, Shelduck, and Eider). Curiously, Prince *et al.* (1970) found that the range of emergence times of the eggs of a clutch of the Mallard, when incubated artificially, increased from 6 to 16 hours going from a semiwild starting population to the incubator-raised F_2 generation, a finding consistent with the hypothesis that in nature some selection pressure tends to stabilize incubation period. Beyond this, however, an active synchronizing process is involved. Faust (1960) observed the incubation habits of the Common Rhea (*Rhea americana*) in captivity. In this species, only the male incubates; the female continues to add eggs to the nest after incubation has started. Remarkably, despite differences in the age of the eggs of up to 9 (commonly) and 12 (exceptionally) days, hatching was highly synchronous; the entire clutch emerged within 2–3 hours. During this study, incubation period (time of laying to time of hatch) was found to vary through extremes from 27 to 41 days. This was strong evidence of an active synchronization process, and Faust suggested that acoustic signals exchanged between the members of the clutch prior to the start of hatching might be the synchronizing mechanism, but he was unable to follow up this lead. It was Vince, who, in a series of elegant experiments, built a very complete picture of the nature of the synchronizing influence, based on the investigation of a variety of galliforms. The summary that follows is based largely on her recent review (Vince, 1969).

Some time after respiratory movements first begin, there is a period during which a peculiar sound, termed clicking, accompanies breathing in all species investigated so far. This sound is audible when the egg is held close to the ear and, presumably, is audible from one egg to another when they are in contact. Since synchronization does not occur when the eggs do not touch, and acceleration of hatching can be obtained when eggs are subjected to stimulation by sound or vibration at about the rate of naturally occurring clicks, it seems clear that clicking plays an important part in this phenomenon. Figure 8 shows the time course of events in the synchronized hatch of two clutches of the Bobwhite. It will be noted that the pip–click interval is the variable part of the scheme, which would suggest that loud clicking in one egg might trigger clicking in another, i.e., act as an accelerator (note that even a 24-hour gap is bridged in this way), but once clicking is under way, a chain of events is activated that culminates in hatching after a given period of time. A closer look at what adjustments in timing are being brought about is afforded in Figure 9, showing evidence of both acceleration as well as retardation effects of proximity of other embryos. Provisionally, a low-frequency sound produced in the "silent" phase of respiration before clicking commences has been ascribed a retarding function; that is, the commencement of clicking in advanced embryos will be somewhat retarded as long as other embryos are producing the low-frequency sound. Quantitatively, acceleration of the late embryos is the more important phenomenon and provides a very neat system for achieving a close synchrony in time of hatch within the clutch—a fine adjustment that is superimposed on other factors (such as the total amount of heat received by each of the eggs, differences in genetic constitution, and length of time elapsing between laying and commencement of incubation).

Summing up, synchronization of hatch that involves an active adjustment in the timing of individual eggs as a result of mutual vocal stimulation in the hours preceding has been demonstrated by incubator trials in waterfowl [*Anas discors* and *Aythya americana*, Fisher (1966); *Anser indicus*, Würdinger (1970 and in preparation)], one

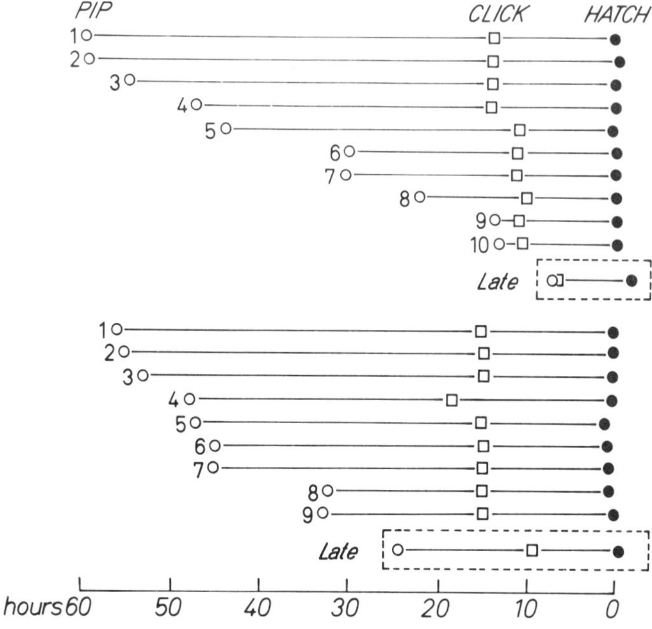

FIGURE 8 Chronology of hatch of individual eggs (numbered in order of pipping) in two clutches of the Bobwhite, *Colinus virginianus*, plotted in relation to mean time of hatch of the entire clutch (= 0 hours). Open circles show time of first pip; squares, time of first click; and solid circles, time of emergence. The eggs of each clutch were incubated in contact to one another. In each clutch, one egg was placed in the incubator 24 hours later than the rest (marked "late"), but nevertheless hatched at about the same time. Note the shortening of the pip–click interval in late-pipping eggs. From Vince (1969), courtesy of Cambridge University Press.

ratite [*Rhea americana*, Faust (1960)], and several gallinaceous birds [*Colinus virginianus, Coturnix coturnix*, and, possibly the domestic fowl; see Vince (1970)]. In all of these species, the brood leaves the nest simultaneously, shortly after hatching. In certain waders [*Limosa limosa*, Lind (1961), *Vanellus malarbaricus*, Jayakar and Spurway (1965)] and in *Fulica americana* (Fisher, 1966), there are indications of a similar process. The last eggs to pip appear to be accelerated by stimulation from the earlier eggs, in that the time they require to emerge is shorter; despite this, however, they do not manage to hatch simultaneously with their siblings. That this speeding-up effect nevertheless has survival value has been stressed by Lind. The suggestion that the Skylark, *Alauda arvensis*, also shows synchronization of this active type (Vince, 1969) requires confirmation by experiment, since consultation of the original source (Delius, 1963) indicates that the range of hatching times reported (4–15 hours) is consistent with the observation that incubation is steady only after completion of the clutch.

THE GASEOUS ENVIRONMENT

Although optimal levels for water vapor (generally measured as relative humidity), oxygen, and carbon dioxide can be accurately specified for conditions of artificial incubation (see summary in

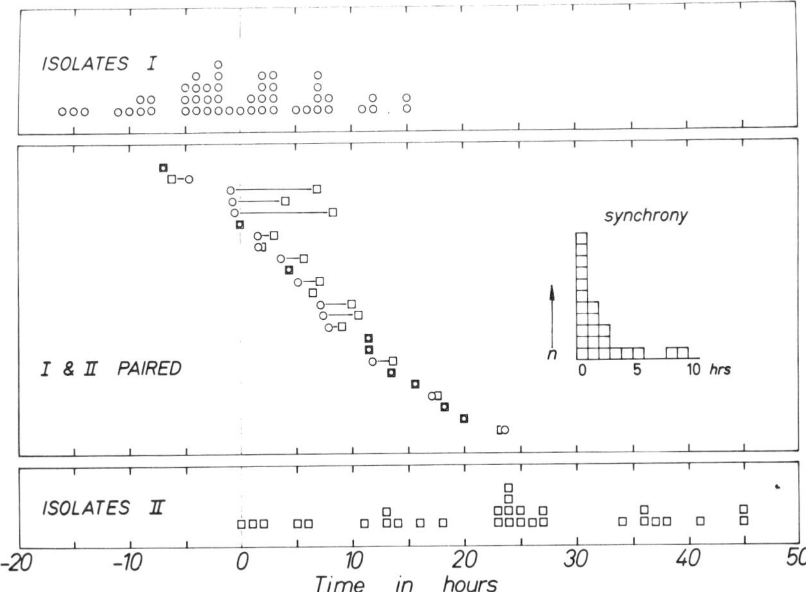

FIGURE 9   Effect of contact between eggs of the Bobwhite on hatching time. Circles show hatching times of individual eggs where incubation started at time I; squares give hatching times of eggs placed in the incubator 24 hours later (time II). The point of the experiment is to show an effect on hatching time when I eggs and II eggs are incubated in pairs, the eggs touching (starting 48 hours before the expected hatching time of the group I eggs), as compared to I eggs and II eggs incubated in isolation. (As expected, II isolates hatch about 24 hours later than I isolates.) All times are plotted with respect to mean hatching time of the I isolates = 0 hours. For paired eggs, hatching times reflect a compromise: Group I eggs in the pair hatch on average 7 hours *later* than their controls and group II eggs in the pair 14 hours *earlier* than their controls. Acceleration of late eggs is thus quantitatively stronger than retardation of advanced eggs. Inset graph shows how closely members of a pair synchronized their hatch. Adapted from Vince (1968), by courtesy of Ballière, Tindall, and Cassell.

Lundy, 1969), we know little of conditions in the nest during natural incubation and absolutely nothing of a possible parental influence on these factors. Several attempts have been made to measure the humidity under the sitting domestic fowl (Burke, 1925; Chattock, 1925; Koch and Steinke, 1944; Kaltofen, 1961), but the more sophisticated equipment now available should be applied to this problem. Indirectly, information on humidity can be obtained from study of weight loss of the eggs: Since the gaseous exchange of the embryo involves only a negligible change in weight, this measure is in fact a reliable indication of water loss from the egg (see discussion in Drent, 1970). Data should be reported as grams lost per day and should include measurement of fresh-egg weight and, preferably also, egg volume, from which the surface area of the egg can be computed. Weight loss per unit surface area per unit time is the measure that would reflect humidity conditions most accurately, although the effect of temperature must also be considered as well as characteristics of the shell. The data available are crude, but at least allow the specification of the normal range in water loss for an egg of given weight (Figure 10). The possibility that in a given study low hatching success is related to humidity conditions could therefore in a preliminary way be checked by collecting data on water loss in the period of linear weight loss before lung ventilation sets in and by comparing this to what is known from previous studies. The tradition of reporting results in terms of percent weight loss during the whole of incubation should be abandoned. Because of this tradition, many investigators have failed to include the basic information on egg weights that is essential if the data are to have a comparative value.

Measurements of carbon dioxide levels in the nest of incubating fowls have been reported by Lamson and Edmond (1914) and Burke (1925) and, as might be expected, are found to lie below the concentrations that have been demonstrated to depress hatchability. There are many shortcomings in the experimental data on carbon dioxide concentrations in relation to hatching success, however, as is clear from the critical discussion of Lundy (1969).

Recently, Baltin (1969) has reported oxygen and carbon dioxide levels in the nesting mounds of *Alectura lethami*, one of the megapodes. Here, as a consequence of the fermentation of the vegetable content of the mound, the principal heat source for incubation in this species, carbon dioxide levels, as measured near the developing eggs, were extremely high (often above 10 percent), whereas oxy-

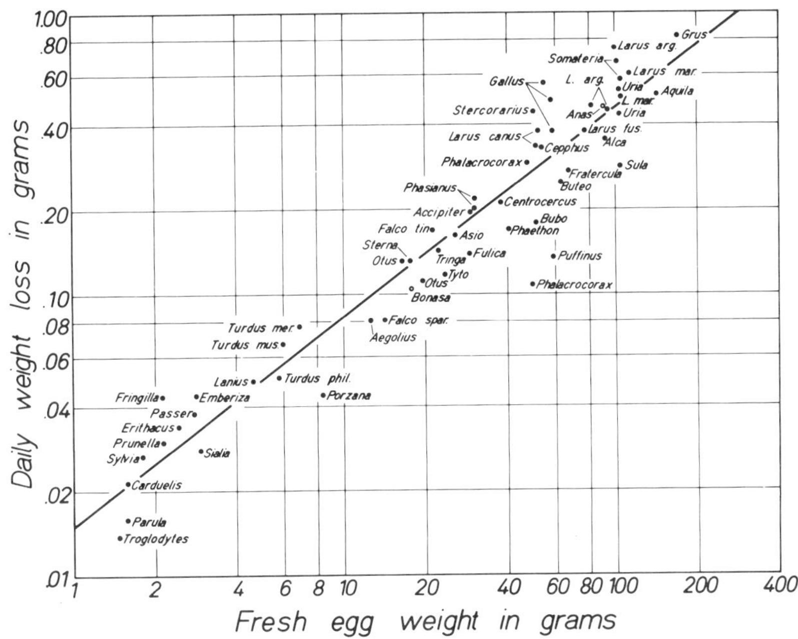

FIGURE 10   Relation between daily weight loss during natural incubation (in grams per egg, ordinate) and initial egg weight (abscissa) in 57 studies on 46 species of birds. The open circles for *Anas* and *Bonasa* are incubator values and have not been included in the calculation of the regression line, which is $Y = 0.0150\,X^{0.742}$ where Y = daily weight loss in grams per egg and X = fresh egg weight, also in grams. From Drent (1970).

gen levels were low (in extreme cases as low as 7–8 percent). Baltin suggests that the mound-tending behavior of the male parent is aimed specifically at counteracting these tendencies toward oxygen deficit and carbon dioxide accumulation, but whether in fact the digging activities of the male are influenced by gas concentrations in some way, or whether gas concentrations are varied as a concomitant of digging aimed at regulating the temperature of the mounds, cannot be decided on the basis of the observations of Baltin. Direct experiments will be required to show whether in this, or indeed in any, species there is a parental control of gas concentrations in the nest as such.

In summary, our knowledge of the gaseous environment of natural incubation is fragmentary to say the least. Progress will depend on sophisticated experimentation, taking care to vary only the parameter under consideration or, alternatively, working with adequate controls. All too often past experiments have involved simultaneous alteration of a host of factors, and many of the conclusions drawn appear premature.

THE THERMAL ENVIRONMENT

Representative measurements of the temperature at which developing embryos are held during natural incubation are presented in Figure 11. The empirically based optimal temperature for the development of eggs of the domestic fowl in modern incubators is also shown and is identical to the internal egg temperature during natural

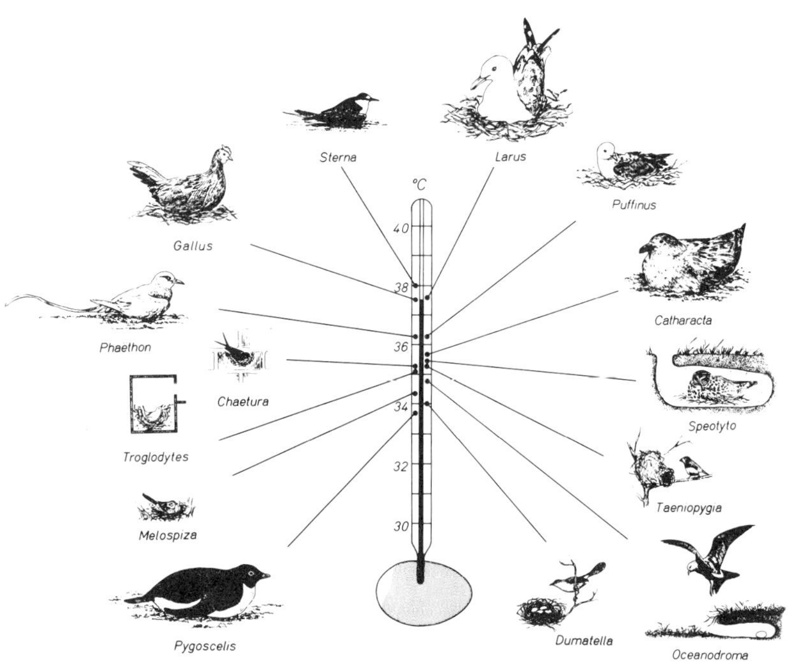

FIGURE 11   Internal egg temperature, measured during undisturbed incubation by telemetry: *Pygoscelis adéliae* (Eklund and Charlton, 1959), *Melospiza melodia* and *Chaetura pelagica* (Huggins, 1941), *Troglodytes aëdon* (Kendeigh, 1936), *Phaëthon rubricauda* (Howell and Bartholomew, 1962a), domestic fowl (Süchting in Groebbels, 1937), *Sterna fuscata* (Howell and Bartholomew, 1962b), *Larus argentatus* (Drent, 1970), *Puffinus pacificus* (Howell and Bartholomew, 1961), *Catharacta skua* [Stonehouse (1956); Spellerberg (1969) reports 36.4 deg for *Catharacta maccormicki*], *Speotyto cunicularia* (Howell, 1964), *Taeniopygia castanotis* (El-Wailly, 1966), *Oceanodroma leucorhoa* (Drent, unpublished), *Dumatella carolinensis* (Huggins, 1941).

incubation in that species. Values from other species cluster close to this level, an expression of the dominance of the body temperature of the sitting bird over the influence of the environment (it should be realized that an extreme range of external conditions is represented in the figure), and of the fact that body temperature within the class Aves is rather uniform. This is not to say that the incubation temperature for eggs of the domestic fowl is to be considered a universal value for all birds; the work reviewed by Rolnik (1970) shows how necessary it is to establish the optima for each species individually. There is no doubt that passerine embryos generally experience lower temperatures than do those of, say, the Herring Gull.

In selecting the data given in Figure 11, only measurements made internally on natural eggs by means of telemetry during undisturbed incubation have been considered, a point deserving emphasis. Huggins (1941), for example, devoted considerable effort toward determining internal egg temperatures typical of natural incubation, but in an effort to include a wide spectrum of species did not generally continue records at any one nest for a long enough time to obtain values indicative of the equilibrium conditions of undisturbed incubation. For several species, however, he obtained records for a span of 6 hours or more, and these values have been entered in the figure, except for the House Wren, where Huggins' work has been superceded by Kendeigh's (1963) more comprehensive data.

Before examining how the parent bird controls or regulates egg temperature, it is useful to consider the situations that the embryo may face when egg temperature is altered. The pattern for the domestic fowl, which is the best documented (Lundy, 1969), is indicative. The optimal temperature for incubation is 37–38 °C, and percentage hatch falls rapidly as temperatures deviate from this. No eggs at all survive continuous exposure to temperature in excess of 40.5 °C, or below 35 °C. The significance of short-term deviations beyond these limits depends to a great extent on age, but overheating of the egg is a real danger at any time; lethal internal temperature varies from 42.2 to 48.3 °C, depending on age. When egg temperature is lowered, difficulties also arise. If the egg is maintained for any length of time in the range between the optimum and the point of no development—the "physiological zero temperature," which has been determined as approximately 25–27 °C—various anomalies such as disproportionate growth of the heart may occur. Below physiological zero, there is a wide range of relative safety, however, at least for moderate periods of exposure. This region of suspended develop-

ment is bordered by temperatures slightly below O °C, which lead to
the formation of ice crystals in the egg that cause irreversible damage. Summing up, we can expect that, in general, the range for optimal development is narrow; death through overheating is uncom-
fortably close at all times; a slight fall in internal temperature, if long
maintained, will lead to abnormal development; and an internal egg
temperature between physiological zero (no development) and freezing can be tolerated for prolonged periods, the most important limitation being that tolerance declines with age. Bearing these constraints in mind, we will now examine the time course of temperatures in the nest of the Herring Gull as a first step toward understanding the role of the parent.

A summary of temperature changes in the course of incubation is
presented for the Herring Gull (*Larus argentatus*) in Figure 12. Tem-

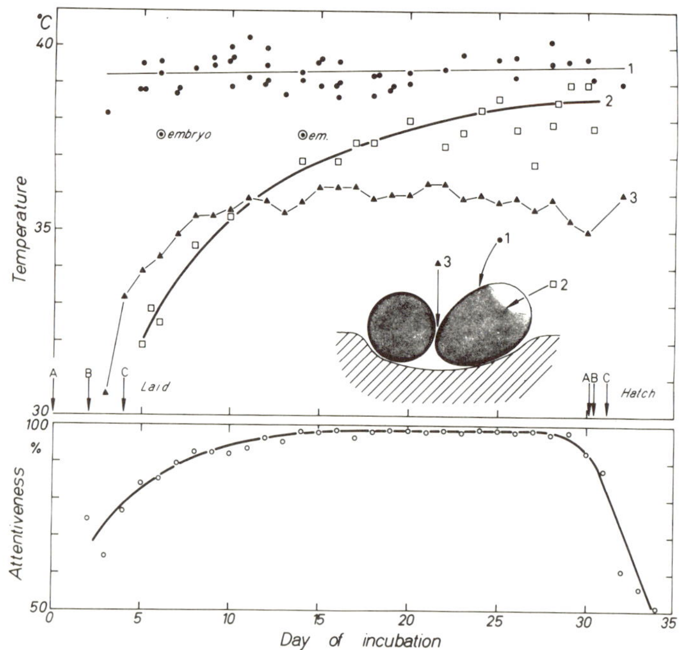

FIGURE 12 Nest and egg temperatures, measured during natural incubation
in the Herring Gull, *Larus argentatus*, (above), compared to constancy of incubation (= attentiveness, below). Sites of temperature measurement indicated
in diagram; in addition, probes were inserted adjacent to the embryo on day 6
and 14. The average time scale for laying and hatching of the three eggs is
shown. Compiled from Drent (1970).

perature, as measured at the top of the eggs, shows a plateau value averaging 39.5 °C, and is virtually constant at least from the time of clutch completion onward. Similar constancy at this site, reflecting a constant skin temperature of the sitting bird, has been reported in most other species investigated and implies that brood-patch defeatherization is complete by the time the clutch is complete [*Troglodytes aedon*, Baldwin and Kendeigh (1932); *Parus major* and *P. ater*, Haftorn (1966); *Branta canadensis*, Kossack (1947); domestic fowl, Burke (1925); *Phasianus colchicus*, Westerskov (1956); *Charadrius alexandrinus* and *C. dubius*, Walters (1958); *Larus canus*, Barth (1949, 1955)]. In *Megadyptes antipodes*, however, Farner (1958) reported that the definitive level was not reached until the fifteenth day of incubation; in fact, the brood-patch of this penguin was found to undergo increased vascularization in this period.

Returning to the figure, air temperature between the eggs is still rising in the first week of incubation, at a time when the egg/brood-patch interface is being held steady. Of major influence here is the pattern of attentiveness; the relatively large eggs of the Herring Gull are slow in warming up and at the beginning of incubation, equilibrium values are rarely reached due to frequent interruptions. Some impression of the gradual change in incubation behavior is given in the bottom graph, depicting the fraction of the time the eggs were actually covered, and the correlation of this curve with that for nest air temperature will be apparent. Beyond this, however, it would be worthwhile to investigate the possible influence of an increasing vascularity of the brood-patch. From the description of Jones (1969), for instance, it is clear that in *Lophortyx californicus* full vascularity of the brood-patch is not reached until well into incubation. Measurements should be made of the heating power of the brood-patch as a function of day of incubation, rather than simply the surface temperature, which is but one of the parameters involved.

Internal egg temperature, as measured centrally, is lower than the surrounding air at first and rises above it from about day 14. This is the direct result of the rise of embryonic heat production, which surpasses the evaporative heat loss of the egg after approximately 10 days of development [as determined in a respirometer (Drent, 1970)], and is responsible for the steadily increasing gradient thereafter. Similar findings have been reported for the domestic fowl (Romanoff, 1941; Romijn and Lokhorst, 1956) and domestic Mallard (Kashkin, 1961). Bearing in mind that the embryo is a relatively tiny structure at first, representing only 1 percent of the total egg weight by the end of its first week and located on the upper surface of the yolk sphere close beneath the egg shell, it is clear that central egg temperature will not adequately reflect the temperatures actually experienced by the embryo in the early period of its life. Accordingly, thermistors were inserted to lie beside the embryo, and measurements on "C" embryos in this early period averaged 37.6 °C (Figure 12). In fact, therefore, the embryo of the Herring Gull experiences an average temperature from 37.6–39.0 °C during steady incubation. By suitable calculations with the data on internal egg temperature (4,779 readings in all), the limits of temperature within which the egg remained during 95 percent of the time can be given for the whole of incubation schematically (see Figure 13) and, aside from the overnight absence of the parent during the laying period, the record compares favorably to that of a modern incubator.

How is this stability of internal egg temperature achieved? It may be surmised that, depending on the environmental circumstances, compensatory behavior of the parent will be called for and, in fact, a great deal of effort in field research has been devoted to this problem (Kendeigh, 1952). The basic pattern of response depends on whether

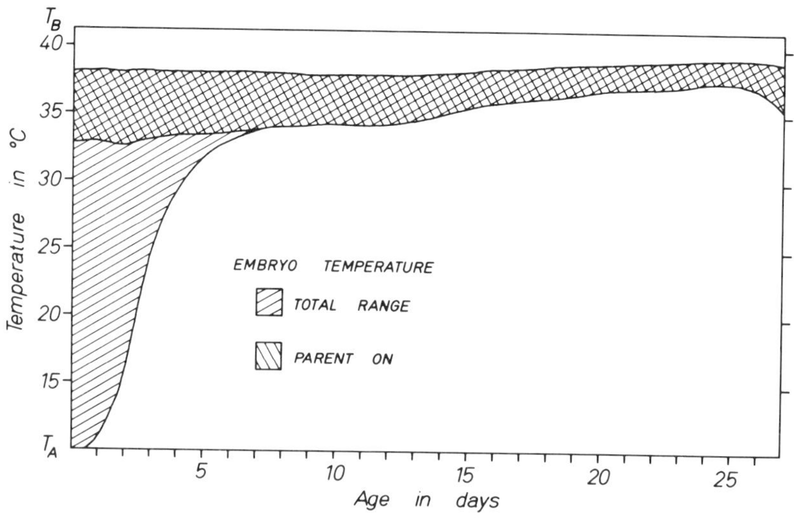

FIGURE 13 Stability of the thermal environment of the developing Herring Gull. During the laying period (up to day 4), the clutch is regularly abandoned at night, and the embryo cools to ambient levels (T_A = 10 °C). Otherwise, the range of temperature experienced by the embryo is quite restricted [the limits of temperature within which the embryo remains 95 percent of the time computed from extensive field measurements in Drent (1970)] but several degrees lower than the body temperature of the sitting parent (T_B = 41 °C).

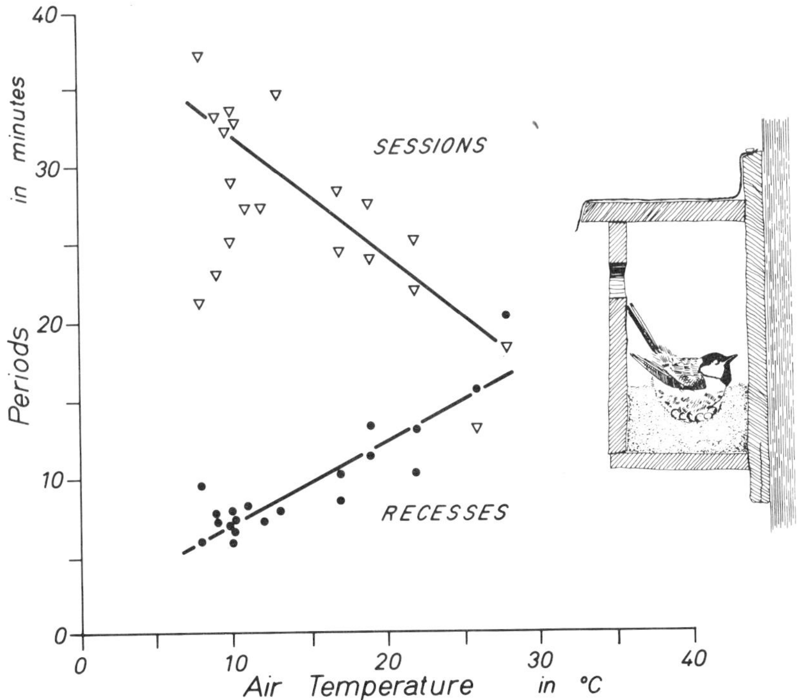

FIGURE 14   Incubation rhythm of the Great Tit, *Parus major*, as registered by activity recorders mounted in nest-box entrance, in relation to external air temperature. Sessions =
sitting bout; recesses = absences; each point is the mean of determinations for one day.
From Kluijver (1950).

the nest is exposed, with direct danger of overheating if the parent
departs (in an abandoned nest of the Herring Gull, lethal internal egg
temperatures were reached within 2 hours on a sunny morning at a
shade air temperature of 18 °C), or sheltered, in which case parental
behavior can show greater variation. Constancy of incubation in the
Great Tit, for example, a species nesting in the shelter of a hole (nowadays often a nest-box) bears a direct relation to air temperature, as
shown in Figure 14 taken from Kluijver's (1950) study. In this case,
the decline in time devoted to covering the eggs as the air temperature rises, results from a simultaneous decrease in length of sessions,
with increase in length of the intervening recesses, but all possible patterns of response occur, some species varying only the sessions, others
only the recesses (von Haartman, 1956). That the variations in the
parental pattern of incubation are, in fact, triggered by temperature

stimuli was shown by von Haartman (1956), who experimentally heated the nest of the Pied Flycatcher, *Ficedula hypoleuca*, another hole-nester, for a 12-hour period and compared behavior of the incubating female with that shown during a 12-hour session of normal temperature the following day. The results were clear-cut, and demonstrated a shortening of sessions during the experimental period (33 °C nest temperature) compared to the control period (nest temperature 16 °C). On both days, the outside air temperature was similar (16–17 °C). It would be very profitable to continue this type of work, to discover whether these effects are due to nest-air temperature directly, or whether they are in fact mediated by differences in egg temperature (for example, the egg temperature measured upon return from a recess).

In the response of the incubating bird to air temperature, we have a mechanism that apportions the time spent covering the eggs according to the needs of the embryo for heat in addition to that provided by the environment. At high air temperatures, the parent is free to devote time and energy to other activities, such as food-getting. In birds whose nests are exposed, protection of the eggs from insolation or predation may make practically continuous coverage a necessity. Even so, some degree of compensatory behavior of the parent is required, for in most cases the heat loss of the eggs—and hence the heat input by the parent needed to stabilize temperature—will be influenced by the external air temperature. In periods of continuous incubation, therefore, we should look for variations in the heat production of the sitting bird, as representing what might be termed a fine adjustment in contrast to the crude adjustment afforded by variations in the time pattern of egg coverage. In fact, data on this point have recently become available.

The data presented in Figure 15 for the Great Tit show the relation between parental heat production and external air temperature for the nighttime period, when incubation by the female is continuous. Mertens (1967), from whose study these data are taken, managed to measure the metabolism of the sitting bird by tapping off air from the nest-box and bypassing this through gas analyzers that allow computations of heat production on the basis of indirect calorimetry. Since nest-box temperatures remained within the thermoneutral zone of the parent throughout the range of external air temperature depicted, the linear relation that heat production shows in relation to external air temperature must be viewed as an

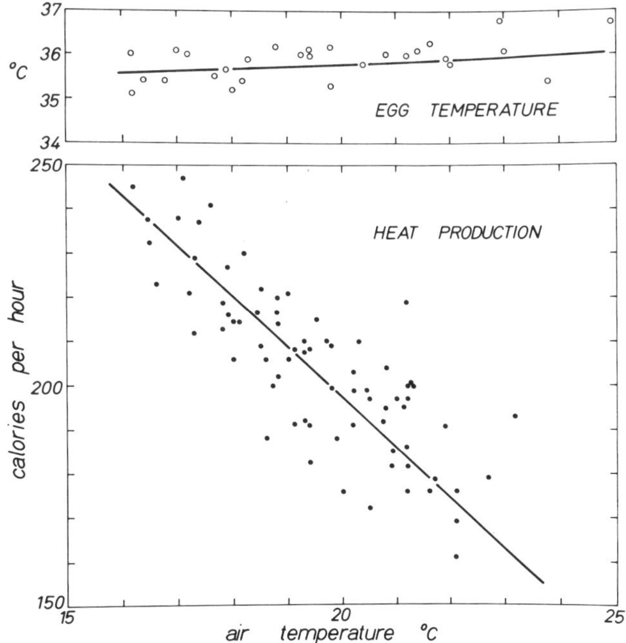

FIGURE 15   Egg temperature (above) and heat production of the incubating female (below) of the Great Tit, *Parus major*, in relation to external air temperature. Each point is based on an hourly determination during the night when the bird sat continuously. Based on data in Mertens (1967); from Drent (1972).

adjustment to compensate for varying amounts of heat loss from the clutch, determined primarily by the gradient between the nest chamber and the outside world. That parental compensation was not complete is revealed by the slight decline in egg temperature with a fall in external air temperature (Figure 15, top). This same relation has been shown in field data for other passerine species (Huggins, 1941; Kendeigh, 1963) and is similar to the registration of rectal temperature with respect to external air temperature in a homiothermal animal (e.g., Aschoff, 1970). In fact, the heat relations between the parent and its eggs, as revealed by this experiment, can be expressed as follows: The parent treats the eggs as an extension of its own body core. The crucial role that the brood-patches play in acting as a heat shunt will be appreciated and, although in a general way we are informed about the vascular networks that develop during the incubatory phase [reviewed by Jones (1971)], detailed study combining physiological experimentation with functional morphology would be most welcome.

Direct experimental alteration of egg temperature is also revealing. Franks (1967) replaced the clutch of the Ring Dove (Barbary Dove, *Streptopelia risoria*) with copper eggs, through which water of carefully regulated temperature could be pumped. The frequency of various behavior patterns, as determined during short-term exposures to egg temperatures altered from –4 to 62°C, was compared to the control situation of normal egg temperature (38–39 °C); selected examples are plotted in Figure 16. The immediate reaction to an abnormal egg temperature was often the adjustment of the contact between the eggs and the ventral body skin, by means of trampling, resettling, and egg-shifting behavior. After several minutes, these preliminary adjustments would be followed by changes in some activities, unmistakably aimed at regulating the body temperature of the sitting parent [Figure 16, top: gular flutter, shivering, and elevation of the feathers; these doves, as confirmed by the work of McFarland and Baher (1968), do not appear to ruffle the plumage at high temperatures as do many other birds]. A generally similar pattern of response has been found in experimentation with above-normal and below-normal egg temperature tests in the Herring Gull (Drent *et al.*, 1970). Of the functionally relevant changes in behavior, only the adjustment of the egg/brood-patch contact can be considered as aimed at regulating the egg temperature directly. Other actions are aimed at safeguarding the body temperature of the sitting bird and, as a corollary thereof, the stabilization of the egg temperature; as stressed earlier, the parent and its eggs act as a unit.

Continuous application of the brood-patch is the only way of correcting a subnormal egg temperature. Similarly, overheated eggs in extreme cases, where the surrounding air temperature exceeds the optimum for incubation, can only be returned to normal by continuous contact with the brood-patches; the blood of the parent in this case carries off the excess heat. In considering the ability of the parent to deal with extremes, therefore, it is pertinent to examine how much time the parent is able to devote to incubation in the strict sense, i.e., unbroken egg contact. The lower part of Figure 16 is meant to convey this. The sharp decrease in time available for egg contact as egg temperature rises above the normal level emphasizes how precarious correction of overheating really is. The strategy of incubation adaptations must be to prevent overheating from occur-

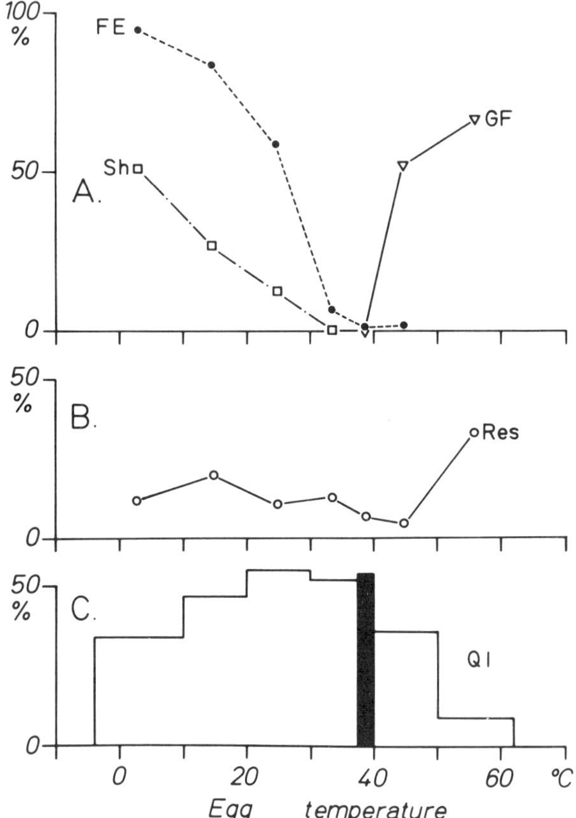

FIGURE 16   Activities of the Barbary Dove, *Streptopelia risoria*, while incubating copper eggs whose internal temperature was varied by the experimenter. Ordinates are percent of total observation minutes in which the activity occurred: FE, feathers elevated or fluffed; SH, shivering; GF, gular flutter; Res, resettling on the eggs; QI, quiet incubation, eyes open but no other activity. Normal egg temperature, 38–39 °C, is shown by the shaded bloc. From Franks (1967).

ring in the first place, rather than correcting aberrations afterward. We will return to this point later.

Several recent field studies describe the reactions of incubating parents to extreme heat at exposed nests, and these support the interpretation of the experimental studies as given above. Figure 17, based on Howell and Bartholomew (1962b), shows *Sterna fuscata* at the nest on a hot day and illustrates the exploitation of various heat-

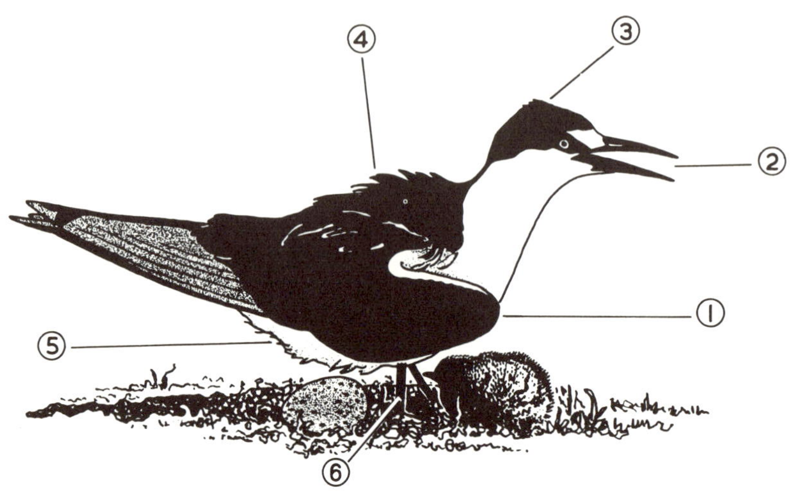

FIGURE 17   Sooty Tern, *Sterna fuscata*, at the nest on a hot day, showing heat dissipating mechanisms: 1, bend of wing exposed; 2, panting; 3, crown and, 4, dorsum feathers ruffled; 5, evaporation from abdomen, wetted periodically by skimming over the sea; 6, legs exposed. Based on photograph in Howell and Bartholomew (1962b); from Drent (1972).

dissipating mechanisms at a level of air temperature that makes direct incubation of the egg unnecessary, while requiring the continuous presence of the parent to prevent exposure to the sun with consequent overheating. At still more extreme temperatures, the parent will be forced to settle on the eggs again and the problem of the regulation of its own body temperature becomes critical. Maclean (1967) observed the Double-banded Courser, *Rhinoptilus africanus*, in the Kalahari and found that the parent frequently shaded but did not incubate the egg at air temperatures between 30 and 36 °C, whereas at temperatures in excess of this the egg was always incubated. At these levels, convective heat loss of the egg will be inadequate to prevent a rise in temperature, so contact with the brood-patch, which now functions to withdraw heat, becomes necessary. Russell (1969) who has studied the White-winged Dove, *Zenaida asiatica*, in the deserts of Arizona, points out that on hot days when ambient air temperatures reach 45 °C in the shade, overheating of the eggs can only be prevented by continuous contact with the brood-patch, which is maintained at approximately 38.5–39.5 °C. The reaction of these birds is to settle more closely on the eggs at high air temperatures; and their effectiveness in coping with these conditions is illustrated in Figure 18. How the parent is able to maintain thermal balance in these extreme conditions, without even showing gular flutter, is an unsolved problem deserving further study.

To summarize this discussion, the regulation of egg temperature in natural incubation is achieved by adjusting the time pattern of egg/brood-patch contact and by adjusting the heat transfer during contact, as though the eggs were part of the core of the parent. The main problem faced by the parents is the maintenance of its own body temperature in the face of environmental stresses imposed while incubating. Finally, the behavior of the parent must be anticipatory, in the sense of taking action before internal egg temperature has reached dangerous extremes, since the ability of the parent to bring aberrant temperatures back to normal is limited. The question of what clues the parent might make use of in adjusting its behavior to the expected needs of the embryo will be touched on in the final section of this paper.

Before leaving this topic, brief mention must be made of the possible role that the embryo might play in improving its thermal environment. From time to time, the suggestion that the embryo,

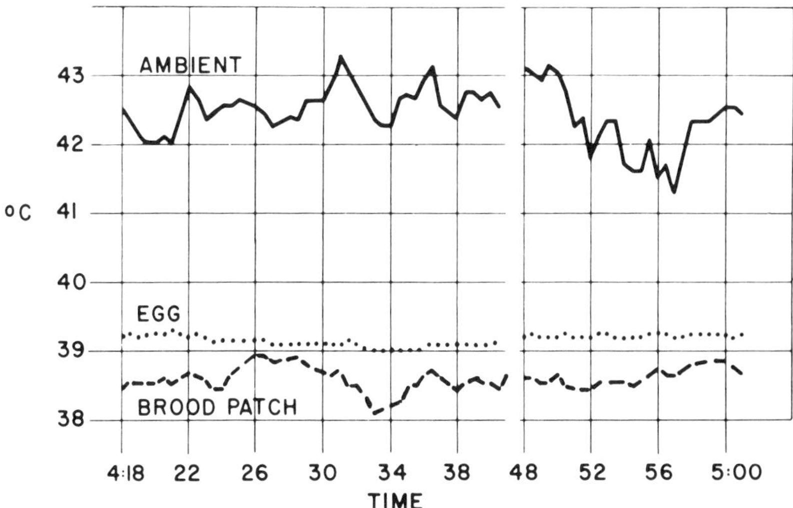

FIGURE 18 Temperatures of ambient air, egg center, and brood-patch as measured at the fully exposed nest of a White-winged Dove, *Zenaida asiatica*. From Russell (1969), by courtesy of the University of New Mexico Press.

through increased heat production, might maintain temperature on its own account crops up in the literature. Evidence for effective compensatory behavior of the embryo when the departure of the parent is simulated experimentally, however, has not been forthcoming (Romijn and Lokhorst, 1955; Kashkin, 1961; Drent, 1970). Indeed, it is unlikely that energy reserves in the egg would make such action feasible, especially in view of the critical role of the yolk remnant as an emergency ration for the immediate posthatch period (Kear, 1965; Reid and Bailey, 1966).

THE PARENT AS AN INCUBATOR

Besides actively influencing the physical environment of incubation the parents also must meet other needs, such as protection of the clutch from predation, and at the same time reach some compromise between duties toward the eggs and meeting their own nutritional demands.

We have seen that data on parental behavior permit a definition of an optimum for egg temperature for incubation and, in a similar way, one can speak of an optimum for egg number. The data assembled in Figure 19 are taken from a study of the Herring Gull by Baerends and his associates (1970) and illustrate clearly that when the egg number deviates from three, the normal clutch size in this species, the amount of interruption in incubation increases dramatically. That such behavioral measures have significance for the eventual outcome of incubation is shown in the top section of the figure, where information on predation rate is presented (in these experiments, the number of eggs was manipulated, i.e., all birds originally had a typical three-egg clutch). In some way, therefore, the Herring Gull is geared to a three-egg clutch, in correspondence to its three discrete brood-patches. Can we in fact speak of a template, or inner expectation value, independent of individual experience? The work of Beer (1965) on the Black-billed Gull, *Larus bulleri*, is of great interest in this connection. This New Zealand species only rarely lays a clutch of three at present, and more usually two, although it still possesses the three brood-patches typical of *Larus* gulls. Curiously, these birds were also found to incubate with least interruption on clutches of three, when the range from one to four was examined by experiment, indicating that we can indeed speak of this "optimum number" as an abstraction.

Similarly, there is evidence (Baerends, in preparation) for an opti-

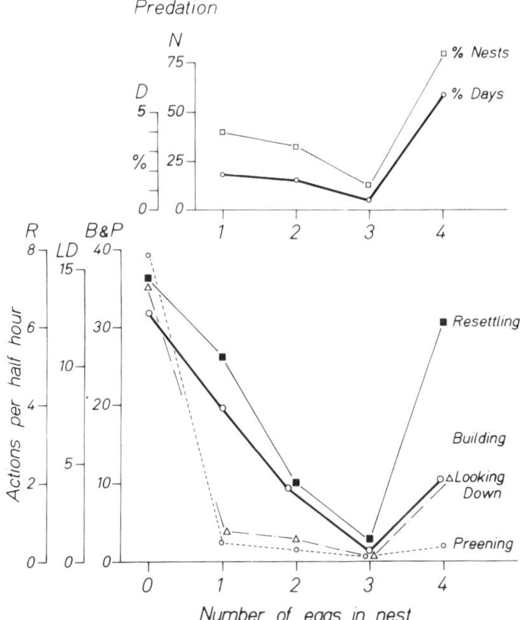

FIGURE 19   Behavior of the incubating Herring Gull, as influenced by experimental alteration of clutch (below: mean
frequencies of interruptive activities) compared to predation
rate (top: as measured by percent days on which predation
occurred, as well as total percent nests predated). The
clutches of 1, 2, and 4 are experimental; 3 is the normal
number. From Baerends *et al.* (1970).

mum in egg size. Incubation shows the least interruption and allows
the least predation with eggs of approximately normal size, but superimposed on this there is a tendency for large eggs to elicit more effective incubation behavior than small ones. This finding may help to
explain the trend in hatching success according to egg size, established
by Parsons (1971) for the same species. Figure 20 shows a bias in
favor of large eggs—a bias that is not explained by the rate of embryonic death + infertility, grouped together as "addled" (see lower
graph). Clearly, differential predation is responsible, but, of course,
these observations concerning naturally occurring variations do not
allow a definite decision on the factors underlying this greater risk of
the smaller eggs. Besides the influence of size per se, it is probable
that we also are dealing with differences in the age and, hence, of the
experience of the parents; there is some evidence that, in the Herring

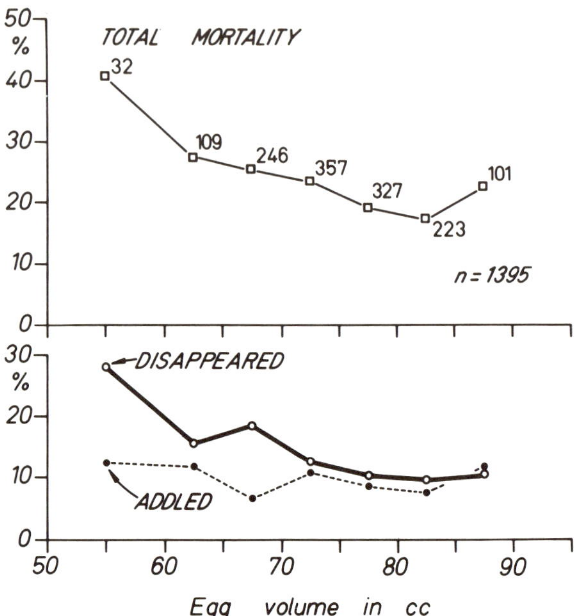

FIGURE 20   Fate of eggs of the Herring Gull in relation to volume, from field data presented by Parsons (1971). The bias against small eggs is mainly due to disappearance (= predation), addled (= infertility + embryonic death) playing only a minor role. Total mortality is equivalent to total failure to hatch.

Gull, egg size tends to increase with age (Andersen, 1957) and we know that age (experience is interwoven) has a profound effect on breeding behavior (Lehrman and Wortis, 1967; Coulson, 1972). As a rule, older birds are better parents.

This kind of information can be assembled in a diagram intended to show the functional structure of incubation behavior (Figure 21). We can theorize that there is an inner template or expectation for a variety of parameters and that information on the current status can be obtained either just prior to taking place at the nest, or during the act of sitting itself. Subsequent actions of the bird will depend on the outcome of the comparison between the moment-to-moment measurement of the variable and the inner expectation for that variable. What variables are in fact involved? Experiments (Baerends *et al.*, 1970) with a variety of egg models, some of which were transparent and possessed no visual value (or perhaps a slightly negative one), have demonstrated the importance of tactile feedback from the brood-patches in the regulation of incubation behavior. That temperature also plays a role has already been mentioned. The brood-patches, together with the legs and probably the visual impression the bird receives before settling, are involved in measuring the position of the eggs in the nestbowl. The incubating bird strives to combine a perfect match of the eggs with the brood-patches, with the direction of sitting preferred because of such reasons as wind and the location of the main route of access. As shown by extensive experimentation with painted wooden egg models (not yet published in detail), information on a large number of visual attributes of the egg also influences incubation behavior. Goethe (1953, 1955) has commented on the probable significance of the vocalizations of the embryo during approximately the last 2 days of incubation, and this view is confirmed by Tschanz's (1968) demonstration of the influence of vocalizations from the egg on the incubation behavior of the parent *Uria aalge*. It is likely that other variables will be identified as work progresses.

There is no more than an analogy to a physical control system, since even with a completely specified sensory input, the reaction of the incubating bird is predictable only in a statistical sense. To incorporate this element of uncertainty in the model, the template or inner expectation is allowed to vary under the influence of motivational factors, as suggested by Baerends (1970). Other solutions, however, are feasible. For instance, in another connection, Jander (1968) introduces variability by interposing a source of "noise" between the outcome from the comparator and the effector organs, and there are doubtless other ways to incorporate this feature. As for the ways in which the output from the comparator is translated into action, a distinction has been made between physiological and behavioral mechanisms, but these are interwoven, as shown in the figure. Physiological mechanisms refer primarily to adjustments made to serve temperature regulation. For example, the signal "internal egg temperature below normal" might be expected to lead to increased blood flow through the brood-patches and to increased heat production by the parent (shivering) in order to maintain a uniform blood temperature, despite the drop experienced as the blood courses through the patches transferring heat to the eggs. The same "cold egg" signal might elicit behavioral mechanisms also, namely, the adjustment of the egg/brood-patch contact via quivering, resetting, trampling, and so forth. Physiological mechanisms can also be brought into play as a by-product of behavioral actions, in addition to direct elicitation by a discrepancy in the incubation comparator, as just discussed. The main reason for making this somewhat artificial separation of mechanisms in Figure 21 is to emphasize the importance of the body temperature control loop in the proper functioning of incubation behavior.

Finally, conflicting demands made on the bird are represented in the category "other systems." Prime among these is the need of the sitting bird to obtain the energy requisite to keep going itself during incubation. Some species can subsist on body fat during incubation, or are fed by the mate and thus normally do not leave the nest, but in all cases the nature of the adjustments is delicate. In birds of prey, for instance, where the male habitually forages for both himself and his partner, the eggs are abandoned when food-getting is difficult [Cavé (1968) and Southern (1970), who observed parental behavior

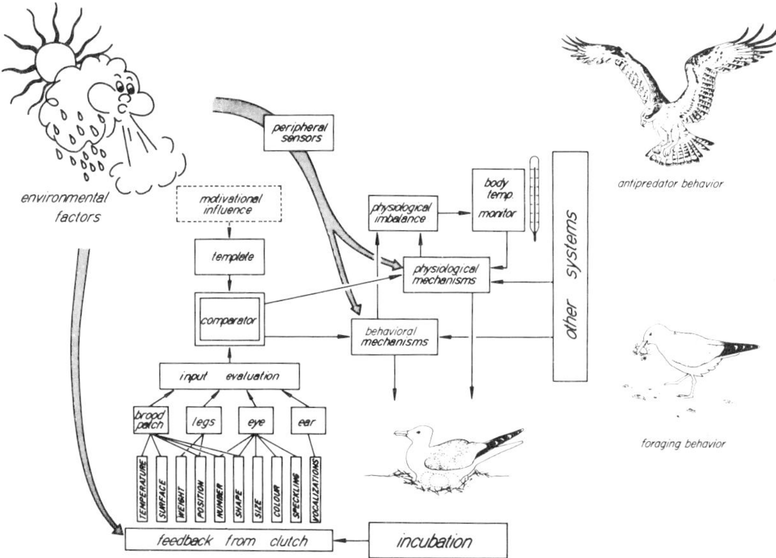

FIGURE 21   Functional control of incubation in the Herring Gull, based mainly on Baerends (1959, 1970). Commencing at lower left, sensory input during incubation constitutes feedback on at least ten characteristics. The measured state of these variables is compared to the template (expectation or "optimum") and mismatch will result in commands activating either physiological mechanisms (primarily those concerned in body temperature control), or behavioral mechanisms, or both. Events are not completely predictable, hence the template is here placed under the influence of motivational factors. Other needs of the animal may modify or interrupt incubation ("other systems"). Environmental events influence incubation directly eliciting anticipatory action ("feed-forward") or via a change in feedback state of the clutch.

and measured food supply at the same time, both reported these breakdowns as happening not uncommonly in *Falco tinnunculus* and *Strix aluco*, respectively]. Desertion of the nest in the face of impending starvation has also been observed in species in which the partners alternate at the nest. When survival of the parent is pitted against survival of the progeny, the former prevails. Nelson (1969) has provided a beautiful example from his work on the Red-footed Booby (*Sula sula*) on the Galapagos, where nearly 70 percent of the eggs laid were subsequently abandoned. This species is a highly specialized feeder that faces long travels to and from the foraging areas and Nelson argues that most desertion occurs because the off-duty partner is unable to return on schedule. This species appears to possess only a precarious margin of time uncommitted to feeding, compared to the other members of the group (Nelson, 1970). In a totally different situation, Murton (1965) has evidence that in times of food shortage incubation may be abandoned in the Wood Pigeon, *Columba palumbus*. Such drastic effects are not known in gulls, but the foraging rhythm of the Herring Gull on the intertidal flats at low water has been shown to influence the incubation rhythm, and the delayed return of the off-duty partner leads to a deterioration of incubation constancy.

It is pertinent now to ask in what ways the demands of incubation may modify the feeding rhythm. For example, if a given species requires, say, 1 hour per day to satisfy its demands for food and water and only one sex incubates, what strategy of nest visitation will evolve? Will the bird subdivide its feeding time and make many brief excursions from the nest, or will it tend to take one long recess? Obviously, a great many factors will enter into the picture, but it may be worth considering the theoretically most efficient way of maintaining the egg at the temperature requisite to development. It is a matter of observation that when eggs are exposed upon departure of the parent, their rate of cooling exceeds the rate of heating when the parent returns. This relation is diagrammed in Figure 22 (top). In other words, the time required to return the eggs to normal temperature will exceed the time of absence. In terms of energy, the parent is penalized for every absence, since returning the egg to normal temperature costs energy, and, until the embryo is again metabolizing normally, there is no return on this energy. Now as the egg cools and the gradient between the egg and its surroundings becomes progressively less, the rate of cooling declines until it becomes hardly perceptible (Figure 22). To avoid unnecessary bouts of warming,

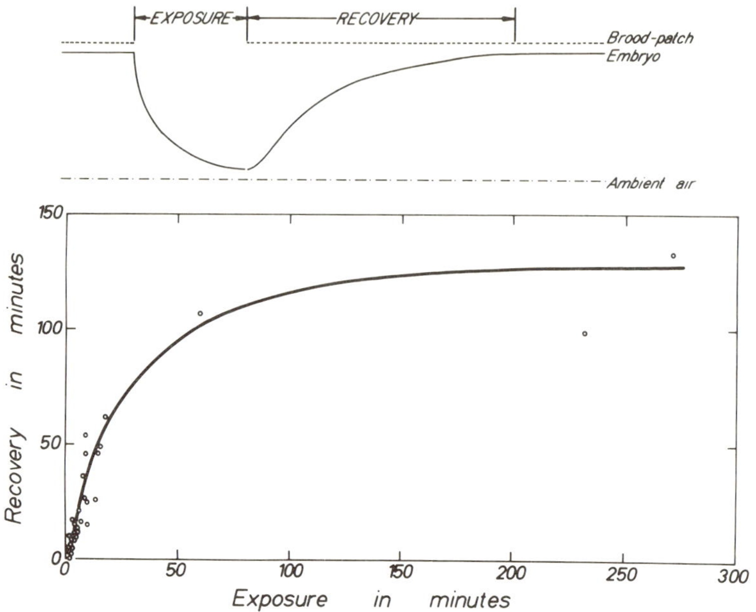

FIGURE 22   Relation between cooling and heating rate of an egg in the nest of the Herring
Gull. A generalized curve of internal egg temperature is shown at top; parental absence
is termed *exposure*, and the time required to return the egg to the initial equilibrium temperature, *recovery*. Field data showing recovery time in relation to the exposure that preceded it are presented in the main graph, based on 83 determinations. Absences of more
than a few moments incur very long recovery times (= energetically wasteful delays in incubation). In undisturbed conditions, absences are usually less than 1 minute in this species.

absences might be timed to exploit this circumstance. The only data
available to show how this might work are from the Herring Gull,
showing how long it took to return the egg to normal temperature,
as a function of the previous absence or exposure (the external air
temperature was approximately 11 °C during these determinations).
For an egg in this size range (100 g), the curve (Figure 22) indicates
that either absences should be avoided altogether, in view of the very
unfavorable relation between exposure time and time required afterward before egg temperature is restored or, if absences cannot be
avoided and must total 1 hour, as in our example, it would be less
disastrous to leave the nest once for 60 minutes (total lost time approximately 60 + 110 = 170 minutes) than to leave ten times for 6
minutes each (total lost time 60 + 10 X 35 = 410 minutes). It would
be of interest to construct "recovery X exposure" curves for eggs of

other weights and to see if this line of reasoning could be applied to the observed incubation rhythm. At the moment this is only a suggested procedure, albeit an intriguing one.

Various antipredator measures may also interfere with incubation—both those aimed at protecting the eggs and those aimed at protecting the parent. Tinbergen and his co-workers have studied antipredator behavior in great depth in the Black-headed Gull, *Larus ridibundus* (Tinbergen *et al.*, 1962; Kruuk, 1964; Patterson, 1965), and argue that the main defense of the eggs is camouflage. Spacing of the eggs is achieved by territorial behavior, and the difficulty of finding the cryptically colored eggs is further enhanced by harassment of potential predators by the nest-owners; the gulls take to the air at a distance that depends on the species of predator. In dealing with some predator species this aerial attack is effective, in others it is scarcely a deterrent, but in the present context it is sufficient to note that the approach of a certain category of animals, either seen at first hand by the sitting gull or signaled by the alarm cries of its neighbors, will bring about the temporary abandonment of the nest. Removal of eggshells, designed to remove a clue that might alert predators to the presence of newly hatched young, is another behavior pattern that will interfere with incubation (more than 24 hours elapse between the hatching of the first and third chick).

On the question of protection of the parent, the appearance of various birds of prey in the gullery, especially if they are large, leads to widespread panic flights in the colony (the Osprey, which takes gulls occasionally, is such an example). Of special relevance to incubation is the nighttime vulnerability of the diurnally active Herring Gull. Outside the breeding season, gulls are extremely wary at night and prefer to sleep either on isolated stretches of flat beach, or out on some protected body of water where they roost in dense rafts. Commencement of incubation brings the adult into an area of high risk; because of the demands of camouflage of the eggs, and later of the chicks, a complex terrain is chosen for the nest, where the parent is at a disadvantage and can be surprised by nightly forays of foxes, for example. What in fact happens is that nighttime incubation only sets in gradually and in relation to the density of incubating neighbors (and hence in relation to the number of potential warning devices, so to speak). Birds isolated from their fellows have been found to abandon the clutch regularly at night, during a period up to 16 days after clutch completion.

Antipredator strategies differ widely from one species to another, as will be clear by considering one further example. The Common Eider, *Somateria mollissima*, in which only the cryptically colored female incubates, lacks a mobbing response toward predators. Instead, the female sits tight and in many cases the nest presumably will be overlooked by predators. If approach is very close, the female flushes at the last moment and, as discovery of the light-colored eggs nestled in a mass of down will now be inevitable, a second line of defense is brought into action—the female defecates over the eggs as she flies off. The fluid ejected at this time is particularly noxious, quite different from normal fecal material, and Swennen (1968) has experimentally demonstrated the repellent effect this has on potential predators. It is not necessary to show that protection is complete, of course, only that the probability of the eggs being taken is reduced.

Turning, finally, to the influence of external weather conditions, we can distinguish two modes of action (see Figure 21). On the one hand, external conditions can influence the parent through a change in the feed-back from the clutch. In the case of the female Great Tit incubating at night (Figure 15), the reaction to lowered environmental temperature was probably triggered by a decline in egg temperature (leading to an increased output of heat from the brood-patches) rather than by the fall in external air temperature *per se*. On the other hand, anticipatory action of the parent is also possible. A clear example is provided by the reaction of many open-nesting birds to the advent of a heavy shower. Skutch (1962) and Rittinghaus (1961) report that birds they were watching foraging away from the nest rushed back to resume incubation when a driving rain commenced. Doubtless many other "predictive clues" are used by the parents in adjusting their behavior; Lind (1961) has shown that *Limosa limosa* covers the eggs at night in the laying phase, particularly at low air temperatures when there is danger that the embryos might freeze. Since the parents do not otherwise visit the nest at this time, this must be a reaction to the fall in air temperature itself. Earlier, in the discussion on the regulation of egg temperature, it was argued that anticipatory action is vital to successful incubation, i.e., prompt action to prevent the consequences of external stresses from having an effect on the egg. This we can call "feed-forward," using McFarland's (1971) terminology, in distinction to the "feedback" from the clutch; far more attention ought to be directed toward finding out just what clues are utilized. We do not know, for example, exactly how a rise in nest temperature leads to a change in the pattern of attentiveness. Is this a reaction to air temperature itself, or to a change in cooling rate of the eggs? Further experimentation along the lines pioneered by von Haartman (1956) would certainly be profitable.

A point not covered in our general scheme of incubation behavior is the importance of nest-building activity. Nests may function importantly in concealing the contents from predators, as has been argued convincingly by Riehm (1970) in her study of the Long-tailed Tit, *Aegithalos caudatus*. Nests of this species are closed structures and, having an outer layer of lichens, tend to blend in with the trunk or branch of the tree in which they are built. Conspicuousness is further reduced by behavior; the parent closes the nest opening with a feather and disguises visits to the nest by an imitation of insect-hunting tactics. Despite all of these measures, however, more than half of all nests studied were destroyed by jays. Other nests may yield physical protection from predation, as has been argued for the domed nest of the Black-billed Magpie, *Pica pica*, by Erpino (1968). The case of the hornbills (Bucerotidae), where the female is sealed in the nesting hole by a layer of mud (Thomson, 1964, p. 378), can be interpreted in this same way. Nests function importantly in temperature regulation as well, acting as an extension of the shell of the parent, so to speak. The Palmgrens (1939) emphasized the significance that insulative value of the nest might have for the distributional ecology of the species and presented data on a number of passerines by measuring the rate of cooling of a flask of water encased within two nests of the species in question. A number of correlations were found among the 15 species investigated. For instance, the difference in insulative value of the nests of *Fringilla coelebs* and *F. montifringilla* was in the direction to be expected on the basis of their distribution. In her study of the survival value of the nest of the Long-tailed Tit, Riehm (1970) also paid attention to this aspect and provides data showing the insulative function of the nest, due primarily to the astonishingly high number of feathers that are worked into its structure (ordinarily 1,500–2,000). This feature must make an important contribution toward reducing the energy requirement for incubation (see Kendeigh's discussion, p. 311), as in this species the female incubates alone and must maintain the temperature of an egg mass equal to her own body weight of 8–9 g [computed from figures given by Heinroth and Heinroth (1926), using mean clutch size as observed by Riehm].

Variations in the nest in accord with the local circumstances also have been noted. Krüger (1965) observed the nesting of Partridge

(*Perdix perdix*) in captivity and found that during the laying phase the female carried out protective measures adapted to the circumstances prevailing at the time of her inspection visits (adding straw, or an earth layer, opening the nest to cool at night, etc.). This calls to mind the intricate behavior of the male Mallee Fowl, *Leipoa ocellata*, that serves to regulate the temperature of the nesting mound, as revealed by the research of Frith (1962). Varying the location of the nest may also serve temperature regulation; the heat of the sun may be utilized (Riehm, 1970), or avoided (Maclean, 1970), or the effect of cooling winds may be exploited (Ricklefs and Hainsworth, 1969). Clearly, study of incubation should include examination of the ways that the parent bird may enhance the microclimate in which the eggs develop by choice of nest site (e.g., Horváth, 1964), the orientation of the nest, and techniques of construction.

CONCLUSION

Although incubation in birds has an obvious function and although a staggering amount of time has been spent by untold observers watching at the nest, our knowledge of the functional aspects of natural incubation is far from impressive. It is the intent of this review to emphasize how badly we still need empirical data on incubation performance before more sophisticated questions on the nature of the control systems involved can be posed and answered. On certain topics enough observations are now at hand to make simple experimentation profitable. The importance of further field investigations will be underscored for each.

MECHANICAL ENVIRONMENT

In only two species are we well informed regarding egg position, which is a very slender base on which to generalize. Although the tendency toward restriction of egg position as incubation proceeds, described for *Limosa limosa* and *Larus argentatus*, may hold for other species as well, precise measurement on individual eggs of other species is urgently needed. One of the consequences of the continued egg-turning behavior of the parent is to bring the developing embryo into a predictable relation to the horizontal plane, i.e., to a part of the egg that can be designated as lying uppermost in the nest the majority of the time. It is tempting to regard the positioning of the embryo in relation to gravity, as well as to the temperature gradient through the egg, as being functionally relevant. One observed effect is that, during hatching, the first hole in the shell is uppermost. That this has something to do with facilitating hatch is at present only a postulate, but relatively simple experiments could be devised to test it.

The problems faced by the embryo in escaping from the shell are primarily the province of laboratory experiment, but the experimental demonstration (R. Faust, M. A. Vince) that the hatching eggs can influence one another in the nest so as to decrease markedly the spread of hatching times is an interesting adaptation about which much can still be learned from field observation. In particular, the functional significance of this mutual adjustment of hatching timetable can only be properly evaluated in the field. By judicious egg transfer experiments in cases in which many nests can be kept under observation, it should be possible to manipulate the spread of hatching times within the clutch and to collect data on the repercussions thereof on survival.

GASEOUS ENVIRONMENT

Knowledge of levels of carbon dioxide and water vapor in the nest during natural incubation is sketchy; an investigator with a flair for instrumentation can make many discoveries here. Study of the ways in which the parent influences these factors, though technically far from simple, is nevertheless of great interest. The next step would be to decide whether there is in fact a parental control or regulation over any of these factors, as distinct from effects coincident with other parental activities.

THERMAL ENVIRONMENT

Specification of the temperatures at which avian embryos are maintained during natural incubation is of intrinsic interest, and the advent of relatively inexpensive but accurate temperature-sensing equipment (thermistors and thermocouples), together with recorders suitable for fieldwork, opens this topic to any interested investigator. The most revealing datum is the embryo temperature, which can be measured in the second half of incubation, without sacrificing the egg, by inserting a temperature probe into the air space and ensuring contact of the sensitive junction with the inner-air-space membrane (i.e., the probe lies against the embryo itself). Eggs so treated, if handled carefully, will hatch normally and will provide useful information on environmental and parental influences on the temperature of the intact embryo. Early in incubation figures representative for the embryo can only be obtained by thrusting the probe into the egg in such a way that the sensitive junction comes to lie where previous study leads one to expect the embryo (in species without heavy pigmentation of the shell, candling can be a useful aid in positioning the probe correctly). Embryos are unlikely to survive this treatment for more than about 24 hours, thus eggs must be sacrificed if measurements are to be made in this phase of development. Although a substantial number of species have been investigated in this way on a spot-check basis (see Figure 11), an indication of the time course of embryo temperature throughout incubation is available for only three species (Kendeigh, 1963; Spellerberg, 1969; Drent, 1970). Empirical knowledge of how closely the parent manages to regulate embryo temperature during natural incubation is still sketchy, and further investigations of the type mentioned would be most welcome. It is understandable, but unfortunate, that many studies have depended on registration of the temperature of dummy eggs filled with paraffin or some other unnatural material. Although eggs so treated are robust, the measurements obtained are useful only as indicators of parental activity. Where feasible, it would be preferable to use real eggs or water-filled dummies that approximate normal conditions more closely.

Besides extending the observational approach to other species, it would certainly be rewarding to carry out simple experiments to identify the temperature clues that trigger parental behavior. We have seen that the time pattern of incubation is varied according to environmental conditions, but a more precise formulation can only be made when various components have been isolated experimentally. We do not know, for example, what influence egg temperature may have independently of air temperature. This is an extremely important question, and one that is amenable to simple field experiment.

In terms of energy, it is not so much the temperature but rather the heat flow from the parent to the clutch that is important. In view of the increasing effort devoted by ecologists to estimating the energy budgets of wild birds we can expect applications of the laboratory techniques devised by El-Wailly (1966) and Mertens (1967) to the field situation. In these studies, the energy demand imposed by incubation was estimated by the increment in food uptake, or the increment in gaseous metabolism. Very appealing for the field-worker is the much less demanding method developed by Kendeigh (1963), where straightforward temperature measurements, coupled with study of the cooling rate of the eggs of the species in question, can be used to arrive at incubation cost indirectly (see Kendeigh, this volume, p. 311).

The significance of the nest as an extension of the shell of the parent deserves careful study. Again, simple experiments can clarify the role that nest structure plays in shielding the eggs from extremes. Such features of the position of the nest as orientation with regard to sun or wind also should receive attention and the possible functional significance thereof tested by experimental alterations. To take a specific example, Orr (1970) indicates that the Desert Lark, *Ammomanes deserti*, reduces the risk of overheating at midday by locating its nest in the shade sector of an overhanging bush or rock and by building a pebble wall around the nest. Although the temperature measurements made in the nest cup that Orr presents are of considerable interest, the specific role of orientation or nest structure could be clarified by varying these features and noting the result, i.e., by removal of shade bush, alterations to pebble wall, and so forth.

THE PARENT AS AN INCUBATOR

Successful incubation in the field hinges on many factors in addition to meeting the physical requirements for embryonic development. Antipredator measures must be taken (camouflage, nest defense, or a combination of these), and the work of Tinbergen and his group on the Black-headed Gull shows how fruitful such a line of field experimentation can be. The incubating parent must also achieve some compromise between tending the eggs and leaving the nest in order to feed. Much is still to be learned about the ways foraging routines influence incubating rhythms. A further aspect of incubating rhythms that has not been touched upon here is the relation these bear to the underlying circadian activity pattern. Hoffmann (1969) recently has summarized what is known of incubation rhythm in pigeons and doves and points out that the time of exchange between the sexes is in accord with generalizations about the endogenous circadian rhythm as formulated by Aschoff. Hoffmann's paper is especially illuminating because it combines an experimental laboratory approach involving manipulated light–dark cycles with direct field observation, supported by a literature survey. Hoffmann also outlines how further field observation can provide a test of the currently accepted hypotheses in rhythms research.

ACKNOWLEDGMENTS

I am indebted to the many people who discussed their investigations with me, discussions leading in most cases to the adaptation of some of their material as illustrations in this review (G. P. Baerends, R. Faust, J. A. L. Mertens, R. W. Oppenheim, J. Parsons, S. M. Russell, M. A. Vince, and I. Würdinger). Mr. L. Hoekstra deserves my thanks for the great trouble he took to provide the drawings of the birds.

REFERENCES

Andersen, F. S. 1957. Egg size and the age composition of bird populations. Vidensk. Medd. Dansk Naturhist. Foren. 119:1–24.

Aschoff, J. 1970. Temperature regulation, p. 43–116. *In* O. H. Gauer, K. Kramer, and R. Jung [ed.] Physiologie des Menschen. Vol. 2. Urban & Schwarzenberg, München, Berlin, Wien.

Baerends, G. P. 1959. The ethlogical analysis of incubation behaviour. Ibis 101:357–368.

Baerends, G. P. 1970. A model of the functional organization of incubation behaviour. Behaviour Suppl. 17:265–312.

Baerends, G. P., R. H. Drent, P. Glas, and H. Groenewold. 1970. An ethological analysis of incubation behaviour in the Herring Gull. Behaviour, Suppl. 17:135–235.

Baldwin, S. P., and S. C. Kendeigh. 1932. Physiology of the temperature of birds. Scientific Publications of the Cleveland Museum of Natural History. Vol. 3. Cleveland Museum of Natural History. 196 p.

Baltin, S. 1969. Zur Biologie und Ethologie des Telegalla-Huhns (*Alectura lathami* Gray) unter besonderer Berücksichtigung des Verhaltens während der Brutperiode. Z. Tierpsychol. 26:524–572.

Barth, E. K. 1949. Redetemperaturer og rugevaner. Naturen (Bergen) 73:81–95.

Barth, E. K. 1955. Egg-laying, incubation and hatching of the Common Gull (*Larus canus*). Ibis 97:222–239.

Beer, C. G. 1965. Clutch size and incubation behavior in Black-billed Gulls (*Larus bulleri*). Auk 82:1–18.

Bjärvall, A. 1967. The critical period and the interval between hatching and exodus in mallard ducklings. Behaviour 28:141–148.

Burke, E. 1925. A study of incubation. Mont. Agr. Exp. Sta. Bull. 178. 44 p.

Cavé, A. J. 1968. The breeding of the Kestrel, *Falco tinnunculus* L., in the reclaimed area Oostelijk Flevoland. Netherlands J. Zool. 18:313–407.

Chattock, A. P. 1925. On the physics of incubation. Philos. Trans. Roy. Soc. (London) Ser. B 213:397–450.

Coulson, J. C. 1972. The pair bond and breeding success in the Kittiwake Gull (*Rissa tridactyla*), p. 424–433. *In* Proceedings of the XV International Ornithological Congress. E. J. Brill, Leiden.

Delius, J. D. 1963. Das Verhalten der Feldlerche. Z. Tierpsychol. 20:297–348.

Drent, R. H. 1970. Functional aspects of incubation in the Herring Gull. Behaviour, Suppl. 17:1–132.

Drent, R. H. 1972. Adaptive aspects of the physiology of incubation, p. 255–256. *In* Proceedings of the XV International Ornithological Congress. E. J. Brill, Leiden.

Drent, R. H., K. Postuma, and T. Joustra. 1970. The effect of egg temperature on incubation behaviour in the Herring Gull. Behaviour Suppl. 17:237–261.

Eklund, C. R., and F. E. Charlton. 1959. Measuring the temperature of incubating penguin eggs. Am. Sci. 42:80–86.

El-Wailly, A. 1966. Energy requirements for egg-laying and incubation in the Zebra Finch, *Taeniopygia castanotis*. Condor 68:582–594.

Erpino, M. J. 1968. Nest-related activities of Black-billed Magpies. Condor 70:154–165.

Fargeix, N. 1963. L'orientation dominante de l'embryon de la caille domestique (*Coturnix coturnix japonica*) et la règle de von Baer. C.R. Soc. Biol. 157:1431–1434.

Farner, D. S. 1958. Incubation and body temperatures in the Yellow-eyed Penguin. Auk 75:249–262.

Faust, R. 1960. Brutbiologie des Nandus (*Rhea americana*) in Gefangenschaft. Verhandl. Deutsch. Zool. Gesellsch. 42:398–401.

Fisher, H. I. 1966. Hatching and the hatching muscle in some North American ducks. Trans. Ill. State Acad. Sci. 59:305–325.

Franks, E. C. 1967. The responses of incubating Ringed Turtle Doves (*Streptopelia risoria*) to manipulated egg temperatures. Condor 69:268–276.

Frith, H. J. 1962. The Mallee-fowl. The bird that builds an incubator. Angus & Robertson, Sydney. 136 p.

Goethe, F. 1953. Experimentelle Brutbeendigung und andere brutethologische Beobachtungen bei Silbermöwen (*Larus a. argentatus* Pontopp.). J. Ornithol. 94:160–174.

Goethe, F. 1955. Beobachtungen bei der Aufzucht junger Silbermöwen. Z. Tierpsychol. 12:402–433.

Groebbels, F. 1937. Der Vogel. Bau, Funktion, Lebenserscheinung, Einpassung. II. Geschlecht und Fortpflanzung. Bornträger, Berlin. 547 p.

Haftorn, S. 1966. Egglegging og ruging hos meiser basert på temperaturmålinger og direkte iakttagelser. Sterna 7:49–102.

Hamburger, V., and R. Oppenheim. 1967. Prehatching motility and hatching behavior in the chick. J. Exp. Zool. 166:171–204.

Heinroth, O. 1938. Aus dem Leben der Vögel. Springer, Berlin. 165 p.

Heinroth, O., and M. Heinroth. 1926. Die Vögel Mitteleuropas. Vol. 1. H. Bermühler, Berlin–Lichterfelde. 339 p.

Hoffmann, K. 1969. Zum Tagesrhythmus der Brutablösung beim Kaptäubchen (*Oena capensis* L.) und bei anderen Tauben. J. Ornithol. 110:448–464.

Holcomb, L. C. 1969. Egg turning behavior of birds in response to color-marked eggs. Bird-Banding 40:105–113.

Horváth, O. 1964. Seasonal differences in Rufous Hummingbird nest height and their relation to nest climate. Ecology 45:235–241.

Howell, T. R. 1964. Notes on incubation and nestling temperatures and behavior of captive owls. Wilson Bull. 76:28–36.

Howell, T. R., and G. A. Bartholomew. 1961. Temperature regulation in nesting Bonin Island Petrels, Wedge-tailed Shearwaters, and Christmas Island Shearwaters. Auk 78:343–354.

Howell, T. R., and G. A. Bartholomew. 1962a. Temperature regulation in the Red-tailed Tropic Bird and the Red-footed Booby. Condor 64:6–18.

Howell, T. R., and G. A. Bartholomew. 1962b. Temperature regulation in the Sooty Tern *Sterna fuscata*. Ibis 104:98–105.

Huggins, R. A. 1941. Egg temperatures of wild birds under natural conditions. Ecology 22:148–157.

Jander, R. 1968. Über die Ethometrie von Schlüsselreizen, die Theorie der telotaktischen Wahlhandlung und das Potenzprinzip der terminalen Cumulation bei Arthropoden. Z. Vergleich. Physiol. 59:319–356.

Jayakar, S. D., and H. Spurway. 1965. The Yellow-wattled Lapwing *Vanellus malabaricus* (Boddaert) a tropical dry-season nester. J. Bombay Nat. Hist. Soc. 62:1–14.

Jones, R. E. 1969. Hormonal control of incubation patch development in the California Quail, *Lophortyx californicus*. Gen. Comp. Endocrinol. 13:1–13.

Jones, R. E. 1971. The incubation patch of birds. Biol. Rev. 46:315–339.

Kaltofen, R. S. 1961. Het keren der eieren van de hen tijdens het broeden als mechanische factor van betekenis voor de broeduitkomsten. Instituut voor de Pluimveeteelt "Het Spelderholt" Beekbergen, Mededeling Nr. 88. 104 p.

Kashkin, V. V. 1961. Heat exchange of bird eggs during incubation. Biophysica (English language version) 6:57–63.

Kear, J. 1965. The internal food reserves of hatching Mallard ducklings. J. Wildlife Manage. 29:523–528.

Kendeigh, S. C. 1952. Parental care and its evolution in birds. Ill. Biol. Monogr. 22:1–358.

Kendeigh, S. C. 1963. Thermodynamics of incubation in the House Wren, *Troglodytes aedon*, p. 884–904. *In* Proceedings of the XIII International Ornithological Congress. American Ornithologists' Union, Ithaca, New York.

Kluijver, H. N. 1950. Daily routines of the Great Tit, *Parus m. major* L. Ardea 38:99–135.

Koch, A., and L. Steinke. 1944. Über Temperatur und Feuchtigkeit bei Natur- und Kunstbrut. Arch. Kleintierz. 3:153–202.

Kossack, C. W. 1947. Incubation temperatures of Canada Geese. J. Wildlife Manage. 11:119–126.

Kovach, J. K. 1970. Development and mechanisms of behavior in the chick embryo during the last five days of incubation. J. Comp. Physiol. Psychol. 73:392–406.

Krüger, P. 1965. Über die Einwirkung der Temperatur auf das Brutgeschäft und das Eierlegen des Rebhuhnes (*Perdix perdix*). Acta Zool. Fenn. 112:1–64.

Kruuk, H. 1964. Predators and anti-predator behaviour of the Black-headed Gull (*Larus ridibundus* L.). Behaviour Suppl. 11:1–130.

Lamson, G. H., and H. D. Edmond. 1914. Carbon dioxide in incubation. Storrs Agr. Exp. Sta. Bull. 76:215–258.

Landauer, W. 1967. The hatchability of chicken eggs as influenced by environment and heredity. Storrs Agr. Exp. Sta. Monogr. 1 (revised). 315 p.

Lehrman, D. S., and R. P. Wortis. 1967. Breeding experience and breeding efficiency in the Ring Dove. Anim. Behav. 15:223–228.

Lind, H. 1961. Studies on the behaviour of the Black-tailed Godwit [*Limosa limosa* (L.)]. Meddelelse fra Naturfredningsrådets reservatudvalg Nr. 66. Munksgaard, Copenhagen. 157 p.

Lundy, H. 1969. A review of the effects of temperature, humidity, turning and gaseous environment in the incubator on the hatchability of the hen's egg, p. 143–176. *In* T. C. Carter and B. M. Freeman [ed.] The fertility and hatchability of the hen's egg. Oliver & Boyd, Edinburgh.

Maclean, G. L. 1967. The breeding biology and behaviour of the Double-banded Courser *Rhinoptilus africanus* (Temminck). Ibis 109:556–569.

Maclean, G. L. 1970. The biology of the larks (Alaudidae) of the Kalahari Sandveld. Zool. Afr. 5:7–39.

McFarland, D. J. 1971. Feedback mechanisms in animal behaviour. Academic Press, London. 279 p.

McFarland, D. J., and E. Baher. 1968. Factors affecting feather posture in the Barbary Dove. Anim. Behav. 16:171–177.

Mertens, J. A. L. 1967. Verslag betreffende het onderzoek over de temperatuur-regulatie en de energiebalans van de Koolmees (*Parus major* L.) in 1965–1966. Oecologisch Instituut, Arnhem. Unpublished report. 21 p.

Mertens, J. A. L. 1970. Verslag van de werkzaamheden van het oecofysiologisch laboratorium in 1969. Oecologisch Instituut, Arnhem. Unpublished report. 10 p.

Murton, R. K. 1965. The Wood Pigeon. Collins, London. 256 p.

Nelson, J. B. 1969. The breeding ecology of the Red-footed Bobby in the Galapagos. J. Anim. Ecol. 38:181–198.

Nelson, J. B. 1970. The relationship between behaviour and ecology in the Sulidae with reference to other sea birds. Oceanogr. Mar. Biol. Ann. Rev. 8:501–574.

Oppenheim, R. W. 1970. Some aspects of embryonic behaviour in the duck (*Anas platyrhynchos*). Anim. Behav. 18:335–352.

Oppenheim, R. W. 1972. Prehatching and hatching behaviour in birds: A comparative study of altricial and precocial species. Anim. Behav. 20.

Orr, Y. 1970. Temperature measurements at the nest of the Desert Lark (*Ammomanes deserti deserti*). Condor 72:476–478.

Palmgren, M., and P. Palmgren. 1939. Über die Wärmeisolierungskapazität verschiedener Kleinvogelnester. Ornis Fenn. 16:1–6.

Parsons, J. 1971. The breeding biology of the Herring Gull, *Larus argentatus*. PhD thesis. University of Durham. 219 p.

Patterson, I. J. 1965. Timing and spacing of broods in the Black-headed Gull *Larus ridibundus*. Ibis 107:433–459.

Prince, H. H., P. B. Siegel, and G. W. Cornwell. 1970. Inheritance of egg production and juvenile growth in mallards. Auk 87:342–352.

Ragosina, M. N. 1963. Die Entwicklung des Haushuhn-Embryos in seinen Beziehungen zum Dotter und zu den Eihäuten. J. Ornithol. 104:82–84.

Reaumur, R. A. F. de. 1751. Konst om Tamme-vogelen van Allerhandesoort in alle Jaartyden Uittebroeijen en Optebrengen zo door 't Middel van Mest als van 't Gewone Vuur. Translated from the original French edition of 1749. Pieter de Hondt, 's Gravenhage. Vol. 1. 384 p.

Reid, B. E., and C. Bailey. The value of the yolk reserve in Adélie Penguin chicks. Rec. Dominion Mus. (Wellington, N.Z.) 5:185–193.

Ricklefs, R. E., and F. R. Hainsworth. 1969. Temperature regulation in nestling Cactus Wrens: The nest environment. Condor 71:32–37.

Riehm, H. 1970. Oekologie und Verhalten der Schwanzmeise (*Aegithalos caudatus* L.). Zool. Jb. Syst. 97:338–400.

Rittinghaus, H. 1961. Der Seeregenpfeifer. A. Ziemsen, Wittenberg. 126 p.

Robertson, I. S. 1961a. Studies of chick embryo orientation using X-rays. I. A preliminary investigation of presumed normal embryos. Brit. Poultry Sci. 2:39–47.

Robertson, I. S. 1961b. The influence of turning on the hatchability of hens' eggs. I. The effects of rate of turning on hatchability. J. Agr. Sci. 57:49–57.

Robertson, I. S. 1961c. The influence of turning on the hatchability of hens' eggs. II. The effect of turning frequency on the pattern of mortality, the incidence of malpositions, malformations and dead embryos with no somatic abnormality. J. Agr. Sci. 57:57–69.

Rolnik, V. V. 1970. Bird embryology. Israel Program for Scientific Translations, Jerusalem. 379 p.

Romanoff, A. L. 1941. Development of homeothermy in birds. Science 94:218–219.

Romijn, C., and W. Lokhorst. 1955. Chemical heat regulation in the chick embryo. Poultry Sci. 34:649–654.

Romijn, C., and W. Lokhorst. 1956. The caloric equilibrium of the chicken embryo. Poultry Sci. 35:829–834.

Russell, S. M. 1969. Regulation of egg temperatures by incubating White-winged Doves, p. 107–112. *In* C. C. Hoff and M. L. Riedesel [ed.] Physiological systems in semiarid environments. University of New Mexico Press, Albuquerque.

Skutch, A. F. 1962. The constancy of incubation. Wilson Bull. 74:115–152.

Southern, H. N. 1970. The natural control of a population of Tawny Owls (*Strix aluco*). J. Zool. (London) 162:197–285.

Spellerberg, I. F. 1969. Incubation temperatures and thermoregulation in the McCormick Skua. Condor 71:59–67.

Stonehouse, B. 1956. The Brown Skua *Catharacta skua lönnbergi* (Mathews) of South Georgia. Scientific Reports of the Falkland Islands Dependencies Survey 14. 25 p.

Swennen, C. 1968. Nest protection of Eiderducks and Shovelers by means of faeces. Ardea 56:248–258.

Thomson, A. L. [ed.]. 1964. A new dictionary of birds. Nelson, London. 928 p.

Tinbergen, N. 1948. Dierkundeles in het meeuwenduin. Levende Nat. 51:49–56.

Tinbergen, N. 1953. The Herring Gull's world. Collins, London. 255 p.

Tinbergen, N., G. J. Broekhuysen, F. Feekes, J. C. W. Houghton, H. Kruuk, and

E. Szulc. 1962. Eggshell removal by the Black-headed Gull, *Larus ridibundus* L.; a behaviour component of camouflage. Behaviour 19:74–117.

Tschanz, B. 1968. Trottellummen. Die Entstehung der persönlichen Beziehungen zwischen Jungvogel und Eltern. Z. Tierpsychol. Beiheft 4:1–103.

Vince, M. A. 1968. Retardation as a factor in the synchronization of hatching. Anim. Behav. 16:332–335.

Vince, M. A. 1969. Embryonic communication, respiration and the synchronization of hatching, p. 233–260. *In* R. A. Hinde [ed.] Bird vocalizations, their relation to current problems in biology and psychology. Cambridge University Press, London.

Vince, M. A. 1970. Some aspects of hatching behaviour, p. 33–62. *In* B. M. Freeman and R. F. Gordon [ed.] Aspects of poultry behaviour. British Poultry Science, Edinburgh.

von Haartman, L. 1956. Der Einfluss der Temperatur auf den Brutrhythmus experimentell nachgewiesen. Ornis Fenn. 33:100–107.

Walters, J. 1958. Über die Brutflecktemperaturen bei See- und Flussregenpfeifern, *Charadrius alexandrinus* und *dubius*. Ardea 46:124–138.

Westerskov, K. 1956. Incubation temperatures of the Pheasant, *Phasianus colchicus*. Emu 56:405–420.

Wetherbee, D. K., and L. M. Bartlett. 1962. Egg teeth and shell rupture of the American Woodcock. Auk 79:117.

Würdinger, I. 1970. Erzeugung, Ontogenie und Funktion der Lautäusserungen bei vier Gänsearten (*Anser indicus, A. caerulescens, A. albifrons* und *Branta canadensis*). Z. Tierpsychol. 27:257–302.

DISCUSSION

S. Charles Kendeigh

In an earlier discussion, I indicated how the amount of energy required for incubation may be measured by the increased food consumption of the incubating adult birds. I would like now to discuss how it may be measured indirectly, using another set of parameters that perhaps may be obtained more easily. The basic point here is that, if eggs are maintained at a relatively constant high temperature during incubation, as they certainly are, then the incubating adult bird must apply heat to the eggs at the same rate at which heat is lost from the eggs to the environment. The question consequently becomes the measurement of the cooling rate of the eggs in the nest.

I have suggested the following equation for these calculations (Kendeigh, 1963):

$$\text{Kcal/bird-day} = \frac{n \cdot w \cdot h \cdot b \cdot (t_e - t_{na}) \cdot i \cdot (1 - c \cdot a)}{1,000},$$

where n = number of eggs in the clutch, w = mean weight of the eggs, h = specific heat of the egg (0.80 g-cal/°C), b = cooling rate of the eggs, t_e = egg temperature maintained in the nest, t_{na} = nest-air temperature surrounding the eggs, i = interval in hours (24, if computations are made on a daily basis), c = percentage of the total surface of the egg covered by the incubating bird, a = percentage of the time interval that the bird is sitting on the eggs, and 1,000 = factor for converting gram-calories to kilogram-calories.

Since eggs lose weight during incubation as a result of evaporation of water, it is desirable to use the mean weight for the entire incubating period. The specific heat of bird eggs has not often been measured and may well vary among species and progressively during incubation, depending on the relative proportion of solid material and water present. The cooling rate of eggs should be the average for the incubation period. If fresh eggs are used for measuring the cooling rate, then allowance must be made for the heat contributed by the embryo, which is especially important in precocial species. The egg temperature should be that of the inside of live eggs. Huggins (1941) found it to average 34 °C in 37 species in 11 orders, but there is some variation among orders. In my experience, 35 °C is a better figure to use for passerine species; Dr. Drent found it to be 37.5 °C in Herring Gulls, and this is probably true also for grouse. It would be a refinement here to obtain the cooling rate of the entire clutch of eggs, which may well be slower than for single eggs. Nest-air temperature is difficult to determine accurately, as it varies among the top, bottom, and sides of the eggs. When the incubating bird is on and off the eggs at frequent intervals, a mean value can only be approximated. The temperature maintained in the nest is also influenced by the effectiveness of the insulation supplied by the nest itself—more on that point later. The percentage of the egg covered by the brood-patch of the incubating bird is important, since this decreases the amount of the surface through which heat is lost from the egg. I estimated it to be about 25 percent in the House Wren (*Troglodytes aedon*). Dr. Drent found the average size of the brood-patch in the Herring Gull to be 18 percent of the average surface area of the egg. Doubtless, this value will vary among species. For those species that do not sit continuously on the eggs, the percentage of time that the birds are attentive must be determined by observation or by an automatic activity recorder. This value, multiplied by the percentage of the egg surface covered, subtracted from 1, gives a correction factor to the overall cooling rate of the clutch.

The cooling rate of eggs has been determined for a few species by inserting temperature probes either into the egg mass itself or into the air space at the large end of the egg in such a way that the sensitive element presses against the egg mass. The rate at which the suspended or fully exposed egg cools at various air temperatures is a constant, in accord with Newton's law of cooling, and is expressed as degree change in egg temperatures per hour per degree difference between egg and ambient temperature. The cooling rate of eggs of different species varies in proportion to the egg mass or weight (w) and surface area and can be calculated from Meeh's formula, $10w^{2/3}$. Thus if the cooling rate of one species is known (b_1), then the cooling rate of another species (b_2) may be calculated from the equation:

$$\frac{b_2 \cdot w_2 \cdot h}{b_1 \cdot w_1 \cdot h} = \frac{10w_2^{2/3}}{10w_1^{2/3}}.$$

The factor h for specific heat cancels out, and the equation simplifies to:

$$b_2 = b_1 \left[\frac{(w_1)^{1/3}}{(w_2)^{1/3}} \right].$$

That is, the rate of cooling is proportional to the cube root of egg weight (Kleiber 1961, p. 143).

The usefulness of this equation may be verified by comparing calculated values with measured values in species that have eggs of different weight (Table 1). Using values for the House Wren egg, the calculated rate for the Dunlin is very close to the measured rate, but for species with much larger eggs the calculated values are too high. Similarly, using values for the large egg of the Herring Gull, calculated values are close to actual for the large eggs of the domestic duck and Jungle Fowl but too low for the two species with much smaller eggs.

TABLE 1   Rates of Cooling for Eggs of Different Species in Degrees Centigrade per Hour per Degree Centigrade Difference between Egg Temperature (t_e) and Nest-Air Temperature (t_{na}) ($b = {}^\circ C/{}^\circ C_{t_e - t_{na}}$-hour)

Species	Weight of Egg	Measured Rate (b_1) (°C/°C-hour)	Calculated Rate (°C/°C-hour)		Reference
			$b_1 = 5.2$ $w_1{}^{1/3} = 1.118$	$b_1 = 1.09$ $w_1{}^{1/3} = 4.55$	
House Wren, *Troglodytes aedon*	1.4g	5.2°[a]	—	4.44	Kendeigh, 1963
Dunlin, *Erolia alpina*	11.2	2.56[b]	2.59	2.20	D. W. Norton, personal communication
Jungle Fowl, *Gallus gallus*	53	1.21[c,d]	1.55	1.32	R. H. Drent, personal communication
Domestic duck	78	1.16[e]	1.36	1.16	Kashkin, 1961
Herring Gull, *Larus argentatus*	94	1.09[d]	1.28	—	Drent, 1967
Gray Lag Goose, *Anser anser*	180	0.73[d]	1.03	0.88	R. H. Drent, personal communication

[a] Average for the incubation period.
[b] Data from 2 eggs; one with embryo heat killed, the other infertile.
[c] Calculations of the cooling rate of eggs of the domestic fowl measured by Moreng and Shaffner (1951) gives 1.20.
[d] Data obtained from fresh eggs.
[e] Calculated from Kashkin (1961, Figure 4) for egg on first day of incubation.

Heating engineers point out that for a given surface area, heat loss from steam pipes of small diameter is greater than for pipes of large diameter—the same may well hold true for eggs of different size. Furthermore, there may be differences in rate of heat conductance through shells of different thickness, and large eggs have thicker shells. The emissivity of heat radiation from the shell surface may vary with the amount of pigment present and with other characteristics. Doubtless the above equation could be refined by including factors for each of these parameters, but this issue will require more study. Meanwhile, I suggest that cooling rates for eggs of different species can be at least approximated by using as reference the cooling rate and the $w_1^{1/3}$ values of that species' egg in Table 1 that most closely agrees with the weight of the particular species under consideration.

One must be careful, however, in using these data and in obtaining new rates of cooling for eggs of other species. In personal communications, W. R. Siegfried reports a measured cooling rate for the 66-g eggs of the Ruddy Duck (*Oxyura jamaicensis*) of 2.46 °C and J. A. L. Mertens reports a rate for the 1.2-g eggs of the Blue Tit (*Parus caeruleus*) of 7.55 °C/°C$_{t_e - t_{na}}$-hour. Both of these rates are high compared to the data for the other five species listed in Table 1 and are very likely due to inconsistencies in the manner in which the measurements were made. There is need to standardize measurements of cooling rates with respect to position of temperature probe in the egg, which probe should register the temperature of the egg mass rather than the air space; to air movements around the egg, which should be none; to relative humidity, which ideally should be the mean relative humidity in the nest (although very little information is available on this parameter); to radiation gradient from egg to surrounding objects; and to the manner in which the egg is suspended or supported (here again perhaps it would be best to support the egg on a platform of material similar to that in the nest lining). I would encourage further study of this problem.

The nest-air temperature may be measured by means of thermocouples (Kendeigh, 1963), or thermistor probes (Drent, 1967), placed in various positions in the nest. In our work with the House Wren, the mean nest-air temperature, combining all attentive and inattentive periods throughout the day and night, averaged about 0.5 °C lower than on the nest bottom or floor. The nest-floor temperature was registered continuously throughout incubation by means of a recording potentiometer. Determination of average nest-air temperature is somewhat simplified in those species that incubate the eggs continuously, as in the Herring Gull, once the gradient from the brood-patch to the eggs to the nest wall is established. Drent (1967) gives a table of nest-air temperatures for several nonpasserine species. We need monographic treatments of the thermodynamics of the incubating process in different species. Temperature gradients doubtlessly vary depending on the insulating characteristics of the nest. Hence, once this relation is established, it may be possible to estimate nest-air temperatures from the type of nest that is built.

The effectiveness of nest insulation is evident from the extent to which nest-air temperature varies at different ambient temperatures (Figure 1). As Drent (1967) points out, the temperature of the Herring Gull nest varies insignificantly with changes in ambient temperature between 12 and 28 °C. The nest is on the ground, from which the eggs are separated by a layer of dried grass. The surrounding grassy vegetation protects the ground from direct insolation during the daytime such that fluctuation in ground temperature is largely dampened and insignificant. Ground temperatures are well below mean air temperatures. Likewise, the 98 percent attentiveness of the parents, who alternate in incubating the eggs, maintains a nearly constant nest-air temperature. Constant incubation and a stable ground temperature therefore eliminate fluctuations with air temperature,

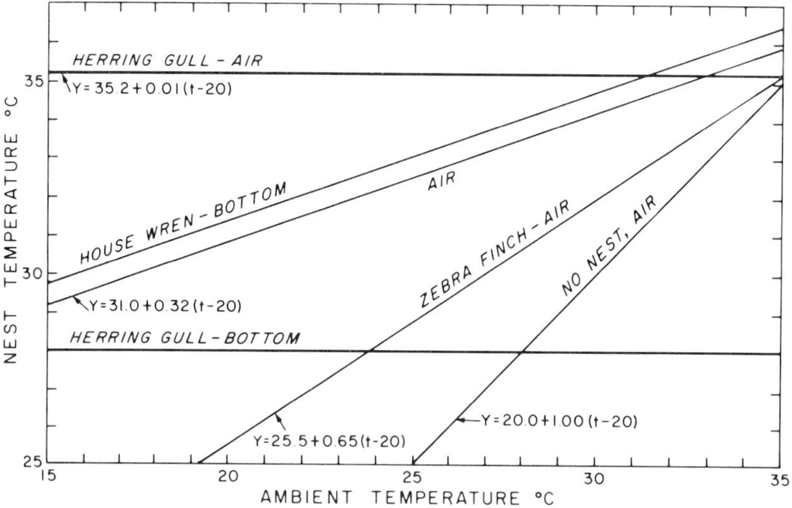

FIGURE 1   Nest temperature (bottom and nest-air) for three species as a function of environmental temperature.

but the low ground temperature produces a steep gradient in temperature between the top and bottom of the eggs. The steep gradient means that the rate of heat loss from the eggs is high, which loss must be compensated for by a high rate of heat input from the incubating birds.

The nest of the House Wren is made of sticks, the nesting cavity is lined with dried grass and feathers, and the whole thing is in a cavity—in this case a wooden box. The good insulation furnished by the feathers, nest, and wooden box permits only a small difference between the nest-air and nest-bottom temperatures. Heat liberated from the brood-patch is used with but little waste. However, since the nest is off the ground and the adult bird incubates only about 75 percent of the time, there is room for considerable response to variations in air temperature.

The nests of the Zebra Finch (*Taeniopygia guttata*) studied by El-Wailly (1966) were made of shredded excelsior and chicken down-feathers that he provided. The nests were partially domed, but were loosely made and directly exposed to room temperatures. Nest-air temperatures were not measured directly but, knowing the energy cost of incubation (kcal/bird-day), the above equation was employed to derive nest-air temperature at various room temperatures. Incidentally, the cooling rate of the Zebra Finch egg, as calculated by El-Wailly, is incorrect. It should be $5.87°C_{(t_e - t_{na})}$-hour, not 4.04 °C. Although the adult birds sit on the eggs 96 percent of the time, with both male and female sharing the duties, the regression line of nest-air on ambient temperature is much steeper than in the House Wren. This reflects the poorer insulation of the nest.

Insulation is usually calculated as the reciprocal of the measured rate of heat conductance from the center of a body through to the surface, from which point the heat is lost through radiation. It would be desirable to measure the rate of heat conductance from the brood-patch of the incubating bird, through the eggs and nest, to the surrounding air. Until this can be done, perhaps we can take the reciprocal of the slope of the regression lines in Figure 1 as a relative index of the insulative properties of the nest together with the incubating bird. These are as follows:

Herring Gull	100
House Wren	3.1
Zebra Finch	1.5
Air alone	1.0

The energy cost of incubation depends on the amount of heat that the incubating bird must supply the eggs to maintain them at the proper incubation temperature. The incubation egg temperature varies slightly between species, but a more striking difference between species that affects energy costs is the insulation characteristics of the nest. The incubation cost in the Herring Gull is high (Table 2), especially during the peak period (fourteenth to twenty-first day of the 30 days that the adults cover the eggs) because of the rapid rate of heat conductance to the ground. However, this is alleviated by the fact that the egg temperature is not raised to the full incubation temperature until about day 14 and to production of considerable heat by the embryo itself after about day 21. Actually, the average cost of incubation for the entire period is about 25 percent of the productive energy the adults have available.

By nesting off the ground and taking advantage of the higher air temperatures, the House Wren and Zebra Finch reduce the energy cost of incubation at the prevailing high late spring and summer air temperatures. It is of interest that it is about the same per bird at 21–22 °C in the two species. However, being thus exposed to air temperatures, the incubating birds are more sensitive to fluctuations in air temperature than are ground-nesting species. Air temperatures are consequently more critical to these species in controlling the onset of nesting activities in the spring, limits of breeding distribution, etc.

TABLE 2 Cost of Incubation in Relation to Productive Energy Available[a]

Species	Ambient Temperature (°C)	Sex That Incubates	Percent of Productive Energy Available	
			Per Individual of Pair	Per Individual If Alone
Herring Gull	12–28	M, F	33[b]	66
			25[c]	50
House Wren	14	F	–	100
	17		–	51
	22		–	*17*
Zebra Finch	14		40	80
	21	M, F	*18*	36
	24		11	21
	29		6	12
	34		0	0

[a] The italicized figures are most representative for natural conditions.
[b] At peak.
[c] Average total period.

I have already mentioned the interrelations between nest insulation, energy cost of incubation, and breeding behavior, as shown by these three species. It is advantageous to the species for breeding behavior to minimize the energy drain of incubation as much as possible. This is done by restricting breeding to a season when ambient temperatures are favorable and by adjustments in actual nesting behavior. It is necessary for both adult Herring Gulls to share incubation, since the cost to a single individual to do it alone would be excessive (50–66 percent). It is also necessary to maintain a constant nest-air temperature because of the steep gradient involved. If the eggs were intermittently allowed to cool because the incubating bird was inattentive, the cost of rewarming the eggs would be much greater than is the cost of maintaining them at the high incubation temperature (Kendeigh, 1963).

The situation with the House Wren is different. Nest insulation is so excellent that the female can meet the energy cost of incubation by herself. Inattentive periods occur frequently during the daytime but are of short duration (8 minutes) so that the drop in egg temperature is slight and quickly restored in the next following attentive period (15 minutes). Freeing the male from incubating duties allows him time to establish and defend a sizeable territory and frequently to become polygynous.

The nests made by the Zebra Finches within the cages may have been somewhat less well insulated than natural nests of the species in the field (Immelmann, 1962). However, there can be no doubt that energy requirements to maintain incubating egg temperatures in this species are greater than in the House Wren. At 21 °C, if only one bird were to incubate, incubation costs would equal 36 percent of its productive energy. By sharing incubation, however, the cost to each adult bird is approximately equal to that of the female House Wren (Table 2). It is also significant that colonial nesting occurs in both the Zebra Finch and the Herring Gull, where the males are continuously busy with incubation and do not have time to defend extensive territories.

I have tried to show a definite link between breeding behavior and reproductive physiology and that the nesting behavior of birds is conditioned to a very large extent by energetics as demonstrated by incubation costs, nest insulation, nest location, ambient temperature, season, and geographic location.

REFERENCES

Drent, R. H. 1967. Functional aspects of incubation in the Herring Gull, *Larus argentatus* Pont. E. J. Brill, Leiden.

El-Wailly, A. J. 1966. Energy requirements for egg-laying and incubation in the Zebra Finch, *Taeniopygia castonotis*. Condor 68:582–594.

Huggins, R. A. 1941. Egg temperatures of wild birds under natural conditions. Ecology 22:148–157.

Immelmann, K. 1962. Beiträge zu einer vergleichenden Biologie australischer Prachtfinken (Spermestidae). Zool. Jb. Abteil. Sys. Okol. Geograph. Tiere. 90:1–196.

Kashkin, V. V. 1961. Heat exchange of bird eggs during incubation. Biophysics 6(1):57–63.

Kendeigh, S. C. 1963. Thermodynamics of incubation in the House Wren, *Troglodytes aedon*, p. 884–904. *In* Proceedings, XIIIth International Ornithological Congress. American Ornithologists' Union, Louisiana State University, Baton Rouge.

Kleiber, M. 1961. The fire of life. John Wiley & Sons, New York. 454 p.

Moreng, R. E., and C. S. Shaffner. 1951. Lethal internal temperatures for the chicken, from fertile egg to mature bird. Poultry Sci. 30:255–266.

INFORMAL DISCUSSION

BEER: I would like to make two comments on this question of synchrony in incubation or hatching that Doctor Drent discussed. I think he may have given the impression that the main factor in effecting synchrony is communication between the eggs, as demonstrated by Margaret Vince. In fact, there is some evidence that the behavior of the parents contributes, too, at least in some species. I think it was Ytreberg* who first showed this in Black-headed Gulls by taking eggs as soon as they were hatched, placing them under birds already incubating, and then returning them to the original parents, so that those eggs all began development as soon as they were laid. For those eggs, the hatching intervals were virtually the same as the laying intervals, suggesting that in the normal case—because the parent delays the onset of effective incubation until the clutch is complete—it, in part at least, contributes to the synchrony in hatching that one finds in those gulls.

The other point that I wanted to make is that synchrony in hatching is not

*Ytreberg, N. J. 1956. Contribution to the breeding biology of the Black-headed Gull (*Larus ridibundus* L.) in Norway. Nytt Mag. Zool. 4:5–106.

universal. There are some birds, particularly the owls and hawks, in which hatching intervals are virtually the same as laying intervals. David Lack* some years ago suggested that this exception might be an adaptation to the widely fluctuating availability of food for owls and hawks, particularly in the Arctic, from season to season.

NORTON: First a brief point again on synchrony. It may not need pointing out, but the value to those species that do indeed depend on a synchronous development is fairly easy to demonstrate if one takes the time to produce asynchronous hatchings as I have succeeded in doing, again with Arctic shorebirds. I have been able to produce a very adversely affected growth rate among chicks by destroying the synchrony of hatching within a clutch by cooling some of the eggs during the incubation period.

Reverting to Doctor Kendeigh's discussion of cooling rate, I think there is some evidence of importance here. Among the shorebirds there is a very stable pattern of a four-egg clutch, which is found across several major taxonomic divisions—the phalaropes, the plovers, and the sandpipers. I have wondered why clutches have no fewer than four eggs. Even when we see ever-increasing evidence of species that lay two successive clutches (or possibly more), one to be incubated by the male and the other to be incubated by the female, why would the establishment of a second nest not bring about a situation in which two eggs are laid in the first nest and two in the second nest, rather than doubling up all of the numbers to maintain a four-egg, four-egg repetition?

A possible explanation for this stable clutch size lies with the cooling rate. By subtracting one egg from a normal four-egg clutch and observing the cooling rate of the resulting experimental three-egg nest, I found a great increase of the cooling rate of an individual egg, which I suspect is due to the simple physical fact that a greater absolute amount of surface area was exposed after subtraction of that one egg.

ABBOTT: I also wish to comment on synchrony. My personal experience is with artificial incubation with both domestic and game species. I wish to reinforce the point about the age of the egg and preincubation effects as important contributors to synchrony.

I would also like to point out that both the total length of the incubation period, as well as the interval between the first pip and hatching, may be influenced by nutrition, genetics (especially inbreeding), oxygen availability, general activity, or egg position, as well as temperature. So, although, this observation on clicking or snoring—or whatever it is called—does seem to have some interest, I think that if one goes into a detailed analysis with large numbers that many of these other factors should be kept in mind as well.

ORING: One additional factor that, on numerous occasions, has struck me as being rather important with regard to synchrony in hatching of shorebird eggs has to do with change in incubation behavior during the last day of incuba-

---

*Lack, D. 1954. The natural regulation of animal numbers. Clarendon Press, Oxford. 343 p.

tion or during the time of hatching. At about the time the first egg begins to
pip, attentiveness to the nest begins to drop. Once the first egg is hatched, in
fact, attentiveness to the nest has dropped remarkably, bringing on quite a bit
of cooling to the later chicks to hatch and, according to Lind's* hypothesis,
thereby increasing their pipping rate and decreasing the length of time to
hatch.

DRENT: Let me make just a summary statement about hatching synchrony, a
topic that seems to have struck the imagination. I did not want to suggest that
"clicking" is the only thing leading to synchrony. What I wanted to say is that
this is a fine mechanism that facilitates minute to minute synchrony. Of
course, these minor adjustments are superimposed on the total developmental
time, which, as Beer pointed out, relates to the time when the bird starts to
apply heat to such an extent that the embryo temperature will surpass developmental thresholds. I hope I have not confused people by simplifying this
story, and I must point out that chickens do not hatch synchronously so,
of course, we cannot disprove a phenomenon on a bird that does not show it.
Gulls do not show synchrony of hatch, either.

*Lind, H. 1961. Studies on the behavior of the Black-tailed Godwit (*Limosa limosa* L.).
Medd. Naturfredningsrådets Reservatudvalg 66:1–157.

Behavioral Components in the Reproductive Biology of Birds

COLIN G. BEER

The relationships between behavior and physiology are several. All behavior is a matter of events in nerves, muscles, and sense organs, and many people consider that the most profound accounts that can be given of behavior have to be in terms of such events. Reductionism of this sort echoes nineteenth century slogans about the nature of mind: "No psychosis without neurosis"; "Ohne Phosphor, kein Gedanke." But physiology has yet to absorb the whole of psychology and the larger part of ethology still treats behavior as more than an epiphenomenom.

Behavior enters as cause and consequence in processes described at the level of functional systems. Cybernetics has shown how such description can be made without reference to material particulars. But even this very general and abstract way of dealing with relationships between behavior and physiology has its limitations. There are behavioral phenomena that remain virtually out of reach of any physiological account, at least so far as present thinking goes. The behavior has been described in terms for which there are no counterparts in physiological language and for which there appear to be at present no alternatives. Discussion of such behavior can proceed, at least in the meantime, only within the universe of discourse constituted by the kinds of terms in which it has been described.

It is therefore prudent, when considering a topic like behavioral components in the reproductive biology of birds, to keep in mind the possibility that these components may be heterogeneous in ways that make it difficult to be both comprehensive and coherent. Comprehensiveness may require bringing together facts, concepts, and theories among which the empirical and logical relationships are obscure; a unified account may be possible only for selected cases.

Here, I attempt neither a comprehensive nor a unified account of our subject. Instead I have chosen to dwell on the heterogeneity I find in it, to recognize which I think is important in appreciating the range of problems posed by the breeding biology of birds. I shall begin with some examples of ways in which behavior and physiology have been found to be linked causally in the breeding biology of birds; I shall then discuss some aspects of behavior that seem to me to be recalcitrant to the dominion of physiological concepts, yet study of which has contributed to our understanding of breeding in birds.

CAUSES IN PATTERNS

Although the relationships between behavior and physiology are so various as to defy tidy classification, it is useful to adopt some system of distinctions. I shall utilize the familiar distinction between external and internal influences (cf Marler and Hamilton, 1966) and some further subdivisions that can be based on it.

Von Holst (1954; von Holst and Mittelstaedt, 1950) drew attention to two kinds of sources of stimulation immediately affecting behavior. First, there is stimulation that originates from a change in the environment and with respect to which the animal is a passive recipient. A cloud moving across the sun would give rise to stimulation of this sort for such an animal as a diurnal songbird. In von Holst's terms, such stimulation constitutes "exafference." The second type of stimulation is that which arises as a consequence of an animal's own movements. For example, if a songbird turns its head, it alters its field of view and thus causes a change in the visual stimulation it receives. Stimulation caused in this way was called "reafference" by von Holst, who showed how it can be involved in the control of movement.

This distinction between passive and active modes of stimulation can be extended to less immediate influences affecting physiology and behavior. Gradual or long-term effects on an animal can result from environmental changes that take place of their own accord, or from changes in the environment that are effected by the animal's activity. The passive type of influence is exemplified by the role of changes in photoperiod in the reproductive cycling of temperate zone birds. Another example is the effect of the emergence of green grass, following heavy rainfall, on the initiation of breeding activity in Red-billed Diochs (*Quelea quelea*) in tropical Africa (Marshall and Disney, 1957). The self-induced type of change in external influences is exemplified by the substitution of one habitat for another that results from migration and by the more specific effects of presence of a nest and eggs on the physiological state and behavior of breeding birds.

In the latter instance, we have reciprocal relations between physiology and behavior; a change in physiological state—perhaps induced by an external influence such as photoperiod—brings about a change in behavior that, in its turn, produces effects on the physiological state. Such patterns of cause–effect relationships fall into three classes that I shall refer to as the conservative, the radical, and the progressive.

The conservative case is one in which the effect tends to negate its cause and thus promotes preservation of the *status quo*; a deviation from a given initial state of affairs causes activity that counteracts the deviation and thus restores the original state of affairs. In short, we have to deal with what has come to be called negative feedback, but I hesitate to use that term lest it be taken to imply that we can give a rigorous mathematical formulation to all such relationships when we perceive them. The conservative pattern is likely to appear in any system that involves maintenance of a steady state. It is familiar to us in the phenomenon of homeostasis, i.e., the regulation of body temperature, hunger, thirst, etc. In endocrinology, it is exemplified by relationships between secretion of tropic hormones by the pituitary and hormone secretion by its target organs. For instance, the amount of thyroid hormone circulating in the blood is regulated by secretion of pituitary thyrotropic hormone, which is increased by a drop in circulating thyroid hormone and decreased when the circulating thyroid hormone rises above a certain concentration.

The secretion of hormones by the gonads also is governed by tropic hormone secretion of the pituitary, but here the relationships are more complicated than they are in the case of the thyroid gland. The stability that is the virtue of homeostasis has to be limited if anything as innovative as reproduction is to take place. In fact, the conserva-

tive effects tend to be transitory and embedded in larger patterns involving progressive change in the reproductive biology of birds. The
incubation period is the time when change is least apparent, but even
here one has to abstract to find anything like a clean example of
negative feedback at work in the control of behavior. Certain features of incubation behavior can be so construed. Because of effects
of preceding events on the reproductive cycle, an incubating gull
appears to be "set" to receive a specific pattern of somaesthetic stimulation from sitting on its clutch. When some feature of this pattern
is lacking, the gull will repeatedly perform the kinds of actions that
normally establish the pattern, i.e., shifting the eggs and settling on
them (Baerends, 1959; Baerends and Drent, 1970; Beer, 1961). In
this and other ways a gull keeps the temperature of the eggs more or
less constant and thus regulates the supply of heat to the eggs in line
with requirement (Baerends and Drent, 1970). Once the nest is built,
a bird may repair damage to it, with the result that the form of the
nest is maintained. A student of mine, Steven Chernesky, has observed what happens when one-half the rim of a nest of a Laughing
Gull (*Larus atricilla*) is removed. He has found that the bird performs
building behavior that is directed almost entirely at the missing portion of the nest and continues until it is restored. Thus even the simple form of a gull's nest can be viewed as the result of deliberate action, guided by something like comparison between conception and
perception. However, whether such a view can be translated into
strict physiological terms is open to question, so I shall turn aside
before I get ahead of myself.

Radical cause–effect relationships consist of cases in which an effect tends to add to the cause that gave rise to it and thus to contribute to its own augmentation. It is familiar in control theory as positive feedback, but, when I use that term here, I do so loosely, without implying the rigorous mathematical connotation it has in that
context. In contrast to negative feedback, positive feedback connections, as long as they persist unchecked, carry a system progressively
farther from its starting point; they are inherently unstable rather
than stable, radical rather than conservative. Hence their operations
are typically transitory and are kept in bounds by limiting conditions.
For example, the hormonal conditions that give rise to certain phases
of reproductive behavior, in turn, are promoted by involvement in
the behavior itself; but this mutual augmentation may be constrained
by the competing demands of functions other than reproduction. In
any case, it is impossible for a bird to behave reproductively for more

than 24 hours a day. In general, in the reproductive biology of birds, radical effects are incorporated within a larger control system in such ways as to play constructive roles, and they may even be utilized in maintenance of a steady state. For example, there is evidence from work on Ring Doves (*Streptopelia risoria*) that prolactin secretion sustains incubation behavior and is itself caused by the stimulation received from incubation (summarized in Lehrman, 1965). The concentration of circulating prolactin mounts as the incubation period proceeds, but incubation behavior, once it has become fully established during or shortly after laying, remains virtually constant in form and temporal distribution until hatching begins, although there may be some increase in the resistance a dove puts up to being dislodged from the nest.

The progressive case is like the conservative in that the effect tends to negate the conditions that gave rise to it, and it is like the radical in that it moves the system away from its initial conditions; instead of restoring the *status quo*, the feedback here brings into being new conditions that produce new effects and thus carries the system from one phase into another. Each of the sharp transitions in the reproductive cycle of birds can be viewed in this light, for both the laying of eggs and the hatching of chicks are consequences of the preceding behavior of the breeding birds and give rise to conditions that change the behavior. The more gradual transitions within the major divisions of the cycle are also effected by such interplay between physiology, behavior, and external stimulation. Courtship sets up the conditions for mating and nest-building; these, in turn, contribute to the process that culminates in egg-laying. Moving to a finer time scale, one can see the pattern at work in social interactions; the behavior of one bird may affect the behavior of another in ways that cause change in the behavior of the first. For example, in the prelaying period of gulls, food-begging by a female can induce her male to regurgitate food. This courtship feeding typically leads into copulation attempts, apparently increasing the sexual arousal of the male and the receptivity of the female. Within the act of copulation itself, there is a sequence, the progress of which is effected by reciprocal influences between the behavior of the two birds. For example, unless the tail-wagging movements and other actions of the mounted male induce the female to cooperate by lifting her tail, the male cannot achieve cloacal contact, and cloacal contact is necessary if insemination is to occur.

The specific kinds of physiological processes mediating these

various effects of behavior on behavior, of course, will differ according to which aspect of the reproductive cycle one is looking at and
how closely. In general, they are only partially understood, but
enough is known to tell us at least what to expect. Among the closest
approaches to an integrated picture are two investigations of the
onset of incubation behavior—one carried out by Hinde and his
group at Cambridge on the female canary the other carried out by
Lehrman and his group at Newark on the Ring Dove. I shall describe
some of this work in sufficient detail to illustrate how the various
cause–effect patterns I have mentioned combine to effect the reproductive process. In both investigations, the strategy has been to
manipulate the external stimuli available and the endocrine state in
various combinations and sequences to test ideas about how they influence and are influenced by the behavioral transitions that lead up
to laying and incubation.

CANARIES

Investigations on canaries have emphasized the point that, as a rule,
the causal factors underlying behavior and the causal consequences
of behavior are multiple and complexly interrelated. Hinde (1956,
1959, 1960, 1970) has questioned the utility of concepts of drive
or instincts in the causal explanation of behavior. Whether his scep-
ticism of old doctrine was a consequence of his research, or provided
motive for his research, I shall not venture to speculate, but that his
work has set what appears to be insuperable obstacles to any currently held theory of drives seems to be beyond question. In studying
the reproductive behavior of canaries, he and his colleagues have
shown that the probability of occurrence of a certain behavior pattern may be influenced by both external and internal factors, the
effects of which may be both short term and long term, peripheral
and central, excitatory and inhibitory, the balance between these
several influences shifting as their consequences take effect.

The reproductive cycle in a female canary begins with environmentally induced hormone secretion (estrogen, consequent to FSH),
which, apparently by acting on certain areas of the brain, causes the
bird to be attracted by the singing and other courtship behavior of
a male (which has been similarly aroused by environmentally timed
production of androgen). The stimulation that the female receives
through thus attaching herself to a male augments her gonadotropin

secretion and hence her ovarian development and secretion (Warren and Hinde, 1961a). Apparently, because of the consequent increase in circulating estrogen, the female becomes receptive to the male's attempts at copulation and is also impelled to perform nest-building behavior. At first this nest-building behavior consists of desultory picking-up and carrying of nest material, but these activities very soon become centered on a nest site where a third category of movements—those involved in arranging the material into the form of a nest—is added.

From close analysis of the sequential patterning of nest-building behavior and the distribution of time between its three components—gathering, carrying, and building—Hinde has concluded that transition from one activity to the next is determined by three interacting influences; the antecedent activity provides stimulation for the subsequent activity (gathering provides the bird with material to carry, carrying brings the bird to the nest site with material to build into it); after a time, performance of an activity has a self-suppressing effect; and there are differences in the levels of hormones necessary to support the behavior—building requires higher levels than carrying, and carrying requires higher levels than gathering (Hinde, 1958). As a consequence of performing these activities, a nest cup is constructed; this provides stimulation that has both short- and long-term effects on the behavior. For a bird in the cup, it elicits and orients the building movements. When the bird leaves the cup, the stimulation it has received there affects how soon it will return and what sort of material it will select to bring back. Given a choice of grass or feathers, the bird will select grass as long as it is engaged in building the outer structure of the cup; it will then switch to feathers for the lining. The transition is effected by stimuli from the grass cup (Hinde, 1958; Hinde and Steel, 1962). The long-term effects of stimuli from the nest involve endocrine changes—addition of progesterone and prolactin to the already circulating estrogen—that hasten the onset of egg-laying and also induce the formation of a brood-patch on the ventral surface.

Brood-patch formation consists of four components: loss of feathers, increase in vascularity, increase in somaesthetic sensitivity, and development of edema. The increased vascularity is a peripheral effect of estrogen acting alone; defeathering is a synergistic effect of estrogen and prolactin; and the increased sensitivity is a synergistic effect of estrogen and progesterone (however, the experiments also indicated that progesterone and prolactin can substitute for one

another in these functions). The edema appears to be a more direct
effect of tactile stimulation from the nest cup and possibly also
from eggs. Stimulation from a nest under construction thus leads to
increased sensitivity to such stimulation (Steel and Hinde, 1963,
1964; Hinde *et al.*, 1963; Hinde and Steel, 1964).

The stimulation from a completed nest tends to inhibit further
nest-building activity, however, and the changed endocrine state
consequent to nest construction also decreases nest-building tendency
(Hinde and Warren, 1959; Warren and Hinde, 1959, 1961b). The
arrival of eggs in the nest, together with the endocrine state present
at laying, induce incubation behavior, stimulation from which leads,
in turn, to further endocrine change that puts a stop to laying (thus
determining the size of the clutch) (Eisner, 1960) and sustains incubation. The events leading up to incubation in the reproductive
cycle of the female canary thus consist of changes in the external
stimuli present to the bird, changes in her hormonal condition,
changes in her behavior, and complex and multiple interrelationships
between these changes [see Figure 26.2 in Hinde (1970)].

RING DOVES

Although there are substantial differences in detail, the story that
has been worked out for the Ring Dove follows the same general pattern as that of the canary. Progress through the reproductive behaviour cycle is effected by changes in endocrine state consequent,
at least in part, on changes in stimulation to which a bird is exposed,
some of which are imposed by the environment or the mate and some
of which result from the bird's own activity.

If a reproductively mature Ring Dove, either male or female, with
or without previous breeding experience, is taken without preparation
and placed alone in a cage containing a nest and eggs, it will not
incubate. A male and female placed together in such a cage begin
incubating after about a week, during which time they engage in
courtship and nest-building behavior. A male and female that have
been together for a week prior to being placed in the cage containing
the nest and eggs begin incubating in about a day. If nest material
has been available to the birds during the week of prior association,
they begin incubating as soon as they are placed with the nest and
eggs. From these observations, it was concluded that association with
the mate and with nest material causes physiological changes that

underlie readiness to incubate (Lehrman, 1958a,b; Lehrman *et al.*, 1961b).

That gonadotropin secretion is involved in this process was indicated by the fact that differences between the experimental groups with respect to readiness to incubate were paralleled by differences in the amounts of oviduct growth and the incidence of ovulation in the females (Lehrman, 1961; Lehrman *et al.*, 1961b). These indices of the effects of prelaying stimulation were used to gage the influence of courtship behavior. When females were presented for 7 days with a view of a male behind a glass partition, they showed less oviduct growth, and fewer ovulated, if the male was a castrate (and hence performed no courtship) than if the male was intact (Erickson and Lehrman, 1964; Lott *et al.*, 1967). Variation in the amount of time a female was exposed in this way to a male and variation in the amount of courtship during a standard exposure time were also reflected in different degrees of oviduct enlargement (Barfield, 1971). Hence, it has been concluded that, "in the female Ring Dove, secretion of gonadotropin is graded in accordance with the amount of courtship behavior perceived by her" (Barfield, 1971, p. 305). However, there is also evidence of what might be called a "momentum effect." Once stimulation from the male and from nest material has started the process leading to ovulation sufficiently along its way, the stimuli can be removed and the process will continue to run its course (Lehrman *et al.*, 1961a). Experiments also have shown that the effect of courtship stimulation is augmented by background auditory stimulation from the rest of the colony (Lott *et al.*, 1967)–a fact that may be relevant to how the social behavior of colonial birds is implicated in the timing and synchrony of their breeding (Darling, 1938; Coulson and White, 1956). However, although such auditory stimulation by itself can induce a certain degree of ovarian development (Lehrman and Friedman, 1969), it is usually insufficient, without the support of at least the visual stimulation from a mirror, to cause a female to reach the stage of egg-laying (Lott and Brody, 1966).

Experiments involving administration of hormones have indicated that full growth of the oviduct prior to ovulation, in female Ring Doves, is brought about by the combined effects of estrogen and progesterone. Estrogen comes first, laying the ground work, as it were, and then acts synergistically with progesterone to give the finishing touches (Lehrman and Brody, 1957). There is also evidence that progesterone is involved in the process that triggers ovulation.

Later experiments compared the effects of treatments with estrogen and progesterone and prolactin on readiness to incubate. Progesterone treatment was the most effective; most of the doves thus treated sat almost immediately when presented with a nest and eggs. Estrogen-treated birds were slower; most took between 1 and 3 days to begin incubation, and some took more than 11 days (Lehrman, 1958b). Results with prolactin, although positive for more than half the birds when the dosage was very high, indicated that this hormone is probably not responsible for the initiation of incubation under normal circumstances. The amount of hormone necessary to induce incubation was so high as also to induce enlargement of the crop—a condition that is not present at the beginning of the incubation period in normally breeding doves, but that does arise toward the end of the period (Lehrman and Brody, 1961). That progesterone invokes its incubation-inducing effect by acting on specific sites in the brain has been demonstrated in a study in which minute quantities of the hormone were placed at various points in the brains of doves that were then tested for incubation (Komisaruk, 1967).

Putting all these and other results together the following story emerges (Lehrman, 1958a,b, 1965):

1.   In the normal reproductive cycle of female Ring Doves, participation in courtship results in stimulation that, acting via the hypothalamic control of gonadotropin secretion, promotes ovarian development and production of estrogen.

2.   The estrogen so produced causes the birds to perform nest-building behavior, stimulation from which changes the pattern of gonadotropin secretion, thus bringing progesterone into circulation.

3.   The progesterone, in addition to its involvement in maturation of the reproductive tract and laying, acts on parts of the brain in such a way as to make the birds respond to eggs by sitting on them.

4.   Stimulation from sitting on the eggs promotes prolactin secretion, which sustains incubation and also induces formation of crop milk in preparation for feeding of the squabs.

The story for males is not so fully worked out. According to Stern and Lehrman (1969), castrated males will not incubate, even when treated with progesterone; they can be induced to incubate only if they are primed with androgen prior to progesterone administration. These results suggest that incubation, like the male courtship display (the "bowing-coo"), is androgen-dependent and that the effect of androgen differs according to the physiological context. For incubation to be induced, androgen requires the addition of progesterone, a condition that inhibits the courtship display (Erickson *et al.*, 1967). However, the doves used in this study were kept in isolation racks until the day of testing. More recent work (Silver, in preparation) has shown that, if males are tested for incubation at regular intervals after being placed with a female, they respond positively after the usual period of 7 or 8 days, even when there is neither androgen nor progesterone present in their bodies.

It also has been found that the effect of a hormone can be influenced by social factors. The probability that a progesterone-treated dove, either male or female, will incubate is significantly greater if the bird is tested in company with a mate than if tested alone; even with a mate present, the bird is more likely to incubate if the mate does also (Bruder and Lehrman, 1967).

From these and other studies, it is apparent that the roles of hormones in the reproductive cycle of Ring Doves is not simply a matter of a different hormone or hormone combination providing the necessary and sufficient conditions for each phase. The production and effects of the hormones depend on a complex of factors, both behavioral and physiological, both past and present. The point has been underlined by a study of the importance of experience in the development of incubation behaviour. Since Ring Doves breeding for the first time in their lives almost always incubate the eggs that they produce and since there is good evidence that progesterone is the hormone involved in the initiation of incubation behavior, it might be expected that reproductively naive doves can be made to incubate simply by injecting them with progesterone and then presenting them with a nest and eggs. A proportion of doves so treated did perform as expected but in a manner that differed from the performance typical of progesterone-treated experienced doves. Whereas experienced doves start incubating within half an hour of being presented with eggs, the naive birds that sat did not do so until more than a day had passed, and among the rest were several that apparently did not even locate the eggs (Lehrman and Wortis, 1960).

A more recent study has shown that reproductively experienced doves are more efficient breeders than are inexperienced doves (Lehrman and Wortis, 1967), but the differences are not very great. In the normal course of events, birds breeding for the first time usually manage better than the tests of progesterone-treated naive birds would lead one to expect. The explanation for this discrepancy appears to lie in effects of experience gained in the beginning of a normal breeding cycle. As a consequence of having participated in courtship and then nest-building, even for the first time in its life, a dove is oriented toward and attached to the place where the eggs appear. Michel (cited in Lehrman, 1971) has allowed naive doves to progress to different points in the reproductive cycle before removing them from the breeding situation, holding them in isolation for several weeks, and then testing them for incubation after treatment with progesterone. He found that doves allowed to go as far as nest-building were not far short of the performance of fully experienced doves in the incubation tests.

In addition to orientation to the nest site, the fact that a bird has formed a pair-bond and hence is partnered by a mate whose physiological state and behavioral orientations have, as a consequence of the behavioral interactions of courtship and nest-building, been meshed with its own, is another consequence of the prelaying experience that probably affects how well a dove performs reproductively. The effects of experience of such partnership persist beyond completion of a single breeding cycle. Morris and Erickson (1971) recently have reported that, when a Ring Dove is given the opportunity to choose a mate from among a number of birds, including one with which it has bred before, it re-establishes the old pair-bond, even if the birds have been apart from one another for as long as 7 months.

GULLS

With talk of pair-bonding and orientation to the environment, we approach matters about which physiology has little helpful to say. We have seen how behavioral relationships such as these can have physiological consequences, and have considered some of the physiological conditions that are necessary to them. But, when we come to look at such relationships in detail, focusing on particular rather than general aspects, the available physiological information proves to be either too coarse in texture for the resolution sought or irrelevant to the questions posed. For the time being at least, study of these aspects and these questions is bounded by behavioral and ecological concepts and methods. Nevertheless, such study can contribute usefully to the understanding of reproductive biology.

Consider the matter of individual attachments. Coulson (1966, 1972) has found a correlation between breeding success and the durability of pair-bonds in the Black-legged Kittiwake (*Rissa tridactyla*). Birds that keep their mates from one breeding season to another are likely to breed earlier, to lay larger clutches, and to raise more young to fledging than are birds that do not. In Kittiwakes, as in other species of gulls, the male and female share incubation. Each pair works out its own way of distributing time on the nest, and the schedules differ considerably from pair to pair. In some cases, the birds fail to achieve a working arrangement and this is one of the major causes of brood loss. Such birds usually "divorce." Successful pair-bonding is thus a matter of finding a compatible mate and then forging a lasting relationship by mutual adjustment of behavior. Precisely how such adjustment is achieved remains to be determined. Before progress can be made on resolving the issue, individual differences in the relevant behavior will have to be subjected to fine-grained analysis. But it is already clear that knowledge of such differences illuminates important aspects of the breeding biology of Kittiwakes. The same is no doubt true for many other species as well.

One obvious way in which individual differences in behavior may contribute to the breeding process is by providing characteristics upon which individual recognition can be based. Whenever one is dealing with a social system involving individual social relationships, individual recognition between the animals thus related must be assumed to exist. Only recently, however, has the basis of such recognition become a subject of direct investigation. In bird studies, most of the work thus far has been directed to recognition by ear. Two approaches have been followed: analysis of the characteristics of vocalizations of different individuals and playback experiments in which observations are made of the responses of birds to recorded calls of different conspecific individuals. For a number of species there is now evidence of individual recognition of voice and some information about the characteristics upon which it is based [see Beer (1970a) for a review]. A brief account of one study will illustrate the insights into social dynamics to which such investigation can lead.

Laughing Gull chicks as young as 24 hours posthatching can discriminate the voices of their parents from those of other gulls (Beer, 1970c).* However, such young chicks respond strongly, with ap-

---

*There is even some evidence suggesting that such discrimination may develop in Laughing Gull chicks prior to hatching, as B. Tschanz (1968) found to be the case in Guillemots (Common Murres), *Uria aalge* (M. Impekoven, personal communication).

proach and vocalization, to playback of an appropriate selection of calls of any adult. Such behavior is most strongly stimulated at this stage by a call I refer to as "crooning." This type of call appears to lack individual characteristics distinct enough to permit individual recognition, and day-old chicks tested with recordings of this call alone failed to discriminate that of the parent from that of a neighboring bird, either in successive or in simultaneous presentations (Beer, in press). In contrast to crooning, the "long-call" possesses characteristics that differ distinctly from one individual to another, so much so that an observer in the field can use them to identify individual gulls by ear. When day-old chicks were tested with recordings of this call alone, they were able to discriminate the call of the parent from the call of a neighbor (Beer, in press). For chicks of this age, it appears that the attracting qualities of crooning can override the recognition of long-calls and other identifying calls that might occur in sequence with crooning. When the chicks are a week old or more, however, they will respond positively only to playback of calls of their own parent (Beer, 1969, 1970b,c). But now the filial behavior of the chicks is elicited by "ke-hah" calls and long-calls of the parents—calls that have marked individual characteristics—and crooning has an effect only if preceded by such identifying calls (Beer, in press).

This shift in a Laughing Gull chick's responsiveness to the calls of adults is associated with enlargement of the range of social contacts to which the chick is exposed and enlargement of the range of circumstances in which the chick has to recognize and locate its parents. During the time that crooning dominates a chick's filial behavior, the chick is attended by one or the other of its parents almost continuously and, since its movements are restricted to the proximity of the nest, it seldom has any close contact with adults other than its parents. By the time ke-hah and long-calling take over, the chick is being left without parental company for varying periods and has extended its range of movement far beyond the vicinity of the nest. Both at home and abroad it must cope with the threat of hostile encounters with adults other than the parents and with the fact that, when one of its parents returns, the location of the parent may be hidden from the chick by vegetation. Thus when the changes in a chick's social life have reached the point where the chick must be able to recognize and locate its parents from a distance, its filial approach behavior has come under the governance of calls that possess the requisite properties, in contrast to the calls that governed this behavior earlier.

The shift in responsiveness of the chicks to the different types of adult calls appears to be a consequence, at least in part, of changes in the behavior of the parents. During the first week or so of a chick's life after hatching, the parents progressively add components to the sequences that terminate in feeding of the chick. To begin with, a parent utters only crooning in regular contiguity with feeding. By this means the parent can augment the visual stimulation that induces the chick to peck at the bill tip and to take food presented there— a function that is apparently prepared for by the chick's auditory experience prior to hatching (Impekoven, 1971). At a later stage the parent regularly begins the sequences with ke-hah calling, often at a distance from the chick, and reverts to crooning only after the chick has begun to approach. Later still, long-calls are added, particularly when a sequence begins with the arrival of a parent after an absence. This progression of events suggests that the parents shape the behavior of the chick by response chaining, the food reward providing the terminal reinforcement. Some support for this suggestion has come from observations of hand-raised chicks (Beer, 1970b, 1972). This, however, is certainly not the whole story, for it leaves certain features of the individual recognition unaccounted for.

An unexpected but intriguing result has been obtained in a study, not yet completed, of the responses of older chicks to playback of the various calls. Some of the chicks were tested with long-calls that had been recorded from their parents on the same day or the day before; other chicks were tested with long-calls that had been recorded about 3 weeks earlier—a day or two after hatching. The chicks tested with the contemporaneous recordings all recognized the calls of the parent and responded positively to them. Most of the chicks of the other group barely gave a hint of discrimination and did not react positively to the calls, in spite of the fact that they had recognized these same calls as those of the parent when tested as nestlings and, in some cases, had reacted positively toward them. It is possible that physiological changes that take place in the parents during the posthatching period affect certain of the identifying features of the long-call, such as its timbre. One's own ear can detect such change in certain individuals, but usually not until later in the season. However, to me, the most distinct identifying characteristics of long-calls—the number, durations, and rate of repetition of notes in the "short note" phase of the call (Beer, 1970c)—appear to remain constant during the period in question here. Another and perhaps even more interesting possibility is that the adults can vary certain features of the long-call in ways that alter what it signifies and that what is indicated by a long-call uttered during the early post-hatching period—when the call is mostly directed at the mate or to other adults and not the chicks—is different from what is indicated when the call is directed by a parent at its own 3-week-old chick (as was the case with most of the contemporaneous calls used in these tests). The recordings have yet to be analyzed to see whether differences can be found that might correlate with their differing effectiveness in the tests. The possibility is not a remote one, for there is already some evidence of such variation associated with differences in the contexts in which long-calls occur. For example, the number of "long notes" included in the call is greater on the occasion of meetings between mates or prospective mates during the courtship phase than is the case in most of the other situations in which the call occurs (Beer, 1970a).

CONTEXTS AND CODES

The range of contexts in which Laughing Gulls utter long-calls is broad, including territorial agonistic interactions, courtship interactions, nest reliefs, parent–chick interactions, adult–chick agonistic interactions, and various interactions in flocks both on the ground and in the air. The same is true of a number of the other calls and display postures in the repertoire of Laughing Gulls—e.g., crooning, choking, head-tossing. The effects of such signals on the recipients vary from context to context. This variability in the eliciting circumstances and functions of a display raises a problem for any simple stimulus–response concept of social communication.

Smith (1965, 1968) has argued that where such variability exists, the effect of a signal on a recipient (its "meaning," in his terms) is determined by both the signal itself and the context in which it occurs, the context including anything present along with the signal, including the recipient animal's physiological state, sex, or social status. Territorial song thus will repel a rival male but attract an unpaired female. As far as the signaling animal is concerned, according to Smith, a common factor will be found in all the contexts in which such a signal occurs. This common factor is what is encoded in the signal (in Smith's terms it is the "message" of the signal); it corresponds to ". . . some aspect(s) of the state (or central nervous state) of the communicator" (Smith, 1968, p. 48). Smith (1966) has applied this concept of communication in interpreting the social behavior of tyrannid flycatchers.

The role of context in effects produced by displays is well exemplified in the social behavior of gulls. For example, in an unpublished investigation of Black-headed Gulls (*Larus ridibundus*), Manley (1960) studied postures that occur in both agonistic and pair-formation contexts and found that whether a posture was interpreted as a hostile signal or a courtship signal was determined by the sequential pattern in which it occurred, its orientation, and subtle details of wing and bill position. Some of these accompaniments were supplied by the displaying bird, however, so that if we assume that a specific aspect or state of the central nervous system is represented by the posture, then it seems that we also have to say that the form of the representation is affected by accompanying aspects or states— which correspond to the contextual details supplied by the displayer— and, hence, that the display is context-determined as well as context-interpreted.

But I do not think that neurophysiological talk is very helpful here. In the absence of any relevant neurophysiological facts, one is liable to fall into circular argument. Perhaps even more serious is the tendency for neurophysiological concepts to exert certain constraints on thinking and to rule out, as inadmissable, ways of viewing the phenomena that may, at least in the short run, be more heuristic. For example, the proposition that each type of signal behavior can encode but a single message will tend to be accepted without question if one is thinking about social communication in terms of physiological causes and effects and so blind one to the possibility that the same behavior might be used to say different things at different times. Attempts to reduce human language to such terms have been dismally unenlightening. I venture to suggest that analogies from language might prove to be more productive of new understanding of social communication in birds, than are any currently held physiological conceptions. One simple illustration might indicate the potential.

I have talked blithely of displays and signal behavior as though they are simply given to observation without any difficulties about distinguishing and classifying them. In fact, selection of units for description is one of the most crucial and challenging steps involved in the analysis of behavior. In any case the plausibility of Smith's conception depends on the units to which it is applied. In the communication behavior of Laughing Gulls, there are certain call notes, postural elements, and movements that enter into various combination and sequences. According to Smith, each of these must reflect a specific physiological state that is peculiar to it but that is the same irrespective of context. The variety of combinations and sequences, however, tends to be such that belief in a common physiological state underlying each and every occurrence of spreading the tail, for example, seems implausible unless one restricts one's reference to the lower levels of the motor system involved. But then it will no longer be appropriate to ascribe a "message" to such a feature. One is more comfortable with the idea that the feature can enter into various combinations and sequences to convey various messages. The analogy that suggests itself is the distinction between phonemes and morphemes in language, i.e., units of sound production, on the one hand, and the units of sense that arise from their combination, on the other. Analysis of the communication behavior of gulls into equivalents of phonemes and morphemes could prove useful in getting at the structure of the behavior and how it is used (cf Beer, 1971). The same could be true of other species.

EVOLUTION

Finally, I shall mention briefly two classes of question that are at the farthest remove from physiological help: questions as to the survival value of behavior and questions about the evolution of behavior. Although physiological considerations are sometimes relevant to interpretation of the adaptive correlations of reproductive behavior, discovery of such correlations by observation and comparison and their testing, where possible, by experiment are primarily matters involving ecological orientation. Physiology has even less to offer the student of behavioral evolution. This method depends heavily on the comparative tradition in zoology, including the old concept of homology. Behavioral homology is anything but a clear and precise concept, either in itself or in its relation to structural homology. The hope that neurophysiology might bring precision and integration to these concepts has so far been unrealized (Atz, 1970). But for those to whom the notions of ritualization, emancipation, and so forth mean something, and for whom questions about phyletic affinities

and the evolutionary antecedents of behavior patterns are worth
pursuing, no better concepts or methods are available than those of
comparative ethology.

CONCLUDING COMMENT

Many scientists question the value of an evolutionary study of behavior, because the conclusions of such study can never be defended
by anything better than indirect evidence and can never be tested by
experiment. Moreover, it is sometimes argued that, to the extent such
study concerns itself with particular historical sequences, it cannot be
science, since the aim of science is supposed to be the discovery of
universal laws underlying phenomena and the particular is considered
interesting to the scientist only insofar as it illuminates the universal.

Something like this attitude has been nurtured in the study of animal behavior. In our laboratories and classrooms, we have a tendency
to regard variability as a nuisance—as noise in the system—to be compensated for by appropriate experimental design, sampling techniques
and statistical analysis. "The mean's the thing wherein to catch the
essence of life's fling," and all deviations must at least be rendered
standard.

But animals are individuals, each with its personal developmental
history, its personal relationships to places and other individuals. It
is sometimes useful to look at them from this point of view when we
are studying their behavior. I believe that there are aspects of social
and reproductive behavior, at least in birds, that demand this viewpoint if they are to be discerned at all—yet the discernment of which
is required for a full understanding of reproductive processes. Physiology deals with the general rather than the particular in behavior.
Therefore I argue that the ideographic viewpoint (Windelband, 1921)
must be allowed a place alongside the physiological in our studies of
behavioral components in the reproductive biology of birds.

Again we have an echo of a nineteenth century attitude. John
Stuart Mill (1875) proposed that the scientific study of mind should
be divided into two approaches: search for the universal laws governing mental life and investigation of the uniqueness of the individual
mind—the development of personal character. Mill suggested that
this "science of character" be called ethology. I should like to suggest
that something like Mill's science of character has a place in ethology
as we know it today.

ACKNOWLEDGMENTS

The work on Laughing Gulls referred to in this paper was supported by Grant
No. MH 16727 from the U.S. Public Health Service. It was carried out in the
Brigantine National Wildlife Refuge. I thank the U.S. Fish and Wildlife Service
for permission to work in the Refuge and to use a federally owned building there
as a field station. I also thank the Refuge manager and his staff for their hospital-
ity and cooperation.

REFERENCES

Atz, J. W. 1970. The application of the idea of homology to behavior, p. 53–74.
    In L. R. Aronson, E. Tobach, D. S. Lehrman, and J. S. Rosenblatt [ed.] De-
    velopment and evolution of behavior. W. H. Freeman and Company, San
    Francisco.
Baerends, G. P. 1959. The ethological analysis of incubation behaviour. Ibis
    101:357–368.
Baerends, G. P., and R. H. Drent. 1970. The Herring Gull and its egg. Behaviour
    Suppl. 17. 312 p.
Barfield, R. J. 1971. Gonadotrophic hormone secretion in the female Ring Dove
    in response to visual and auditory stimulation by the male. J. Endocrinol.
    49:305–310.
Beer, C. G. 1961. Incubation and nest building behaviour of Black-headed Gulls.
    1. Incubation behaviour in the incubation period. Behaviour 18:62–106.
Beer, C. G. 1969. Laughing Gull chicks: Recognition of their parents' voices.
    Science 166:1030–1032.
Beer, C. G. 1970a. Individual recognition of voice in the social behavior of birds.
    Adv. Study Behav. 3:27–74.
Beer, C. G. 1970b. On the responses of Laughing Gull chicks (*Larus atricilla*) to
    the calls of adults. 1. Recognition of the voices of the parents. Anim. Behav.
    18:652–660.
Beer, C. G. 1970c. On the responses of Laughing Gull chicks (*Larus atricilla*) to
    the calls of adults. II. Age changes and responses to different types of calls.
    Anim. Behav. 18:661–677.
Beer, C. G. 1971. Diversity in the study of the development of social behavior,
    p. 433–455. *In* E. Tobach, L. R. Aronson, and E. Shaw [ed.] The biopsychol-
    ogy of development. Academic Press, New York.
Beer, C. G. 1972. Individual recognition of voice and its development in birds,
    p. 314–331. *In* Proceedings of the 15th International Ornithological Congress
    (The Hague, 1970). E. J. Brill, Leiden, The Netherlands.
Beer, C. G. A view of birds. Minn. Symp. Child Psychol. 7. (In press)
Bruder, R. H., and D. S. Lehrman. 1967. Role of the mate in the elicitation of
    hormone-induced incubation behavior in the Ring Dove. J. Comp. Physiol.
    Psychol. 63:382–384.
Coulson, J. C. 1966. The influence of the pair-bond and age on the breeding biol-
    ogy of the Kittiwake Gull, *Rissa tridactyla*. J. Anim. Ecol. 35:269–279.

Coulson, J. C. 1972. The pair bond and the breeding success in the Kittiwake, *Rissa tridactyla*. *In* Proceedings 15th International Ornithological Congress (The Hague, 1970). E. J. Brill, Leiden, The Netherlands.

Coulson, J. C., and E. White. 1956. A study of colonies of the Kittiwake, *Rissa tridactyla*. Ibis 98:63–79.

Darling, F. F. 1938. Bird flocks and the breeding cycle. Cambridge University Press, Cambridge. 124 p.

Eisner, E. 1960. The relationship of hormones to the reproductive behaviour of birds, referring especially to parental behaviour: A review. Anim. Behav. 8:155–179.

Erickson, C. J., and D. S. Lehrman. 1964. Effect of castration of male Ring Doves upon ovarian activity of females. J. Comp. Physiol. Psychol. 58:164–166.

Erickson, C. J., R. H. Bruder, B. R. Komisaruk, and D. S. Lehrman. 1967. Selective inhibition by progesterone of androgen-induced behavior in male Ring Doves (*Streptopelia risoria*). Endocrinology 81:39–44.

Hinde, R. A. 1956. Ethological models and the concept of drive. Brit. J. Philos. Sci. 6:321–331.

Hinde, R. A. 1958. The nest-building behaviour of domesticated canaries. Proc. Zool. Soc. London 131:1–48.

Hinde, R. A. 1959. Unitary drives. Anim. Behav. 7:130–141.

Hinde, R. A. 1960. Energy models of motivation. Symp. Soc. Exp. Biol. 14:199–213.

Hinde, R. A. 1970. Animal behaviour. 2nd ed. McGraw-Hill, New York. 876 p.

Hinde, R. A., and E. A. Steel. 1962. Selection of nest material by female canaries. Anim. Behav. 10:67–75.

Hinde, R. A., and E. A. Steel. 1964. Effect of exogenous hormones on the tactile sensitivity of the canary brood patch. J. Endocrinol. 30:355–359.

Hinde, R. A., and R. P. Warren. 1959. The effect of nest building on later reproductive behaviour in domesticated canaries. Anim. Behav. 7:35–41.

Hinde, R. A., R. Q. Bell, and E. A. Steel. 1963. Changes in sensitivity of the canary brood patch during the natural breeding season. Anim. Behav. 11:553–560.

Impekoven, M. 1971. Prenatal experience of parental calls and pecking in the Laughing Gull chick (*Larus atricilla*). Anim. Behav. 19:475–480.

Komisaruk, B. R. 1967. Effects of local brain implants of progesterone on reproductive behavior in Ring Doves. J. Comp. Physiol. Psychol. 64:219–224.

Lehrman, D. S. 1958a. Effect of female sex hormones on incubation behavior in the Ring Dove (*Streptopelia risoria*). J. Comp. Physiol. Psychol. 51:142–145.

Lehrman, D. S. 1958b. Induction of broodiness by participation in courtship and nest-building in the Ring Dove (*Streptopelia risoria*). J. Comp. Physiol. Psychol. 51:32–36.

Lehrman, D. S. 1961. Hormonal regulation of parental behavior in birds and infrahuman mammals, p. 1268–1382. *In* W. C. Young [ed.] Sex and internal secretions. Williams & Wilkins, Baltimore.

Lehrman, D. S. 1965. Interaction between internal and external environments in the regulation of the reproductive cycle of the Ring Dove, p. 355–380. *In* F. A. Beach [ed.] Sex and behavior. Wiley, New York.

Lehrman, D. S. 1971. Experiential background for the induction of reproductive behavior patterns by hormones, p. 297–302. *In* E. Tobach, L. R. Aronson, and E. Shaw [ed.] The biopsychology of development. Academic Press, New York.

Lehrman, D. S., and P. Brody. 1957. Oviduct response to estrogen and progesterone in the Ring Dove (*Streptopelia risoria*). Proc. Soc. Exp. Biol. Med. 95:373–375.

Lehrman, D. S., and P. Brody. 1961. Does prolactin induce incubation behaviour in the Ring Dove? J. Endocrinol. 22:269–275.

Lehrman, D. S., and M. Friedman. 1969. Auditory stimulation of ovarian activity in the Ring Dove (*Streptopelia risoria*). Anim. Behav. 17:494–497.

Lehrman, D. S., and R. P. Wortis. 1960. Previous breeding experience and hormone-induced incubation behavior in the Ring Dove. Science 132:1667–1668.

Lehrman, D. S., and R. P. Wortis. 1967. Breeding experience and breeding efficiency in the Ring Dove. Anim. Behav. 15:223–228.

Lehrman, D. S., R. P. Wortis, and P. Brody. 1961a. Gonadotropin secretion in response to external stimuli of varying duration in the Ring Dove (*Streptopelia risoria*). Proc. Soc. Exp. Biol. Med. 106:298–300.

Lehrman, D. S., P. Brody, and R. P. Wortis. 1961b. The presence of the mate and of nesting material as stimuli for the development of incubation behavior and for gonadotropin secretion in the Ring Dove (*Streptopelia risoria*). Endocrinology 68:507–516.

Lott, D. F., and P. N. Brody. 1966. Support of ovulation in the Ring Dove by auditory and visual stimuli. J. Comp. Physiol. Psychol. 62:311–313.

Lott, D. F., S. D. Scholz, and D. S. Lehrman. 1967. Exteroceptive stimulation of the reproductive system of the female Ring Dove (*Streptopelia risoria*) by the mate and by the colony milieu. Anim. Behav. 15:433–437.

Manley, G. H. 1960. The agonistic behaviour of the Black-headed Gull. PhD thesis. Bodleian Library, Oxford, England.

Marler, P., and W. J. Hamilton. 1966. Mechanisms of animal behavior. Wiley, New York. 771 p.

Marshall, A. J., and H. J. de S. Disney. 1957. Experimental induction of the breeding season in a xerophilous bird. Nature 180:647–649.

Mill, J. S. 1875. A system of logic. 9th ed. Longmans, Green Reader & Dyer, London. 622 p.

Morris, R. L., and C. J. Erickson. 1971. Pair bond maintenance in the Ring Dove (*Streptopelia risoria*). Anim. Behav. 19:398–406.

Smith, W. J. 1965. Message, meaning and context in ethology. Am. Nat. 99:404–409.

Smith, W. J. 1966. Communication and relationships in the genus *Tyrannus*. Publication of the Nuttall Ornithological Club, No. 6. 250 p.

Smith, W. J. 1968. Message-meaning analysis, p. 44–60. *In* T. A. Sebeok [ed.] Animal communication. Indiana University Press, Bloomington.

Steel, E. A., and R. A. Hinde. 1963. Hormonal control of brood patch and oviduct development in domesticated canaries. J. Endocrinol. 26:11–24.

Steel, E. A., and R. A. Hinde. 1964. Effect of exogenous oestrogen on brood patch development of intact and ovariectomized canaries. Nature 202:718–719.

Stern, J. M., and D. S. Lehrman. 1969. Role of testosterone in progesterone-induced incubation behaviour in male Ring Doves (*Streptopelia risoria*). J. Endocrinol. 44:13–22.

Tschanz, B. 1968. Trottellummen. Z. Tierpsychol. Suppl. 4. 103 p.

von Holst, E. 1954. Relations between the central nervous system and peripheral organs. Brit. J. Anim. Behav. 2:89–94.

von Holst, E., and H. Mittelstaedt. 1950. Das Reafferenzprinzip. Naturwissenschaften 37:464–476.

Warren, R. P., and R. A. Hinde. 1959. The effect of oestrogen and progesterone on the nest-building of domesticated canaries. Anim. Behav. 7:209–213.

Warren, R. P., and R. A. Hinde. 1961a. Does the male stimulate oestrogen secretion in female canaries? Science 133:1354–1355.

Warren, R. P., and R. A. Hinde. 1961b. Roles of the male and the nest-cup in controlling the reproduction of female canaries. Anim. Behav. 9:64–67.

Windelband, W. 1921. An introduction to philosophy. T. Fisher Unwin, London. 659 p.

DISCUSSION

Donald A. Jenni

In recent years, we have made tremendous progress in understanding the role played by the hypothalamo–hypophysial system. Control of reproductive hormones by this system is especially important in the initiation and termination of reproductive behavior (Farner, 1967). The next step,–i.e., to learn what role, if any, the neuroendocrine system has in controlling specific behavioral elements in, for example, a courtship sequence–is exceedingly difficult. Beer has described how the performance of specific acts by a bird can have feedback effects on its physiology. We quickly move to a level where endogenous and exogenous factors interact. In some cases, the endogenous factors probably play more of a permissive role, and the exogenous factors are involved in the timing and orientation of the behavioral acts—perhaps in the selection of which particular acts are performed. These exogenous factors and behavioral acts, in turn, affect the internal state. From Beer's review we see that progress has been made, but only a beginning; most of the work and most of the discoveries must lie ahead.

Physiological considerations appear most helpful in understanding the breeding behavior of birds when we consider fine-grain analyses of behavior as Beer did in his review of Hinde's work with canaries and Lehrman's with Ring Doves. The need to analyze at this fine level draws our attention away from problems of social organization that seem to be best understood on a broader basis or on a coarser analysis. Social organization is the sum of the interactions between the individual members of the species. These interactions have important consequences for the spatial relationships of the individuals. Thus, social organization and spatial relationships may be viewed as consequences of behavior, but they could just as well be viewed as part of social behavior. Recent reviews of social organization in birds have been especially thorough (Crook, 1965; Lack, 1968; Orians, 1971). By emphasizing the adaptiveness of various types of social systems, these authors reflect the ecological emphasis that we find in the literature on social organization.

In a review of breeding behavior, however, we must consider social organization during the reproductive season. There are at least two ways to approach the subject. One can emphasize spatial relationships, but perhaps it is more direct to emphasize basic breeding units. It is necessary to describe the behavioral interactions between the individual members of the breeding units. In the ornithological literature these relationships have been classified according to the ratio of sexes, and sometimes according to permanence. The basic mateship systems are as follows:

• Monogamy, in which one male and one female consort with each other for part or all of one or more breeding seasons and who exclude other sexually active individuals from their immediate space. According to Lack (1968), approximately 91 percent of the bird species are monogamous.

• Polygamy, in which one individual consorts with two or more individuals of the opposite sex. Polygamy may take one or two forms; the more common is polygyny, in which one male consorts with two or more females; the other is polyandry, in which one female consorts with two or more males. If polyandry is to be functionally adaptive, at least some reversal of the typical behavioral roles of the sexes must occur. The supposition that reversal of the roles of the sexes is evidence of polyandry has resulted in many erroneous reports thereof.

• Promiscuity, in which no consort relationships are formed.

The other approach to describing social organization during the breeding season is to start at the population level and to emphasize the spatial relationships between the breeding units. The spatial distributions of these basic breeding units range all the way from highly clumped to widely dispersed. At the one extreme, we find colonial seabirds; on the other, songbird territories of the type we now call classical. What causes breeding units of a particular species to be distributed across the landscape in a particular pattern seems quite intractable to physiological analysis, to an examination of internal causal factors. Ecological examination of the adaptiveness of the space relationships seems more appropriate. Although ecologists have considered a variety of potentially adaptive pressures, the way in which the resources required by a particular species are distributed in space often seems to correlate well with the way the breeding units are distributed (Crook, 1965). In some, the particular space distribution appears to be related more toward countering predator pressure (e.g., Kruuk, 1964). However, in all such analyses, it is difficult to relate proximate factors, be they physiological or ecological, from the more remote ultimate factors. It is, of course, invalid to assume that present selective forces for which the social organization (or any other characteristic of a species) is adaptive are the same selective forces that operated on the evolution of the system. However, we must study how these characteristics are adaptive if we are to understand what is responsible for a particular pattern of distribution (Tinbergen, 1965).

Whenever we attempt to classify the diversity of mateship systems or spatial arrangements into rigid categories, we are apt to find nature more flexible than we are. For example, one species may include both monogamous and polygamous mateship groups. For how many hours or days must a male consort with each individual female before we consider him no longer promiscuous, but serially monogamous instead? Several species of grouse (Tetraonidae) are considered promiscuous, and indeed the males may be so, but in many the females apparently mate only once. Identifying a species as monogamous, territorial, or altricial does not explain causation, function, or evolution; it is only a shorthand form of description. This is an example of the kind of trap we can create for ourselves in trying to understand a social system when we are overeager to classify and do not gather data adequate to describe the precise relationships between the individuals.

In the final analysis, it is necessary to include both approaches. A description of the social organization of an avian species must include the spatial relationships between the breeding units as well as the behavioral relationships between the individuals composing the breeding units.

Investigation into the causation of mateship systems may at first seem as intractable to physiological explanation as does the spatial distribution of the breeding units. But I would like to pursue this idea briefly. Beer pointed out that the studies on the roles of hormones throw little light on the mechanisms of pair formation. How these hormones act in determining the basic breeding units remains unknown. It is possible that they play no direct role, but rather play permissive, priming, or triggering roles.

The polyandrous mateship system of the American Jacana (*Jacana spinosa*) is a good example of the problems that we encounter in attempting to understand the causation of mateship systems. The behavioral roles of the sexes in the American Jacana are the reverse of what we consider "normal" for birds. The male builds the nests, incubates the eggs, and cares for the young; the female indulges in some desultory nest-building but she neither incubates nor cares for the young. The male is strongly attached to a piece of real estate and the question of which female becomes his mate is an affair settled among the females. In a matriarchial system where the females weigh 75 percent more than males, the male has little choice but to go along with the arrangement worked out among the females. A female most often maintains simultaneous pair-bonds with, and defends the territories of, two males (9 of 19 cases), less often with three males (6 of 19 cases) or four males (1 of 19 cases). Occasionally, the females are monogamous (3 of 19 cases). Why do some females have more males than other females? Although more data are needed, it appears more fruitful to analyze exogenous factors than to look for internal differences in the females. But the possibility must remain that the females differ internally. The American Jacana is the only species of bird in which polyandry has been proven in the wild by marking individual birds (Jenni and Collier, 1972). This discovery generates a number of intriguing questions. What is the physiological mechanism that regulates this system? One hesitates to hypothesize a polyandry hormone, but one does face the enigmatic problem: What role do hormones play in the regulation of such a system? One would hope that it is not necessary to assign causation to the black box called higher centers. The most direct way to study this problem would probably be through a comparative study. An equally significant line of questioning is more ecological. What is the adaptive significance of this unique form of social organization? How has it evolved? How is such a system maintained? Questions on adaptiveness, as difficult as they are, seem more tractable than questions on physiologcial control. Critical experimental proof of the adaptive value of any social system, to my knowledge, has never been offered; however, the descriptive and comparative data in some cases are overwhelmingly convincing (e.g., Crook, 1964; Tinbergen, 1967).

The question of how this polyandrous system is maintained can be studied and described ethologically. Given time, it should be possible to discover its adaptive value and, perhaps, even gain some insight into the circumstances leading to its evolution. Given answers to those intriguing questions, we are still left with the dilemma of explaining the role of internal mechanisms in determining or permitting polyandry in the American Jacana and monogamy, or other systems, in closely related groups. The work of neurophysiologists does not hold much promise. Key factors appear to be the small size of male territories and the large size of female territories. The role that neural and endocrine systems play in determining territory size, perhaps through control of aggressive levels, remains unknown. Is it practical to look inside the animal for causal factors of different social systems?

The phalaropes are classic examples of polyandry, yet recent critical work has failed to confirm polyandry in any of these species (Johns, 1969; Höhn, 1971). Nevertheless, these birds have extreme reversal of the roles, similar to the reversal seen in the American Jacana. It is apparently this characteristic that is responsible for the erroneous assumption that they are also polyandrous. A series of studies into hormonal regulation in the Wilson's Phalarope (*Steganopus tricolor*) by E. W. Pfeiffer and his collaborators at the University of Montana suggest that the physiological approach to this problem is practicable. Preliminary experimental work showed that testosterone- and prolactin-induced brood-patch development in both male and female phalaropes (in the absence of any stimuli from the nest) even though females do not normally develop a brood-patch. Neither testosterone, prolactin, nor estradiol, when tested alone or in any combination other than testosterone and prolactin, induced brood-patch formation (Johns and Pfeiffer, 1963). Further experiments showed that testosterone caused development of nuptial feathers in both sexes (Johns, 1964). Although male Wilson's Phalaropes typically lack the bright nuptial plumage of the female, there is considerable variation. These findings led to the hypothesis that the females lacked adequate prolactin levels to stimulate development of the brood-patch. Measurement of the prolactin in the pituitaries of males and females collected during the breeding season in Montana revealed that, prior to the development of brood-patches by the males, they lacked prolactin, whereas the pituitaries from the females contained low levels of the hormone. Later, when incubation patches had developed, the level of prolactin in the females was double that of earlier level, but the males had 3–4 times the amount of prolactin found in the females at the same time (Nicoll *et al.*, 1967). These results were interpreted as supporting the earlier hypothesis that the low level of prolactin is responsible for the failure of the brood-patch to develop in females and may be largely responsible for the reversal of sexual roles in these species. However, the possibilities of feedback from nest construction behavior, or stimuli from the nest and eggs, were not explored. In light of Beer's lucid description of the way behavior and sensory input can modify the bird's physiology, these conclusions must be regarded as premature. The most significant value of this work may be that it is a tentative step toward taking physiological analysis into the field. However, failure thus far to continue this very promising line of research may be indicative of the difficulty of doing so. This research also stands as a plea for the application of the team approach.

It is difficult to overemphasize either the importance of careful descriptions and analyses of social systems or the danger of generalizing from one species to another, even closely related species. There has been, at least in the past, a tendency to make the generalization that, on the northern breeding grounds, the two most abundant groups of ducks in North America, the surface-feeding ducks (tribe Anatini) and the pochards (tribe Aythyini), each has a basic form of reproductive social organization. The dabbling ducks, typified by the Mallard (*Anas platyrhynchos*), were all thought to be basically monogamous or monogamous–promiscuous. The pairs space themselves out over the available habitat and the males defend loafing and feeding areas that the females use prior to and during nest-building and egg-laying. Sometime during incubation the male quits defending this area, abandons the female, joins other postbreeding males, and undergoes a simultaneous molt of the flight feathers. It was felt that other members of the genus *Anas*, as well as the other dabbling ducks, followed such a pattern. On careful analysis, it turns out that this assumption is invalid. Earlier in this volume, McKinney carefully reviews the social organization—behavior of two puddle ducks that have widely different systems; his material must come as a surprise to many. The time-honored method of censusing breeding puddle ducks by counting pairs, and counting each loafing but unaccompanied male as a pair, simply cannot be expanded to include any species unless its basic social organization during the reproductive season is known.

The pochards have a disparate sex ratio, and it has long been recognized that the presence of a male does not necessarily indicate the presence of a female. The status of such a species on the breeding grounds can best be determined by counting the number of females or, perhaps more precisely, the number of broods. Yet this group includes the Redhead (*Aythya americana*), a species that is partly parasitic (Weller, 1959) in that females deposit many eggs in the nests of other waterfowl and may never nest or incubate their own eggs. If we measured Redhead production only by counting the number of females with broods, we might reach very different conclusions than we would if we counted the number of ducklings.

For biologists involved in the management of avian species, extrapolation from other species, even congeners, does not often work. With this diversity in the social organization of ducks in mind, one cannot but wonder at the significance, let alone the precision, of such statements as Gavin's (1971) that waterfowl populations average 2.8 pairs per square mile on that part of Alaska's North Slope under consideration for oil development. How can one interpret the significance of a single male, or a group of three males of a particular species? Clearly, the interpretation must be different for different species. Examples taken from waterfowl seem especially appropriate. I can think of no group of animals so important, about which so much has been written, and that is so well known behaviorally. Yet McKinney's review and the examples above show how little we know.

Microanalytical procedures for the assay of hormones are now available. The time seems ripe for critical field testing of hypotheses regarding the roles of these hormones. This research absolutely demands the team approach—close collaboration between the ethologist and the endocrinologist. I hope that the results of such interdisciplinary research will have the kind of synergistic effects I sense this symposium has had and will lead to increased collaboration and that this collaboration will itself become an example of positive feedback.

ACKNOWLEDGMENT

The research on the American Jacana discussed here has been supported by NSF Grant GB-21279.

REFERENCES

Crook, J. H. 1964. The evolution of social organization and visual communication in the Weaver Birds (*Ploceinae*). Behaviour, Suppl. 178 p.

Crook, J. H. 1965. The adaptive significance of avian social organizations, p. 181–218. *In* P. E. Ellis [ed.] Social organization of animal communities. Academic Press, London and New York.

Farner, D. S. 1967. The control of avian reproductive cycles, p. 107–133. *In* Proceedings of the XIV International Ornithological Congress. Blackwell's, Oxford and Edinburgh.

Gavin, A. 1971. Alaska—Oil and the environment, p. 235–241. *In* Transactions of the Thirty-Sixth North American Wildlife and Natural Resources Conference. Wildlife Management Institute, Washington, D.C. 534 p.

Höhn, E. O. 1971. Observations on the breeding behaviour of Grey and Red-Necked Phalaropes. Ibis 113:335–348.

Jenni, D. A., and G. Collier. 1972. Polyandry in the American Jacana (*Jacana spinosa*). Auk 89:743–765.

Johns, J. E. 1964. Testosterone-induced nuptial feathers in phalaropes. Condor 66:449–455.

Johns, J. E. 1969. Field studies of Wilson's Phalarope. Auk 86:660–670.

Johns, J. E., and E. W. Pfeiffer. 1963. Testosterone-induced incubation patches of phalarope birds. Science 140:1225–1226.

Kruuk, H. 1964. Predators and anti-predator behaviour of the Black-billed Gull (*Larus ridibundus* L.). Behaviour, Suppl. 129 p.

Lack, D. 1968. Ecological adaptations for breeding in birds. Methuen & Co. Ltd., London. 409 p.

Nicoll, C. S., E. W. Pfeiffer, and H. R. Fevold. 1967. Prolactin and nesting behavior in phalaropes. Gen. Comp. Endocrinol. 8:61–65.

Orians, G. 1971. Ecological aspects of behavior, p. 513–546. *In* D. S. Farner and J. M. King [ed.] Avian biology. Vol. 1. Academic Press, New York.

Tinbergen, N. 1965. Behavior and natural selection. *In* Ideas in modern biology. Proc. XVI Int. Congr. Zool. 6:521–539.

Tinbergen, N. 1967. Adaptive features of the Black-headed Gull (*Larus ridibundus* L.), p. 43–59. *In* Proceedings, XIV International Ornithological Congress. Blackwell's, Oxford and Edinburgh.

Weller, M. W. 1959. Parasitic egg laying in the Redhead (*Aythya americana*) and other North American Anatidae. Ecol. Monogr. 29:333–365.

DISCUSSION

K. Homma

Doctor Beer placed special emphasis on the benefits of the ethological approach in investigating reproductive and associated behavior in birds. He has pointed out that numerous terms commonly used in behavioral sciences are difficult to translate into appropriate physiological terms because of the basic difference in concepts between the two domains. In the former, deviation from the average is considered to be important, whereas in the latter only generalizations are regarded as valuable. In my own view, these differences are rather superficial so long as we are dealing with the same object. Significant differences might arise in the ways of evaluating results. In the behavioral sciences, there are no approved universal units, but physiology, in a broad sense, may be thought of as a science that deals with biological phenomena in standardized units. It is clear, however, that either attitude by itself is inadequate to describe such an extremely complicated thing as the brain and its functional manifestations.

I hope to bridge the gap between the various biosciences related to reproduction, rather than to widen the gap by criticism. Although construction of a durable bridge may require more rocks of evidence than those now available, it does seem possible to suggest where to make the bridge, how to design it, and what materials should be sought.

I comment briefly here on the characteristics of sensory and endocrine systems of the birds and then point out several practical problems that seem to require the cooperative study of behaviorists and physiologists to obtain a relevant solution.

VISUAL ACUITY AND COLOR SENSE IN BIRDS

The ability to recognize forms and colors is not merely a function of the eye but also of the central nervous system. Nevertheless, an understanding of the basic character of the avian retina is a prerequisite for complete behavioral study. Incidentally, the fact that birds can distinguish different colors to the same extent we can was first proven through behavioral studies.

Birds eyes, except those of a few nocturnal species, are well equipped optic systems. In nearly all avian species, cones are the predominant retinal photoreceptors. The distribution of photosensitive elements in the avian retina is much denser than that in the human retina and the ratio of photosensitive elements to optic nerve fibers in birds is higher than in humans. These are regarded as the reasons why birds have better visual acuity than humans. In contrast to the acuity, the ability of the birds to integrate visual information is poor when judged by responses to various visual stimuli.

Birds often ignore static objects, but readily respond to the same object when it moves. Some investigators hold that this character is related to a peculiar structure in the bird's eye (the pecten), though evidence supporting this hypothesis is not conclusive.

The mechanism underlying such responsiveness to moving objects seems to play a role in certain aspects of reproductive behavior; for instance, major elements contained in the courtship behavior are sets of monotonous repetitive movements seldom seen during non-breeding season. Significance of special combinations of image, color, and motion of the parents in feeding behavior to the young have been studied extensively by behaviorists, but similar reports on reproductive behavior seem to be scarce.

Special senses, beyond those of human perception, are sometimes hazards in our conclusions on the behavioral study of birds. Encephalic photoreception is undoubtedly participating with photo-gonadal reflex and certain behavior of normal birds, but the precise mechanism and its functional significance await future elucidation (Benoit and Assenmacher, 1959; Menaker, 1971).

ROLE OF TOUCH OR TACTILE STIMULATION IN AVIAN REPRODUCTION

The neural network around the body surface of birds is not evenly distributed. The bill, crown, foot, and cloacal area are the parts where Pacinian (Herbst's) corpuscles are densly distributed. Such touch-sensitive locations seem to be important during mating in inducing special postural reflex. For example, when the neck or back of a female bird is seized by the bill of the male, the seized female will assume a posture to accept the service of the male. This reflex can be induced experimentally in either sex by pinching the appropriate area with a paper clip. It is surprising that the unexperienced male seizes the proper site of the female at the first copulation, as if he knew the reflex pattern.

Behaviorists tend to overlook simple reflexes that may assemble into a complicated behavioral pattern. Evaluation of such reflexes from an evolutionary viewpoint seems attractive.

ENDOCRINOLOGY OF AVIAN SPECIES

Recent progress in the endocrinology of avian reproduction, especially as it concerns the role of the hypothalamo–hypophysial axis was presented earlier in this volume (see Follett, p. 209, and Assenmacher, p. 158). It is clear that the hypothalamus is the site where information from the higher levels of the brain is transformed into molecular signals and conveyed to the adenohypophysis via the portal vessels. The hypothalamus is endowed with a considerable degree of autonomy. It is capable of controlling most gonadal activity, as indicated by isolation techniques in the rat; this has not been well established in birds experimentally. Also, the use of stereotaxic instruments and appropriate brain atlases has facilitated the investigation in mammals, but is still limited in birds.

The prevailing attitude among investigators in this field has been physiological. Much effort was devoted to establish clear point-to-point correspondence, either functional or structural, and the results thus obtained still contain many contradictions. Extending this approach into future research does not seem to be effective in elucidating the exact picture of any extensive regulatory mechanism in reproduction that includes sensory systems and other neural and humoral components.

Steroid feedback to the hypothalamus has attracted the attention of many investigators. At present, it seems extremely difficult to decide whether it is possible to designate the target of the negative feedback system in a limited area of the hypothalamus. Action of the steroid itself should be considered more carefully, since steroid actions on the central nervous system differ with dose level. It is known that progesterone and some other steroids are potentially anesthetics at high dose levels, whereas the same steroids stimulate the central nervous system at low dose levels; in addition, the threshold value may fluctuate with age and other factors. Information about the effects of recently developed antisteroidal substances on the central nervous system would provide a better understanding of the real mechanism of the steroid on the central nervous system, especially on the hypothalamus.

The action of steroids on the central nervous system of birds is somewhat different from mammals as regards the mode of action and age dependency. In the domestic fowl, for instance, administration of small amounts of androgen causes permanent impairment of the mating behavior of the male without any detectable damage to gametogenesis, except when the hormone is given before the critical thirteenth day of embryonic life (Wilson and Glick, 1970).

The negative feedback of steroid hormones on the avian central nervous system becomes less effective with age, and at sexual maturity—at least in the domestic fowl and Japanese Quail—hens are practically insensitive to physiological doses of exogenous estrogen. Apparently, this is correlated with the production of egg-yolk protein by the liver, which requires a high estrogen level.

In general, a negative feedback through humoral means in bird reproduction does exist but does not seem to be a potent regulatory system. On the contrary, regulation by such external environmental factors as photoperiod governs the major epochs in reproduction.

EFFECTS OF STRESS ON REPRODUCTION

We often have great difficulty in raising wild avian species in captivity. In most of such cases neither inadequate nutrition nor poor environmental control of temperature or photoperiod can account for the reproductive failure. As a consequence, we suspect that psychological stress associated with deprivation of freedom in the limited space imposed by the artificial environment is the main reason. If this is true, we have no way to clarify the real stressors through environmental studies in the laboratory. Deprivation or blocking of special senses, alone or in combination, would give much information about the nature of the factors that disturb normal reproduction. Research of this sort is scarce in the behavioral literature. However, poultry men are aware from experience that transportation of breeding stock at night markedly reduces the incidence of reproductive failure as compared to transport in day light. Similar strategy appears to be beneficial in handling wild avian species.

SYNCHRONIZATION OF RELEASE OF GAMETES

For successful reproduction of higher forms of life, gonadal maturity
and synchronization of release of gametes by the two sexes are
essential. In birds, sychronization may differ from 1 to 3 weeks,
since avian sperm may survive for a time in the female genital tract
after copulation.

In nature, the mechanisms responsible for sexual arousal and
synchronization appear to involve a variety of information and components. Some species utilize visual signals; others utilize photoperiod
as the main cues for synchronization. The role of vocal signals,
however, seems to be nothing more than accessory to reproduction
in a strict sense. Among the species that utilize photoperiodic information, there is considerable variation in the mode of responses.
Through such a variety of responses, established by generations of
adaptation, different avian species can share limited natural
resources.

REFERENCES

Benoit, J., and I. Assenmacher. 1959. The control by visible radiation of the
gonadotropic activity of the duck hypophysis. Rec. Prog. Hormone Res.
15:143–164.
Menaker, M. 1971. Rhythms, reproduction, and photoreception. Biol. Reprod.
4:295–308.
Wilson, J. A., and B. Glick. 1970. Ontogeny of mating behavior in the chicken.
Am. J. Physiol. 218 (4):951–955.

INFORMAL DISCUSSION

DRENT: Doctor Beer, I understand, of course, that you had to make a selection
of what you were going to tell us. To make a pun, your problem was to quote
or not to quote. But your problem of the identification of the individual parent call must imply some kind of matching between the members of the pair.
I wondered whether you would like to comment on whether birds mate with

others who have similar voices, or whether they learn to match, as you might surmise from that 3-week difference, and there must be all kinds of nuances that the individuals can use.

Doctor Jenni, even if with sophisticated hormone aliquot plasma methods, and so forth, you show that the female in these phalaropes species has a low level of prolactin, and therefore does not form the brood-patch, is it still only a mechanism? I want to know what approach you would suggest to determine whether we should accept the argument that a hormone aberration had led to the very peculiar breeding arrangement, or that there are other reasons for the breeding arrangement, and that in order to achieve this other role of the sexes, the mechanism has been changed in this way. I am reminded very much of the kind of arguments that Murton used earlier concerning the photoperiodic sensitivity of geese, when he mentioned the one Himalayan species that showed such an aberration. Is that the reason this goose has a different breeding range, or is this just the adaptive mechanism that fits the goose to that particular niche?

BEER: That question helps me to explain why I am harping on individuality. Some years ago I did some playback experiments in an attempt to find out whether birds of a pair could recognize one another by ear. I got successful results and found that a bird on the nest would respond to the call of its mate, but not to the call of anybody else. When I examined the calls that I was using, I found considerable similarity among a number of the birds as to those characteristics of the long call that appeared to convey identity. I immediately made a general rule that the long-calls of mated birds tend to be more similar to one another than to the calls of birds at large.

Then the question was posed whether birds match their calls after they are mated, or do birds with initially similar calls tend to mate? In attempting to answer this question, I collected many recordings of both the males and females of different pairs. What I find is that in some cases the calls are very similar and in other cases they are not similar at all. So my answer is that, in some pairs, the calls are similar in those characteristics that convey identity, and in others they are not. I thus taught myself a little lesson about generalizing prematurely. I have become very much aware of the idiosyncracies between individuals and between pairs in the birds I have examined.

The young seem to be perfectly well able to distinguish and recognize the calls of both parents, even when the long-calls are markedly different from one another. Of course, this has been found in other birds as well. I think Tschanz* was one of the first to experiment with the possibility of individual recognition by ear in his studies of the Guillemot. He showed experimentally that a Guillemot chick, even before it hatches, can learn to distinguish the voices of both of its parents, even when they are quite different from one another, from the calls of any other adults.

*Tschanz, B. 1968. Trottellummen. Z. Tierpsychol. Suppl. 4. 103 p.

JENNI: Perhaps I should let Doctor Murton answer the question, but I would like to answer it myself. If we do find that the proximal factors controlling polyandry and/or reversal of the role of the sexes is hormonal, we will have identified the proximal mechanism whereby the system is maintained. The proximal mechanism is not necessarily related to the original evolutionary causation or to the adaptive significance of the system. Polyandry and the physiological mechanism for regulating it on a proximal basis have evolved together, not in sequence.

DYER: Are there any dissimilarities in birds that show divorce?

BEER: You are referring to the Kittiwake work. Really I cannot answer the question. You would have to put it to Coulson. As I recall from his published work, he found that, in many cases, divorce ensued when the birds of the pair failed to mesh their schedules, so that the eggs were not effectively incubated. As a consequence of this, many eggs failed to hatch. Exactly how the birds communicate with one another and what individual characteristics are involved there, I don't think he has the information. He may, but if he does, I don't know about it.

HAILMAN: I have a question about the possible generality of the elegant studies of Hinde and Lehrman. The general picture of hormonal behavioral interaction seems to be that the stimuli from the mate or the stimuli from immediate environment brought about by the behavior of the mate (say, the nest situation) are the controlling factors in subsequent timing of the reproductive behavior. But it occurs to me that the investigated animals (canaries and Ring Doves) are both very highly domesticated and tested in a laboratory situation.

Now, if we look at relatives of these two species in North America, for instance, the canary's congener, the American Goldfinch (*Spinus tristis*), it is a very late breeder. It breeds, at least in latitudes where I am familiar with it, after the summer solstice, so that the "classical" picture we have been talking about earlier in the symposium of increasing day length effects would seem to require additional mechanisms and environmental triggering in order to bring about reproduction.

Or consider Ring Doves. One of its relatives here, the Mourning Dove (*Zenaidura macroura*), breeds in Maryland in every month of the year, so that it seems to be a quite aperiodic breeder, although I am sure there is emphasis on breeding in the spring.

I just wonder what the generality of these laboratory birds could be with regard to more typical species that are tropical in origin, that come to the northern climates to breed and go away again, where the whole reproductive cycle may be much more tied to environmental changes and to migratory patterns.

BEER: I could not agree more. I wonder about the generality of this work, too. But I think perhaps two things ought to be distinguished here. One is, again, the attempt to find out whether there are general rules and, perhaps, how they might have to be qualified from one species to another, just what are the particulars, and so forth. The other point is that, whether the rules are general, they tend to be far from simple. My aim in choosing those two examples was to illustrate the varieties of ways in which behavior and physiology can be related causally. Even though the conclusions may not have general application, I think they do illustrate what we might expect to find wherever we look: that you will find short-term and long-term effects, peripheral and central effects, positive feedback, negative feedback, and something that I call progressive feedback. That was really the point I was trying to make—to illustrate the diversity, where one can find it, of the patterns of relationship between behavior and physiology when they can be seen to be causally linked.

About the generality, I think it is an open question. In a study, one is always constrained by the techniques that one uses. There is an inevitable relativity in any scientific method, and one must be very careful to keep this in mind when tempted to make sweeping generalizations from particular species studied in particular situations with particular methods. Every method has its limitation and its way of selecting. In short, it imposes selection.

FOLLETT: A primary rationale for developing methods to measure hormones in small volumes of plasma has been to apply these techniques to the field situation. However, an unexpected difficulty has arisen in that much individual variation exists in the plasma concentration of luteinizing hormone and testosterone. It is thus difficult to interpret much of the data. The reasons for the variation are unclear but it may reflect the facts that the birds are altering hormone output minute by minute (there is evidence for this in domestic mammals) and that each bird has a stable level, with individuals varying in the position of this set point. Clearly, these problems must be resolved.

KERN: I would like to comment on the canary studies of Hinde. By way of preface, let me say that canaries build a nest in two stages. They first construct an outer frame of "grass" and then line this with feathers. On the basis of some elegant studies,* Hinde postulates that canaries show an increasing preference for feathers late in the building period and cease to build shortly before egg-laying, because of information that they receive about the nest by way of the incubation patch. Preference for feathers is thought to require a small nest bowl or one constructed of coarse nest material. Cessation of building requires both.

I am now examining these postulates. I modify the information available to the female canary through her patch by covering it with a pliable plastic diaper or denervating it surgically. Diapered hens can feel the bowl (i.e., assess its size), but can no longer determine its texture. If Hinde is correct, one would expect a diapered hen to show an increasing perference for feathers, but *not* to stop building before she lays her clutch. She does show the increasing preference for feathers. However, she also ceases to build shortly before egg-laying. Denervated birds can no longer perceive either the size or the tex-

---

*Hinde, R. A. 1965. Interaction of internal and external factors in integration of canary reproduction, p. 381–415. *In* F. A. Beach [ed.] Sex and behavior. John Wiley & Sons, New York.

ture of the nest through the patch. If Hinde is correct, one would expect these females to show no preference for feathers and to continue to build up to and beyond the time when they lay eggs. However, they too show an increasing preference for feathers and they too stop building one to several days before egg-laying. In short, diapered and denervated females build and behave like normal females.

These results suggest that the canary is receiving information about the nest from more than one source. She has two perfectly good eyes as well as the tactile receptors in the incubation patch and she can assess the condition of the nest with either. I suggest that when one source of sensory information is absent, she relies upon another.

Medway* came to the same conclusion in his earlier studies of pigeons with denervated venters (pigeons lack incubation patches). He found that denervated hens would incubate eggs in a lit room, but often did not incubate in the dark. They did crush their eggs, which says something about the role of pressure receptors in the venter. At any rate, Medway's conclusion, like my own, was that more than one kind of receptor is involved here; if sensory information sent to the brain by one receptor is interfered with, the bird relies on another for information about nest and clutch.

IMMELMANN: I would like to refer to the question of the similarity or dissimilarity of the calls of the Laughing Gull. Doctor Beer, do you have any indications that perhaps the pairs with similar calls had been mated for a longer period than those with more dissimilar calls? I am referring here to the work by Mundinger, who found that, in certain species of Cardueline finches, the partners of a pair come to share one or more flight call patterns and that several other calls are also modified by vocal imitation. Perhaps the Laughing Gull pairs with similar calls just had more time to match their calls. Do you have any indication?

BEER: My answer is quite brief. No, I have no indication one way or the other. It is very difficult to keep track of individual birds, even within one season, in the gullery in which I have been working. We have nothing like the beautiful situation that Coulson has, where he can get at his birds easily and follow them from one year to the next. So, really, at the moment it is an open question as to what extent, on the one hand, birds can modify their calls to achieve some kind of match, to what extent birds with initially similar calls are more likely to mate than birds with dissimilar calls, and as to what the effects are—we have, in fact, no information at all about divorce and fidelity in the Laughing Gull. We should like to have it, but the situation is such that as yet we have not been able to collect it.

HAILMAN: I want to respond a little to Immelmann's question about the study done at Rockefeller by Mundinger. If you examine his paper closely, he did several experiments in which he recorded sonograms of birds' calls as a function of the time they spent together—this was in heterosexual and homosexual

*Medway, L. 1961. Domestic pigeons: The stimulus provided by the egg in the nest. J. Endocrinol. 23:9-18.

groups. But it is very interesting that the degree of familiarity is not the only thing that changes during the time of his experiments. In fact, sexual behavior appears, even in homosexual groups. As he shows, they become homosexual in the reproductive sense as well, and all birds come into breeding condition. So one wants to know whether the observed change is due to increasing familiarity of the individuals or to a change in the physiological condition of the animals.

Now the calls do become similar, but it is interesting to note that from group to group, even in separate experiments, the calls all become similar to some archetype. So one possible interpretation is that the birds in breeding condition all give a very similar species call that has nothing to do with the phenomenon of the individuals' learning one another's call at all.

ORIANS: I have a question that stems from some comments that Doctor Jenni made concerning our inability at the moment to deal in terms of physiological mechanisms with such acts as territorial behavior or responses to multiple mates. We do seem to have reasonable physiological methods for dealing with responses such as gonadal responses to hormones. However, if we became interested in, for example, the physiological basis of territoriality, where would we begin to look? What sort of neural bases or hormonal bases might one begin to test?

FARNER: As for methodology, certainly techniques for the placement of lesions and implants of various sorts in the central nervous system are developing at a rate that promises excellent possibilities for obtaining useful information on the components of systems that are involved in behavior.

I share with Beer the frustration, which I believe to be ephemeral, that comes with seeking physiological explanations of very complex elements of behavior, such as defense of territory and interaction between mates. I think that this reduces itself to thermodynamic considerations in reference to extremely complex systems with very low entropy for which what we have to construct "explanations" are very unsatisfactory bits and pieces, the total entropy of which is high. Certainly, the approach that must come, if we wish to have physiologic explanations, is a resort to careful and systematic reductionist procedures. I am optimistic that these can be effected. I will not, however, argue that the explanations thus obtained will be worth the effort, but I think that they can be obtained.

To say that, from the aspect of behavior, one might find more satisfaction in looking at the problems of linguists suggests to me that one certainly will find similar problems and that it may be interesting and comfortable to compare them to the problems of "explaining" behavior. But I submit, if I may make an overly simplistic analogy, that this is akin to two people with fever. They have a few problems in common and can engage in sympathetic discussions, but they solve nothing. Had each, instead, consulted a physician who could have applied systematic diagnostic procedures, they would have at least taken the first step toward solving their problems.

To assert that it is impossible or even unlikely, that there is a physiological resolution or explanation for complex behavioral patterns is really to assert that the second law of thermodynamics is invalid, or that there is a host of Maxwell's demons who need pay no attention to it! I think that the real point here reflects our experience with scientific research in general, which is that one begins with a black box and attempts to understand why it reacts as it does and why it has the properties that it does. One eventually opens the box, only to find that the "explanation" consists of a number of smaller black boxes each with its peculiar characteristics. One must then decide whether he has an adequate "explanation" or whether he must open the still smaller boxes to seek the "explanation." So, at this point one arrives not at the question of whether or not it is possible to find physiological bases for very complex behavioral problems, but whether finding it through a whole series of successively smaller black boxes is, in fact, the way that one wishes to go.

BEER: To deal adequately with Professor Farner's remarks would require more time than we have available here. I shall limit myself to a few brief comments.

Professor Farner may be right in his belief that, if we wish to have physiological explanations of complex behavior, our only resort is to turn to "careful and systematic reductionist procedures." I question whether such procedures are at present available to us for certain of the kinds of questions upon which I touched. However, Professor Farner seems to concede that the approach is yet to come. In the search for it I think a number of points should be borne in mind. First, the logic of reductive explanation is simple in only the small proportion of cases that tend also to be the least interesting. It is a matter about which philosophers of science are divided. Second, I doubt whether conceptions at the physiological or physicochemical level are so well established that it is a closed question to what kinds of terms complex behavioral phenomena should be reduced. At one time, the aim of behavioral analysis was held to be the reduction of all behavior to reflexes or stimulus-response links—an atomistic approach against which there has been reaction in both psychology and behavioral physiology that has tended to emphasize more holistic conceptions of organization, integration, and control. A similar swing has occurred in genetics, where atomistic conceptions of the gene and of gene action have been opposed by the conception of the genome and the gene pool as organized and integrated systems. Like other kinds of explanation in science, reductionism is attempted within a context involving models or theories—a "paradigm" as Thomas Kuhn has described it. As science advances, models and theories come and go. At one time the telephone switchboard was the model of nervous system function; nowadays, it is the computer; tomorrow, who knows?

It is at least conceivable that new concepts of physiological functioning will be forthcoming from the attempt to reduce behavior to physiology; as Piaget would have it, physiology may not be able to assimilate behavior without itself undergoing accommodation. Indeed, such accommodation could even involve a reversal of reductionism: explanation of the molecular in terms of the molar. Reversal of this sort is not without precedent in the history of science. Professor Farner's reference to thermodynamics reminds me that the development of this science involved a transference of attention from the mechanics of individual particles in motion to the statistical properties of populations of such particles.

Such changes of direction in science have often involved the recognition of new questions or the separation of questions that had hitherto been confused. One of the crucial steps in Darwin's thinking was his separation of what is implied by the existence of heritable variations in natural populations of organisms from the question of the origin of such variations (a step, incidentally, that some present-day biologists, such as Waddington, believe that we now have to try to take back). In my formal presentation I tried to argue that behavior raises a diversity of questions, investigation of which requires that the distinctions be marked and respected if progress is to be achieved, given our present state of knowledge and theory. I am reminded of an incident that occurred to me when I was a student. I arrived at the university one morning to find a crowd of people gazing at an unmentionable bedroom utensil perched on the topmost point of the clock tower. The bewilderment of the people was divided between two questions: How was the object placed where it was? Why should someone have undertaken the risk and difficulty of putting it there? The first question was a matter of engineering; the second a matter of motive. Some of the people were interested in one of the questions more than the other, but nobody suggested that there was only one question or that the answer to one of the questions would dispose of the other as well. Professor Farner's parable of the fever victims seems to imply that all questions about behavior are ultimately physiological and hence that the physiologist is the authority to consult if we want to make headway toward their solutions. Given present knowledge, the first proposition can only be an article of faith. Whatever the ultimate case may turn out to be, I repeat that, for several of the kinds of question that observations of behavior raise, we are now at a loss to know how they might be expressed in physiological terms. The physiologist, it seems to me, has no more authority than the ethologist in this sort of situation. Indeed, it is my opinion that he may be, and has been, more hindrance than help. The model of the reflex arc has been a mixed blessing in psychology, and I have mentioned what I think has been a retarding influence of physiology on the progress of thinking about animal communication. To turn Professor Farner's parable to my purpose, if the cause of the fever were an ailment unknown to medical science, the sufferers could well be better off consulting one another rather than a doctor who assumes his knowledge covers the case. However, the situation I have in mind with respect to animal communication and linguistics may be more like the situation of someone involved in a divorce suit: For certain aspects of the case at least, the person will need to consult a lawyer, not a physician. To suppose that there is only one legitimate approach to the understanding of the reproductive biology of birds is an attitude like that attributed to Benjamin Jowett, the Master of Balliol College, Oxford, during a large part of the nineteenth century. Jowett is supposed to have said: "I am the Master of this College. What I don't know isn't knowledge."

Fecundity, Mortality, and Avian Demography

ROBERT E. RICKLEFS

One of the objectives of population biology is to predict changes in population size as the result of observable changes in primary demographic parameters—fecundity and mortality—whether these changes are caused by natural environmental factors or by man. Implicit in this aim is to determine to what extent populations can adjust to changes in their environment and remain viable.

This paper concerns the effect of variation in demographic parameters on the growth potential of populations. We cannot equate growth potential with the actual growth rate of a population because density-dependent factors absorb surplus productivity and maintain populations close to constant levels. Growth potential represents the ability of a population to recover from decimation and also indicates the rate at which a population can be exploited under controlled management practices.

The "means" of population study frequently are to reduce demographic phenomena to a convenient set of numbers on paper that can be subjected to mathematical analysis and statistical scrutiny much more readily than their counterparts in the natural populations that are the object of the study. All too often, however, the "means" become the end, and the mere existence of the numbers on paper, rather than their validity, becomes the object of population studies.

We must constantly remind ourselves that the mathematical analysis
of population is valuable only in that it can tell us something of the
behavior of populations in nature. As Hickey (1955) points out,
"bird populations have a field reality and a paper existence," a distinction that must be kept in mind.

DEMOGRAPHIC PARAMETERS IN NATURAL POPULATIONS

A brief review of our knowledge of population parameters will serve
both to introduce the basic factual knowledge (and lack of it) on
which a practical understanding of avian demography can be based
and to outline the population parameters that form the basic life
equation of the species.

The basic components of the population equation (the complete
quantitative description of the demography of a population) include
all of the birth and death statistics of the population that, for birds,
may be conveniently broken down into the following components:
the number of eggs laid per clutch, the frequency at which clutches
are laid, the survivorship of eggs and young and to the age at first
reproduction, and the subsequent survival of the adults throughout
their lifetime.

CLUTCH SIZE

The number of eggs that birds lay has been the subject of a formidable literature, much of it controversial. The topic has been reviewed
extensively, beginning with David Lack's (1947, 1948b, 1954, 1968)
pioneering studies and augmented by many more (e.g., Moreau 1944;
Skutch, 1949, 1967a; Cody, 1966; Klomp, 1970; von Haartman,
1971), not to mention numerous studies on the evolution of clutch
size.

In passerines and other small land birds that feed their young,
clutch size varies between two and perhaps a dozen eggs in the average clutch of most species. Clutches tend to be larger at high latitudes than in the tropics by a factor of two or three (Cody, 1966);
hole nesters lay more eggs than do species with open nests; birds lay
fewer eggs in coastal than in continental regions; at the same latitude,
clutch size is low in habitats with low productivity (for example, in
our southwestern deserts), although the opposite trend occurs in the
tropics (Lack and Moreau, 1965; Marchant, 1960). Clutch size tends
to increase with altitude in temperate regions (e.g., Johnston, 1954),

but not in tropics (Skutch, 1967b). Most raptors lay fewer eggs than most passerines, and clutches are smaller in progressively larger species (Ricklefs, 1968b).

Species that do not feed their young lay more eggs than passerines by a factor of two or three, but variations in clutch size are thought to parallel those observed in passerines (von Haartman, 1971, p. 425). Some precocial species, particularly charadriids, have seemingly sacrificed clutch size for the sake of increasing egg size and typically lay four or fewer eggs per clutch (Lack, 1968).

Oceanic Procellariiformes and Pelecaniformes rarely lay more than one egg and, moreover, are not capable of raising more than one young in experimental situations (e.g., Rice and Kenyon, 1962; Harris, 1966), although such inshore feeding species as penguins, cormorants, and gulls usually lay between two and four eggs.

Although only seabirds with single-egg clutches have a good mathematical reason for lacking variability in clutch size, several groups of nidicolous land birds that typically lay larger clutches also do so without variation, even within an entire family. All hummingbirds and most pigeons and doves lay two eggs, regardless of the time of the season, quality of the habitat, or geographical locality. The clutches of most other nidicolous species are sensitive to environmental variations, at least to some extent (Lack, 1954; Klomp, 1970).

As banding programs and more refined methods of determining age of birds in the field have been developed, it has become increasingly apparent that the number of eggs laid increases with age in many species. Average clutch size of the Black-legged Kittiwake (*Rissa tridactyla*) increases from 1.8 during the first breeding season to 2.4 after the third (Coulson and White, 1961); in the Netherlands, Starlings (*Sturnus vulgaris*) lay an average of 4.9 eggs during their first season and 5.9 thereafter (Kluijver, 1935). Richdale (1957) found that in Yellow-eyed Penguin (*Megadyptes antipodes*) one-egg clutches comprises 38 percent of all clutches during the first breeding season, only 6 percent during the second, and less than 2 percent thereafter (the normal clutch is two). While these do not represent major age-specific trends in fecundity, such information should nonetheless be incorporated into demographic calculations when available.

LENGTH OF BREEDING SEASON

The number of clutches that a female lays each year is an integral part of the annual fecundity of the individual, as is the size of the clutch itself, and depends in large part on the length of the season that is suitable for reproduction. All habitats are seasonal to a greater or lesser degree, and the timing of reproductive activities strongly reflects this. Breeding seasons are shortest at high latitudes and where rainfall is sparse and highly seasonal (Lack, 1950; Skutch, 1950; Ricklefs, 1966; Serventy, 1971; Immelmann, 1971). The period over which eggs are laid varies from 1 to 2 months in the Arctic, 3 to 5 months in the temperate regions, up to 7 months in subtropical regions, and up to 10 months (rarely 12) in the tropics. Within the tropics, dry climates may have breeding seasons on the order of 4 months (Marchant, 1959). Some tropical areas are characterized by a double breeding season (e.g., Miller, 1965) that reflects two peaks in the seasonality of rainfall.

Within a region, specific breeding seasons correspond to the availability of the particular food resource used and may vary considerably. Birds that specialize on adult insects (flycatchers, swallows, etc.) tend to nest late and for relatively short periods, as do species that feed their young seeds (goldfinches, grassquits, etc.). Species with varied and flexible diets, particularly such omnivorous birds as pigeons, have long breeding seasons. Hole-nesters tend to nest early and for relatively short periods, although the reason for this is not clear. Raptorial birds must lay eggs long before other land birds to accommodate their long incubation and nestling periods.

The trends in breeding seasons exhibited by land birds are closely paralleled in seabirds. Whereas boreal species always raise but a single brood, initiated during a narrowly defined season, some tropical seabird colonies may have eggs laid in every month of the year (Ashmole, 1971). Breeding conditions are so nearly uniform in some tropical areas that seabirds renest immediately after fledging a brood, generally less than a year from egg-laying, rather than wait for some external seasonal stimulus to initiate reproduction on an annual basis. On Ascension Island, the Black Noddy Tern (*Anous tenuirostris*) has a reproductive period of about 45 weeks that is well synchronized within the colony. Although terns exhibit highly seasonal breeding, reproductive activity peaks occur at a different time each year (Ashmole, 1963).

NUMBER OF BROODS

While the overall length of the annual reproductive period is relatively easy to determine for populations of most species, few data are available on the number of clutches laid, or broods raised, by multiple-brooded species. There are two major reasons for this. First, relatively few marked populations of birds have been followed in sufficient detail to evaluate annual reproductive performance directly. Second, the time required for reproductive cycle, including the time between broods and time wasted on unsuccessful breeding attempts, is too poorly known for most species to calculate annual fecundity indirectly from the length of the breeding season and the time required for each brood. Despite all the discussion of clutch size in the literature, the annual fecundity of birds is hardly understood at all.

In some species, the nesting cycle is so long that no more than one brood could possibly fit into a single season, and failures may not even be replaced unless the loss occurs very early in the season. Because their nesting cycles are longer than a year, King Penguins (*Aptenodytes patagonica*) nest twice every 3 years (Stonehouse, 1960); condors, albatrosses, and large eagles can nest only once every other year (Amadon, 1964; Rice and Kenyon, 1962).

In the restricted breeding seasons of temperate and arctic regions, even small land birds with relatively short breeding cycles may nest only once; in the Arctic, they may not even attempt to replace losses. Many hole nesting species, such as the Great Tit (*Parus major*) and the Pied Flycatcher (*Ficedula hypoleuca*), are single-brooded in the northern parts of their ranges, but frequently replace lost broods (Lack, 1966; von Haartman, 1971). In Pennsylvania, Starlings commonly attempt a second brood after successfully raising their first, but relatively few of the second broods fledge young (although this varies from year to year). Many temperate-zone species with longer breeding periods may attempt more than two broods. Wood Pigeons (*Columba palumbus*) usually lay two or three sets of eggs during a season (Murton, in Lack, 1966). Nice (1937) found that in one year of her study on Song Sparrows (*Melospiza melodia*), 12 pairs made three nesting attempts and 4 pairs made four attempts; only 18 of the 42 clutches were successful (one each for 11 pairs and two each for 4 pairs). Lack (1966, p. 123) gives several cases of extreme high productivity in the European Blackbird (*Turdus merula*), citing one pair that raised 16 young in five broods and another that raised 17 young in four broods within a single season. But Snow (1958) found that in the Botanic Garden at Oxford, where nesting success was higher than usual, about 20 percent of pairs raised no young, and the average number of young was only slightly over four. Snow (1962) found that the Black-and-white Manakin (*Manacus manacus*) usually lays three clutches per season in Trinidad, and some individuals attempt as many as five. Other small passerines probably attempt two to four broods there. Even over the geographical range of a species, the number of broods typically increases toward lower latitudes. The Song Thrush (*Turdus ericetorum*) is single-brooded at the northern end of its range (65–70°N), but attempts two to three broods, and perhaps four, at the southern end of its range (45–50°N) (Siivonen, in von Haartman 1971, p. 428). The Rough-winged Swallow (*Stelgidopteryx ruficollis*), however, appears to be single-brooded throughout its range from southern Canada to Argentina (Ricklefs, 1972).

Little or no renesting has been reported for most gallinaceous birds in the United States (Hickey, 1955), but there are some exceptions. The proportion of female Ring-necked Pheasants (*Phasianus colchicus*) that are successful, judged on the basis of individuals with broods, was 1.5 to 2 times greater than the proportion of successful nests in one study, which suggests that renesting was prevalent (Hickey, 1955).

It is evident that direct estimates of annual fecundity will be difficult to obtain, at least for many small, potentially multiple-brooded species. The simplest indirect estimate is to divide the total length of the season by the time required for the completion of a successful brood, including the delay before renesting. This will underestimate the actual number of nesting attempts, however, because nest failures take less time than successful broods; the method will overestimate the number of successful broods because it indicates the maximum possible number of successful broods. This indirect technique will be refined later to provide a more accurate estimate (see p. 401), but I will make a preliminary calculation at this point. Elsewhere, I have summarized the length of the nesting cycles for temperate and tropical regions (Ricklefs, 1969b). The average length of a successful nest cycle (egg to egg) is about 2 months in the tropics and $1\frac{1}{3}$ months in temperate regions. Thus, in the tropics, if eggs are laid by a population over a period of 8 months, an individual could squeeze in four full nest cycles and lay a fifth at the end; four successful broods could be produced within a 6-month egg laying season. For a typical temperate-zone species, egg laying periods of 2, 3, and 4 months would permit the raising of two, three, and five successful broods, respectively. Thus, there is reason to suspect that some tropical species may attempt a half-dozen or more clutches during a season, while three or four would appear within the reach of many temperate birds.

NESTING SUCCESS

Nesting success has been reviewed many times (e.g., Kalmbach, 1939; Kendeigh, 1942; Lack, 1954; Hickey, 1955; Nice, 1957; Skutch, 1966; Ricklefs, 1969a). Among passerine birds in temperate regions, the proportion of eggs that eventually become flying young varies widely between about 30 and 80 percent, although some species may be subjected to decimating failures of reproduction in some areas in some years (e.g., Kale, 1965). Nests in marsh habitats are least secure, and success progressively increases in ground-nesting, tree-nesting, and hole-nesting species. A strong latitudinal gradient is also evident among passerines with open nests, decreasing from about 60 percent in the Arctic to 47 percent in temperate regions and to 32 percent in the humid tropics (Ricklefs, 1969a). Nesting success of birds in dry tropical habitats (44 percent) is comparable to success in temperate regions and to one semiarid temperate locality in Arizona (40 percent) (G. Austin, unpublished data).

Increasing size among altricial land birds is associated with decreased daily nest mortality rates, presumably because large species can better defend their nests against predators and the young can better withstand inclement weather because of their large size and energy reserves. Because nest periods are much longer in raptorial birds than in smaller species, however, the overall survival of eggs and nestlings (40–80 percent) is in the same range as for small land birds. The nesting success of gallinaceous birds (to hatching, which requires nearly as long as the combined incubation and nestling periods of most small land birds) is similar to that for ground-nesting passerines (20–60 percent) (Hickey, 1955). The Tetraonidae are somewhat more successful than the Phasianidae.

Whereas daily mortality rates during the nest period are fairly low among seabirds, especially during the nestling period, the young remain in the nest for such a long period that fledging success is often quite low, varying from about 5 percent in boobies (*Sula*) on Ascension Island (Dorward, 1962) to 30–60 percent in two tropic birds, a storm petrel and several cormorants (Ricklefs, 1969a). The gulls, terns, and alcids, with higher daily nest mortality but shorter nest periods, have comparable overall rates of nest success (6–60 percent, half of which are between 26 and 44 percent). Shorebirds (most of which nest in the Arctic) and other precocial water birds (including grebes, ducks, and rails) exhibit relatively high egg success (most studies exceed 50 percent).

Predation is by far the most important mortality factor for small land birds. Nolan (1963) found that about 88 percent of all nesting failures of birds in a deciduous scrub habitat were due to predation. Lack (1954) has blamed predators for about three quarters of the nest mortality in Britain, and Nice (1957) has estimated predation as 40, 43, 55, and 73 percent of losses in four species. Snow (1962) stated that at least 86 percent of nest failures of the Black and White Manakin in Trinidad were attributable to predators, mostly snakes. The intensity of predation on passerine birds apparently decreases as one moves out of the tropics into higher latitudes and may even be negligible in the Arctic (Ricklefs, 1969a). Jehl (1971) has indicated, on the contrary, that predation is the major cause of egg mortality in arctic water birds and shorebirds, even though average success rates are in excess of 75 percent. Other factors, such as starvation, inclement weather, desertion, disease, and brood parasitism, influence nesting success to a greater or lesser extent, depending on the geographical region and nesting habitat (Ricklefs, 1969a). In raptors and seabirds, which are relatively immune to predation, many losses are caused by starvation of the young and by "internal" factors (Pettingill, 1939) attributable to intraspecific strife in crowded nesting grounds.

POSTFLEDGING SURVIVAL

The survival of young birds after they leave the nest is difficult to assess because direct observation is rarely possible, and the indirect methods used to calculate the survival of adult birds are not applicable. For this reason, the data presented below include most of the information that is available for this population parameter.

Mortality may be estimated in three ways: by direct observation of the fate of young birds after fledging (or hatching in the case of precocial young); by changes in the ratio between juvenile and adult birds in populations (which gives an estimate of survival relative to that of the adults); and, when adult mortality rates and the production of fledglings are known, by calculating the survival of young needed to replace adult losses. The latter method assumes that the population is stable and that all young enter the breeding population.

[Rates determined by this method cannot validly be used to construct life tables unless independent data demonstrate the population is stationary and has a stable age distribution (see p. 397).] In many marine birds whose young do not reach adulthood for several years, conventional banding techniques may be used to obtain the annual mortality of immature birds.

Several studies based on direct observation of small numbers of fledglings have provided estimates of postfledging mortality in small songbirds. Survival usually can be assessed only until the young become independent of parental care and leave the nesting area, usually up to a week or two after fledging. These data are as follows: Wood Pigeon, 40 percent to 30 days (Murton, quoted by Lack, 1966); European Blackbird, 81 percent to 5 days and 66 percent to 15–20 days (about 62 percent of young that survive to independence also survive to the following breeding season, about the same as the adult survival rate) (Snow, 1958); Ovenbird (*Seiurus aurocapillus*), 56 percent to 30–35 days (Hann, 1937), Cactus Wren (*Campylorhynchus brunneicapillus*), 87 percent to 21 days (Ricklefs, 1968a); Black-capped Chickadee (*Parus atricapillus*), 82 and 97 percent (2 years) to 21–28 days (Smith, 1967); European Tree Sparrow (*Passer montanus*), 62 percent to 1 month (Pinowski, 1968); Skylark (*Alauda arvensis*), 22 percent to 25 days (Delius, 1965). In general it appears that the postfledging survival rate in those species with the longest nestling periods (i.e., whose young fledge in a more developed state) is highest.

The survival of precocial young of gallinaceous birds is somewhat higher than that of the altricial passerines mentioned above, but this is not surprising in view of their larger size and more advanced state of development. Typical data (Hickey, 1955) are Prairie Chicken (*Tympanuchus cupido*), 50 percent to 4 weeks; Ring-necked Pheasant, 70 and 88 percent to 8 weeks; Bobwhite Quail (*Colinus virginianus*), 71 percent to 8 weeks; and Sage Grouse (*Centrocercus urophasianus*), 54 percent to 9 weeks.

It has often been stated that mortality rates of young birds increase sharply for a short period after the young become independent of their parents. However, I know of no direct evidence bearing on this point, and I see no particular reason to hold this view, unless one assumes that after the young leave the protective mantle of parental care they are thrown into competition with the general population, in which they fare poorly compared to more experienced adults.

Just how much experience the young need to attain adult behavioral and physiological capabilities (and thus, adult survival rates) is open to question. Lack (1946) has stated that the survival rates of first year European Blackbirds, Song Thrushes, and Starlings reach adult levels by January 1, although Kluijver (1951) found the December–May mortality in the Great Tit to be 43 percent for first-year birds, compared to only 25 percent for adults. The prolongation of parental care in seabirds, especially in the tropics, suggests that a long period is required for the attainment of proficient hunting skills (Ashmole and Tovar, 1968). Direct observations by Orians (1969) on the feeding success of Brown Pelicans (*Pelecanus occidentalis*) of different ages support this view.

Survival rates of first-year birds are known to increase during the first winter in quail (see Figure 1), Common Cormorants (*Phalacrocorax carbo*) (Kortlandt, 1942) and the Herring Gull (*Larus argentatus*) (Paynter, 1947). The survival of many large species with delayed reproduction continues to increase for several years before attaining adult levels. For example, immature Herring Gulls sustain a 20–25 percent loss during their first winter (September–March, 3–4 percent

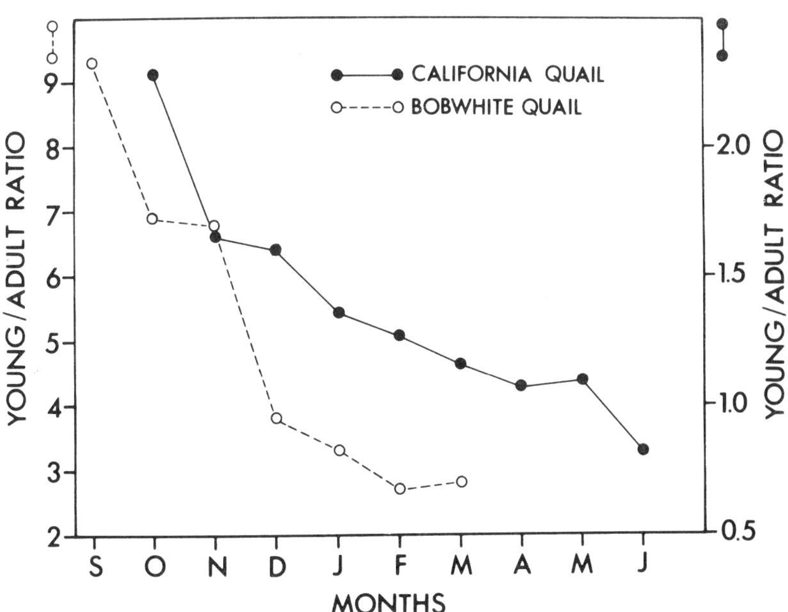

FIGURE 1 Ratio of immature to adult individuals in populations of the Bobwhite (Robel and Fretwell, 1970) and the California Quail (Emlen, 1940) from fall to spring.

per month), a 20–22 percent loss during the ensuing year (just under 2 percent per month), and 4–9 percent annual mortality as adults (about 0.5 percent per month) (Kadlec and Drury, 1968). The survival or Caspian Terns (*Hydroprogne caspia*) increases during their first 3 years from 56 to 74 to 82 percent (Hickey, 1952).

A complicating factor for demographic calculations is that the survival of the young is often not independent of other population parameters. Perrins (1965) has shown that postfledging survival of the Great Tit varies both with the size of the brood from which the young fledged (see also Lack, 1948a; Pinowski, 1968) and with the age of the female parent. Snow (1958) observed a similar trend in fledgling European Robins (*Erithacus rubecula*) in the Botanic Garden at Oxford. Of the young from first breeding attempts, 56 percent survived after leaving the nest until independence. Seventy percent of the fledglings of older birds survived the same period.

AGE AT FIRST REPRODUCTION

The age at which birds are recruited into the breeding population has a major influence on the growth potential of the population (Cole, 1954; Cody, 1971). Age at first breeding has been summarized recently by Lack (1968) and Cody (1971) for birds generally, and by Ashmole (1971) for seabirds. Virtually all terrestrial species first breed in the season following their birth, with the exception of the swifts (2 years), many parrots (2–3 years), and raptorial birds (3 or more years). Among passerines, the males of some species are known not to breed until their second year [e.g., the Starling (Coulson, 1960) and, occasionally, the Red-winged Blackbird, *Agelaius phoeniceus* (Orians, 1961)] and it is likely that the onset of reproduction is delayed, perhaps in both sexes, in many tropical species. Many have a first-year male plumage. Van Tets (1968) found that the mortality rate of first-year (nonbreeding) gulls of several species was lower than adult mortality rate during the breeding season but was usually higher during some portion of the nonreproductive season.

Among water birds (herons, ducks, waders, rails, etc.), and particularly among seabirds, the onset of sexual maturity is delayed—until as late as 9–12 years in large albatrosses and condors (Amadon, 1964). The age of onset of reproduction closely parallels trends in the annual survival of adults (Figure 2), and delayed reproduction presumably has the effect of restricting the recruitment of young into the breeding population to a level that is compatible with adult losses. The

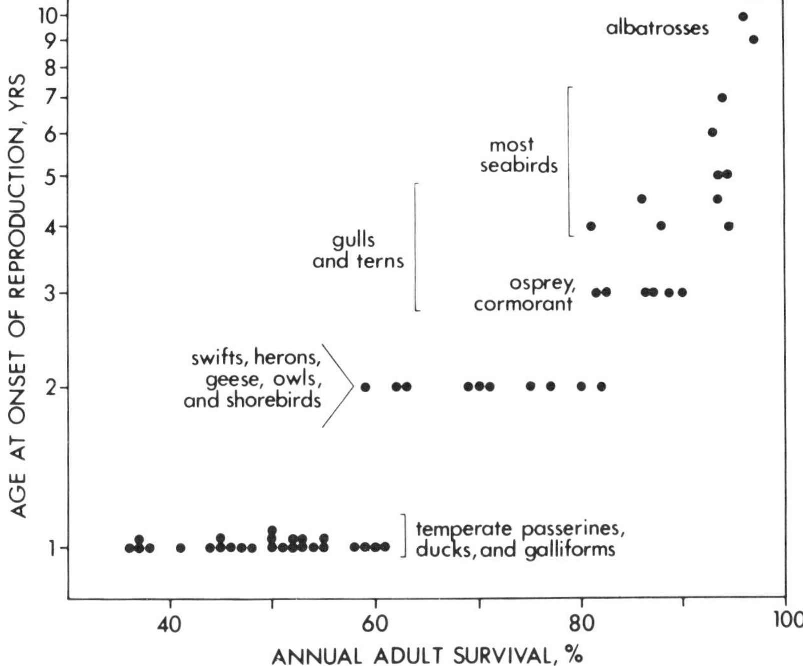

FIGURE 2 Relation between annual adult survival and the age at first reproduction in birds (from various sources).

adaptive significance of delayed maturity is not well understood, however, and has been the subject of considerable disagreement (Wynne-Edwards, 1955, 1962; Amadon, 1964; Lack, 1966, 1968).

The age at first breeding varies considerably in some populations. Favorable conditions for reproduction usually promote the early onset of maturity (references in Cody, 1971, p. 468). Females often breed at an earlier age than males (e.g., the Yellow-eyed Penguin; see p. 407), possibly because females suffer greater mortality (see p. 378), which opens more places in the breeding population each year. One might also argue that early death resulting from high mortality selects for earlier onset of reproduction; the causality is not at all celar. The White Stork (*Ciconia ciconia*) apparently has an inherent variation in age at first breeding (Lack, 1966, p. 221). Among pairs for which data were obtained, 12 percent bred for the first time at 3 years, 40 percent at 4 years, and 48 percent at 5 years. Reproductive success at first breeding increased with age (53, 60, and 78 percent, respectively).

A serious demographic complication related to the delay of reproduction is that some populations contain "reserves" of nonbreeding birds, usually of undetermined age (von Haartman, 1971, p. 433–435). Many of these individuals are probably first-year birds that are unable to obtain territories (e.g., von Haartman, 1951), but it is also possible that many birds that survive to older ages are reproductively "dead" because they cannot secure territories or attract mates and, thereby, do not contribute demographically to the population. Either way, the nonbreeding fraction of the population complicates the population equation.

The extent of "floating" populations is difficult to evaluate because nonbreeding birds often cannot be distinguished, or even observed. Estimates of the proportion of nonreproductive adults in populations are on the order of 20 percent in the Herring Gull (Kadlec and Drury, 1968), 10 percent in the Song Sparrow population of Mandarte Island, British Columbia (Tompa, 1964), and 10 percent in the British Skylark population studied by Delius (1965). Such values do not seem overly worrisome, but experiments in which breeding male populations of songbirds were removed repeatedly from a local area (Stewart and Aldrich, 1951; Hensley and Cope, 1951) suggest that population reserves may be greater than commonly believed [see von Haartman (1971), p. 434, for additional references]. Carrick (1963) has described the very unusual social organization of the Australian Magpie (*Gymnorhina tibicen*) in which large breeding flocks defend group nesting territories and exclude the majority of the population (73 percent) from suitable breeding habitats.

ADULT SURVIVAL

A considerable wealth of data on the survival of adult birds has accumulated [reviews by Hickey (1952, 1955) on gallineaceous birds; Farner (1955) and Boyd (1962) on waders; Lack (1954) and Ashmole (1971) on seabirds; see also von Haartman (1971), p. 441–449, and Cody (1971), p. 488–491]. Survivorship is generally low among small land birds (40–60 percent in most temperate passerines and 20–50 percent in gallinaceous species), but is much higher in swifts (about 80 percent). Tropical land birds apparently have much higher survival rates than do their high latitude counterparts (80–90 percent) (Snow, 1962; Willis, 1961). Extensive banding programs have not been carried out in the tropics until recently. Preliminary analyses based on returns of small numbers of banded birds over several years in forests near Belem, Brazil (T. Lovejoy, unpublished data) indicate annual survival rates of 63 percent for the Bare-eyed Antbird (*Phlegopsis nigromaculata*), 55 percent for the White-throated Thrush (*Turdus albicollis*), and 51 percent for the Pectoral Sparrow (*Arremon taciturnus*). Adult survivorship in raptors generally ranges between 60 and 80 percent, but must be substantially higher in some of the larger species that have not been studied.

Among water birds, the Anatidáe have rather low adult survival (40–60 percent). Herons, some gulls, and most waders fall in the 60–80 percent range, above which one finds most of the seabirds; the most long-lived seabirds (large albatrosses) have annual survival rates in excess of 95 percent.

Sexual differences in survivorship have been noted for several species on the basis of skewed sex ratios in populations. In most cases, the survival of the female is lower, presumably because of the strain imposed by reproductive activities. Coulson (1960) reported that the survival of European Starlings during the first year of life (after August 1) was 30 percent in females, and 61 percent in males, which do not breed during their first year. The annual survival of female Shags (*Phalacrocorax aristotelis*) (80 percent) is about 5 percent lower than that of the males (Potts, 1969). Perrins (1965) similarly found that female Great Tits enter the breeding population at an earlier age than the males and have a greater adult mortality rate.

In many dimorphic species, the easier role of the male in reproduction appears to be more than offset by his greater conspicuousness to predators (and hunters), and perhaps to deaths either directly or indirectly related to aggressive encounters with other males. Among gallinaceous species with little sexual dimorphism (quail, partridge), Hickey (1955) found that the sex ratio in autumn hunters' bags favored males (58–66 percent in 8 cases; samples of young shot during the same period contained between 47 and 54 percent males in 7 cases). Bags of the highly dimorphic Ring-necked Pheasant contained only 26 percent males, but a rapidly growing population of Protection Island, Washington, where hunting was prohibited, had a sex ratio *favoring* males by almost 2:1 after 3 years (Einarsen, 1945).

The annual survival rate of birds after attaining maturity is essentially constant (Deevey, 1947) [cf Caughley (1966) for mammalian data]. If birds are maintained in caged conditions free of most natural mortality factors, their survival is generally higher and also shows a conspicuous decline with age (Comfort, 1962). Thus, birds in natural populations do not usually survive long enough to show signs of senescence. This is consistent with current theories on the evolution of senescence (Williams, 1957; Hamilton, 1966). Berndt and Sternberg (1963, in Lack 1966) reported the remarkable finding that the survival rate in natural populations of the Pied Flycatcher (*Ficedula hypoleuca*) in Lower Saxony decreased with age from 74 percent during the first adult year to 65, 41, 30, and finally 0 percent (no returns of banded females). If these data are an unbiased representation of survivorship in the population, the situation is probably highly unusual.

On the other hand, there are instances in which survivorship increases with age after the onset of reproduction. For example, Morel (1966) reported that the survival rate of the Red-billed Firefinch (*Lagonosticta senegale*) increases during the first 4 years of life from 27 to 39, 50, and 60 percent, although these figures seem so unusual for a songbird that they should be seriously questioned. The general relationship between survival during the first year of life (usually measured from an intitial late summer date) and adult survival is presented in Figure 3. First-year survival varies about an average trend of 50–60 percent of adult level, irrespective of the absolute adult survival rate or whether the onset of reproduction is deferred beyond the first year.

Unfortunately, few of the data in Figure 3 pertain to small passerines. Additionally, the values do not give a true estimate of survival during the first year of life out of the nest because the initial dates of most banding programs occur during the late summer or early fall, well after most young have fledged. As a result, many of the values shown in Figure 3 are too high. The actual survival of young birds from fledging until they begin to reproduce can be crudely estimated by dividing the adult mortality rate (i.e., the number of openings in the breeding population that must be filled each year) by an estimate of the number of fledglings per adult. Nice's (1937) classic study of the Song Sparrow will serve to demonstrate how such calculations are made. The annual productivity of fledglings is estimated from the following data:

$$\frac{4.05 \text{ eggs/clutch} \times 47.4\% \text{ egg success} \times 3.33 \text{ nesting attempts}}{2 \text{ adults/pair}}$$
$$= 3.2 \text{ fledglings/individual/year}$$

The adult mortality is probably on the order of 50 percent, and so fledgling survival to breeding required to balance the loss of adults is

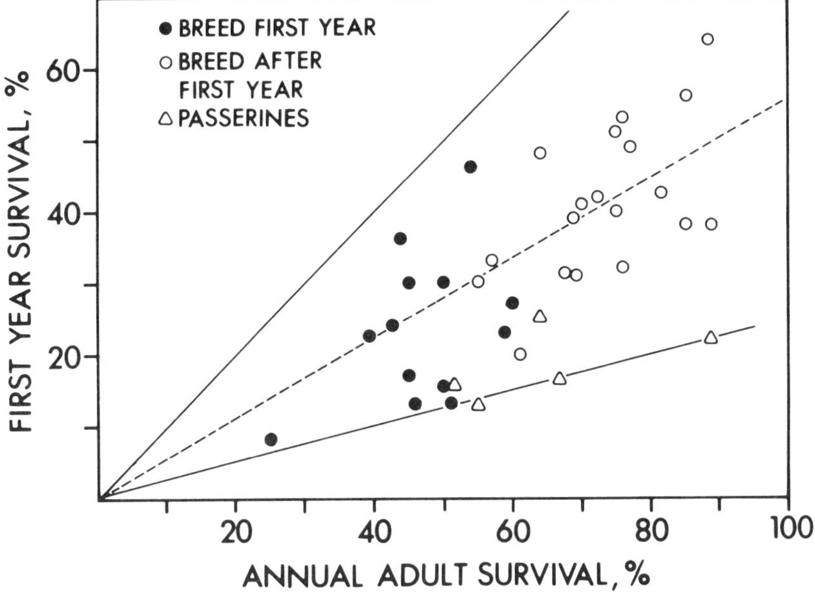

FIGURE 3　Relation between annual adult survival and survival during the first year of life.

0.5/3.2 = 15.6 percent. Similar calculations for two other temperate-zone passerines give 12.9 percent for the Great Tit and 16.1 percent for the European Blackbird. (The adult survival of the latter in most habitats is probably lower than the 67 percent for the Botanic Garden used in these calculations, but so too is nesting success.) Snow (1958) found by direct observation of banded young European Blackbirds, however, that the survival of fledglings to the beginning of the next breeding season was about 42 percent. The survival of fledgling Wood Pigeons to adulthood was calculated to be 25 percent (adult survival is 64 percent). For the tropical Black-and-White Manakin, calculated first year survival is 22 percent, although Snow (1962) directly observed the survival of 5 of 15 (33 percent) fledglings to the beginning of the subsequent breeding season. Whatever lengthens the life-spans of adult birds in the tropical environment also appears to enhance the survival of immatures. (An estimate of first year survival for the Firefinch *Lagonosticta senegale* is 37 percent, but since this is higher than the reported annual survival of first-year adults, both figures should probably be discarded). In general, the survival of small land birds from fledging to breeding is about one quarter of the annual adult survival (Figure 3).

THE GROWTH POTENTIAL OF POPULATIONS

The principal objective of demographic analysis is to determine the effect of changes in population parameters on the growth rate of the population or, alternatively, on the rate at which the population may be exploited. The following discussion sets out some basic features of population growth, both in an unrestricted phase that represents the full biotic potential of the population and under regulation by density-dependent factors. A few examples of population increase primarily of introduced species are presented to provide some concrete figures for the full biotic potential of birds.

UNRESTRICTED (EXPONENTIAL) GROWTH

Because population growth is based upon a multiplicative process, the increase in size of a population is self-accelerating (young, in turn, produce young) and tends to follow an exponential rate of increase. That is to say, the population increases by a constant factor during each time interval, rather than by the addition of a constant increment. A simple mathematical formulation of exponential growth relates the rate of increase in the population with respect to time (dN/dt) to a constant (the exponential growth rate of the population, r) times the size of the population (N). Thus

$$dN/dt = rN \qquad (1)$$

Equation (1) may be integrated over time (t) to obtain the growth of the population over that time interval $[N(t)]$ as a function of the initial population size $[N(0)]$ in the following form:

$$N(t) = N(0)e^{rt}, \qquad (2)$$

where e is the base of the natural logarithms.

DENSITY-DEPENDENT REGULATION

The exponential growth rate of a population cannot remain constant indefinitely. As the population increases, it places greater strain upon its resources and becomes more vulnerable to predation and disease, with the result that reproduction and survival fall to the point that recruitment just balances mortality. Thus populations are regulated by density-dependent effects on fecundity and mortality that greatly restrict the biotic potential of the population—a point of view that has been eloquently argued by Lack (1954, 1966). Because the fecundity of most populations greatly exceeds the required annual recruitment of new birds, density-dependent factors absorb much of the variation in demographic factors and buffer the population against change in size.

In spite of the general acceptance of the notion of density-dependent regulation of populations, density-dependence has actually been difficult to pin down in bird populations. This is partly because bird populations are relatively stable compared to most other animal populations and so they do not provide opportunities for the observation of demographic responses to a wide range of densities. One can most readily demonstrate density-dependent effects in introduced populations that rapidly increase to stably maintained levels from a small number of colonists. Einarsen's (1945) study of the introduced Ring-necked Pheasant population on Protection Island (Figure 4) is exemplary. In 5 years, the spring population increased from 8 (6 females and 2 males) to 1,325, but the rate of increase dropped steadily throughout the period. Most of the density-dependent response of

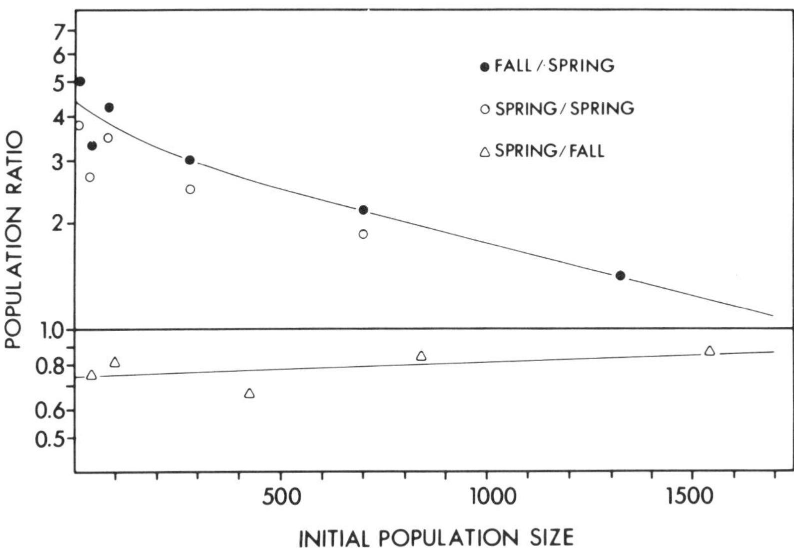

FIGURE 4   Relation between change in the introduced Ring-necked Pheasant population on Protection Island and the population density, based on data in Einarsen (1945).

the population occurred in relation to reproduction (spring to fall increase) rather than changes in winter mortality. Einarsen suggested that nesting success dropped at high density because of male interference in nesting activities.

Analysis of long-term population fluctuations also provides evidence of density-dependent change (e.g., Errington, 1945, for the Bobwhite). Lack (1966) has identified density-dependent responses in the clutch size and fledging success of Great Tits in deciduous forests near Oxford. An interesting experimental demonstration of density-dependence has been provided by Kluijver (1966) with the Great Tit on an isolated Dutch island in the North Sea. During several experimental years, the number of fledglings in the population was artificially reduced by 40 percent. In the wake of reduced competition from young birds, survival rates about doubled, to 20 percent in juveniles from 11 percent during initial control years and to 54 percent from 26 percent in adults.

BIOTIC POTENTIAL AND EXPLOITATION

Bird populations are generally maintained (often loosely) about some equilibrium size. Thus, we are most interested in a population's ability to withstand disturbances that alter demographic properties, to recover from decimation by environmental perturbations, and to sustain exploitation by predators (including hunters).

Because density-dependent factors act most strongly on dense populations, the growth potential of a population is least restricted when its numbers are most reduced, which is also the time at which it faces the greatest probability of extinction by random accident. For example, the reproductive rate of Ring-necked Pheasants on Protection Island was greatest when their population was small, and it steadily decreased as the population grew. Thus, because of the properties of population-regulating mechanisms in most vertebrates, the extinction of a population in a local area should be a fairly rare occurrence. Diamond (1969) found, however, that between 17 percent and 62 percent of the bird populations present in 1917 on the several Channel Islands in California had disappeared by 1968 and were replaced by other species largely at random. Because island populations are not fed by continual immigration from outside, they are more susceptible to extinction than are local mainland populations.

The relationship between capacity for population increase and population density, which will generally carry a population through most of its vicissitudes, may be completely inadequate in the face of an efficient predator (or hunter). If a population is to maintain its numbers, predators cannot remove any more than the reproductive surplus produced each year. The surplus is the difference between the mortality of adults and the recruitment of new birds into the breeding population (Figure 5). We will assume for the moment that adult mortality rate is independent of population density (it probably increases somewhat with density in nature), and so the mortality curve in Figure 5 is a linearly increasing function of density. If most of the burden of population regulation is borne by prereproductive individuals (Tanner, 1966), the annual recruitment rate per adult will fall off with increasing population density, and the curve for total recruitment into the population will level off above a certain density and may even begin to decline.

The hypothetical population depicted in Figure 5 has an annual adult mortality of 40 percent. At low population density, the recruitment rate is 0.8 individuals per adult (or 80 percent) and so the population initially can grow at a rate of 40 percent per year (recruitment minus adult mortality). As the population's density increases, however, the recruitment rate falls off until at a density of 125 individuals it equals the adult mortality rate. Above that point, the population

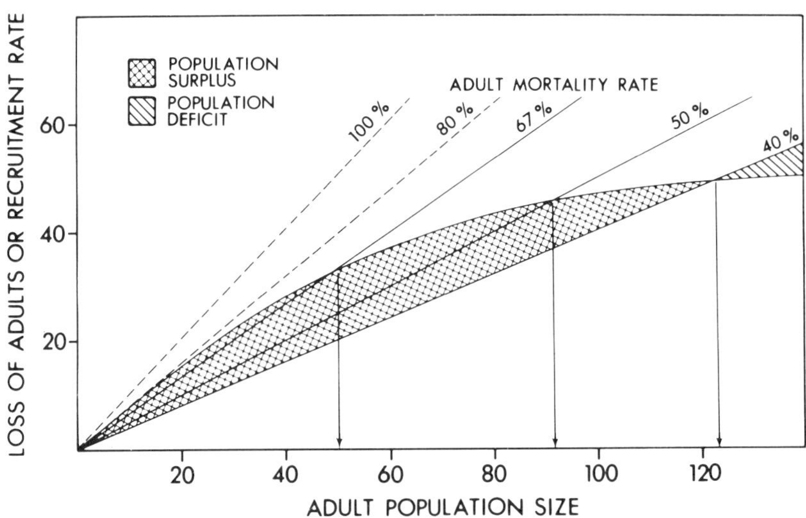

FIGURE 5 The relationship between adult mortality and recruitment as a function of population density. Various levels of adult mortality are indicated. Above a level of 80 percent the population will decrease to extinction at the level of recruitment given.

runs an annual deficit (hatched area) and the population declines. At lower population densities, the surplus of recruitment over mortality (crosshatched area) varies, being greatest (33 − 20 = 13) at a density of 50 individuals.

The effect a new predator would have on the demography of this population would depend on how efficiently it exploits its prey. If the predator removed about 10 percent of the adult population over and above its natural mortality of 40 percent, recruitment would balance the 50 percent adult mortality rate at a density of 92 individuals. Predators usually prey on the young to a greater extent than on adults (although this is not the case for hunting mortality). Increased mortality of prereproductives could be incorporated into Figure 5 either by translating them into adult equivalents or by reducing the height of the recruitment curve. The removal of a prereproductive bird does not have the same dampening effect on population growth as the removal of an adult. Normally, many prereproductives die before reaching maturity, so they are generally more expendable in terms of the population equation. It should also be pointed out that some additional predation would remove individuals that might have died by other causes. Thus some of the predator-caused deaths are absorbed by other density-dependent factors. A 20 percent increase in predation might raise the total mortality by only 10 percent.

A predator's maximum harvest from the hypothetical population in Figure 5 would be 13 individuals at a population density of 50 (a predation rate of 27 percent and a total mortality of 67 percent). At higher predation rates, the equilibrium population size would be reduced further, and the return to the predator would diminish. If the predator's exploitation rate exceeded 40 percent (total mortality of 80 percent), the population could no longer sustain the removal and would become extinct. The hypothetical population's ability to recover from random fluctuations, or to sustain predation, is greatest above some intermediate population level (50 individuals in this case) and steadily decreases below that point. It is likely that, when the prey population became rare, the predator would experience increased difficulty of finding individual prey and the level of predator exploitation would decrease. Thus, it is not likely that a predator could directly cause the extinction of its prey (unless it was incredibly efficient), but predators may reduce prey populations to the point that they are more vulnerable to extinction by random environmental factors.

I have calculated a mortality-recruitment curve for the Ring-necked Pheasant from Einarsen's data (Figure 6). Unfortunately, the censuses were interrupted by World War II before the population had reached an equilibrium level. Whereas the recruitment rate initially stood at more than 300 percent, the population showed signs of leveling off before the end of the observation period. The population probably would have been able to sustain an exploitation rate of about 600 individuals per year above its normal adult mortality at a population of between 600 and 1,000 individuals (a predation rate between 60 percent and 100 percent). But managing such a population to obtain the maximum rate of harvesting is not simple. No population can sustain an adult mortality of 100 percent all at once, unless there are abundant potential recruits in the population.

THE GROWTH POTENTIAL AND COMPARATIVE STABILITY OF POPULATIONS

The growth potential of a population determines how rapidly it will return to its equilibrium from a reduction in size. Maximum growth potential is not readily determined for natural populations. It is not

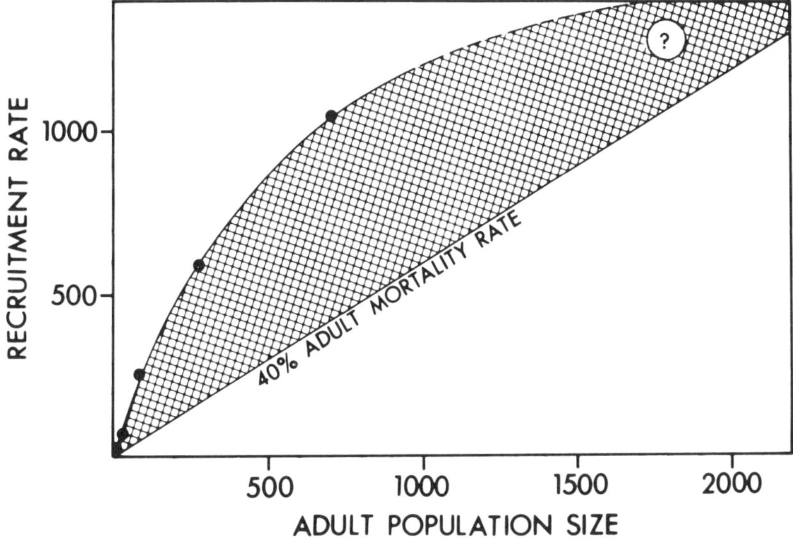

FIGURE 6 The relationship between mortality and recruitment in the Ring-necked Pheasant on Protection Island as a function of population size.

valid to cite the fecundity of a pair as the growth potential because many of the eggs die before reaching adulthood, even under optimal conditions. Some of the studies on introduced birds, whose populations have increased rapidly, give an indication of the potential growth of avian populations. The Ring-necked Pheasant on Protection Island initially increased with a recruitment rate of more than 300 percent annually, almost one half of the maximum possible, assuming a clutch size of 12 and an even sex ratio among chicks. The European Starling has exhibited a spectacular increase in numbers since its introduction to the United States in 1890. Peterson (1948) had estimated that within 60 years, the population had increased a millionfold over the original 120 birds introduced, a doubling every 3 years. This, of course, is much less than the Starling's theoretical maximal annual recruitment of 400–500 percent, but is an average (23 percent) that includes the leveling of populations in many areas. The Shag has apparently been increasing at the rate of 11 percent per year in the North Atlantic (Potts, 1969).

Only a few other detailed local accounts of introduced populations have been published; Gill and Lanyon (1965) followed the increase of an introduced House Finch (*Carpodacus mexicanus*) population on Long Island, New York. The number of breeding birds at the Kalbfleish Station increased from 2 in 1960 to 5, 13, and 22 in subsequent years, indicating an initial annual rate of increase $[(1/N)(dN/dt)]$ of 0.9 (90 percent). The number of individuals trapped during winter increased from 1 in 1958 to 12 and 73 in the next 2 years, giving an annual rate of increase of about 260 percent. The buffering effects of breeding territories could account for the discrepancy between breeding and winter populations, and the winter capture rate is probably a better index of overall population density. One must remember that the House Finch has greatly expanded its range in the eastern United States and this geographical expansion could absorb much of the local productivity.

The same is true of the Starling population, whose increase was documented in Ohio, Massachusetts, and New York by Davis (1950). Davis' population growth curves indicate much greater statewide annual rates of increase and the values from Davis' study represent geographical expansion more than local population growth.

I have summarized data on the increase of the Cattle Egret (*Bubulcus ibis*) population in southern Florida from Christmas Bird counts in five localities. The total population increase was about 100-fold in 10 years $[(1/N)(dN/dt) = 0.46]$, varying for each of the areas

(0.33 at Coca, 0.84 at Fort Myers). I assume that these values are a fairly good estimate of the growth potential of the Cattle Egret population, which is consistent with the low reproductive rate of the egrets compared to pheasants and House Finches.

Lack (1966) has presented considerable data on year-to-year variation in populations of birds, which give some indication of the maximum rates of population increase over the range of densities normally encountered in nature. The Common Heron (*Ardea cinerea*) attained an annual growth rate of about 15 percent during several mild years following a population low. White Storks, which have been censused in the province of Oldenberg, Germany, for many years, attained rates of increase of 27 percent and 30 percent during 2 years. Some smaller species with higher reproductive rates exhibited somewhat higher rates of increase during their 2 best years (Great Tit 69 percent and 104 percent; Rock Ptarmigan, *Lagopus mutus*, 55 percent and 102 percent, but the Pied Flycatcher in the Forest of Dean, England, exhibited smaller increases (30 percent, 31 percent; its maximum decline, however, was also lower: −13 percent and −18 percent, compared to −26 percent for 2 years for the Rock Ptarmigan and −45 percent and −50 percent for the Great Tit. Hickey (1955) summarized data on the rates of increase in gallinaceous bird populations obtained from fluctuations in established populations. The Bobwhite exhibited maxima of 74 percent and 166 percent, and increases for several species in Wisconsin were generally below 100 percent. All of the above data, based on intensive censuses of local areas, are complicated by the immigration of birds into areas whose populations have been subjected to local decline. But the growth potentials of bird populations may be summarized as being on the order of 10 percent to 30 percent in large species with low reproductive rates and between 50 and 100 percent during favorable years in smaller species with higher reproductive potentials.

How do these rates compare with declines in natural populations? We have seen above that maximal declines for three populations of small land birds were between −18 percent and −50 percent. Bad years for White Storks were −28 percent and −31 percent and for the Common Heron, about −45 percent and −65 percent. Hickey (1955) suggested that on good range, annual declines in gallinaceous species rarely exceed −50 percent. He also stated that declines were largely caused by reproductive failure rather than by abnormally high adult losses, unless local declines greatly exceeded −50 percent. For example, 2 years of large population declines in the Ruffed Grouse (*Bonasa umbellus*) were −45 percent and −66 percent, which was not much greater than the fall to spring decline in adults (40 percent) during years of population increase.

In general, most population declines—except locally disastrous years (for example, for seabirds when large areas of ocean freeze over) (von Haartman, 1971, p. 408)—can be recuperated during a single good year of population growth. For this reason, fluctuations in the populations of most birds are short term and are quickly reversed. Again, exceptions occur among species whose habitats are changing or that are engaged in predator–prey cycles, as some gallinaceous species appear to be (Keith, 1963). However, von Haartman (1971, p. 409–410) has referred to studies by Beven and by Yapp that indicate that some passerines undergo fairly long-term cycles of abundance in local areas.

THE LIFE TABLE

The life table is a summary of the vital statistics for each age group of a population under a given set of conditions. From the basic life table parameters (age-specific fecundity and age-specific survivorship), one can determine the growth rate of the population and its age structure under constant conditions and assess the effect of a change in one of the life table parameters on the growth performance of the population.

THE LIFE TABLE PARAMETERS

Age-specific fecundity (b_x), generally the number of female eggs laid per female* each year, is determined independently for each age group. In most birds, fecundity is constant after a certain age, often the first breeding season.

The age-specific survivorship (l_x) is the proportion of individuals that survive to age x. Determining l_x for natural populations can be

---

*Life tables are usually based upon the statistics of females. For most monogamous birds in which both sexes care for the young, the distinction is unnecessary, although it will be followed here for consistency. In the case of polygamous and promiscuous species, it would be almost impossible to determine the fecundity of males. Also, to base the life table on the female sex is helpful when the survival rates of the sexes differ (see p. 378). Unless it is known to be otherwise, one assumes that the sex ratio of young birds is evenly balanced.

difficult because the survivorship to any given age is a function of the survival rates during all previous age intervals. We have seen that it is often nearly impossible to estimate survival prior to the attainment of adulthood, and yet this statistic is as vital as any other.

One can but rarely follow directly the survival of individuals to each age to obtain survivorship data, particularly in long lived species. So, rather than construct "dynamic" life tables based on the survival of a cohort (a group of individuals that begins life together), one usually constructs a "time-specific" life table based on data gathered during a short period (Deevey, 1947; Hickey, 1952; Farner, 1955). The time-specific life table offers the advantage for population studies that it summarizes demographic parameters for the environmental conditions that existed during only one year. Thus, while it does not portray the average dynamic structure of the population, the time-specific table does potentially show how the population responds to environmental fluctuations.

Age-specific survivorship may be calculated from data gathered during an interval of one or more years either from the survival of individuals of known age over a year, or from the number of individuals of each age that are found dead. Frequently, composite life tables are constructed from data for many cohorts or for many years and, thus, incorporate aspects both of dynamic and time-specific life tables.

I have put together a hypothetical life table in Table 1 to illustrate how a life table is constructed from field data. As mentioned above, fecundity is measured directly, although there are some indirect methods of calculation for multiple-brooded species that will be discussed later (see p. 401). Suppose that the original survivorship data consists of the number of birds of each age that die during a year, or any other interval (Column A). Of the dead birds recovered, 47 were less than 1 year old, 30 between 1 and 2 years old, and so on. From these data, we can assume that all 137 of the recovered birds were alive when they were banded as immatures in the fall (Column B), that 90 were alive at the end of their first year in the following fall (137 minus 47 that died during the first year), 60 were alive at the end of the second year (90 minus 30), and so on. We divide each of these values by the number of birds alive at age zero (137) to obtain the age-specific survivorship (Column C). Note that if the original data had been the age-specific annual survival rates (S_x) given in Column D, we would have arrived at the same survivorship data by progressively multiplying the survivorship to each age by the survival

TABLE 1   Life Table for a Hypothetical Population Based on Recoveries of Birds of Known Age (Column A).

	A	B	C	D	E	F	G	H
Age	Recoveries	Age Structure	Survivorship (s_x)	Annual Survival (s_x)	Annual Mortality (m_x)	Fecundity (b_x)	$l_x b_x$	$x l_x b_x$
0	47	137	1.000	0.66	0.34	0	0	0
1	30	90	0.657	0.67	0.33	1.2	0.79	0.79
2	26	60	0.438	0.57	0.43	2.3	1.01	2.02
3	16	34	0.248	0.53	0.47	2.7	0.67	2.01
4	13	18	0.131	0.28	0.72	2.9	0.38	1.52
5	5	5	0.036	0.00	1.00	3.0	0.11	0.55
6		0	0.000					
TOTAL 137		344					2.96	6.89

rate during the following age interval. That is, 66 percent $\times$ 67 percent = 44 percent, 44 percent $\times$ 57 percent = 25 percent, and so on.

In the example in Table 1, the decrease in annual survival rate, from about two thirds during the first 2 years to about one quarter during the fifth, is unusual for birds. But the sample size is small and the drop survival rate may not be significant. The 100 percent mortality during the sixth year is purely an artifact of the small sample size, for undoubtedly some individuals survive beyond that age.

One can estimate the overall annual survival rate of the population from the total number of birds in our sample that were alive at the beginning of the year (137). The overall mortality in the population is 40.2 percent (137/344), which indicates an annual survival rate of 59.8 percent. Expectation of further life after a given age (E_x) is the reciprocal of the mortality rate after that age, but since individuals die throughout the age interval one must subtract about one-half year. Thus at age 0, individuals will live an average of (344/137) − 0.5 = 2.51 − 0.5 = 2.01 years. After the end of one year, E_x is (207/90) − 0.5 = 2.30 − 0.5 = 1.80 years, after the second year (117/60) − 0.5 = 1.95 − 0.5 = 1.45 years, and so on.*

The fecundity in Table 1 (Column F) was determined by direct

---

*The formula for expectation of further life is

$$E_x = \frac{\sum\limits_x xD_x}{l_x} - 0.5, \tag{3}$$

where D_x (the age-specific death rate) is $l_x - l_{x+1}$. The equation can be rewritten

$$E_x = \frac{\sum\limits_x l_x}{l_x} - 0.5, \tag{4}$$

which simplifies calculations. A precise expression for Eq. (4) is

$$E_x = \frac{1}{\log_e (1 - l_x / \sum\limits_x l_x)}. \tag{5}$$

Thus, after 3 years, the data in Table 1 give

$$E_3 = \frac{1}{\log_e (1 - 0.248/(0.248 + 0.131 + 0.036)} = 1.1 \text{ years}$$

compared to an estimate of 1.17 from Eq. (4).

observation of the clutch size of birds of known age. All of the values
are actually half the actual clutch size because only the female component of the population is being considered (a 50:50 sex ratio is
assumed). The low fecundity of females during their first year is, in
this hypothetical case, partly due to the fact that not all females
breed at that age.

Another statistic that can be calculated from the life table, and
that will be used below for other calculations, is the expected number of female progeny for a female during her lifetime (the net reproductive rate, R_0). Since a female can be expected to live to age x
with probability l_x, her expectation of fecundity at age x is $l_x b_x$,
(Column G, and the total expectation of fecundity during her entire
life-span is the sum of $l_x b_x$ terms for each age ($R_0 = \sum_0 l_x b_x$). In this
case, the net reproductive rate is 2.96; this suggests that, if the population is remaining at a constant level, about two thirds of the young
do not survive to the initial date of the banding program, age "0".

ASSUMPTIONS OF THE LIFE TABLE

A life table is always an abbreviation of the vital statistics of a natural
population. To prepare a complete life table, one would have to determine age-specific parameters continuously for all ages, which is impossible. Normally, survivorship data are calculated for yearly intervals
that begin on an arbitrary anniversary date, frequently January 1 but
often in the late summer or early fall. Banding programs are also
often highly seasonal, occurring when birds are easiest to obtain:
during fall migrations or winter feeding congregations for many species or on the breeding grounds for others. The choice of an anniversary date is completely arbitrary; ideally, however, it is as soon as
possible after the end of the breeding season. In this way, one measures more realistically the survival of young to the age at first
reproduction, upon which survivorship to all subsequent breeding
seasons is based.

An equally critical problem is that the realization of fecundity in
natural populations is not achieved instantaneously at the end of the
age interval, as it is on paper. Reproduction occurs over a season of
finite length during which many adult birds die. If both parents participate in nesting, the death of one may not cause complete loss of
the clutch or brood, but the nesting attempt is frequently terminated.
Death also precludes the possibility of further reproduction during
the year. Thus it is difficult to estimate precisely the proportion of

females that survive to breed during their xth year, because breeding does not occur at one point in time.

If the breeding season is relatively short (all the better if only one nesting is attempted), fecundity can often be determined as the number of young fledged per nesting attempt, thus accounting for the death of adults during the nest period. When the breeding season is longer and several broods are normally raised, a simple adjustment is not possible. Also, one cannot validly, or even safely, calculate mortality during the breeding season by prorating the annual mortality rate during the breeding season. Mortality rates are known to vary seasonally and are often highest during the breeding season because the adults are most vulnerable to predators. From Nice's (1937) data on the Song Sparrow, the mortality rate of males during the two months of the breeding season (April to June) was 11.8 percent per month (12.9 percent for females), but decreased during the postbreeding season to 5.5 percent per month and over winter to 4.6 percent per month. Similarly, Leedy and Hicks (1945) found that the mortality of Ring-necked Pheasant hens in Ohio was 42 percent during the nesting season, compared to estimates of 54 percent and 70 percent for the entire year. Van Tets (1968) reported that mortality rates in five of six gull species were lowest during the breeding season.

Most of the potential irregularities in the life table caused by sex-related differences in survival are taken care of by ignoring males. However, Hickey (1955) has pointed out that, if females are the more abundant sex in a population, some will go without mates in a species that adheres to a monogamous mating system. In polygamous species, the nesting of many females around one male when the sex ratio is extremely unbalanced may reduce the nesting success of individual females. The quantitative implications of unbalanced sex ratios for the life table are completely unknown, however.

ASSUMPTIONS ABOUT FECUNDITY

Age-specific fecundity is usually calculated on the basis of the number of eggs laid by females at different ages. From that point on, the fate of the egg in the population life table is followed in the age-specific survivorship column. It is the usual practice to lump the survivorship of eggs of females of all ages together in the same column, and so the influence of parental age on productivity is assumed to cease after the clutch has been laid. Yet, this is demonstrably false in some natural populations. In virtually every case where it has been checked in the field, the survival of young birds is strongly influenced by the age of their parents. This has already been pointed out for the Great Tit (Perrins, 1965; Lack, 1966, p. 45). Similar observations have been reported for the Skylark by Delius (1965). The nesting success of Gannets (*Sula bassana*) during their first breeding season is lower (50 percent) than during subsequent years (79 percent) (Nelson, 1966). More young are raised by experienced Kittiwakes than by new recruits into the breeding population (Coulson, 1966) and the length of the pair-bond additionally influences nesting success independently of breeding experience. Snow (1958) found that the age of parent European Blackbirds influenced both nesting success (3.4 young fledged per pair of year-old birds versus 6.3 for older pairs) and the survival of the young from fledging to independence (56 and 70 percent, respectively, but based on relatively small samples).

Clearly, to list separate survivorship columns in the life table based on the age of the parent would be cumbersome, but extended parental effects on survival should be incorporated into a life table when the data exist. Two approaches are acceptable. First, one may adjust the clutch size or the number of eggs laid during a breeding season to account for parent-determined differences in survival of the young. For example, if the nesting success of first-year birds is only 60 percent that of older pairs, one could simply multiply the number of eggs by 0.6 during the first year. A less cumbersome but demographically less satisfactory approach is to calculate fecundity as the production of immature birds, taking into account clutch size, nesting success, and survival to independence where possible, all of which may vary with the age of the parent. Then, the initial age for the life table (age "0") would be the age at which immatures become independent of parental care. Differential survival of young to the anniversary date because of differences in the date of fledging (e.g., Perrins, 1965; Lack, 1966, Figure 13) adds a further complication to life table calculations, but it will not be dwelt on here.

THE CALCULATION OF POPULATION GROWTH

Provided the conditions for which the life table parameters are defined remain constant, the population will assume a stable age distribution (all age groups increase or decrease at the same rate) and the annual growth rate of the population $[(1/N)(dN/dt)]$ will assume a constant value (r), which can be calculated from the life table.

The net reproductive rate (R_0) represents the number of times that a female replaces herself in the population by her female offspring. If the R_0 is positive, let us say 1.7, each female can be expected to leave 1.7 female offspring and the population will increase by approximately 70 percent in one generation. Knowing the generation time (T), therefore, one may make a reasonable estimate of the annual growth rate of the population. Remember that

$$\frac{N(t)}{N(0)} = e^{rt} \tag{6}$$

from Eq. (2). If we now replace t with the generation time T, we know that $N(t)/N(0) = R_0$; therefore

$$R_0 = e^{rT}, \tag{7}$$

which can be rearranged in the following form to solve for

$$r = \log_e (R_0)/T. \tag{8}$$

If $R_0 = 1$, that is, if a female replaces herself exactly once, $\log_e (R_0)$ is 0 and the population size remains constant. If R_0 is less that 1, its logarithm is negative and the population decreases, whereas, if R_0 exceeds 1, its logarithm is positive and the population increases. Note that the growth rate of the population is more sensitive to changes in the average generation time (T) than in the net reproductive rate (R_0). For example, if the net reproductive rate is doubled, r increases by 69 percent. ($\log_e 2 = 0.69$), whereas if the mean generation time is halved, r increases 100 percent. Thus changes in the age at reproduction have more effect than seemingly equivalent changes in fecundity.

The mean generation time of a population may be estimated from the life table as the mean age at which eggs are laid by an individual. Thus T is the sum of the $l_x b_x$ terms weighted by the age x, divided by the net reproductive rate

$$T = \frac{\sum\limits_{0} x\, l_x b_x}{\sum\limits_{0} l_x b_x} \tag{9}$$

In the life table provided in Table 1, the mean generation time is calculated as 6.89 (the sum of Column H) divided by 2.96 (the sum of Column G), or 2.33. Note that because fecundity increases with age, the mean generation time is longer than the mean expectation of life (E_0), which is 2.01 years. From the above value of T, r may be estimated as $\log_e (2.96)/2.33 = 0.47$. Equation (9) provides a reasonably

accurate estimate of mean generation time if the population is not
growing rapidly, but overestimates the actual generation time in a
rapidly growing population (underestimates T in a declining population) because the generations of parent and offspring overlap
(Laughlin, 1965). Before a female has completed her life-span, some
of her oldest progeny will have offspring, which, in turn, may have
bred, and so on. To correct for the problem of overlapping generations, Lotka (1925) derived an equation for the exact calculation of r
from the life table:

$$1 = \sum_0 e^{-rx} l_x b_x \ . \tag{10}$$

For a particular set of data, the equation must be solved by the trial
and error substitution of r values until the sum adds to 1, which can
amount to a tedious bit of calculating. Computational aspects of
Eq. (10) will not be pursued further here, primarily because the
equation is rarely applied in avian demography [cf Mertz's (1971)
paper on the California Condor, *Gymnogyps californianus*, and
Leslie's (1966) paper on the Guillemot (Common murre), *Uria aalge*].
To show the sort of error that results from using Eq. (8) to estimate
r, the value obtained from Eq. (10) lies between 0.53 and 0.54, compared to 0.47.

The calculations of r from time-specific life tables do have valuable
potential application to the problem of predicting the growth of
populations under constant conditions that may obtain for short
periods. This approach has been extensively employed by economic
entomologists (e.g., Birch 1948, 1953).

THE DEMOGRAPHIC CONTRIBUTIONS OF BIRTH AND DEATH

Birds are rather unusual in that fecundity and mortality are on the
whole relatively constant during the adult life-span. This simplifica-
tion of the life table allows us to determine the relative contributions
of changes in fecundity and mortality to the growth rate of the population from Eq. (10). If we assume that b_x is some constant B and
that survivorship l_x is a function of some constant mortality m
(which is a rate of exponential decrease), we may substitute $l_x =
E^{-mx}$. Equation (10) now becomes

$$1 = B \sum_0 E^{-(r+m)x}, \tag{11}$$

which may be solved for

$$r = B - m. \tag{12}$$

The growth rate of the population is equal to the birth rate minus the death rate. Thus, an increase of 0.1 new recruits per individual into the breeding population has the same enhancing effect on the growth rate of the population as reducing the annual adult mortality by 10 percent. The additive relationship between B and m simplifies the task of predicting the effect of a change in any one demographic parameter on the population growth rate. For example, if the productivity of a game-bird population dipped by 25 percent one summer, the population could be expected to sustain only three-quarters the normal mortality without declining.

RECRUITMENT REQUIREMENTS OF A POPULATION

Life table information is frequently used to determine the recruitment rate required to maintain a population at a constant level. The basic recruitment can be stated as

$$\text{recruitment} = \frac{\text{mortality rate of adults}}{\text{survival of young to reproduction}}, \tag{13}$$

where the age from which the survival of young to reproduction is measured is the age at which recruitment is specified. The recruitment ($\overline{m}$) (Henny *et al.*, 1970) is expressed with mathematical symbols as

$$\overline{m} = \frac{1 - s}{s_0}, \tag{14}$$

where s is the average survival rate of adults, and s_0 is the proportion of young that survive to reproduction. Henny *et al.* (1970) provide a comprehensive summary of computations of recruitment requirements for the more complicated situation in which the annual survival of adults increases with age. For example, if the survival rate is lower during the first year of adult life than it is thereafter, the equation for the recruitment requirement is adjusted in the following way:

$$\text{recruitment} = \frac{\text{mortality rate of adults } (1-s)}{\left(\begin{array}{c} \text{survival of the young} \\ \text{to reproductive age,} \\ s_0 \end{array} \right) - \left(\begin{array}{c} \text{loss of potential breeding} \\ \text{adults before second year} \\ \text{due to reduced survival,} \\ s_0(s-s_1) \end{array} \right)},$$

where s_1 is the survival of adults between the first and second breeding seasons. For example, if values of s_0, s_1, and s were 0.15 (from the egg), 0.50, and 0.70, the necessary recruitment would be $(1 - 0.70)/[0.15 - 0.15(0.7 - 0.5)] = 0.3/(0.15 - 0.03) = 2.5$ eggs. This value must be multiplied by two because the eggs are approximately half male. Henny *et al.* (1970) provided tables of recruitment requirements as a function of first-year and adult survival rates for several special cases including delayed onset of sexual reproduction. In most cases, the authors consider the "first year" as after the first anniversary date, so survival of eggs to the anniversary is considered to be 1, and s_1 is calculated for the following year. Recruits are counted on the anniversary date. For example, survival rates for the Canvasback Duck (*Aythya valisineria*) would be $s_0 = 1$, $s_1 = 0.23$, and $s = 0.59$ (Henny *et al.*, 1970). From these data, we calculate the recruitment requirement as $(1 - 0.59)/[1 - (0.59 - 0.23)] = 0.41/0.64 = 0.64$ young surviving to their first anniversary date per adult.

Estimated recruitment requirements based on adult survival rates have also been used as a check on the known productivity of a population to determine whether survival rates obtained from banding programs are realistic. According to Farner (1955), such calculations are usually reasonable for most passerines, but estimates of required recruitment are often too high for most seabirds, probably because band loss makes mortality rates seem too great (bands may "die" while their owners survive). As Farner points out, one should always try to balance population parameters against each other to see if they make up a reasonable population equation.

THE ESTIMATION OF DEMOGRAPHIC PARAMETERS

The data required to construct a life table are straightforward in their meaning and interrelationship, but their estimation is almost always difficult. Some of the parameters, particularly clutch size and nesting success, are relatively easy to obtain because the focus of life-history events during breeding is stationary. The only major assumptions involved in the measurement of clutch size and nesting success, other than that the sample is representative, are that clutches are complete and that nests are found prior to the beginning of incubation. Exceptions to the latter assumption may be corrected by Mayfield's (1961) "exposure" method.

On the other hand, estimation of the annual productivity of the pair (number of fledglings produced or clutches laid), or the survivorship of individuals, involves greater potential pitfalls because one usually bases calculations on indirect evidence of the parameter in question. Survivorship estimates are rarely based on the survival of individuals, but are usually derived from the effect of survival on the age structure and death rate of the population. Most of the numerical techniques involved in the calculation of mortality from banding data are covered adequately by Farner (1955) and references therein.

FECUNDITY

Fecundity and clutch size are not equivalent when the breeding season is sufficiently long that replacement clutches or multiple broods can be produced. When pairs tend to return to the same nesting site (or, better yet, the same nest box) for subsequent nestings, the annual fecundity of a nest site may be equated to the annual fecundity of an adult. Even if birds do not renest in exactly the same spot, but remain within the same general area, a diligent worker can follow the reproductive outcome of marked birds, as Nice (1937) did with Song Sparrows. But the amount of work involved in nest-finding, particularly in the many tropical birds whose successive nests may be spaced months apart, largely precludes direct methods of estimating fecundity for an adequate sample of individuals.

Fecundity may be calculated indirectly when other reproductive parameters are known well enough for the species as a whole. This approach is particularly useful to compare life tables among species or between geographical areas.

The number of nestings attempted during a season is roughly the length of the season divided by the time required for each nesting attempt. The length of the combined incubation and nestling periods cannot be used for the length of the average nesting attempt, because nests that fail require less time than broods that succeed and because the initiation of the following brood is usually delayed for a week or more after the young have fledged or the nest has been destroyed. A more precise relationship for the length of the nesting attempt (Ricklefs, 1970) is

$$\overset{*}{T} = P(T + r_s) + Q(T_f + r_f), \tag{15}$$

where

$$\overset{*}{T} \quad = \text{ the average length of a nesting attempt,}$$
$$P \quad = \text{ proportion of nesting that succeeds, } (Q = 1-P)$$

$$T \quad = \text{combined length of laying, incubation, and nestling periods,}$$

$$T_f \quad = \text{the average time to failure in an unsuccessful nest,}$$

$$r_s, r_f = \text{time before initiation of a new nest after a success and a failure, respectively.}$$

The parameters in Eq. (15), including T_f, can be estimated directly by checking nests each day. This may be too time consuming in practice, in which case T_f can be calculated from P and T. If one assumes that the probability that a nest will fail in a given interval is constant throughout the nest period (Ricklefs, 1969a), which is sometimes not the case, time to failure may be calculated by the equation:

$$T_f = \frac{Q - mPT}{mQ}, \tag{16}$$

where m, the daily mortality rate, is calculated by

$$m = \frac{-\log_e P}{T} \tag{17}$$

(Ricklefs, 1970). By way of example, suppose that a species exhibits the following parameters: $P = 0.40$; $Q = 0.60$; $T = 30$ days. In this case, $m = -\log_e (0.40)/30 = 0.031$ per day, and $T_f = 0.60 - (0.031)$ $(0.40)(30)/(0.031)(0.60) = 12.4$ days. Note that T_f is just less than half the length of the nest period, which would be the average time to failure if nest losses were distributed linearly with time rather than exponentially. If P is not too low, T_f may be reasonably estimated as $T/2$. Suppose that in the above example, r_s and r_f were both 10 days. We would now calculate $\overset{*}{T}$ from Eq. (15) as $\overset{*}{T} = 0.40\ (30 + 10)$ $+ 0.60\ (12.4 + 10) = 29.4$ days.

After obtaining the average expected length of a nesting attempt, the number of attempts during the year is calculated as the length of the season divided by $\overset{*}{T}$. However, since the last brood is not followed by a period r, we first subtract time T from the season length (we are only interested in broods that are successful and one started fewer than T days before the end of the season for successful fledgling will not succeed) and then calculate the number of nest attempts for the rest of the season. Suppose that in the above example, the season length (between the first suitable conditions for egg-laying and the last suitable conditions for fledging young) is 200 days. The number of nesting attempts that could be made during the season would be $[(200 - 30)/29.4] + 1 = 6.8$, of which 40 percent (2.7) are successful. Thus, if clutch size was 4 and brood size at fledging was 3, on the average, a female would lay 27 eggs and fledge 8.1 young each year.

Season length is not so sharply defined as I have suggested above. Its beginning and end are usually rather more imprecise. Nesting seasons are most often measured as the period over which clutches are initiated, but the maximum span of laying is not the best estimate of season length because nesting activities are generally curtailed at the beginning and toward the end of the season. MacArthur (1964) chose an index of "equally probable months for nesting,"

$$\text{months} = e^{-\Sigma p_i \log_e p_i} \tag{18}$$

where p_i is the proportion of all nests that are started during time interval i, usually a month. As the name of the index suggests, it gives stronger weight to equitably distributed clutches, as shown in the following tabulation:

| | Proportion of Clutches Laid during | | | | | Season Length |
Sample	Apr.	May	June	July	Aug.	Index (mo.)
A	0.15	0.20	0.32	0.20	0.13	4.75
B	0.08	0.59	0.27	0.04	0.02	2.93

Thus, while eggs are laid over 5 months in both samples, the season length index is probably a much more reasonable estimate of the time available for breeding. When the egg-laying season is known, we simply divide the season length index (converted to days) by $\overset{*}{T}$ and add 1 to find the expected number of nesting attempts.

Another index of the overall production of young is the ratio of immatures to adults at the end of the season. In one sense, this is a better estimate of productivity than the indirect method outlined above because it takes into account the differential survival of young that fledged at different times during the season. It has a marked disadvantage, however, in that one is not sure at what age the ratio is obtained. Also, the number of adults in the population decreases during the breeding season; thus the immature adult ratio could overestimate the fecundity of individual birds.

SURVIVORSHIP

Adult survivorship is usually measured in one of three ways: by intensive studies of survival in local populations, by extensive data on returns of dead birds of known ages, and from ratios of first-year birds to older birds in the breeding population. These methods are discussed in detail by Farner (1955), who also discusses the assumptions made when recoveries of dead birds are used to calculate mortality rates.

Values from intensive studies can only be taken as minimum survival rates because banded birds that emigrate from the study area are counted as deaths. Site specificity varies greatly among species and among sexes (e.g., Lack, 1966, p. 111).

Mortality is calculated from age specific data on the recoveries of dead birds by the equation (Haldane, 1955)

$$m = \frac{N}{\Sigma x D_x,} \tag{19}$$

where N is the total number of recoveries, x the year of death, and D_x the number of recoveries for that age class. Thus for the data presented in Table 1, $M = 137/[47 \times 1) + (30 \times 2) + (26 \times 3) + (16 \times 4) + (13 \times 5) + (5 \times 6)] = 137/344 = 40$ percent. In the denominator of Eq. (19), the number of recoveries of each age class is multiplied by their age to account for the number of years that the bird has survived before the year of its death. That is to say, by determining the year of death, we also know the number of years that the bird survived. If a substantial number of banded birds remain in the population at the end of the recovery program, a correction must be used to avoid the resulting bias toward higher mortality at older ages (Haldane, 1955; von Haartman, 1971, p. 441).

The use of age ratios to determine mortality rates has intrigued me for some time. Although the technique is commonly employed in studies of gallinaceous birds (e.g., Hickey, 1955; Robel and Fretwell, 1970), it has rarely been applied in passerines (cf Snow, 1956), partly because first year birds are difficult to distinguish from adults in some species. I have checked some of the published records of mass bird kills at television towers (Graber and Graber, 1962; Tordoff and Mengel, 1956) for ratios between immatures and adults in fall migration samples. One would not necessarily expect the age ratio to have settled down to its breeding season level by the early fall because of subsequent age-differential survival. An additional bias could be produced by age-differential migration periods combined with limited samples, (e.g., Hussell *et al.*, 1967; Ricklefs, 1972). The result of the studies (Table 2) are not very encouraging; the age ratio in one case suggested higher adult survival than is known to occur, when the

TABLE 2 Percentages of Adults in Samples of Passerine Birds Killed during Fall Migration Compared with Survival Rates Determined from Recovery of Banded Birds

Number of Species[a]	% Adult, or Adult Survival	Range	Reference
Migration kills			
7	67	43–86	Graber and Graber, 1962
7	45	32–68	Tordoff and Mengel, 1956
Recovered birds			
5	51	32–60	Farner, 1955
9	47	37–58	Lack, 1954

[a]Only species represented by 20 or more specimens are included.

discrepancy should have been in the opposite direction. Tordoff and Mengel's sample (Table 2) seems much more in line with survival rates determined by other methods.

The use of museum specimens for the calculation of age ratios may be useful for species that cannot easily be aged in the field (to eliminate collecting bias) but whose age can be distinguished by subtle characters in the museum tray. Snow (1956) applied this method to the Blue Tit and found that among British populations in museum trays, the percent of adults was 20 in October–November, 31 in December–February, and 30 in March–June. These data suggest that the difference in adult and juvenile survivorship disappears by midwinter and also that the adult mortality rate is about 70 percent, which is very high, but not inconsistent with the species' large clutch size (11.6) in Britain.

Immature/adult ratios can also be applied to the question of breeding productivity and differential survival of first-year birds. Provided the necessary assumption be made that a collection is an unbiased sample of a natural population, we may derive an expression for the differential mortality (or survival) rate between the adult and immature segments of the population. The immature/adult ratio (R) can be defined for some initial time (0) as

$$R(0) = \frac{N_{\mathrm{I}}(0)}{N_{\mathrm{A}}(0)}, \tag{20}$$

where N_{I} and N_{A} are the number of immature and adult individuals in the population. After some time period (t), the ratio is $N_A(t)/N_A(t) = R(t)$. From Eq. (2), we know that $N(t) = N(0)e^{-mt}$, where m is the exponential rate of decrease (or mortality) of a segment of the population. Thus,

$$R(0) = R(0)e^{-(m_A - m_I)t}, \qquad (21)$$

which may be solved

$$m_A - m_I = \frac{\log_e [R(t)/R(0)]}{T}. \qquad (22)$$

Applying Eq. (22) to Emlen's (1940) data for immature/adult ratios in quail (see Figure 1), we find that during the 6-month period between November and May, the ratio fell from 1.65 to 1.09 and so

$$m_a - m_I = \frac{\log_e(1.09/1.65)}{6 \text{ months}} = 7 \text{ percent per month}, \qquad (23)$$

which is equivalent to a mortality differential of 83 percent per year. Similar calculations for the Bobwhite from September through March (Figure 1) give a value of 19 percent per month.

The survival of fledglings until they have become a part of the general adult population is probably the most formidable challenge for the field demographer. Direct observation of young after they leave the nest is tedious and necessarily precludes large sample sizes. Also, the young can be observed directly only as long as they remain within the nest area, less than 2 weeks for most songbirds. In such "ideal" species as the Cactus Wren, the young return to roosting nests in their parents' territory for a month or so after fledging, and an accurate picture of fledging survival can be obtained by periodic nest checks. Such opportunities should be followed up wherever they occur. Most of the data in the section on postfledging survival (p. 373) was obtained by direct observation. Extensive banding of nestlings and subsequent recovery involves too many sampling problems to be used reliably. The dispersal of young birds makes impossible local intensive studies on the survival of banded young. Because many young die before they have a chance to disperse widely throughout the population, returns of dead birds during the first few months postfledging will be too low, particularly if nestlings are banded in a few restricted areas but dead birds are recovered over broad areas. Indirect estimates may be made from immature/adult ratios shortly after the breeding season, if the population is single brooded. For example, the I/A ratio of Rough-winged Swallows in the United States sample during July was 1.75, which was only slightly lower than the productivity of fledgings (Ricklefs, 1972), indicating that juvenile survival was very high. In the long run, it is often easier, and probably equally valid, to estimate the juvenile and immature survival as "the part that's left over" after all the other components of the life table have been worked out.

REPRESENTATIVE LIFE TABLES

I have put together tentative life tables for two species, the Yellow-eyed Penguin and the European Tree Sparrow, not because they are "typical," or for that matter even the best documented, but because they demonstrate adequately the construction of life tables and they also differ from each other in the extreme.

The data for the Yellow-eyed Penguin come from Richdale's (1957) classic study on a small colony near Dunedin, New Zealand, which has been summarized by Lack (1966, p. 230–238) (Table 3).

Note that the table is missing a value for survival of the young to the first breeding season. Richdale provided a figure, but for the time being I have assumed that it is questionable and calculated a value

TABLE 3 Basic Reproductive and Survival Parameters[a] as a Function of Age

	Age (Years)				
	1	2	3	4	>4
Age at first reproduction					
males	0	8%	35%	33%	26%
females	0	48%	52%	0	0
Clutch size					
1 egg		38%	6%	1.5%	1.5%
2 eggs		62%	94%	98.5%	98.5%
Hatching success					
first year		32%	70%		
later			82%	87%	92%
Survival of young to eight weeks		33%	84%	89%	89%
Overall fecundity per female at eight weeks	0	0.08	1.24	1.54	1.62
Survival to following year		81%	82%	84.2%	84.2%

[a]Assumed to be constant after 5 years of age.

TABLE 4 Life Table for the Yellow-eyed Penguin (*Megadyptes antipodes*)[a]

Age (x)	Fecundity (b_x)	Annual Survival Rate (s_x)	Survivorship (l_x)	Expected Fecundity ($l_x b_x$)
0	0	0.305[b]	1.000	0
1	0	0.810	0.305	0
2	0.08	0.820	0.247	0.020
3	1.24	0.842	0.203	0.250
4	1.54	0.842	0.171	0.263
5	1.62	0.842	0.144	0.233
6	1.62	0.842	0.121	0.196
7	1.62	0.842	0.102	0.165
8	1.62	0.842	0.086	0.139
9	1.62	0.842	0.072	0.117
10	1.62	0.842	0.061	0.099
10+	1.62	0.842	(0.323)[c]	0.523
				$R_0 = 2.005$

[a] Constructed from data in Richdale (1957).
[b] Condition for a stable population.
[c] Calculated as $l_{10}/(l-s)$; see text.
[d] Number of young alive at 8 weeks of age.

that would be required for the population to remain stable.* The life table in Table 4 is calculated only for females and, because of the complexities of reproductive success during the first few years, fecundity is measured when the young are 8 weeks old. Survivorship is calculated by successively multiplying annual survival rates. Thus, for the third year, $0.305 \times 0.81 \times 0.82 = 0.203$, and so on. Note that one need not extend the life table forever, because, if mortality rate is constant, the expected number of breeding seasons that a newborn individual will survive from age x on is the probability of surviving to age x (l_x) divided by the mortality rate ($l-s$). Thus, the expected number of breeding seasons for a Yellow-eyed Penguin after age 5 is $0.119/0.158 = 0.75 (1/0.158) = 7.3$ for birds that actually survive to 5 years. The required mortality rate between 8 weeks of age and the end of the first year of life is 30.5 percent, which makes the net reproductive rate 2.01 (Table 4). The actual value determined by Richdale from the returns of banded nestlings was a reasonably close 25.7 percent. Birds wandering during their first year to breed in other colonies could account for the difference.

*This can be done by first assuming that $s_0 = 1$ and calculating the net reproductive rate on that basis. The required survival (s_0) is then the reciprocal of the initial net reproductive rate, or $2/R_0$ if eggs of both sexes are counted.

TABLE 5   Life Table for the European Tree sparrow (*Passer montanus*)[a]

Age (x)	Fecundity (b_x)	Annual Survival Rate (s_x)	Survivorship (l_x)	Expected fecundity ($l_x b_x$)
0	0	0.075	1.000	0
Hatching (12 days)			0.787	
Fledging (27 days)			0.688	
Independence (57 days)			0.444	
1	13.1	0.45	0.075	0.983
2	13.1	0.35	0.034	0.445
3	13.1	0.33	0.012	0.157
4	13.1	0.25	0.004	0.052
5	13.1	–	0.001	0.013
				$R_0 = 1.650$

[a]Constructed from data in Pinowski (1968).

The second life table is based on Pinowski's (1968) study on the population dynamics of the European Tree Sparrow near Warsaw, Poland. This sparrow is very productive in the area and frequently fledges three broods in a season (the most frequent clutch is five). The life table (Table 5) summarizes Pinowski's data at face value. The average number of eggs laid per season (13.1) is assumed to be an age-constant fecundity. The net reproductive rate of the population according to the life table is only 0.825 females per female (1.650 ÷ 2), and so either the population is decreasing at the rate of 17.5 percent per generation (highly unlikely!) or one of the survival values is somewhat inaccurate—most likely the survival from independence to first breeding. If s_0 were 0.091 (0.075 ÷ 0.825), the life table would be balanced.

Pinowski mentions that the proportion of first-year birds in the breeding population is about 55 percent, which is consistent with the adult survival rates given in Table 5. The proportion of first-year birds in the population is calculated by the expression $l_x / \sum_1 l_x$, which is identical to that computed on page 393 for the average mortality of adults in the population. For the European Tree Sparrow, the proportion of first-year birds is 0.075/(0.075 + 0.034 + 0.012 + 0.004 + 0.001) = 0.595.

The life tables of the Yellow-eyed Penguin and the Tree Sparrow exemplify extreme life-history patterns in birds: on the one hand, high annual survival, low fecundity, and delayed onset of reproduction; on the other, high annual mortality, high fecundity, and early onset of breeding. These patterns are summarized by their $l_x b_x$ curves in Figure 7. Since the areas of the curves are 1.0 for stable populations, adjustments to life-history parameters are maintained within this constraint by density-dependent factors. If fecundity is reduced, the period of reproduction is extended, and so on.

COMPONENTS OF VARIATION IN THE LIFE TABLE

It seems to me that the principal objective of population study should be not to describe the life table of a species under constant conditions, but to understand the dynamics of populations under changing conditions. One cannot really understand a system until one knows how it will respond to outside perturbations. I will approach this problem by responses of a population to its own density (and pre-

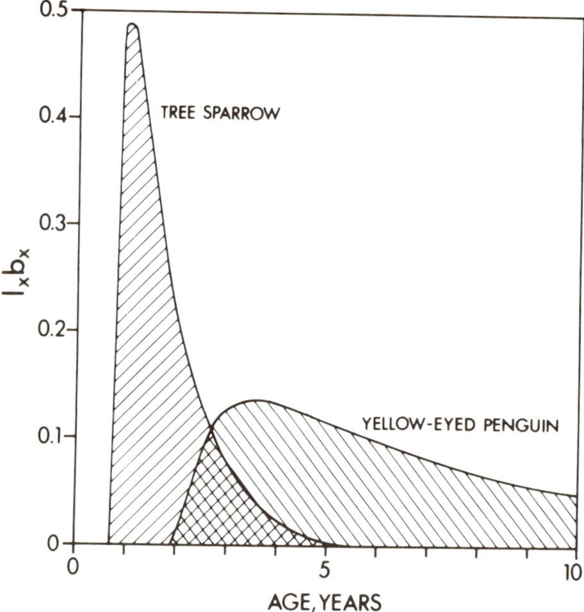

FIGURE 7   The shape of the $l_x b_x$ curves for the Tree Sparrow and the Yellow-eyed Penguin from Tables 4 and 5. The fecundity values upon which this figure is based are half those given in the life tables to discount male offspring. The total areas under the curves are roughly equivalent.

sumably under constant environmental conditions), by variations in population parameters from year to year, and by comparing population parameters among species and geographical regions. Through the first approach one can potentially learn about the regulation of population size; through the second, about the inherent stability of the population and how it is maintained; through the third, about how populations adapt their demography to different conditions.

DENSITY DEPENDENCE

Although we know little about the changes in population dynamics as a function of density in birds, we can make the following general statements. First, population phenomena in birds exhibit classical density dependence (Lack, 1954, 1966) that leads to the regulation of population size about some stable and relatively constant equilibrium. Second, deaths related to population regulation are incurred primarily by the young (Tanner, 1966). Most populations produce far more eggs than are needed to replace adult losses, so mortality of the young is usually high. Because the young are inexperienced, they are not able to compete effectively with adults in times of hardship and their mortality is thus more strongly influenced by density. Also, predation, which is often a density-dependent factor, acts primarily on the young. Any change in the overall predation rate on a population will affect the young most, because predation is a larger percentage of their total mortality than of adult mortality.

We have seen that, during the population increase of the Ring-necked Pheasant on Protection Island, the spring-to-fall increase (production of young) exhibited density dependence, whereas the fall to spring decrease (mortality of older birds) did not. Errington (1945) found a similar pattern in population fluctuations of the Bobwhite. Yet as Kluijver (1966) has shown by experiments on the Great Tit, adults can also participate in density dependent processes when a population is experimentally reduced at the end of the reproductive season by the prior removal of nestlings (see p. 384).

YEAR-TO-YEAR VARIATION

Year-to-year variation in population size can be used as a fairly reliable index of the constancy of a population with respect to environmental fluctuations if one assumes that age distribution remains relatively constant. The variation in individual population parameters also points out what the vulnerable points are in the dynamics of the population. Comparison of the relative stability of birds from the same general region, and thus which are subjected more or less to the same overall climatic fluctuations, may suggest which aspects of a species' ecology act to stabilize its population.

Table 6 summarizes year-to-year variation for several parameters for several species from tables in Lack (1966) and Snow's (1966) study of the European Blackbird and Bergerud's (1970) study of the Willow Ptarmigan (*Lagopus lagopus*). Except for the White Stork (Germany), the Yellow-eyed Penguin (New Zealand), and Bergerud's Willow Ptarmigan study (Newfoundland), all of the studies were carried out in Britain. For each parameter, the following data are provided: years of the study, mean value for the parameter, and its coefficient of variation (standard deviation/mean) in percent. The coefficient of variation is suitable for comparisons among parameters and among species and will be used here as an index of fluctuations.

Population Size

Variation in population size was relatively low (20 percent to 30 percent for breeding birds), but was between 40 percent and 50 percent in the two tits, the Willow Ptarmigan, and the White Stork. The ptarmigan population underwent a long-term cycle, first decreasing steadily from 86 adults to 24 in 7 years and then increasing back to 90 adults in the next 4 years. The White Stork population in the Oldenburg province of Germany underwent a similar long-term cycle and so the coefficient of variation greatly overestimates the actual year-to-year fluctuations in population size by superimposing a longer cycle. Populations of the two species of tits in the woods near Oxford, England, are maintained at unnaturally high levels by artificial nesting boxes. At these densities, the tits may be forced to conform to environmental fluctuations more closely than if their populations were less dense.

Although Lack (1966, p. 16–18) has shown that tit populations in southern England and in the Netherlands fluctuate in parallel from year to year, local fluctuations must be at least partly independent in some species. For example, the variation in the population index of the European Blackbird, which was calculated from banding records over a large area, was only 10 percent, compared to 27 percent and 20 percent for more local censuses. Two statewide indices of Bobwhite abundance had coefficients of variation of 10.9 and 16.7 percent (J. J. Hickey, personal communication).

TABLE 6 Year-to-Year Variation in the Population Parameters of Several Birds[a]

Parameter and Species	Parameter	Years	Mean	CV[b](%)	Table
Population Size					
Great Tit	Number of pairs in boxes	17	36.9	45	2
Great Tit	Total summer population	16	293.8	30	11
Great Tit	Estimated maximum winter population	11	162.1	50	11
Blue Tit	Number of pairs in boxes	17	25.6	42	2
Pied Flycatcher	Approximate number of pairs	16	70.7	21	20
European Blackbird	Pairs in Botanic Garden	5	14.2	20	26
European Blackbird	Singing males in oak wood	15	8.5	27	26
European Blackbird	Population index	8	89.6	10	c
Tawny Owl	Number of pairs	13	24.6	19	27
Wood Pigeon	Nests/acre	5	23.0	22	32
Wood Pigeon	Size of postbreeding population	5	143.4	20	32
Wood Pigeon	Dec. population (maximum)	6	91.5	55	32
Wood Pigeon	Feb.–Mar. population (minimum)	6	52.0	69	32
Wood Pigeon	Apr. population (maximum)	6	65.5	38	32
Wood Pigeon	July population (breeding)	6	57.2	21	32
Red Grouse	Number (Oct.–Jan.)	5	391.2	28	35
Red Grouse	Number (Feb.–breeding)	5	296.2	34	35
Red Grouse	Breeding birds/40 hectares	5	26.2	35	35
Ptarmigan	Number of adults (spring)	12	57.3	45	39
Ptarmigan	Number seen/10 km (summer)	19	21.3	59	39
Ptarmigan	Pairs/km^2	9	1.5	36	d
White Stork	Number of pairs	36	147.7	41	40
Clutch Size					
Great Tit	Mean clutch	18	9.5	11	2
Blue Tit	Mean clutch	18	10.7	11	2
Pied Flycatcher	Mean clutch	16	7.0	5	20
European Blackbird	Mean clutch	15	3.8	2	c

TABLE 6 (*continued*)

Parametor and Species	Parameter	Years	Mean	CV[b](%)	Table
Tawny Owl	Mean clutch	11	2.4	15	27
Red Grouse	Mean clutch	5	7.4	11	35
Ptarmigan	Mean clutch	19	6.4	21	39
Ptarmigan	Mean clutch	10	10.2	7	d
Nesting Success					
European Blackbird	Fledging success (%)	15	33.2	17(8)	c
Tawny Owl	Pairs breeding (%)	12	59.0	45(65)	27
Wood Pigeon	Fledging success (%)	5	56.0	20(25)	32
Red Grouse	Hatching success (%)	5	77.4	15(53)	35
White Stork	Nesting success (%)	36	67.1	22(45)	40
Yellow-eyed Penguin	Fledging success (%)	16	59.7	18(27)	45
Productivity of the Pair					
Great Tit	Young flying/pair	16	6.7	25	11
Blue Tit	Young flying/pair	16	7.6	22	13
Tawny Owl	Young fledged/pair	13	0.77	56	27
Wood Pigeon	Young raised/pair	5	2.9	7	32
Ptarmigan	Average brood (full grown)	16	3.5	52	39
Ptarmigan	Young (10–12 weeks)/adult	19	1.1	73	39
Ptarmigan	Juveniles (Aug./Sept.)/adult	11	2.7	24	d
White Stork	Young/successful nest	36	2.9	9	40
White Stork	Young/pair	36	1.9	27	40
Productivity of the Population					
Great Tit	Total young fledging	16	222.1	26	11
Blue Tit	Total young fledging	16	181.7	40	13
Pied Flycatcher	Total young fledging	16	381.9	19	20
Tawny Owl	Total young fledging	13	17.5	51	27

414

Red Grouse	Number of young (Aug.)	5	437.8	54	35
Proportion of Young in Populations					
European Blackbird	Young in population, % (North)	5	74.2	12	c
European Blackbird	Young in population, % (South)	5	75.6	12	c
European Blackbird	First-year birds (%)	7	37.0	16	c
Wood Pigeon	% juveniles, postbreeding (calculated)	5	59.2	3	32
Wood Pigeon	% juveniles, Feb. (shot)	6	30.0	52	32
Ptarmigan	% juveniles, Oct. (wings)	11	71.1	10	d
Survival of Young					
Great Tit	% recovered after 3 mo.				
	brood < 8	17	8.8	82	8
	brood 8–10	17	9.9	60	8
Red Grouse	Survival to Aug. 1 (%)	5	50.4	31(32)	35
Yellow-eyed Penguin	Survival to breed (%)	14	25.7	42(15)	45
Ptarmigan	Survival to Aug. (%)	11	57.7	21(30)	d
Survival of Adults					
Great Tit	Summer to spring (%)	16	29.0	49(20)	11
Great Tit	Winter to spring, estimated (%)	11	53.5	23(27)	11
Pied Flycatcher	Recovery of breeding birds following year				
	% males	5	40.2	31(22)	20
	females	7	39.3	17(11)	20
European Blackbird	Annual adult survival (%)	7	63.1	7(11)	c
Yellow-eyed Penguin	Annual adult survival (%)	16	85.4	8(48)	45

[a] Data are primarily from Lack (1966: tables indicated), unless otherwise specified.
[b] Coefficient of variation (CV) is the mean divided by the standard deviation; CV for mortality is presented in parentheses.
[c] Data from Snow (1966).
[d] Data from Bergerud (1970).

Brewer (1963) has analyzed population variation in five breeding bird census areas in eastern U.S. hardwood forests. Of the 50 species for which coefficients of variation were calculated, 22 percent were 10–20 percent, 32 percent were 30–40 percent, and 16 percent were over 40 percent. Brewer concluded that the variation of a particular species was lower in habitats that were most suitable (judged by the density of its population). He also found that the variation in the total population for each census was much less than for individual species (indicating relative independence of population fluctuations) and was lower in the mesic lowland and floodplain habitats than in the drier upland forests (Table 7). Thus different habitats may impose different levels of intrinsic variation on the populations of their inhabitants.

In a Wood Pigeon population that was censused several times during the year, variation was greatest during the winter (60 percent in February–March) and then declined to 21 percent in the breeding population in July. These data suggest not only that population changes occur in the fall without regard to the density of the population but also that density-dependent processes operate during the winter and early spring months to reduce the year-to-year variation in the populations.

In the Great Tit, variation is about the same in breeding and winter populations (45 percent and 50 percent), which suggests that density-dependent regulation of the population must occur through reproduction and the survival of young. In fact, variation in the total summer population of the tits (young plus adults) is only 30 percent, indicating a density-dependent response [as shown by Lack (1966, p. 74)] in the production of fledglings per pair.

TABLE 7 Variation in Total Population for Each Breeding Bird Census by Habitat

	Upland Oak	Maple–Elm–Oak	Beech–Maple	Lowland Beech–Maple	Floodplain
Area (acres)	18.5	55.3	65	55	23.4
Number of years	10	8	18	15	7
Mean no. of species	15.6	23.3	24.7	25.8	27.6
Mean no. of pairs	48.7	142.4	138.1	124.8	123.3
Mean coefficient of variation (%)	32	37	33	23	28
Coefficient of variation for total pairs (%)	18	15	14	8	8

Clutch Size

Clutch size is less variable than population size (CV = 2–21 percent). Since clutch size is generally thought to be related to the number of young that parents can nourish (in nidicolous species, anyway), there would be little advantage in varying clutch size unless future conditions could be predicted with accuracy [cf references cited by Cody (1971, p. 469–470)]. Other mechanisms, such as brood reduction, are widely employed (e.g., Ricklefs, 1965) to bring brood size into line with local feeding conditions.

Nesting Success

The coefficient of variation in nesting success can be based on either the percent of success or the percent mortality, depending on one's purpose. If it is to compare the contribution of nesting success to productivity, success should be used as a base (the standard deviation is the same for success or mortality). If one is interested in mortality rates, which provide more valid comparisons between species whose nesting success differs greatly, percent mortality should be used. For example, if nesting success was 60 percent ± 20 percent standard deviation, the coefficient of variation would be 33 percent (20/60) for success and 50 percent (20/40) for failure.

Among the studies presented in Table 6, variation in success ranged between 15 percent and 45 percent (usually 22) and in failures between 8 percent (for the European Blackbird, over a large area) and 65 percent. It is probably significant that the highest value was for the Tawny Owl, which during the summer months feeds primarily on populations of rodents, whose populations fluctuate greatly from year to year. Most of the nesting failure of the owl is attributable to starvation.

Productivity of the Pair

The variation in the number of young produced per pair was very low in the Wood Pigeon (7 percent), suggesting that variation in fledging success per nesting attempt (20 percent) was largely compensated for by renesting. Variation in the tits and the White Stork was between 20 and 30 percent; the value in the stork (9 percent) is lower than the measure of productivity, which includes whole nesting failures. The high variation in the productivity of the Tawny Owl again seems attributable to its variable food supply, whereas that of the Willow Ptarmigan in Britain seems due partly to the fact that productivity was measured after the young had become full-grown and mortality factors had acted for a longer period.

Productivity of the Population

This is an absolute measure of the reproductive performance of a population and can usefully be compared to the variation in the size of the breeding population to determine whether density-dependent factors act on reproduction. Coefficients of variation in the five species varied from 19 percent to 54 percent. Compared to the breeding populations, variation in the number of young produced was lower in the Great Tit (26/45 percent), about the same in the Blue Tit (42/40 percent) and Pied Flycatcher (21/19 percent), and much greater in the Tawny Owl (*Strix aluco*) (51/19 percent) and Red Grouse (*Lagopus scoticus*) (54/35 percent), indicating a decreasing order of density dependence, probably reaching zero in the last two species.

Proportion of Young in the Population

The proportion of young in the total population can be used as a measure of productivity or recruitment, depending on whether it is obtained just after or just before the breeding season. Snow (1966) used banding records for the European Blackbird over much of southern England to show that the variation in the proportion of young in the population after the breeding season was comparable to that of first-year birds in the breeding population (16 percent), which indicates that the mortality of first-year birds and adults roughly paralleled each other during most of the nonbreeding season. In the Wood Pigeon, the proportion of young immediately after breeding had a variation of 3 percent, but this had increased to 52 percent by February. Thus, the mortality rates of first-year birds seem to be more variable than that of the adults.

Survival of the Young

The proportion of young Great Tits banded as nestlings and recovered after 3 months or more had a high variation (60–82 percent), but samples studied were small. Nevertheless, the vulnerable young may be especially sensitive to variations in the environment. Survival of young Red Grouse and of young Yellow-eyed Penguins to their first breeding season has intermediate levels of variation (31 percent, 42 percent). Year-to-year variation in the survival of young Ptarmigan is relatively low (21 percent).

Survival of Adults

Unfortunately, relatively few data have been collected for adult survival over many years. Variation is high for the Great Tit (49 percent), intermediate for the Pied Flycatcher (males, 31 percent; females, 17 percent) and low for the European Blackbird (7 percent) and the Yellow-eyed Penguin (8 percent); but the coefficient of variation appears to be inversely related to survival rate. Variation in the mortality rate is more uniform for the British passerines (11–22 percent), but may be higher for the Great Tit during the winter months (27 percent winter–spring versus 20 percent summer–spring).

VARIATION AMONG SPECIES

We have seen that substantial differences occur between the life tables of species with respect to the production of young and age at first reproduction, but what of the more subtle adjustments between similar species breeding in the same area? Which components of their life histories are subject to change, and how do these affect the overall life equation?

To begin to answer such questions the components of productivity in passerine birds in several localities have been analyzed in detail.* The components of productivity included were clutch size, length of breeding season, and nesting success. In the temperate sample, nesting success was calculated for all species from the relation between nesting success and the length of the nestling period obtained for a few of the species. The number of broods attempted during the year was calculated by the method described in the discussion of fecundity (p. 401).

The variables and their coefficients of variation are presented in Table 8. Briefly, intraregional variation in production was based on different components of the life cycle in different areas, and the overall variability in productivity was greatest in the moist tropical sample.

The temperate sample was based on species for which breeding

---

*These anlyses were performed by George Bloom, as an undergraduate independent research project.

TABLE 8 Components of Variation among Passerine Birds in the Annual Production—Offspring to Fledging

Locality	Number of Species	Component of Variation	Mean	CV (%)	Correlation with Production (r)
Kansas	12	Season length (months)	2.69	21	0.54
(temperate)		Clutch size	4.47	16	0.49
		Nesting success	0.57	25	0.49
		Production (young)	6.14	22	—
Costa Rica	6	Season length	4.12	25	0.63
(wet tropical)		Clutch size	2.25	12	0.20
		Nesting success	0.31	42	0.53
		Production	2.26	35	—
Ecuador	10	Season length	3.59	35	0.82
(arid tropical)		Clutch size	3.30	24	0.57
		Nesting success	0.45	18	-0.11
		Production	4.86	43	—
Arizona	9	Season length	3.09	46	-0.20
(semiarid temperate)		Clutch size	3.17	18	0.55
		Nesting success	0.42	33	0.84
		Production	2.80	59	—

season data were given in Johnston (1964) and included 12 species. The three parameters of productivity varied with approximately the same order of magnitude and contributed about equally to the interspecific variation in productivity. Among components, clutch size and nesting success were postively correlated ($r = 0.61$), but season length was not significantly correlated with clutch size or nesting success (Figure 8).

The pattern is quite different for the tropical samples. In the sample from the wet tropics (6 species), based primarily on Skutch's work (1949–1967), nesting success was the most variable component of production. Nesting success and season length contributed about equally to variation in production. As in the temperate sample, nesting success was negatively correlated both with season length ($r = 23$) and with clutch size ($r = -0.27$), which indicates that interspecific differences tend to involve relatively long breeding season, small clutch size, and reduced nesting success, and vice versa. In the sample

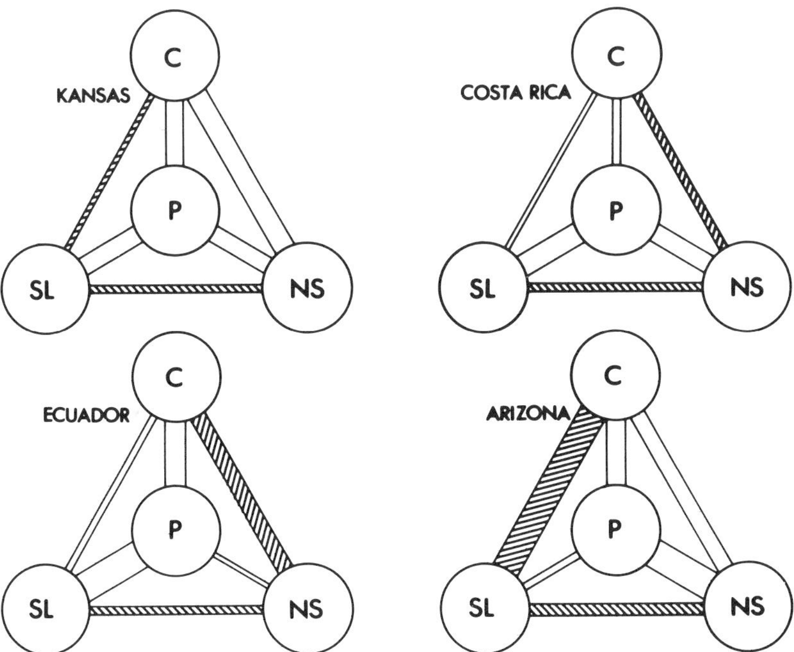

FIGURE 8 Correlations among the components of production. The width of the band connecting the variables is proportional to the correlation coefficient between them.

C, clutch size; SL, season length; NS, nesting success; P, productivity. C–SL and SL–NS correlations are all negative.

of 10 species from the dry tropics, based on Marchant's (1959, 1960) studies in Ecuador, clutch size and season length independently contributed to variation in productivity. The semiarid temperate data (Arizona), provided by George Austin (unpublished), are unique in showing a strong inverse correlation between season length and clutch size and a strong relationship between nesting success and productivity. The general trend among the components is the same for all of the localities, namely, high clutch size, short season length, and high nesting success.

INTERREGIONAL VARIATION

From the mean values presented in Table 8, we can determine which factors are responsible for major geographic trends in productivity. The overall level of productivity was highest in the sample from temperate regions (6.14 fledged young per season), next highest in the dry tropics (4.86), then the semiarid temperate (2.80), and least in the wet tropics (2.26). The length of the breeding season in Costa Rica is very short for tropical localities. This is partly because Skutch lives in a relatively seasonal environment compared to other wet tropical areas, but also because his sample is limited primarily to birds of second-growth habitats, which tend to have shorter breeding seasons than do forest-dwelling species in the same area; the latter are far more difficult to study. Both nesting success and clutch size are important factors in determining the trend in productivity. The length of the nesting season runs opposite to the general trend in productivity.

These comparisons involve different species from one locality to another, but there is no reason to believe that individual species would behave differently. The trends in nesting season, clutch size, and nest success are very general gradients that apply within, as well as between, species (Lack, 1954; Cody, 1971). For example, nest success in the Red-winged Blackbird is about 45 percent in Washington but only 22 percent in Costa Rica (Orians, quoted by MacArthur, 1969). Few data have been gathered for the nesting productivity of a species in different geographical parts of its range, however. Snow's (1956) study of the Blue Tit suggested that variation in productivity was primarily related to changes in clutch size (Figure 9), but his immature/adult ratios were taken during the breeding season and thus are actually a measure of adult survival rate (assuming first-year birds replace dead adults in the population) rather than nesting productivity. I compiled data for the Rough-winged Swallow through-

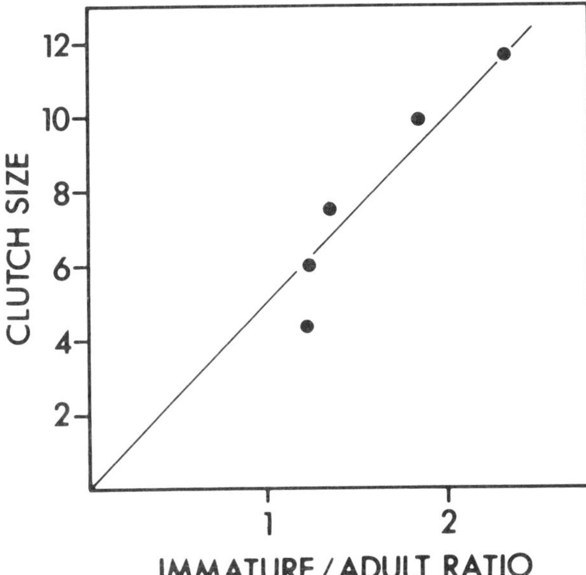

FIGURE 9 The relationship between clutch size and the immature/adult ratio during the breeding season in the Blue Tit; based on data in Snow (1956). The localities are, in order of increasing clutch size, Canary Islands, Iberian Peninsula, North Africa, Northern Europe, and Britain.

out its extensive range (southern Canada to Argentina) based on ratios of immature to adult males in museum collections just after the end of the breeding season (Ricklefs, 1972). Productivity was found to drop markedly from north temperate into tropical regions and then to rise again in the south temperate subspecies. Because the Rough-winged Swallow is probably single brooded throughout its range and because the clutch size of the species does not decline very much in the tropics, nesting mortality must be the principal cause of geographical variation in productivity.

THE ECOLOGY AND EVOLUTION OF LIFE TABLE PARAMETERS

From the standpoint of the theoretical ecologist, the study of demography seems likely to lead to some understanding of the ecological and evolutionary processes of which life table parameters are merely symptoms or indices. Lack (1954, 1966) has applied this approach to the problem of population regulation; my own interest lies in determining the nature of adaptive responses and their constraints.

Demographic parameters are clearly influenced by ecological factors. After all, every bird must die of something! Environmental forces also apply the selective pressure that stimulates evolutionary response. All evolutionary systems tend in the same direction: to make fecundity as high as possible and mortality close to zero. I will consider first the external limits to success in this endeavor—the major environmental factors that control patterns of fecundity and mortality—and then the internal constraints posed by limits to the adapting system—various physiological rates that are beyond the reach of natural selection, problems of allocation that lead to compromising evolutionary strategies, and so on.

ECOLOGICAL INFLUENCES

A few major trends are noted here; others are treated more thoroughly by Lack (1954, 1966, 1968), von Haartman (1971), Cody (1971), and others.

Body size is an important aspect of the environment (large size = a relatively small environment). Large birds typically have low mortality rates both as young and as adults. Because of their favorable surface-volume ratio, they are well buffered against environmental fluctuations. They also are larger than most of the predators that attack smaller birds. The life cycles of large species run more slowly for reasons that are not adequately understood from the physiological standpoint. Nevertheless, large birds consistently have small clutches, slow nestling growth, and delayed maturity.

One of the most striking characteristics of tropical bird populations is that, while adult survival appears to be higher than for their temperate counterparts, eggs and nestlings are less fortunate. How does one explain this seeming discrepancy? Cody (1971) has suggested that because tropical adult populations are very constant from year to year (due to the supposed "stability" of the tropics), a rather constant food source of vulnerable eggs, nestlings, and juveniles is available to predators. This favors increased specialization of the predators and higher exploitation rates on the young. Skutch (1966) states that the increased mortality of the young, in fact, is caused primarily by heavy predation. On the other hand, adults are more likely to succumb to inclement weather (and its effect on their food) and to accidents, rather than to predators. Thus a more favorable tropical climate would reduce adult mortality.

I have suggested elsewhere (Ricklefs, 1969a) that the increase in nest mortality in the tropics is caused by an increased diversity of predators and by the bird's consequent lack of effective antipredator adaptations when beset by many predation strategies. Whether the increased diversity of predators is caused by the stability of tropical populations, or whether it is incidental to other aspects of avian demography cannot be answered at this time. Most of the nest predators in the tropics are probably opportunists rather than specialists, and I doubt very much that breeding populations of birds in the tropics are nearly as constant as is commonly supposed.

Seasonality and year-to-year predictability are important aspects of the environment that have received relatively little attention. In highly seasonal environments, adult populations are usually restricted by the carrying capacity of the environment during the poorest season of the year, so the excess availability of resources over adult requirements during the favorable season will be great, and potential reproductive rate will be high (Ashmole, 1963; Ricklefs, in preparation). This model predicts an inverse relationship between season length and clutch size, which occurs both within populations and between populations (see p. 419).

Year-to-year variation in the environment produces fluctuations in reproductive parameters that test the strength of density-dependent regulation. Populations are relatively helpless in unpredictable environments because the only way to plan for change is to plan for all possible changes. This puts a strain on the allocation of adaptations and resources and probably reduces the productivity of populations compared to those in more predictable environments. In an unpredictable environment, uncertainty increases as an event is increasingly removed in time. In birds, clutch size cannot be adjusted to account for chance variations in the environment that occur within the relatively long period between egg-laying and feeding young. Many species have shortened this time lag between fluctuation and response by adopting a strategy of asynchronous hatching and subsequent selective starvation of the smallest young if feeding conditions should deteriorate after the clutch is laid. (A large clutch is laid to accommodate all possible ranges of feeding conditions.) Thus by "brood reduction," the size of the brood may be quickly brought into line with the resource availability (Ricklefs, 1965).

A tremendous variation in life-history parameters that has been imposed upon the otherwise uniform "song bird" adaptive system attests to the role of the environment in molding the age-specific fecundity and survivorship curves.

EVOLUTIONARY RESPONSES

Evolutionary change is a way by which some organisms surpass others.
Because all populations are pushed by selection toward the same ends
and because they are all about equally successful at any given time
(populations remain relatively constant), there is no way to judge
adaptation in an absolute sense. It is no surprise, then, that most of
the recent discussions on the evolution of demographic strategies have
dealt with compromises between different components of strategy,
particularly reproductive effort and adult longevity (see for example,
Cody 1966, 1971; Williams, 1966; Murphy, 1968; Cole, 1954; Gadgil
and Bossert, 1970; Tinkle, 1969; Tinkle *et al.*, 1970; Hairston *et al.*,
1970; MacArthur and Wilson, 1967; Pianka, 1970; Lewontin, 1965).

One argument goes as follows: If the mortality associated with reproduction is high compared to nonreproductive mortality, it may be
a good strategy to reduce or delay reproduction in order to ensure
longer life (Williams, 1966; cf Gadgil and Bossert, 1970). Life history
parameters of birds are consistent with this model. When adult mortality rates are low, fecundity is also typically low and reproduction
is delayed. But demographic parameters themselves do not constitute
sufficient evidence that adjustments of demographic parameters are
"strategies" over and above the influence of the environment on each
independently. For example, food is critically limiting for the seabirds
because of where and how they feed (Ashmole, 1971). Also, because
there are few predators where seabirds feed and nest and because their
flight powers allow them to outflank storms and to find patchily distributed food, adult and nestling mortality is very low. But is fecundity
reduced below what it could be, or is the onset of first reproduction
delayed beyond what it could be, thus increasing adult life-span? The
importance of such questions is matched only by their inaccessibility
to experimental testing. One might follow up extensive twinning experiments with observations on the survival of adults compared to
those with the normal single young brood. But to demonstrate that
normal reproductive levels increase mortality over nonreproductive
mortality is not enough. This has also been shown for Song Sparrows,
and nobody has seriously suggested that they have evolved reduced
fecundity for the sake of longevity.

All reproductive effort and feeding effort, too, involves risk that
must be adjusted to a tolerable level on the basis of its return. Observations of this kind are the basis for evolutionary demography. One
must be able to relate effort (measured in terms of adult mortality),

to return (productivity of young) before one can specify optimum
demographic strategies and to test demographic hypotheses.

A suitable, though somewhat oblique, test can be made for a related system to demonstrate this point. Skutch (1949) suggested that
tropical birds may have small clutches that reduce activity around
the nest; predators supposedly being attracted by activity around the
nest. One would predict that the rate of nest success (return) would go
down when the clutch size (effort) was increased. Although it is im-
practical to enlarge broods artificially to test this hypothesis in most
tropical species, we also would predict that nest mortality during the
incubation period would be lower than during the nestling period,
because there is relatively little commotion and coming-and-going
during the incubation period. But, in fact, mortality rates calculated
from Skutch's (1966) nest success data are somewhat higher during the
incubation period (Ricklefs, 1969), and it is therefore unlikely that
the relationship between effort and return would make reduced clutch
size a good strategy.

CONCLUSIONS

Variation is the key to all demographic studies. A steady state can
tell us nothing about how a population is organized and regulated
demographically. Changes in demographic parameters through different conditions from year to year provide insight into the stability
and growth potential of the population. In most cases, we must rely
on the vagaries of nature to perform experiments for us because populations are so large that we cannot hope to manipulate them. Small
island populations, however, do provide opportunities for experimentation that could be extremely enlightening (e.g., Kluijver, 1966).
Regardless of the circumstances, more emphasis should be placed on
variation in parameters in relation to environmental factors, rather
than on their means for each species.

Interspecific variation in demographic parameters, especially when
it is matched to suitable environmental gradients, may also be used to
provide information on the organization of population dynamics.
Here one must remember that demographic parameters are not independent. An increase in fecundity must be balanced by an increase in
mortality. Which is driven by gross environmental changes, which by
density-dependent factors? It is often not easy to tell. It is currently
thought that, in birds, mortality is adjusted to changes in fecundity

(Lack, 1966). But if adult mortality was to decrease and a population became stabilized at a higher density, would not intraspecific competition reduce food availability and, hence, clutch size?

Evolution is itself based upon genetic variation, but, whereas ecological variation either occurs spontaneously or may be reproduced experimentally, much genetic variation is irretrievably lost in the course of evolution. Studies on the evolution of demographic parameters require the evaluation of the fitnesses of phenotypes that have long since been eliminated. It is not good enough to reconstruct these phenotypes by conjecture because our knowledge of adaptation is far too limited. We can sometimes reconstruct phenotypes artificially, perhaps by hormonally speeding the onset of maturity. We can also look for systems that involve unambiguous problems in allocation, such as the use of time for prolonged parental care versus laying a new clutch or the production of a large egg at the expense of fecundity (e.g., Tinkle *et al.*, 1970). The costs of intensified or prolonged parental care, in terms of fecundity, are easy to evaluate in these cases, even if the benefits in terms of survival of the young are not.

As we survey natural populations, we cannot help but be impressed with the tremendous uniformity that they possess. Not only are the absolute values, the "means" of population parameters similar on the whole among ecologically related species, but levels of variation are also remarkably uniform among species that otherwise differ greatly. Species that are vulnerable to population fluctuations, either because they inhabit varying environments or are inherently prone to external factors, also respond rapidly to population change. Those species that are slow to change are also slow to respond and recover. In either case, density-dependent responses reduce or eliminate the danger of extinction by random factors unless gross and lasting environmental changes, such as the destruction of habitat or efficient exploitation, occur. Such perturbations can force the decline of a species to a point below which random factors make extinction a probable, or even a likely, event.

Extinction is the ultimate demographic event of the population. The changes in populations that lead to extinction normally take place over a different time scale than we have considered here. Extinction is the antithesis of stable population regulation. But does extinction occur in the absence of gross and persistent environmental changes? Unless one subscribes to the notion of "racial senescence," no. But "persistent environmental changes" may include efficient competitors and parasite and predator pressure. George Cox and I

(1972) have outlined a model to account for the extinction of populations of birds on islands by a progressive weakening of competitive ability through counterrevolution from the island biota. Competitive ability can be altered by very small changes in the demographic parameters of the population, and so the long-term processes that lead to extinction are based on a much smaller scale of demographic change than we witness in the year-to-year variations in populations. It seems ironic to me that the mechanisms of the ultimate demographic event may be nearly imperceptible.

ACKNOWLEDGMENTS

I wish to thank George Austin, Tom Lovejoy, and George Bloom for providing unpublished data. The manuscript benefited greatly from the substantive and editorial comments of Henry Wilbur and Susan White. Parts of this review were supported by National Science Foundation grants GB 12612 and GB 31554X.

REFERENCES

Amadon, D. 1964. The evolution of low reproductive rates. Evolution 18:105–110.

Ashmole, N. P. 1963. The regulation of numbers of tropical oceanic birds. Ibis 103b:458–473.

Ashmole, N. P. 1971. Sea bird ecology and the marine environment, p. 223–286. *In* D. S. Farner and J. R. King [ed.] Avian biology. Vol. I. Academic Press, New York and London.

Ashmole, N. P., and S. H. Tovar. 1968. Prolonged parental care in Royal Terns and other birds. Auk 85:90–100.

Bergerud, A. T. 1970. Population dynamics of the Willow Ptarmigan *Lagopus lagopus alleni* L. in Newfoundland 1955 to 1965. Oikos 21:299–325.

Berndt, R., and H. Sternberg. 1963. Ist die Mortalitätsrate adulter *Ficedula hypoleuca* wirklich unabhängig vom Lebensalter? Proc. Int. Ornithol. Congr. 13:675–684.

Birch, L. C. 1948. The intrinsic rate of natural increase of an insect population. J. Anim. Ecol. 17:15–26.

Birch, L. C. 1953. Experimental background to the study of the distribution and abundance of insects. I. The influence of temperature, moisture and food on the innate capacity for increase of three grain beetles. Ecology 34:698–711.

Boyd, H. 1962. Mortality and fertility of European Charadrii. Ibis 104:368–387.

Brewer, R. 1963. Stability in bird populations. Occasional papers of the C. C. Adams Center for Ecological Studies. No. 7, p. 1–12.

Carrick, R. 1963. Ecological significance of territory in the Australian Magpie,

Gymnorhina tibicen, p. 740–753. *In* Proceedings of the XIIIth International Ornithological Congress. American Ornithological Union, Baton Rouge, Louisiana.

Caughley, G. 1966. Mortality patterns in mammals. Ecology 47:906–918.

Cody, M. L. 1966. A general theory of clutch size Evolution 20:174–184.

Cody, M. L. 1971. Ecological aspects of reproduction, p. 461–512 *In* D. S. Farner and J. R. King [ed.] Avian biology. Vol. I. Academic Press, New York and London.

Cole, L. C. 1954. The population consequences of life history phenomena. Q. Rev. Biol. 29:103–137.

Comfort, A. 1962. Survival curves of some birds in the London Zoo. Ibis 104:115–117.

Coulson, J. C. 1960. A study of the mortality of the Starling based on ringing recoveries. J. Anim. Ecol. 29:251–271.

Coulson, J. C. 1966. The influence of the pair-bond and age on the breeding biology of the Kittiwake Gull *Rissa tridactyla*. J. Anim. Ecol. 35:269–279.

Coulson, J. C., and E. White. 1961. An analysis of the factors influencing the clutch size of the Kittiwake. Proc. Zool. Soc. London 136:207–217.

Davis, D. E. 1950. The growth of Starling, *Sturnus vulgaris*, populations. Auk 67:460–465.

Deevey, E. S., Jr. 1947. Life tables for natural populations of animals. Q. Rev. Biol. 22:283–314.

Delius, J. D. 1965. A population study of Skylarks, *Alauda arvensis*. Ibis 107:466–492.

Diamond, J. M. 1969. Avifaunal equlibria and species turnover rates on the Channel Islands of California. Proc. Natl. Acad. Sci. U.S.A. 64:57–63.

Dorward, D. F. 1962. Comparative biology of the White Booby and the Brown Booby, *Sula* spp., at Ascension. Ibis 103b:174–220.

Einarsen, A. S. 1945. Some factors affecting Ring-necked Pheasant population density. Murrelet 26:39–44.

Emlen, J. T., Jr. 1940. Sex and age ratios in survival of the California Quail. J. Wildlife Manage. 4:92–99.

Errington, P. L. 1945. Some contributions of a fifteen-year local study of the Northern Bobwhite to a knowledge of population phenomena. Ecol. Monogr. 15:1–34.

Farner, D. S. 1955. Bird banding in the study of population dynamics, p. 397–449. *In* A. Wolfson [ed.] Recent studies in avian biology. University of Illinois Press, Urbana.

Gadgil, M., and W. H. Bossert. 1970. Life historical consequences of natural selection. Am. Nat. 104:1–24.

Gill, D. E., and W. E. Lanyon. 1965. Establishment, growth, and behavior of an extralimital population of House Finches around Huntington, New York. Bird-Banding 36:1–14.

Graber, R. R., and J. W. Graber. 1962. Weight characteristics of birds killed in nocturnal migration. Wilson Bull. 74:74–88.

Hairston, N. G., D. W. Tinkle, and H. M. Wilbur. 1970. Natural selection and the parameters of population growth. J. Wildlife Manage. 34:681–690.

Haldane, J. B. S. 1955. The calculation of mortality rates from ringing data, p. 454–458. *In* Proceedings of the 11th International Ornithological Congress, 1954. HRGS. von Adolp Portmann Lund, Ernst Sutter, Basel, Birkhauser.

Hamilton, W. D. 1966. The moulding of senescence by natural selection. J. Theoret. Biol. 12:12–45.

Hann, H. W. 1937. Life history of the Ovenbird in southern Michigan. Wilson Bull. 49:145–237.

Harris, M. P. 1966. Breeding biology of the Manx Shearwater, *Puffinus puffinus*. Ibis 108:17–33.

Henny, C. J., W. S. Overton, and H. M. Wight. 1970. Determining parameters for populations by using structural models. J. Wildlife Manage. 34:690–703.

Hensley, M. M., and J. B. Cope. 1951. Further data on removal and repopulation of the breeding birds in a spruce-fir forest community. Auk 68:483–493.

Hickey, J. J. 1952. Survival studies of banded birds. U.S. Fish Wildlife Ser. Spec. Sci. Rep. Wildlife 15:1–177.

Hickey, J. J. 1955. Population research on gallinaceous birds, p. 326-396. *In* A. Wolfson [ed.] Recent studies in avian biology. University of Illinois Press, Urbana.

Hussell, D. J. T., T. Davis, and R. D. Montgomerie. 1967. Differential fall migration of adult and immature Least Flycatchers. Bird-Banding 38:61–66.

Immelmann, K. 1971. Ecological aspects of periodic reproduction, p. 341–389. *In* D. S. Farner and J. R. King [ed.] Avian biology. Vol. I. Academic Press, London and New York.

Jehl, J. R., Jr. 1971. Patterns of hatching success in subarctic birds. Ecology 52:169–173.

Johnston, R. F. 1954. Variation in breeding season and clutch size in Song Sparrows of the Pacific Coast. Condor 56:268–273.

Johnson, R. F. 1964. The breeding birds of Kansas. Univ. Kans. Publ. Mus. Nat. Hist. 12:575–655.

Kale, H. W., II. 1965. Ecology and bioenergetics of the Long-billed Marsh Wren in Georgia salt marshes. Publ. Nuttall Ornithol. Club 6:1–142.

Kalmbach, E. R. 1939. Nesting success: Its significance in waterfowl reproduction. Trans. N. Am. Wildlife Conf. 4:591–604.

Keith, L. B. 1963. Wildlife's ten-year cycle. University of Wisconsin Press, Madison. 201 p.

Kendeigh, S. C. 1942. Analysis of losses in the nesting of birds. J. Wildlife Manage. 6:19–26.

Klomp, H. 1970. The determination of clutch-size birds. A review. Ardea 58:1–124.

Kluijver, H. N. 1935. Waarnemingen over de levenswijze van den Spreeuw (*Sturnus v. vulgaris* L.) met behulp van geringde individuen. Ardea 24:133–166.

Kluijver, H. N. 1951. The population ecology of the Great Tit *Parus m. major*. Ardea 39:1–135.

Kluijver, H. N. 1966. Regulation of a bird population. Ostrich Suppl 6:389–396.

Kortlandt, A. 1942. Levensloop, samenstelling en structuur der Nederlandse aalscholverbevolking. Ardea 31:175–280.

Lack, D. 1946. Do juvenile birds survive less well than adults? Brit. Birds 39:258–264.

Lack, D. 1947. The significance of clutch-size. I and II. Ibis 89:302–352.

Lack, D. 1948a. Natural selection and family size in the Starling. Evolution 2:95–110.

Lack, D. 1948b. The significance of clutch-size. III. Some interspecific comparisons. Ibis 90:25–45.

Lack, D. 1950. The breeding seasons of European birds. Ibis 92:288–316.

Lack, D. 1954. The natural regulation of animal numbers. Clarendon Press, Oxford. 343 p.

Lack, D. 1966. Population studies of birds. Clarendon Press. Oxford. 341 p.

Lack, D. 1968. Ecological adaptations for breeding in birds. Methuen, London. 409 p.

Lack, D., and R. E. Moreau. 1965. Clutch size in tropical passerine birds of forest and savanna. Oiseau 35(Suppl.):75–89.

Laughlin, R. 1965. Capacity for increase: A useful population statistic. J. Anim. Ecol. 34:77–91.

Leedy, D. L., and L. E. Hicks. 1945. The pheasant in Ohio, p. 57–130. *In* W. L. McAtee [ed.] The Ring-necked Pheasant and its management in North America. The American Wildlife Institute, Washington, D.C.

Leslie, P. H. 1966. The intrinsic rate of increase and the overlap of successive generations in a population of Guillemots (*Uria aalge* Pont.). J. Anim. Ecol. 35:291–301.

Lewontin, R. 1965. Selection for colonizing ability, p. 77–94. *In* H. G. Baker and G. L. Stebbins [ed.] The genetics of colonizing species. Academic Press, New York.

Lotka, A. J. 1925. Elements of physical biology. Williams & Wilkins, Baltimore. (Reprinted as *Elements of Mathematical Biology*, Dover, New York, 1956.) 460 p.

MacArthur, R. H. 1964. Environmental factors affecting species diversity. Am. Nat. 98:387–397.

MacArthur, R. H. 1969. Patterns of tropical communities. Biol. J. Linnean Soc. 1:19–30.

MacArthur, R. H., and E. O. Wilson. 1967. The theory of island biogeography. Princeton University Press, Princeton, New Jersey. 203 p.

Marchant, S. 1959. The breeding season in S. W. Ecuador. Ibis 101:137–152.

Marchant, S. 1960. The breeding of some S. W. Ecuadorian birds. Ibis 102:349–382, 584–599.

Mayfield, H. 1961. Nesting success calculated from exposure. Wilson Bull. 73:255–261.

Mertz, D. B. 1971. The mathematical demography of the California Condor population. Am. Nat. 105:437–453.

Miller, A. H. 1965. Capacity for photoperiodic response and endogenous factors in the reproductive cycles of an equatorial sparrow. Proc. Natl. Acad. Sci. U.S.A. 54:97–101.

Moreau, R. E. 1944. Clutch size: A comparative study, with special reference to African birds. Ibis 86:286–347.

Morel. M.-Y. 1966. Productivité et renouvellement des populations de *Lagonosticta senegale* dans la basse vallée du Sénégal. Ostrich Suppl. 6:435–442.

Murphy, G. I. 1968. Pattern in life history and the environment. Am. Nat. 102:391–403.

Nelson, J. B. 1966. The breeding biology of the Gannet *Sula bassana* on the Bass Rock, Scotland. Ibis 108:584–626.

Nice, M. M. 1937. Studies in the life history of the Song Sparrow. Vol. 1. Trans. Linnean Soc. N.Y. 4:1–247.

Nice, M. M. 1957. Nesting success in altricial birds. Auk 74:305–321.

Nolan, V., Jr. 1963. Reproductive success of birds in a deciduous scrub habitat. Ecology 44:305–313.

Orians, G. H. 1961. The ecology of blackbird (*Agelaius*) social systems. Ecol. Monogr. 31:285–312.

Orians, G. H. 1969. Age and hunting success in the Brown Pelican (*Pelecanus occidentalis*). Anim. Behav. 17:316–319.

Paynter, R. A., Jr. 1947. The fate of banded Kent Island Herring Gulls. Bird-Banding 18:156–170.

Perrins, C. M. 1965. Population fluctuations and clutch size in the Great Tit, *Parus major*. J. Anim. Ecol. 34:601–647.

Peterson, R. T. 1948. Birds over America. Dodd, Mead and Co., New York. 342 p.

Pettingill, O. S., Jr. 1939. History of one hundred nests of Arctic Tern. Auk 56:420–428.

Pianka, E. R. 1970. On r- and K-selection. Am. Nat. 104:592–597.

Pinowski, J. 1968. Fecundity, mortality, numbers and biomass dynamics of a population of the Tree Sparrow (*Passer m. montanus* L.). Ekol. Polska Ser. A 16:1–58.

Potts, G. R. 1969. The influence of eruptive movements, age, population size and other factors on the survival of the Shag [*Phalacrocorax aristotelis* (L)]. J. Anim. Ecol. 38:53–102.

Rice, D. W., and K. W. Kenyon. 1962. Breeding cycles and behavior of Laysan and Black-footed Albatrosses. Auk 79:517–567.

Richdale, L. E. 1957. A population study of penguins. Oxford University Press (Clarendon), London and New York. 195 p.

Ricklefs, R. E. 1965. Brood reduction in the Curve-billed Thrasher. Condor 67:505–510.

Ricklefs, R. E. 1966. The temporal component of diversity among species of birds. Evolution 20:235–242.

Ricklefs, R. E. 1968a. On the survival rate of juvenile Cactus Wrens. Condor 70:388–389.

Ricklefs, R. E. 1968b. Patterns of growth in birds. Ibis 110:419–451.

Ricklefs, R. E. 1969a. An analysis of nesting mortaility in birds. Smithson. Contrib. Zool. 9:1–48.

Ricklefs, R. E. 1969b. The nesting cycle of songbirds in tropical and temperate regions. Living Bird 8:165–175.

Ricklefs, R. E. 1970. The estimation of a time function of ecological use. Ecology 51:508–513.

Ricklefs, R. E. 1972. Latitudinal variation in breeding productivity of the Rough-winged Swallow. Auk 89:826–836.

Ricklefs, R. E., and G. W. Cox. 1972. Taxon cycles in the West Indian avifauna. fauna. Am. Nat. 106:195–219.

Robel, R. J., and S. D. Fretwell. 1970. Winter mortality of Bobwhite Quail estimated from age ratio data. Trans. Kan. Acad. Sci. 73:361–367.

Serventy, D. L. 1971. Biology of desert birds, p. 287–339. *In* D. S. Farner and

J. R. King [ed.] Avian biology. Vol. I. Academic Press, New York and
    London.
Skutch, A. F. 1949. Do tropical birds rear as many young as they can nourish?
    Ibis 91:430–455.
Skutch, A. F. 1950. The nesting seasons of Central American birds in relation to
    climate and food supply. Ibis 92:185–222.
Skutch, A. F. 1966. A breeding census and nesting success in Central America.
    Ibis 108:1–16.
Skutch, A. F. 1967a. Adaptive limitation of the reproductive rate of birds. Ibis
    109:579–599.
Skutch, A. F. 1967b. Life histories of Central American highland birds. Publica-
    tions of the Nuttall Ornithological Club, No. 7.
Smith, S. M. 1967. Seasonal changes in the survival of the Black-capped Chicka-
    dee. Condor 69:344–359.
Snow, D. W. 1956. The annual mortality of the Blue Tit in different parts of its
    range. Brit. Birds 49:174–177.
Snow, D. W. 1958. The breeding of the Blackbird *Turdus merula* at Oxford. Ibis
    100:1–30.
Snow, D. W. 1962. A field study of the Black and White Manakin *Manacus ma-
    nacus*, in Trinidad. Zoologica 47:65–104.
Snow, D. W. 1966. Population dynamics of the Blackbird Nature 211:1231–
    1233.
Stewart, R. E., and J. W. Aldrich. 1951. Removal and reproduction of breeding
    in a spruce-fir forest community. Auk 68:471–482.
Stonehouse, B. 1960. The tropic birds (genus *Phaethon*) of Ascension Island. Ibis
    103b:124–161.
Tanner, J. T. 1966. Effects of population density of growth rates of animal pop-
    ulations. Ecology 47:733–745.
Tinkle, D. W. 1969. The concept of reproductive effort and its relation to the
    evolution of life histories of lizards. Am. Nat. 103:501–516.
Tinkle, D. W., H. M. Wilbur, and S. G. Tilley. 1970. Evolutionary strategies in
    lizard reproduction. Evolution 24:55–74.
Tompa, F. S. 1964. Factors determining the numbers of Song Sparrows, *Melos-
    piza melodia* (Wilson), on Mandarte Island, B. C., Canada. Acta Zool. Fenn.
    109:1–73.
Tordoff, H. B., and R. M. Mengel. 1956. Studies of birds killed in nocturnal mi-
    gration. Univ. Kan. Publ. Mus. Nat. Hist. 10:1–44.
Van Tets, G. F. 1968. Seasonal fluctuations in the mortality rates of three
    northern- and three southern-hemisphere gulls. CSIRO Wildlife Res. 13:1–9.
von Haartman, L. 1951. Der Trauerfliegenschnäpper. II. Populationsprobleme.
    Acta Zool. Fenn. 67:1–60.
von Haartman, L. 1971. Population dynamics, p. 391–459. *In* D. S. Farner and
    J. R. King [ed.] Avian biology. Vol. I. Academic Press New York and Lon-
    don.
Williams, G. C. 1957. Pleiotropy, natural selection, and the evolution of senes-
    cence. Evolution 11:398–411.
Williams, G. C. 1966. Natural selection, the costs of reproduction and a refine-
    ment of Lack's principle. Am. Nat. 100:687–690.

Willis, E. 1961. A study of nesting Ant-tanagers in British Honduras. Condor 63:479–503.

Wynne-Edwards, V. C. 1955. Low reproductive rates in birds, especially seabirds, p. 540–547. *In* Proceedings of the 11th International Ornithological Congress, 1954. HGRS. von Adolp Portmann Lund, Ernst Sutter, Based, Birkhauser.

Wynne-Edwards, V. C. 1962. Animal dispersion in relation to social behaviour. Oliver and Boyd Ltd., Edinburgh and London. 653 p.

DISCUSSION

Lars von Haartman

In spite of the impressive wealth of data on clutch size, we are still not very well informed about a number of basic questions. One of these unsettled questions concerns geographical variation. Two trends have been presumed to exist: a south–north trend, which implies that clutch size, because of the longer day for catching food, rises from the south toward the north, and a west–east trend, which implies that clutch size, for somewhat more obscure reasons, rises in Europe from the west toward the east.

There is strong evidence of an increase in the clutch size from the tropics toward the temperate region and from the Mediterranean area toward central and north Europe. Whether or not there is in temperate Europe a general increase of clutch size toward north is less certain. In two thoroughly studied species, the Great Tit and the Pied Flycatcher more southerly populations may produce larger clutches than more northerly ones, but other species support the argument for a south–north trend (von Haartman, 1967). The alleged west-east trend is even less well supported.

A couple of recent studies on clutch size make it possible to compare clutch sizes in temperate Europe. The south–north trend should show up if we compare the clutch sizes in Switzerland (Glutz von Blotzheim, 1962) and Finland (von Haartman, 1969). In order to make the data more readily comparable, I have, where possible, excluded Swiss data from high altitudes, but considered those from the lowland. In all, 45 species [44 passerines and one near-passerine

(*Jynx torquilla*)] could be compared (Table 1). The result strongly favors the concept of a south–north trend.

The west–east trend cannot, in the absence of adequate data, be tested as effectively. A recent handbook of the birds of Norway (Haftorn, 1971) provides some statistical data that make possible comparisions of ten species (9 passerines and *Jynx torquilla*) in Norway and Finland (Table 1). The data do not support the alleged west–east trend. We must recognize, however, that the comparison concerns only a fragment of the breeding species in both countries.

If we regard the west–east trend as uncertain and accept the south–north trend as a reality, what causes the geographical variation in clutch size? Is it day length, as has usually been assumed? There are facts militating against such an interpretation. For instance, owls show the trend in the same way as diurnal birds (von Haartman, 1971). Therefore, other factors may be considered. I propose two: (1) The short summer season in the north concentrates the production of animal and vegetable food, so that birds enjoy superior food conditions during their breeding season in the north; (2) it is well known that the fauna in more northerly regions is much poorer in species, whereas the abundance of each species may be higher. This may favor the foraging of the birds, making it easier for them to form "searching images" of the food to search for and to learn where to find it.

TABLE 1   Clutch Size Trends in European Passerine Birds

Trend	No. of species
South–North	
Finnish population has larger clutch	23
Clutch size equal[a]	20
Swiss population has larger clutch	2
Mean clutch size in Finland (45 sp.)	5.56
Mean clutch size in Switzerland (45 sp.)	5.28
West–East	
Finnish population has larger clutch	2
Clutch size equal[a]	5
Norwegian population has larger clutch	3
Mean clutch size in Finland (10 sp.)	6.04
Mean clutch size in Switzerland (45 sp.)	6.09

[a] A difference of 0.2 egg or less was accepted as equality.

So far as differences between the dry Mediterranean region and central and north Europe are considered, climatic differences may contribute to the differences in clutch size.

REFERENCES

Haftorn, S. 1971. Norges Fugler. Universitetsforlaget, Oslo. 862 p.
von Blotzheim, U. Glutz. 1962 Die Brutvögel der Schweiz. Verlag Aargauer Tagblatt AG, Aarau. 648 p.
von Haartman, L. 1967. Geographical variations in the clutch-size of the Pied Flycatcher. Ornis Fenn. 44:89–98.
von Haartman, L. 1969. The nesting habits of Finnish birds. I. Passeriformes. Commentat. Biol. 32:1–187.
von Haartman, L. 1971. Population dynamics, p. 391–459. *In* D. S. Farner and R. King [ed.] Avian biology. Vol. I. Academic Press, New York.

DISCUSSION

R. J. Robel

I feel that construction of life tables, be they dynamic, time-specific, or composite, is often regarded as an end in itself. Often only one figure is used from a life table—i.e., the average annual mortality. Quite often, too, life tables are constructed by people not directly involved in the collection of the field data, who may therefore incorporate some major biases in their preparation.

The first is trapping bias. For the table to be sound, large numbers of banded birds must be included in its development. This may lead the investigator to use the most efficient trapping method—most efficient in numbers of birds trapped—rather than the method that contains the least bias or captures birds truly reflecting the composition of the population. Trap-happy or trap-shy birds, or segments of the population, can radically alter the results obtained in life table calculations. Similarly, since large numbers of birds are needed for life table construction, data from birds from all types of habitat are pooled, even from birds captured over a period of several years. What one has then is a mosaic picture, although the results are often quoted later as species-specific characteristics, rather than a reaction of the species to a set of conditions, i.e., food supply, nutrients, predation, climate, disease, parasites, and inter- and intraspecific competition.

The mere appearance of data in a life table is impressive and data contained in them are often accepted without close scrutiny of how, where, and when they were collected.

My second comment deals with failure to consider possible behavioral influences on avian reproduction. We evaluate in depth the influence of clutch size, breeding season length, age-at-first reproduction, density-dependent regulation, evolutionary responses, etc., but seldom look critically at the influence of behavior in unconfined avian populations. Let me just give a couple examples here based on studies on Red Grouse.

Red Grouse are territorial birds. The male establishes and defends a territory on which the female nests and on which both parents feed. The area for territory establishment is limited; birds unable to establish territories cannot reproduce and experience much higher mortality. When food abundance is low, chicks grow slowly; slow-growing chicks are more aggressive than fast-growing chicks and control larger territories the next nesting season. Therefore, nesting population size may be governed not by the food available during a specific summer, but rather by that available the preceding summer. Doctors Jenkins, Watson, Miller, and Moss have published extensively on this and seem to have built a fairly strong case for this hypothesis.

My second example is based on work done in Kansas on Greater Prairie Chickens (*Tympanuchus cupido*). Prairie Chickens are a lekking species: Males gather on display grounds (booming grounds) and perform mating dances there. Females are attracted to these grounds, where mating takes place. On the booming grounds, males establish a social hierarchy; the most socially dominant male controls the most preferred territory and conducts most of the matings. During a 6-year study, 92 percent of all matings on three mating grounds were conducted by the two most socially dominant males on each ground. During that 6-year study, all of the females visiting the ground were successfully mated.

To test the importance of stable social hierarchy in reproductive success, we decided to reduce the stability of the social hierarchy by removing the three most aggressive birds, with this result:

	Female Visits/Year	Agonistic Interactions between Males	Interrupted Copulations	Successful Copulations
Control ground	39	24/morning	2	32
Experimental ground	41	109/morning	28	3

Clearly, our experimental alteration of the social structure of the booming ground did not appear to affect the booming ground's attractiveness to females, increased agonistic interactions among males, increased interruptions of copulations, reduced copulations by about 90 percent. One might now ask: How does nutritional level influence the development of stable social hierarchies, keeping in mind that testosterone production is reduced in undernourished birds?

Since, in Prairie Chickens, social hierarchies are developed in late winter, will birds on low nutritional levels have sufficient "excess" reserve energy for reproductive behavior necessary for social hierarchy development? If males must delay the establishment of a social hierarchy until the hens' visit, reproductive success may be reduced.

As posed previously by Doctor Linder (p. 42), how do sublethal levels of pesticides affect reproductive behavior? What are the effects of sublethal pesticide dosages and nutritional levels on incubation behavior and parental care?

Many questions remain unanswered and we must not overlook the potential influence of bird behavior on reproductive success. Laboratory studies on avian physiology and endocrinology alone will not provide us a full understanding of avian reproduction. We must include field studies of reproductive behavior.

INFORMAL DISCUSSION

KENDEIGH: I do not believe that the larger clutch sizes at northern latitudes in the Northern Hemisphere are due directly to the longer photoperiods. The longer photoperiod may be involved, but the proximate factor is more likely temperature. Birds feed primarily to satisfy their energy requirements, such as loss of weight during the preceding night that needs to be replaced, or by prevailing low temperature during the daytime. At high environmental temperatures (22 °C) in the laboratory, these requirements are mostly met by White-throated Sparrows (*Zonotrichia albicollis*) during the first 2 or 3 hours of the day; they loaf the rest of the light period. Under stress of low temperature or forced exercise (Figure 1), the rapid early morning feeding is prolonged for additional hours and, under extreme stress, is continued throughout the day.* Caged House Sparrows (*Passer domesticus*) and Redpolls (*Acanthis* sp.) even feed at night when under energy stress. Free-living White-crowned Sparrows (*Zonotrichia leucophrys*) have a somewhat different pattern: They feed most rapidly in the early morning and again in the evening, with a loafing period during the middle of the day. But here again, under stress of low temperature or nightly unrest (*Zugunruhe*), they give up their loafing period and feed continuously throughout the day.†

In experiments measuring existence metabolism at two different photoperiods in some 13 species, the amount of food metabolized is always higher on the longer photoperiods, but the amount of increase is never proportional to the increase in number of hours of light. With longer photoperiods, activity and high body temperatures are sustained for a longer period. The extra energy metabolized is simply to meet this extra energy demand, and there is no accumulation of energy reserves in the body.‡

I believe that the size of the clutch is affected in large part by the relation of prevailing temperatures to the temperature at which they can mobilize the largest amount of productive energy. This is at 22 °C for the House Sparrow (see p. 111). The size of the clutch depends on the amount of energy the birds can readily mobilize over and above what they need for existence.

Clutch size varies seasonally in many birds, being smaller in mid-summer than in late spring. This phenomenon is related to the fact that temperatures in mid-summer surpass the point of maximum available productive energy. The same relation may hold latitudinally in that higher temperatures at southern latitudes repress laying of large clutches; this repression is absent at northern latitudes.

HICKEY: Doctor von Haartman has pointed out that it is extremely difficult for the investigator working on a small area to measure the survival of birds returning after migration, particularly after they have been marked as young birds. From what I could find out from five species, the return to the site of hatching follows an exponential relationship in that with each unit of distance away from the site of the nest—1 km, 2 km, and so on—we have concentric

---

*Kendeigh, S. C., J. E. Kontogiannis, A. Mazac, and R. R. Roth. 1969. Environmental regulation of food intake by birds. Comp. Biochem. Physiol. 31:941–957.
†Morton, M. L. 1967. Diurnal feeding patterns in White-crowned Sparrows, *Zonotrichia leucophyrs gambelii.* Condor 69:491–512.
‡Kendeigh, S. C. 1970. Energy requirements for existence in relation to size of bird. Condor 72:60–65.

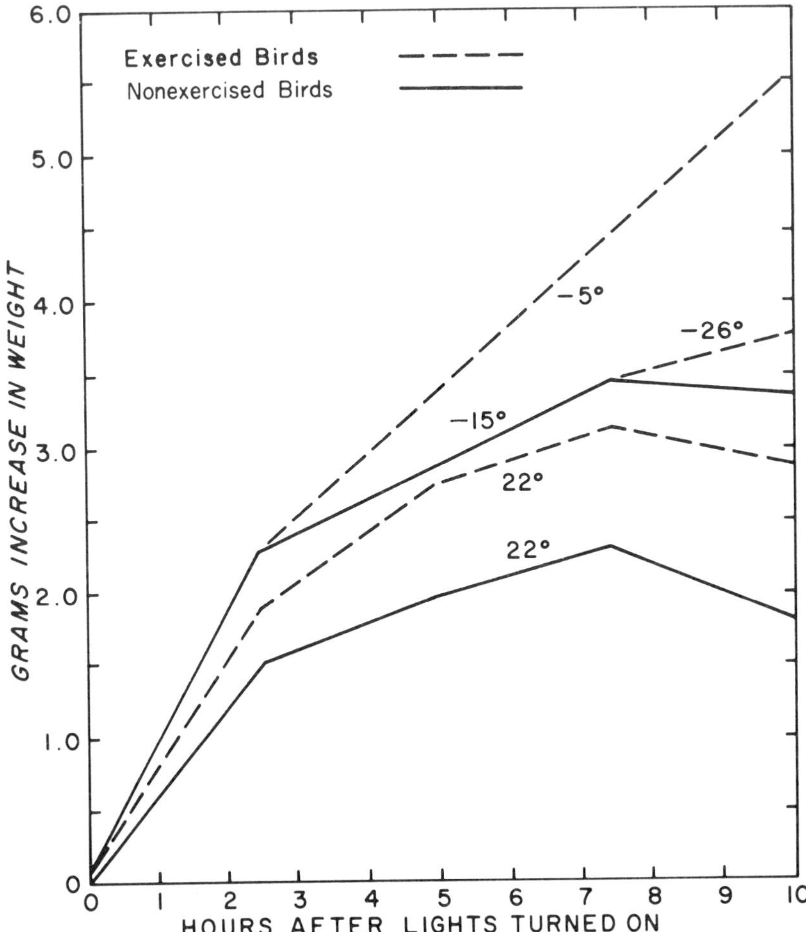

FIGURE 1   Patterns of daytime increases in weight of caged White-throated Sparrows under different degrees of stress (Kendeigh *et al.*, 1969).

circles of different acreages, but the density of returning birds in these circles drops at a constant rate. This means that an investigator like von Haartman will perhaps never be able to find all of his young Pied Flycatchers. There is always going to be some small percentage far off in the distance.

In Ricklefs' paper, I found it confusing to follow the terms growth rate, recruitment rate, and rate of increase. It is a common principle in game biology to recognize that the rate of reproductive gain is inversely proportional to population density in spring. This is something that Errington derived from his own observations and it follows the well-known sigmoid curve of population growth. It is true that we can harvest the maximum in a population by looking at the middle part of this curve. However, in species like the Bob-white, fall populations often are directly related to the fall populations of the preceding year. This inversity principle, while common in terrestrial vertebrates, has rather minor significance in the application of hunting regulations.

However, in looking at population phenomena, I think that we need to stress the point that Ricklefs made about the size of the study area, i.e., local versus widespread populations. In an early review paper.* I tried to integrate population data on the basis of local, regional, and continental phenomena. The coefficient of variation is going to drop as one goes from one level of integration to another.

When David Lack published his book on *The Natural Regulation of Animal Numbers,** I realized that he was not defining a stable population. So I tried to look at some published data on stable songbird populations, and I computed the coefficient of variation for these. Some of these populations include a census by A. B. Williams of a 65-acre maple–beech forest near Cleveland. I came up with the conclusion that a stable bird population was one with a co-efficient of variation in density of 50 percent or less.

At the time, there were not many studies of insect populations available, but I had about three, mostly from Andrewartha and Birch as I recall, and the coefficient of variation on these was simply tremendous. The other group of birds that had a different and high coefficient of variation were cyclic species.

It develops then that in looking at population phenomena, we must become involved with this phenomenon of movement. I believe Kluijver pointed this out very well for the Great Tit, where local populations are a function of natality, mortality, and movement. Every time we begin to deal with a local population, we have this phenomenon of what is going on outside that population.

The experimental pheasant population that was described included two tomcats in the original design. I felt that, while Einarsen was trying to be honest in putting some predation on this island, this perhaps was not enough. Actually, the island itself represents a special case in population dynamics in that there is no ingress or egress. This is very different, therefore, than the local segment of a continental population. In dealing with an island, we are more or less getting over to where the fisheries people are, and we can even begin to talk about biomass.

MURTON: No mention has been made of an extremely important method of analyzing animal population data and one that should be of vital importance to game biologists. This is the technique of "key-factor analysis" developed by G. C. Varley and G. R. Gradwell.† The technique is an advance over the

*1954. Oxford University Press, London. 343 p.

†Varley, G. C., and G. R. Gradwell. 1960. Key factors in population studies. J. Anim. Ecol. 29:399–401; Varley, G. C., and G. R. Gradwell. 1968. Population models for winter moth, *In* T. R. E. Southwood [ed.] Insect abundance. Symp. Roy. Entomol. Soc. London 4:132–142.

method of R. F. Morris,* and its validity has recently been tested and confirmed.† The method consists of calculating the various mortalities k_0 ... $k_1 - k_n$ affecting an animal during a single generation, or year in the case of birds. These mortalities are represented as logarithms so that they add to give the total annual mortality K. Data for several years are plotted, whereupon the individual k best approximating to the changes in total K is seen as the key factor. This is the mortality that is most responsible for changes in population size from year to year. Pest control workers or game preservers need to identify this specific mortality factor if they are to be successful in managing a population. The next stage involves plotting the various k's against the populations upon which they act. Those giving straight lines with significant positive or negative slopes enable the density-dependent or inverse-density-dependent factors that are responsible for regulating the population at a particular level to be recognized. Delayed density dependence can also be discovered.

My second point concerns the question of social hierarchies and the suggestion that they have evolved to regulate population size; similar kinds of argument apply in the case of those hormone mechanisms that lead to stress syndromes. The subject can become "a which came first, chicken or egg" type of controversy. In other words, are social hierarchies the consequence of animals having to live together because this arrangement is the most profitable ecological adaptation, perhaps for obtaining food most efficiently, or have they evolved for the prime purpose of population regulation? I regard social hierarchies as consequences and consider it a sterile hypothesis to make them the central theme. In the same way I believe adrenal stress to be a consequence of the fact that for some species it is ecologically desirable to associate in flocks and herds in order to get food; the flocking habit enables population size to increase and not be depressed. Pertinent field data and experimental details are given in full elsewhere.‡

CONLEY: Regarding Doctor Hickey's comment on ingress and egress, this has been shown to be extremely important in microtine populations; it should not be ignored, although it does not become apparent until you restrict immigration and emigration.

I have two questions for Doctor Robel. What time of year were these three dominant males removed in relation to time required to establish the hierarchy? Second, how successful were copulations of the birds that were lower in the hierarchy? Do you have any indication of a reduced fitness in terms of lower hatchability or survival of these clutches? If so, this would seem to indicate a lower-order function of aggressive behavior as related to individual

---

*Morris, R. F. 1959. Single-factor analysis in population dynamics. Ecology 40:580–588.
†Luck, R. K. 1971. An appraisal of two methods of analysing insect life tables. Can. Entomol. 103:1261–1271.
‡Murton, R. K. 1971. The significance of a specific search image in the feeding behaviour of the Wood Pigeon. Behaviour 40:10–42; Murton, R. K., A. J. Isaacson, and N. J. Westwood. 1971. The significance of gregarious feeding behaviour and adrenal stress in a population of Wood-Pigeons, *Columba palumbus.* J. Zool. London 165:53–84.

fitness, rather than a higher-order population-level phenomenon, involving orderly breeding seasons and so forth.

ROBEL: Regarding the first question, we removed the dominant males just before the females began to come on the grounds, after the social hierarchy had been established. I feel the social hierarchy normally is established in January or February. The females begin to visit the grounds in March. We determined which we thought were the socially dominant males, not by just territory size, but by interactions and so on. Therefore the removal of males occurred before the females arrived, but not long enough before for the males to re-establish an entirely new social hierarchy before the females arrived for mating.

I cannot answer the second question. We have no data on chick survival in relation to which male mated with which female. Regarding hatchability, we have data from about 100 or 110 nests on our study area. As far as we can determine, when the clutch hatches, all of the eggs hatch. The nests that are not successful are normally destroyed by predators or something like that before hatching begins. So it is an all or nothing situation on our study area.

There are some data that indicate that a selection for high aggressiveness in males may be balanced in the females, that is, the more submissive female may be mating with the more aggressive males; if that is the case, you would have a stable level of aggression in the population. We have no hard data to measure aggressive tendencies or to detect social hierarchies in the females yet.

FRETWELL: I would like to comment on this issue of social behavior and population regulation. As soon as we admit that populations can be regulated by social behavior, we allow ourselves to introduce ultimate and proximate factors.

In particular, if a population uses social behavior to regulate its size, it can use that social behavior in anticipation of whatever actually limits the population. I think the best possible example of this is in the Great Tit, as reported in Lack's review in 1966.* I disagree with Murton that Lack was unable to show that food was a limiting factor for this population. The density-dependent mortality in the Great Tit did seem to occur in the juveniles entering the adult ranks and that mortality was correlated rather well with the beech mast crop, a major food for the species. However, the mortality all seemed to occur before the beech mast itself became available as a food supply. That is, it seems that perhaps social behavior was regulating the Great Tit population and was creating density-dependent mortality in the young. It was doing this perhaps in August, September, October, prior to the actual existence of the beech mast, which seems to be at least part of the limiting food resource.

If this interpretation is correct, social behavior was regulating the population. Density-dependent mortality was occurring, and it was occurring in the

---

*Population studies of birds. Clarendon Press, Oxford. 341 p.

juvenile birds, but it was occurring in anticipation of and prior to the actual existence of any food shortage. In fact, Lack shows that, although winter food seems to be regulating these Great Tit populations in the period of greatest food shortages (i.e., December through March), there is no density-dependent mortality.

I think it is worth pointing out that Wynne-Edwards* has argued that this sort of anticipatory regulation is in fact very expedient for a population. That is, if density-dependent mortality is delayed until the food supply becomes actually limiting, the result is perhaps an overutilization of the food resources. One can imagine, for example, too many Great Tits eating all of the winter food supply by February and having nothing to eat through March. The mortality then would be nearly total for the population. Thus, one not only finds this anticipatory social regulation of population but, perhaps, also can expect it in most species.

DYER: My colleagues and I have worked with a "goodness of the environment" concept with the Red-winged Blackbird, and Doctor Orians has really developed this concept with some of his theories on polygyny. I would like to show how demographic and environmental quality considerations can be very important in trying to assess how a population increases or decreases in time on a local or on a continental basis.

For the Red-winged Blackbird, there is a recent association with the agricultural ecosystem, but there are still areas where the species is breeding apart from intensive agriculture in what would be termed a natural environment. I recently have been working with detailed censuses in both of these habitats in a large sector of the Midwest to ascertain the population levels in these two types of habitat from year to year, from area to area, and within areas.

We were very fortunate to conduct these censuses during a period when we encountered major fluctuations over the entire population and in one area nearly a crash. Furthermore, we were able to compare fluctuations of natural breeding populations with those in the agricultural ecosystem. While the population was nearly crashing in areas where it was breeding naturally, there were no changes whatsoever over long periods of time in the agricultural ecosystem. Thus the interpretation is that energy flow in the environment becomes very important in a very general sort of way. This is nothing new, of course, but the degree is exceedingly variable and surprising. Furthermore, we can state that much of the energy from an agricultural ecosystem is being funneled into a Red-winged Blackbird population, energy that is not available to natural populations; we can carry this argument one step further and state that agriculture is actually increasing the population of Red-winged Blackbird in some areas, whereas in other areas natural regulation is allowing the population to fluctuate, but usually at lower levels.

---

*Wynne-Edwards, V. C. 1962. Animal dispersion in relation to social behavior. Oliver and Boyd, Edinburgh. 653 p.

ZWICKEL: I just wanted to say a bit about technique in demographic studies. First, we have heard several comments here about correlations. At the same time, we have seen some very nice experimental work in physiology, some in behavior, etc., but very little, except for that which Doctor Robel mentioned, in demography. Even here, I think, his study was mainly behavioral.

There are techniques for manipulating populations, with a shotgun if need be. One can make a population effectively polyandrous or polygynous by sex-selective removals. At this time, we just are not doing much of this in demographic work. We follow populations, describe them, and say we obtained a correlation with food or correlation with this or that. Some people might argue that there are experiments going on involving removals by hunting, etc., but I suggest that if we are really going to advance our knowledge demographically, we must carry out experiments with good controls.

With respect to one of Ricklefs' comments, I would suggest that there is a way we can start trying to find out what the biotic potential of a species is by manipulating its level in spring. It is rather troublesome to shoot a big brown-eyed hen sitting on a nest, but it can be done.

HOMMA: This question is addressed to the people working in the field. In the case of birds, inbreeding depletion is surprisingly high, especially in the wilder species. If we repeat sibling matings for three to four generations, most birds die out. Part of the trouble is low fertility; part is high mortality. In insect so-cieties—for example, bees—there are several mechanisms working to avoid in-breeding. How are these mechanisms connected with pair-bond matings? If birds repeat sibling matings, it will be very dangerous to their survival. If birds return to the same territory, is there any chance of an increased inbreeding? Also, inbreeding depletion depends on the size of the birds; in the smaller birds, it seems to be very high. In the fluctuation in number of males to fe-males, how is the inbreeding problem involved?

ZIMMERMAN: I think it holds as a generalization that, even if male birds are territorial and show the phenomenon of *Ortstreue*, the females do not. So the probability of brother mating with sister is very low.

HICKEY: Concern about inbreeding in game populations has been prevalent in British and Dutch game circles for a long time, but as we have developed marked populations of songbirds—Song Sparrows, House Wrens, Pied Fly-catchers, Great Tits, and others—the phenomenon of inbreeding has been found to be very, very low. It is of no consequence as a population problem in this group of vertebrates.

JENNI: Doctor Robel, territorial social organization and hierarchial systems seem to be mutually exclusive. I wonder if you could describe briefly the so-cial system that you spoke of as having both of these characteristics? Let me restate the question. A territorial species is dispersed. A species that has a so-cial hierarchy must be a "flocking" species with opportunity for contacts be-tween all individuals within the hierarchy. Yet I think that you said your population has both a social hierarchy and a territorial system. I wonder if you would describe the system for us?

ROBEL: No, I do not think I did. The Prairie Chicken has a social hierarchy on the booming ground, and each of the males controls a territory. This is in contrast to the Red Grouse, which is territorial and, I believe, has no obvious social hierarchy, since they are individuals dispersed throughout the environment and each male does not have contact with a large number of other males to establish a good peck order.

I wanted to make a comment about something that Doctor Zimmerman said regarding inbreeding. You would think that there would be a great deal of inbreeding in Prairie Chickens since very few males are involved in breeding, but as Zimmerman mentioned, this does not hold for females. We have data on renesting by females. In fact, one radio-equipped female renested twice, that is, she had three nesting attempts in one year. That female was mated by a different male prior to each nesting attempt. Moreover, she was mated on three different booming grounds during that period. Thus, the problem of inbreeding in the population is probably minimal.

Changes in the
Breeding Bird Fauna
of North Europe

LARS VON HAARTMAN

The avifauna of north Europe has undergone considerable changes within the period of somewhat more than 100 years over which we can look back. According to a recent handbook on the birds of Finland [von Haartman *et al.* (1963–1972) author of the distribution paragraphs mainly P. Linkola], out of 233 breeding species about 34 percent have enlarged their areas and/or become more numerous, 25 percent have receded, 22 percent have fluctuated in abundance, and only 20 percent seem to have remained fairly stable. Considerable change has taken place in the Norwegian (Haftorn, 1971), Swedish (Curry-Lindahl, 1959–1963), and Danish (Løppenth, 1967; Salomonsen, 1967) avifauna as well.

The causes of this remarkable lack of stability are no doubt complex. In the following discussion I will concentrate mainly on the expanding and/or increasing species. The decrease of the others, as far as it is caused by man, presents a rather different set of considerations (e.g., the biocides) that would call for separate treatment.

The following explanations of the expansion of birds in north Europe have been proposed. It will be apparent that they are in part mutually exclusive, in part not:

- Changes may be apparent rather than real. The strong increase in ornithological activity and the refinement of methods of species identification will tend to reveal the occurrence of rare species in areas where they were formerly overlooked.

• What we witness is the last step in the postglacial reoccupation of the northern area.

• A rise in temperature has made north Europe accessible to a number of thermophilic species, whereas thermophobic species have been forced to withdraw. Changes in humidity have also played a role.

• The climate in the steppe region of the Eurasian continent has become more arid, thereby forcing a number of birds of eutrophic lakes to emigrate.

• The rise in temperature has prolonged spring migration in a number of species.

• Changes in nature caused by man and his domestic animals have affected the bird fauna both favorably and unfavorably.

• The distributional changes may not necessarily be caused by external factors, but by internal ones, i.e., modifications or adaptational processes within bird populations.

APPARENT CHANGES IN THE AVIFAUNA

The enormous postwar increase of bird watchers and the consequent dramatic growth of information has raised the question whether the expansion of many birds is apparent rather than real. This may be illustrated by a simple model. Let a checkerboard represent the area wherein a species occurs and a die be thrown for every square to give a digit varying from 1 to 6. The effect of low-intensity investigation may be shown by accepting only the sixes as a token of occurrence. The species will now appear to have a very restricted distribution. Again, the effect of high-intensity investigation may be represented by accepting all figures, from 3 to 6, for instance, as a token of occurrence. The species will now appear to have spread over almost the entire board.

Is this what has happened in the case of the northern bird fauna? Eriksson (1969a) has studied the increase in numbers of observations of Grasshopper Warblers (*Locustella naevia*) in Finland in 1955–1965 and compared it with the increase in ornithological activity as indicated by the number of birds ringed annually, the number of ringers, and the number of participants at certain ornithological meetings. It was found that "the number of Grasshopper Warblers has increased less than expected, which might indicate that in fact

the species is gradually decreasing, contrary to the general belief"
(Eriksson, 1969a, p. 124). Similar results were obtained with respect
to the River Warbler (*L. fluviatilis*) and Blyth's Reed Warbler
(*Acrocephalus dumetorum*) (Eriksson, 1969a,b). Hyytiä (1970) has
challenged these results, maintaining that the number of records is
not a linear function of the number of observers, because the study
areas of different observers may overlap and interest in publishing
records may decrease as the number of records accumulate. Just how
important these factors are it is impossible to judge. After considering the ratio between positive and negative records (observed or
absent) and long-term figures for certain relatively well-studied areas,
Källander (1970) concludes that the increase of the Grasshopper
Warbler in Sweden is real.

Examples like these may convince us how difficult it is to ascertain whether a species is spreading or not. Continuous observations
made in restricted areas by the same observer are more reliable
evidence of changes in the avifauna than are records stemming from
an inconstant number of observers of varying skill. I have carried out
such observations from 1935 to 1971 in an area in southwest Finland (Lemsjöholm and surroundings) and, for several species, will
rely on these data as ultimate proof of increase.

CHANGES IN THE CLIMATE AND THEIR INFLUENCE ON THE AVIFAUNA

A sample of the 1,400-m-deep land ice at Camp Century, Greenland,
has provided us a picture of the climatic changes that have taken
place in the last 100,000 years. By analyzing the period since 1200
A.D. (Dansgaard *et al.*, 1971), it has been possible to predict the
forthcoming development with a certain degree of reliability. The
amelioration of the northern climate, which has prevailed since about
1870 or 1880 [von Rudloff (1967) dates the beginning of the "true
climatic change" very exactly to 1897] and which reached its peak
in the 1930's, seems to have come to an end and will from now on
probably be replaced by a period of deterioration. This means that
we will soon have a unique opportunity to assess the role of the
climate in the recent changes in bird distribution. In the last 100
years, i.e., within the "historical time" of faunistics, three trends
have been observed, all of which are of interest with respect to bird
distribution: an increase in the numbers of bird watchers, an amelio-

ration of the climate, and the alteration of nature by man. The reversal of one of these trends would greatly facilitate conclusions regarding the causes of faunal changes.

The first to draw attention to the possible effect of the amelioration of the climate upon the avifauna was Jägerskiöld (1919). He found that the distribution of the Thrush Nightingale (*Erithacus luscinia*) had varied considerably. In Linnaeus' time, when temperature was favorable, it occurred as far north as central Sweden. When the climate became less favorable, it receded into the southernmost part of the country and, still later, when the temperature rose again, started reoccupying its old area (cf Bergström, 1940; Curry-Lindahl, 1959–1963; Bjärvall, 1965). In other species, a similar connection was established between their northern limit and the temperature, although their distribution cannot be traced as far back in time as that of the popular Thrush Nightingale.

In 1937, Siivonen and Kalela (cf Kalela, 1938) drew attention to some important new facts regarding the climate and the avifauna, and Kaisila (1962) showed that many Finnish Lepidoptera had expanded their areas in a manner similar to that of the birds. In countries outside north Europe, the bird distribution also seems to have changed synchronously with the climate (summary in Udvardy, 1969).

RESIDENT AND EARLY-ARRIVING BIRDS

As pointed out by Siivonen and Kalela (1937), temperature changes have affected the different seasons in different ways. The winter temperature began to increase in Helsinki as early as about 1880, the spring (April–May) temperature rose after about 1885, and the June temperature only after 1930 (Figures 1 and 2). From 1940 on, the winter temperature dropped again, owing not to a general deterioration of the winter climate, but to an increase in the number of very hard winters (Ångström, 1958). From 1941 to 1968, the average winter temperature was about 1 °C lower than from 1901 to 1930.

Among the species that have enlarged their area are a number of resident and early-arriving species. On the whole, their expansion has started earlier than that of late-arriving "summer birds" (Kalela, 1949). The chronology of the expansion of the avifauna given in Table 1 is based on material the reliability of which cannot be guaranteed for all the species; space does not permit a discussion of individual cases.

It is well known that hard winters may reduce the population of

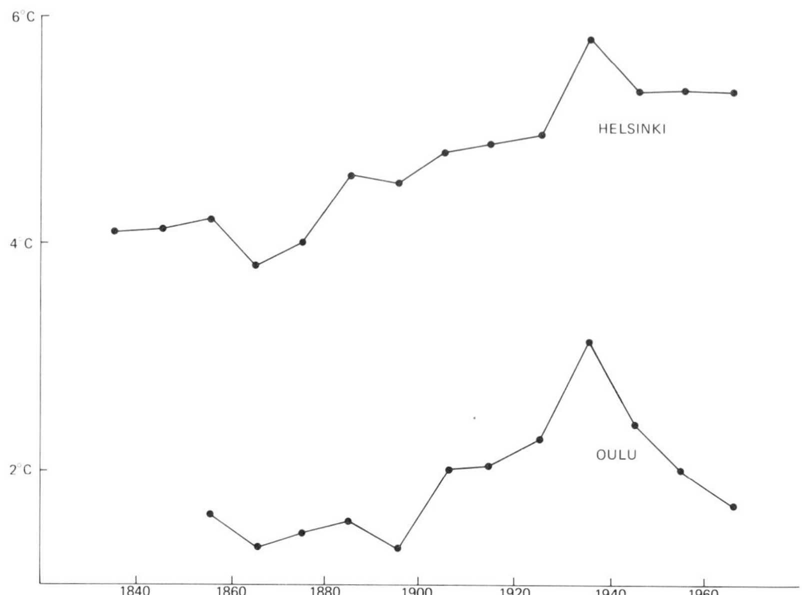

FIGURE 1 Average yearly temperature in south (Helsinki) and middle (Oulu) Finland. The last decade is incomplete, ending 1968. Courtesy of the Meteorological Institute, Helsinki.

nonmigratory birds [e.g., *Regulus regulus* (Palmgren, 1936; Siivonen, 1956), *Parus major* (von Haartman, 1971), but populations of *Turdus merula* are not greatly affected (Linkola in von Haartman *et al.*, 1963–1972)]; the same is true with the effect of cold spells in the spring upon early migrants [e.g., *Vanellus vanellus* (Vepsäläinen, 1968)]. It seems justified to assume that amelioration of the climate has not only an occasional but also a permanent effect on the avifauna.

In a number of species in Table 1 the expansion, however, may have begun too early to be correlated to increasing temperature. This may possibly apply to *Aythya ferina, Fulica atra, Larus ridibundus, Parus caeruleus,* and *Carduelis carduelis. Prunella modularis*, on the other hand, started its expansion well after the peak of the temperature increase. In north Finland, the expansion generally appears to have been much later than the temperature increase, but this may be due partly to the lack of accurate information from this region prior to the end of the nineteenth century.

A number of northern species have been receding simultaneously with the expansion of the southern ones. Merikallio (1951) has

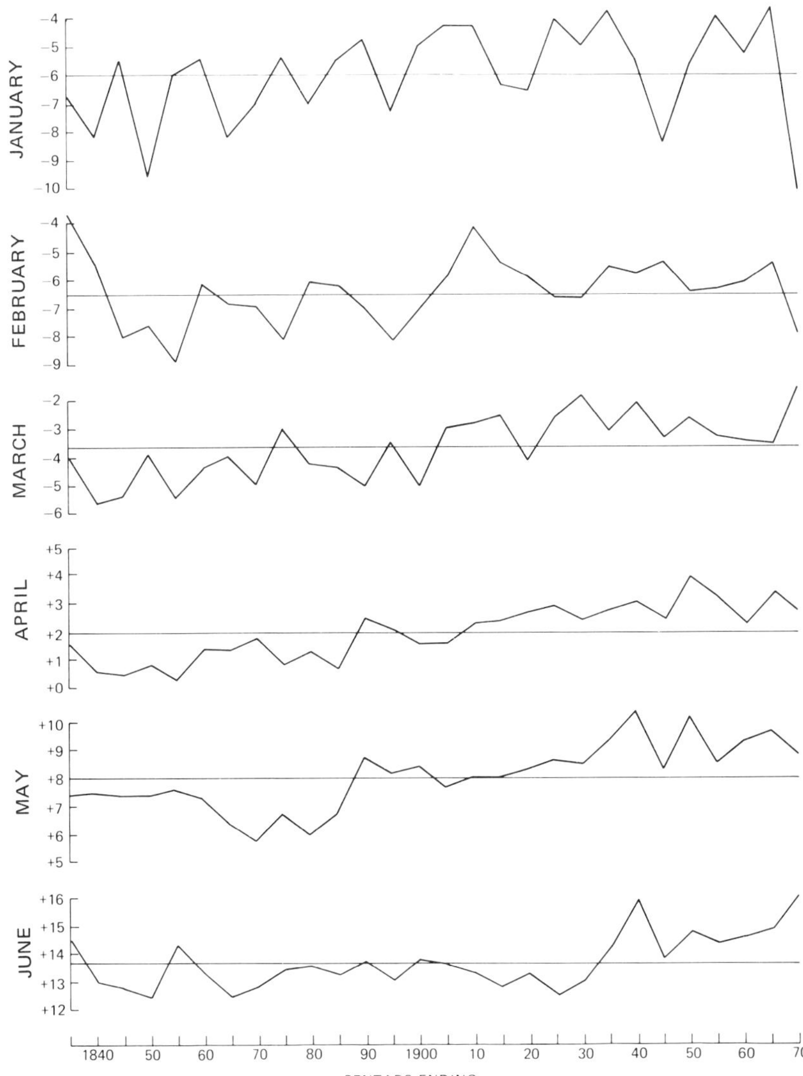

FIGURE 2 Mean temperature of different months in Helsinki. Data from Siivonen and Kalela (1937), modified with data since 1936.

TABLE 1 Expansion or Increase of Avifauna in Finland; Resident or Early-Arriving Species

Species	Increase of Records of Nonbreeding Individuals	First Breeding Established	Increase of Breeding Records
South Finland			
Aythya ferina	1835–53?	1867	1889–1950
Cygnus olor		1934 (intro.)	1934–
Rallus aquaticus	(1915–?) 1947–	1950	1950–
Gallinula chloropus	1880's?	1880	No steady increase
Fulica atra	1846–58,1870–72	1875	(1875) 1890–1930's
Vanellus vanellus			1880–
Larus ridibundus		1811? 1864	1864–1900, 1920–
Streptopelia decaocto	(1953) 1959–	1966	1966–
Strix aluco		1878	1878–
Corvus frugilegus		(1885)	1880–1930
Parus caeruleus		1859	1881–1910, 1930's
Parus cristatus			Åland: 1930's or 1940
Turdus merula	1842–	Before 1859	1910
Prunella modularis			After 1945
Carduelis carduelis	1845–	1860's	1910–55
North Finland			
Corvus cornix			After 1945
Parus major			1930's–
Parus cristatus			(1900) 1930–
Turdus philomelos			1930–
Erithacus rubecula			After 1945
Regulus regulus			1930's–
Carduelis spinus			1920's–
Fringilla coelebs			After 1910
Emberiza citrinella			1930's?–

shown that the Brambling (*Fringilla montifringilla*) and the
Lapponian Tit (*Parus cinctus*) have receded from 1910 to 1949,
whereas their southern counterparts, the Chaffinch (*Fringilla coelebs*)
and Crested Tit (*Parus cristatus*), have enlarged their areas. Whether
the cause is the climate, as Merikallio assumes, or habitat changes
is difficult to judge. In contrast to the Chaffinch, the Brambling
prefers open forests. In the period 1936–1967, which included both
warm and cold springs, there were no Brambling territories at
Lemsjöholm, but a small population settled there in the spring of
1968, after a hurricane in August 1967 had felled a huge number of
trees (see Figure 3).

LATE-ARRIVING MIGRATORY BIRDS

The late-arriving and late-nesting species have generally started to penetrate northward more recently than the non migrants and early-arriving migrants. This is in accord with the fact that the June temperature began to rise as late as about 1930 and supports the view that their expansion was influenced by the climate (Table 2) (Leivo, 1946; Välikangas, 1951).

The expansion of a couple of easterly species [*Phylloscopus trochiloides* (Nowak, 1971) and *Carpodacus erythrinus* (Jozefik, 1960)] has been as much toward the southeast as toward the northwest. Besides, the expansion may be expected to have a northwestward direction in Finland because of the inclination of the isotherms. The expansion of *Emberiza rustica* seems to be directed westward, not northward. *Acrocephalus schoenobaenus* and *Emberiza hortulana* have become more numerous in south Finland, whereas their populations have remained fairly stable in the north. In species like these, it is by no means self-evident that the expansion is caused by amelioration of the climate.

In several of the species listed in Table 2, the chronology of the expansion is less well known than the data may indicate. The first records may have come long after the beginning of the invasion. This is especially true of the night-singers, which have only recently become the subject of systematic study. Eriksson (1969a,b) casts doubt on the expansion of some of these species.

Quite a few of the expanding "summer" birds are Siberian faunal elements, whose glacial refugia were situated far from north Europe. It seems possible that their westward penetration has taken a considerable time. Some of these species may well be examples of a "normal" postglacial reoccupation of the Eurasian arctic region.

BIRDS OF EUTROPHIC LAKES

Lönnberg (1924) assumed that a recent period of aridity in Central Asia has caused desiccation of shallow lakes and so forced their avian inhabitants to emigrate. Kalela (1940) accepted the chief outlines of this hypothesis and presented some documentation concerning the effect of drought in the arid zones of the Soviet Union and as far westward as Hungary. The species in Table 3 may be tentatively listed as birds of eutrophic lakes that have expanded their areas in Finland.

Buchinsky (1963) has surveyed the changes in humidity in the arid steppe region of the Ukrainian Soviet Socialist Republic. Historically, he finds that, "beginning from the fifteenth century, the

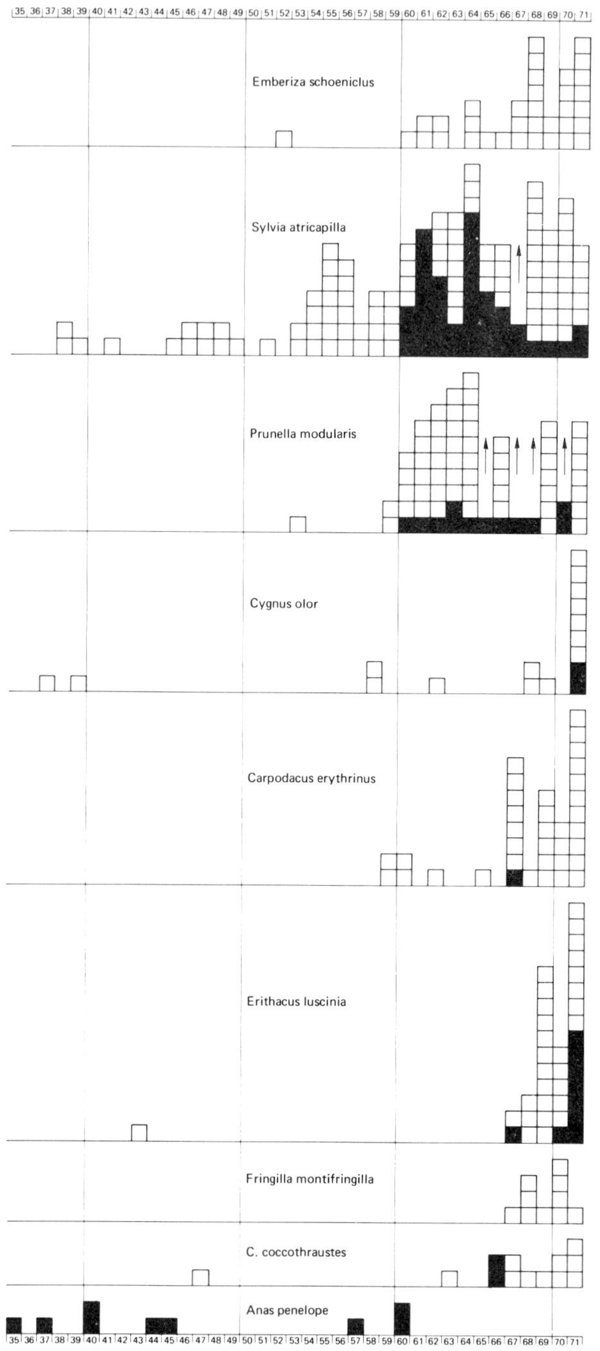

FIGURE 3   Changing abundance of a number of bird species at Lemsjöholm (for *Cygnus olor*, *Anas penelope*, and *Erithacus luscinia* Lemsjöholm and its surroundings) in southwest Finland.

456

TABLE 2   Expansion or Increase of Avifauna in Finland; Late-Arriving Migratory Species

Species	First Appearance	Main Expansion
South Finland		
Locustella naevia		After 1950
Acrocephalus arundinaceus	1930	Still very rare
Acrocephalus scirpaceus	1922	1926–
Acrocephalus palustris	1944	1950–
Acrocephalus dumetorum	1930	1947–
Acrocephalus schoenobaenus		After 1900
Sylvia nisoria		After 1945
Phylloscopus trochiloides	1935	Still rare
Ficedula parva		After 1900?
Serinus serinus		First nest 1967
Carpodacus erythrinus		(1920?) 1945–
Emberiza aureola		1920?–
Emberiza rustica		1910–
Emberiza hortulana		After 1910 or 1920
North Finland		
Sylvia borin		1940's
Emberiza pusilla	1935	Overlooked before?

observed territory experienced increased wetness. Moreover, in describing droughts from the first half of the second millennium, the chronicles refer more often to burning woods, *dried-up rivers, lakes and swamps* than in the second half" (italics mine). From 1841 on, a series of observations from Odessa show that relatively moist cycles alternated with dry cycles. Severe spring droughts occurred in the south Ukraine as follows: 1871–1900, 5; 1901–1930, 4; 1931–1959, 10. There is not much evidence that a catastrophic drought occurred in this region at the time of the recent expansion of waterfowl into north Europe.

In itself, the hypothesis that the expansion represents a forced emigration does not seem very attractive. Vagrant individuals capable of founding new populations have probably always been arriving in north Europe. The main question is why they have been able to settle successfully only from the second half of the nineteenth century on. If conditions in north Europe had not improved, the destruction of the original breeding grounds, though increasing

---

Open squares, singing males or pairs (in *Coccothraustes* and *Cygnus* individuals are indicated separately); filled squares, nests; arrows, the species was not censused, but its number was about as high as indicated.

TABLE 3 Expansion or Increase of Avifauna in Finland; Birds of Eutrophic Lakes

Species	Increase of Records of Nonbreeding Individuals	First Breeding Established	Increase of Breeding Records
Podiceps cristatus			1880–
Botaurus stellaris		1848	End 19th cent.–1930; 1950–
Cygnus olor		1934 (intro.)	1934–
Anas querquedula			(1850–80) 1940's?
Aythya ferina	1835–53?	1867	1889–1950
Circus aeruginosus		1922	1922–
Rallus aquaticus	(1915–?) 1947–	1950	1950–
Gallinula chloropus		1880	No steady increase
Porzana porzana			1860–1905; later decrease
Fulica atra	1846–48, 1870–72	1875	(1875) 1890–1930's
Larus ridibundus		1811? 1864	1864–1900, 1920–
Larus minutus	1860's	1879	1879–90
Chlidonias niger		1960	Very rare
Acrocephalus scirpaceus		1922	1926–
Acrocephalus arundinaceus	1930–40	1942	Very rare
Acrocephalus palustris	1944–47	(1950) 1960	After 1950?
Acrocephalus dumetorum	1930	1947	1947–
Acrocephalus schoenobaenus			After 1900
Locustella naevia			After 1950

the number of vagrants, would have led only to abortive invasions.

MARITIMIZATION OF THE SUMMERS IN NORTHWESTERN EUROPE

Until recently, a period of rainy summers has prevailed in northwestern Europe. Durango (1946, 1950) assumes that the decrease of the Roller (*Coracias garrulus*) and of the Red-backed Shrike (*Lanius collurio*) has been caused by the increasing humidity, which has made their main food, large flying insects, less easily available. The catastrophic decrease of the White Stork (*Ciconia ciconia*) in western Europe has also been attributed to the maritimization of the summers, but this is only one of many tentative explanations of the enigmatic decline of this intensively studied species.

PROLONGED SPRING MIGRATION

The theory that birds may enlarge their breeding range by prolonging their spring migration was proposed by Wallengren (1854–1856) and Palmén (1874). In its modern form, the theory claims that there is a trend among migratory birds to continue in their direction of migration rather than to deviate to the left or right and that the prolongation of migration, as a rule, is released by external stimuli, among which can be cited (Otterlind, 1954) overpopulation, stimulation by high temperature, and stimulation by other individuals in social species.

It is possible, although not fully established, that overpopulation in spring may cause a directed emigration of the surplus individuals rather than a randomly orientated vagrancy (in which case it would be meaningless to speak about prolonged migration). As to the influence of temperature, laboratory tests have confirmed that high temperature releases migratory restlessness in spring. It is possible that prolonged spring migration may speed up the expansion of birds into regions north of their breeding area. In a long-term study at Valassaaret, Hildén (1966), however, failed to find any clear correlation between populations of southerly species and the temperature in spring. Prolonged spring migration, at any rate, cannot explain the success of expansion. Generally, for expansion to succeed, the conditions in the area of immigration must have changed with respect to climate or habitats, or both. Neither do we have any reason to believe that prolonged spring migration is a necessary condition of expansion, as nonmigratory species seem to be as expansive as migratory species.

The role of social stimulation in prolonged migration is obviously of minor importance, though it may have been implicated in, for example, the rapid spread of the Mute Swan, *Cygnus olor*, in southwest Finland. The first Mute Swans were released from a bird park on the Åland islands, but it seems likely that further colonization has included individuals from Sweden, accompanying the Finnish Mute Swans on their spring migration.

CONCLUDING REMARKS

The amelioration of the climate has obviously influenced the recent northward expansion of southerly species. To examine whether the climate has been the main cause of the expansion, we must compare the magnitude of both phenomena.

According to Ångström (1939) the mean yearly temperature in central Finland rose 0.5 °C between 1859–1900 and 1901–1930. This means that the isotherms moved northward about 2/3 deg. of latitude (cf map in Kolkki, 1966, p. 32).

The possibility remains that the rise in temperature has been so much greater in certain key seasons as to explain the large change in bird distribution, If we compare the isotherms for different months during 1881–1890 (Atlas öfver Finland, 1899) and 1931–1960 (Kolkki, 1966), we notice a change, but as a rule the isotherms have moved less than 2 deg. of latitude northward (Figure 4). I am fully aware of the difficulties in drawing the isotherms precisely, especially for an earlier period, but a very large difference between then and now would have shown up more distinctively.

Figures 5–9 show the expansion of some comparatively well-studied birds. In all of them, except *Carpodacus erythrinus* and *Erithacus luscinia*, whose spread has been mainly toward the west, the northward expansion has been greater than the northward movement of the isotherms. We, therefore, will have to search for other explanations.

HABITAT CHANGES AND THEIR INFLUENCE ON THE AVIFAUNA

GENERAL REMARKS

Changes in vegetation caused by amelioration of the climate have been too small to explain the expansion of bird areas. For instance,

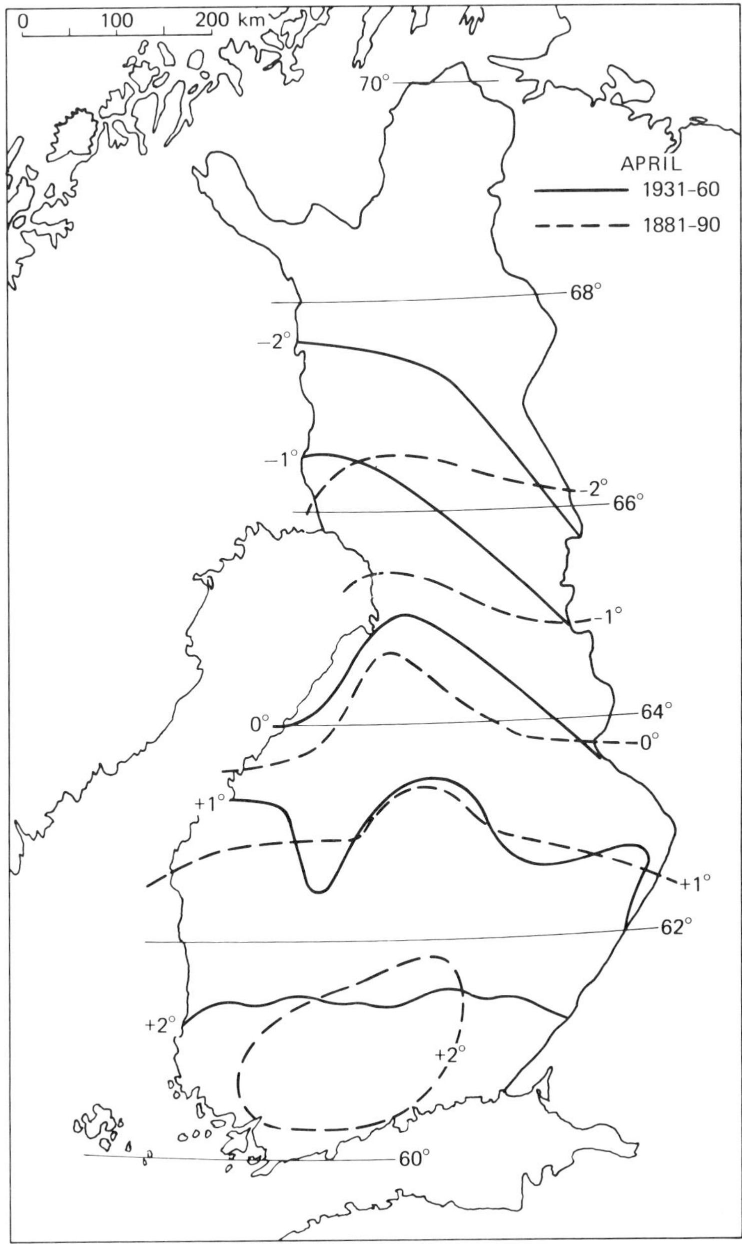

FIGURE 4   April isotherms in 1881–1890 (from *Atlas ofver Finland*, 1899) and 1931–1960 (from Kolkki, 1966).

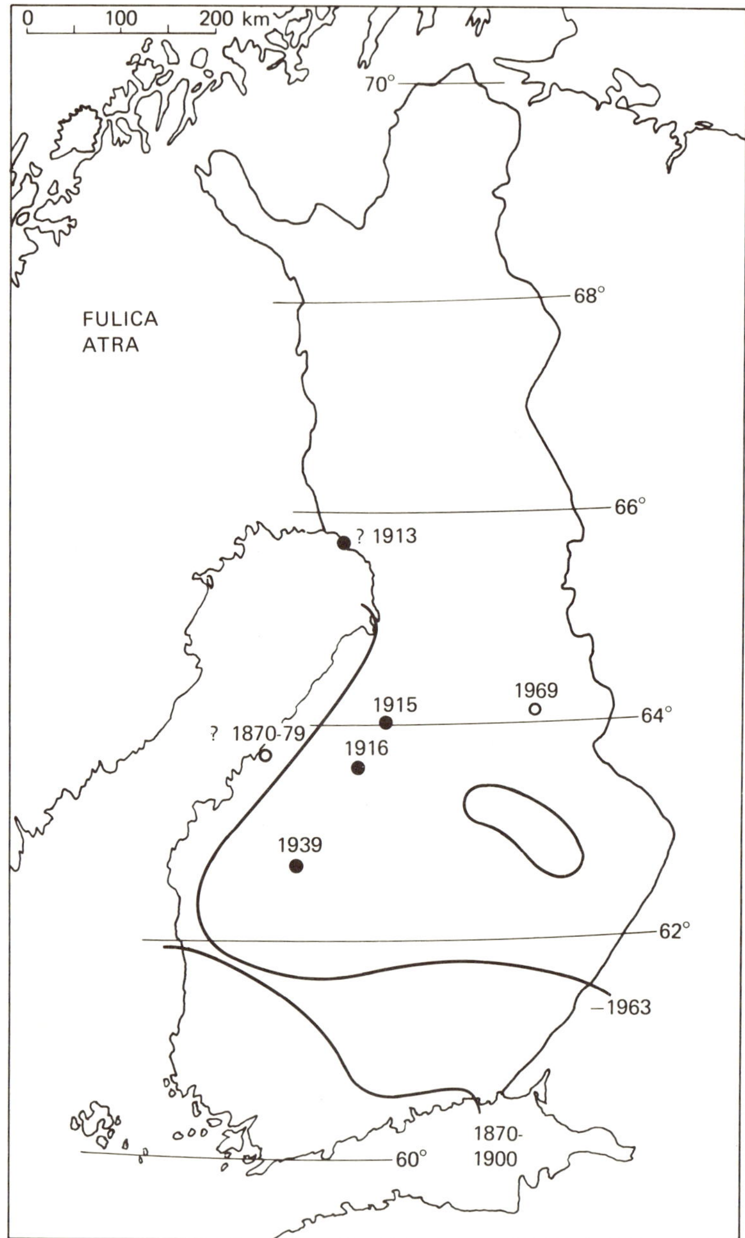

FIGURE 5   Extension of the breeding area of *Fulica atra* in Finland. Data from von Haartman *et al*. (1963–1972) and nest cards of the Finnish Society of Sciences.

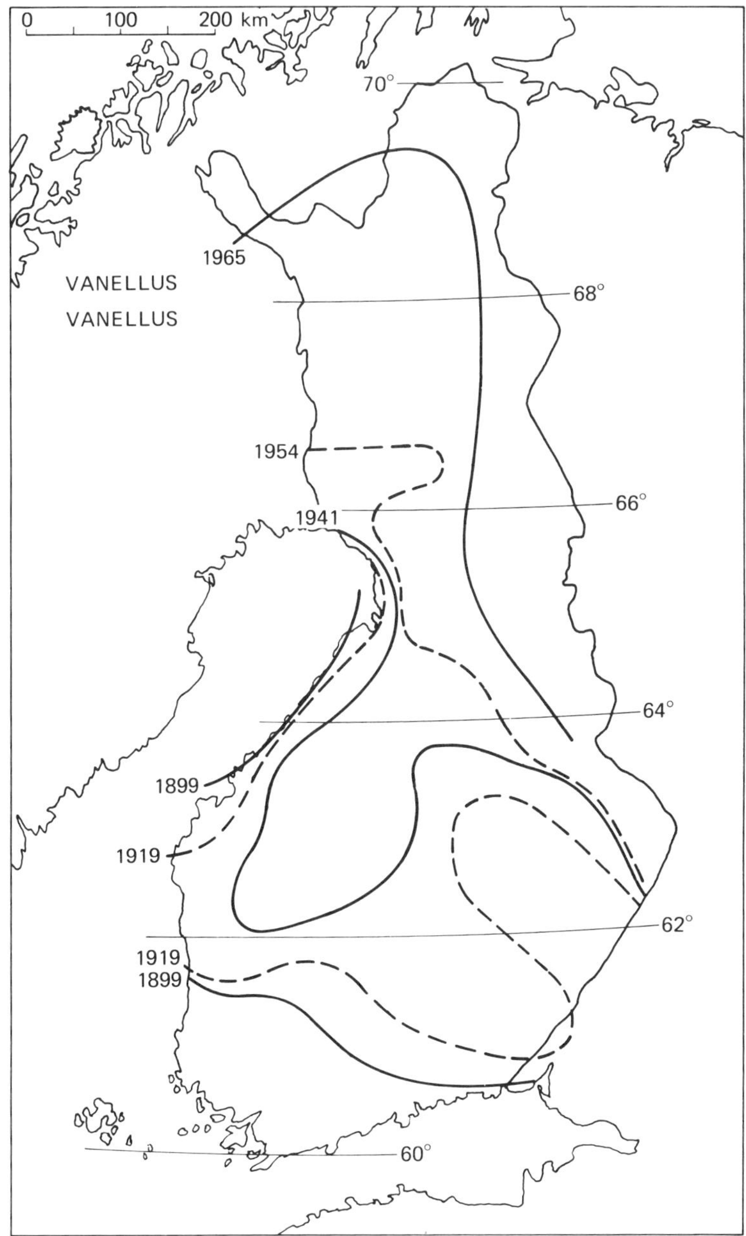

FIGURE 6   Extension of the breeding area of *Vanellus vanellus* in Finland. According to Kalela (1955) (modified) and data in Komonen (1963) and von Haartman *et al*. (1963–1972).

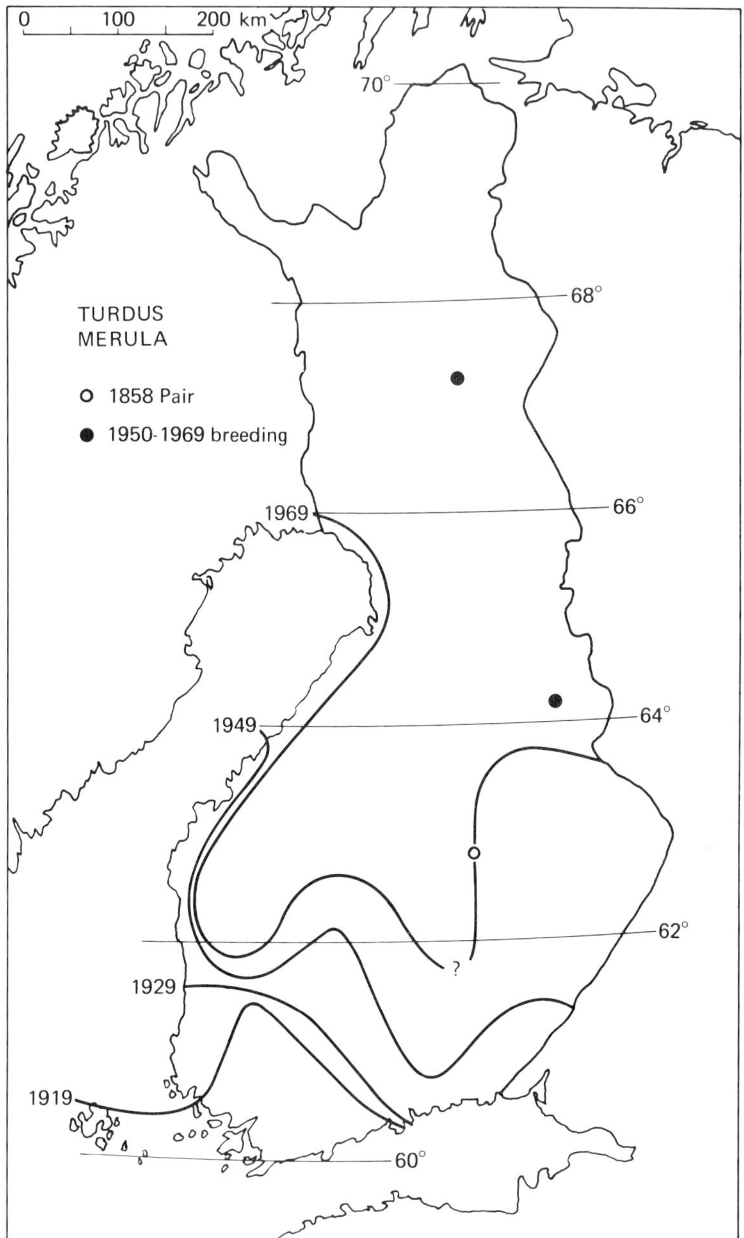

FIGURE 7 Extension of the breeding range of *Turdus merula* in Finland. Data from von Haartman *et al*. (1963–1972) and the nest cards of the Finnish Society of Sciences.

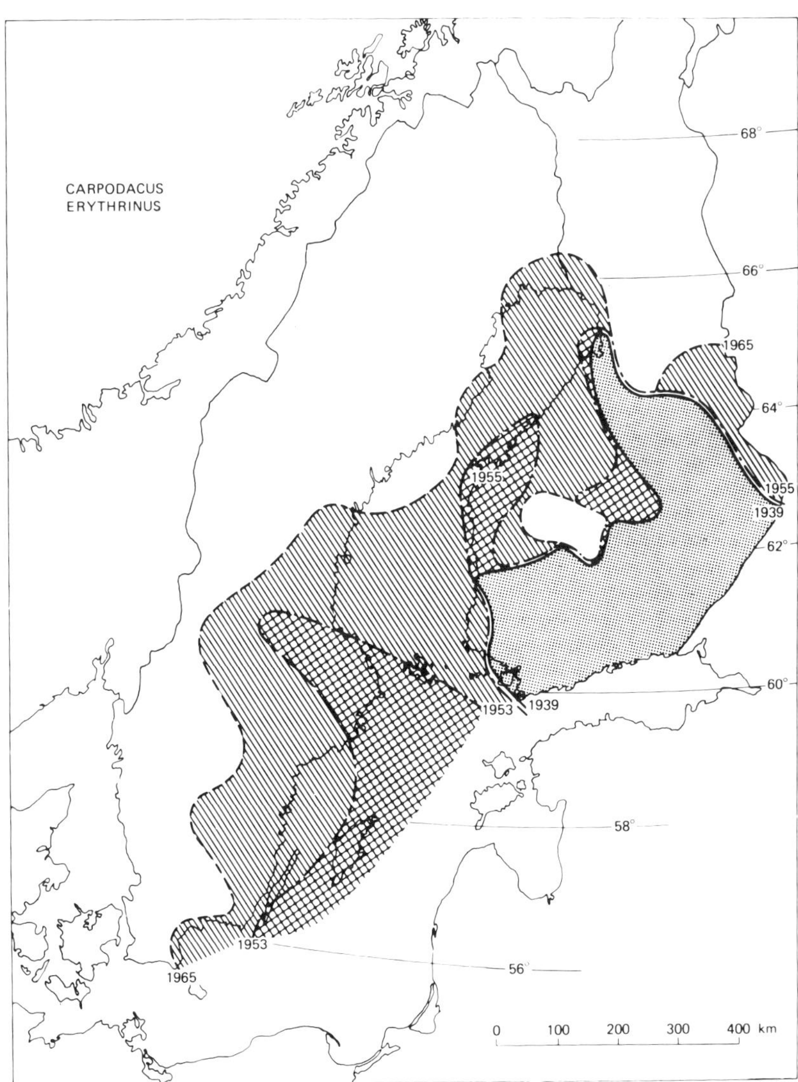

FIGURE 8   Area of regular occurrence of *Carpodacus erythrinus* in Finland and Sweden. Finnish data from Reinikainen (1939), Nordström (1956), and T. Stjernberg (unpublished data). Swedish data from Risberg (1970).

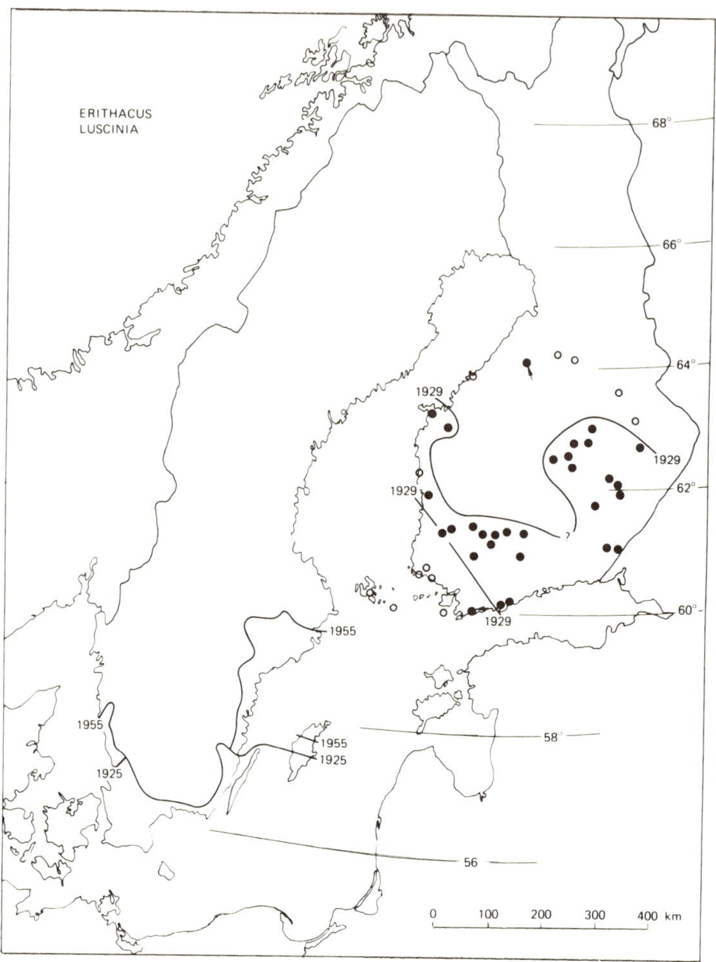

FIGURE 9   Area of *Erithacus luscinia* in Finland and Sweden. Finnish data from Hortling (1929–1931), von Haartman *et al.* (1963–1972), and the nest cards of the Finnish Society of Sciences. Swedish data from Curry-Lindahl (1959–1963). Open circles indicate observations after 1930 in Finland.

the northward movement of the northern limit of the pine (Sirén, 1955) is only a small fraction of the changes in the distribution limits of many birds.

In contrast, the influence of man on nature has been profound. This influence has only recently begun to be fully appreciated, which is not surprising. Whereas temperature and humidity have been recorded with the aid of instruments for the last 300 years, man-made changes are difficult to observe. They usually come too gradually to be striking and are too diversified to be surveyed by any single person. Man even tries to deny or hide his own destructive influence, as in the case of the technician polluting nature or the poacher killing the last eagle.

An instructive example of the complex causal nexus between nature and human activity is offered by the decrease of the Corncrake (*Crex crex*) (Norris, 1945; von Haartman, 1958). The Corncrake was probably not a member of the Finnish avifauna before the agricultural period. It was highly favored by the development of agriculture, but has been disappearing over large areas of Europe since mowing-machines came into use (the first relatively efficient type was developed in England in the 1850's) and because hay harvest comes earlier in the modern farming schedule. Its decline occurred earliest in the areas with the most efficient agriculture, and latest in the most old-fashioned areas. This course of events is hardly compatible with Kalela's (1938) suggestion that the decrease of the Corncrake is due to some change in climate.

The first, or one of the first, to draw attention to the influence of man on birds was Kolthoff (1907). His attitude was rather nostalgic (see especially Kolthoff, 1896), and he stressed the harmful effect of human activity. Later students of the problem have given a less biased picture of man's influence, which has also had many favorable aspects.

FOREST SUCCESSION

At least in southern Finland, nature has been remodeled by man since time immemorial. His influence has varied with the size of the human population and the methods of land use.

In 1858, the German forestry expert von Berg traveled around in Finland on the behest of the government (Kalliola, 1950, p. 374). His description of the state of the forests is depressing. He pointed to three main causes of their poor condition: grazing by cattle, burn-beating, and forest fires. Burn-beating and fires are interconnected since the fire easily gets out of control when burn-beating is practiced.

Burn-beating was carried out about every thirtieth year; rye was cultivated for a few years, until the crop became poor, and then the area was abandoned for another 30 years. The first stage in the forest succession on moist ground is deciduous forest with *Alnus incana*, birch (*Betula* sp.), and aspen (*Populus tremula*). Of the conifers, pine, but not spruce, occur at this early stage. Birch soon becomes dominant, but when birch forest canopy becomes closed, it loses the ability to rejuvenate, as young birches do not grow in shade. At this point, the forest is invaded by spruce and after a century has turned into a mixed forest, wherein the birches will soon start to die off and the spruce will take over. After a few hundred years, the only evidence of the earlier stages of the succession is a few moribund birches. The pines remain for a much longer time, but after about half a millennium, the last ones die, and the climax formation—the spruce taiga—is reached. On dry ground, the succession is different; the climax forest is formed by pine.

In western Finland, burn-beating was abandoned long ago, but in eastern Finland it continued into this century (Heikinheimo, 1915). Consequently, the forests are at different seral stages in the west and the east of the country. In the west, the spruce dominates; in the east, the birch forms an important element of the forest and the spruce plays a minor role. But also in the east the spruce is gaining ground, a process that is speeded up by forestry practice; until recently, the birch was used for little else than fuel, and among Swedish foresters this beautiful tree went under the nickname of the white harlot.

Birch forest is richer in bird species and in individuals than spruce forest, the main reason being the richer invertebrate fauna. The forest succession, therefore, tends to impoverish the avifauna. Soveri (1940) calculated how the numbers of birds would change in the parish of Lammi, south Finland, when the spruce had reached its final abundance (Table 4).

It now seems that Soveri's prediction was too pessimistic, as it did not take into account the favorable influence of human culture. Species like the Great Tit, *Parus major*, and the Pied Flycatcher, *Ficedula hypoleuca*, are restricted to birch forests partly because of the superior nesting facilities that they offer (von Haartman, 1971). It has become the fashion to put up nest-boxes, which probably more than compensates for the loss of terrain suffered by the Great Tit and the Pied Flycatcher as a result of the forest succession. In addition, winter feeding is of the utmost importance for the Great Tit. After the Soviet troops had left the Porkala enclave, and before the Finns returned, there were no Great Tits nesting in the area, except close to its borders (Bergman, 1957). Nowadays, Great Tits are fed at many

TABLE 4   Calculated Change in Numbers of Nesting Pairs in Spruce Forests Versus Birch Forests in Lammi, South Finland[a]

Favoring Spruce		Favoring Birch	
Carduelis spinus	1,900–2,800	*Fringilla coelebs*	23,000–20,000
Pyrrhula pyrrhula	1,000–1,800	*Anthus trivialis*	1,100– 200
Regulus regulus	3,500–6,000	*Parus major*	800– 200
Turdus philomelos	2,300–3,000	*Parus montanus*	5,500– 3,000
Erithacus rubecula	3,500–4,500	*Muscicapa striata*	4,500– 1,000
Parus cristatus	900–1,700	*Ficedula hypoleuca*	3,500– 300
		Phylloscopus trochilus	7,000– 1,300
		Sylvia borin	2,800– 600
		Turdus iliacus	1,400– 700
		Phoenicurus phoenicurus	500– 200

[a]Data from Soveri (1940).

homes, and, with the general rise in living standards, the amount of waste that can be utilized by the tit has probably increased. The Great Tit is one of the species that have spread northward in Lapland. In like manner, feeding stations may well have favored the northward expansion of the Cardinal (*Richmondena cardinalis*) in eastern North America.

All the increasing species of the spruce woods mentioned in Soveri's list (Table 4), except the Bullfinch (*Pyrrhula pyrrhula*), have expanded their areas in north Finland. Linkola (in von Haartman *et al.*, 1963–1972) has correlated the expansion of the Crested Tit with an increase of its population in its old areas. The same may, perhaps, apply to the other birds of the spruce forest.

LAND USE AND THE FORESTS

Forests are now cut at a relatively young age, at the point when the growth of the trees slows down so much that the capital represented by the forest ceases to give a sufficient return. Recently, Finnish forests have been somewhat overcut, among other reasons, because of high prices on the world market.

But the influence of modern forestry has not been merely destructive from the standpoint of nature conservancy. Replanting has become common. Also, there is a recent trend to abandon the cultivated fields of small and unprofitable farms and plant them to spruce, or leave them to natural succession, which starts with a stage of deciduous bushes in which willow (*Salix*) is a dominant species. In Sweden, this trend has become very pronounced; in Finland, it is just beginning.

The deleterious effect of the loss of large trees (which are needed, e.g., by the Black Woodpecker, *Dryocopus martius*, and the birds utilizing its old nest-holes) and the removal of dying trunks have been amply demonstrated by nature conservationists (e.g., Linkola in von Haartman *et al.*, 1963–1972). The difference between managed and unmanaged conifer forests has been studied by Haapanen (1965–1966).

There is, however, a point that seems largely to have escaped the students of the Finnish avifauna and that is the influence of livestock.

In 1900, the number of large domestic herbivores (Figure 10) about equaled the number of people. As horses and cattle are large animals, the total food demands of the herbivores must have exceeded that of man. Of course, man utilizes natural resources for other, partly

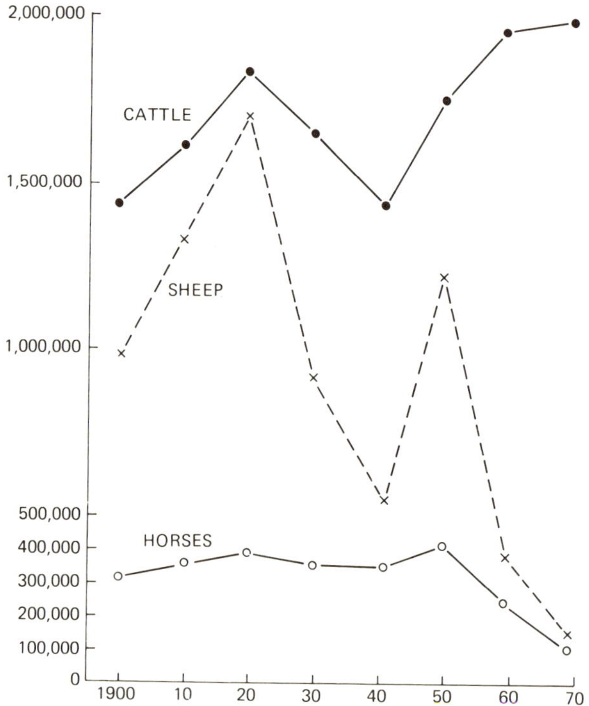

FIGURE 10 The number of domestic herbivores in Finland in the twentieth century.

unnecessary purposes, but at any rate the influence of the large herbivores upon nature has been important.

The number of horses and sheep has become negligible since 1950, but the number of cows has risen slightly. This does not mean that the influence of the cows on nature has remained unchanged. On the contrary, whereas until quite recently cattle were grazed in forests and on shore meadows in summer, this practice has now been abandoned, because of the damage the cattle do to the forest and the fact that high milk production is inconsistent with poor grazing. Today, the cattle feed on cultivated fields.

The effect of cattle on the forest is profound. Not only is the rejuvenation of the deciduous trees impossible but the young spruce also are damaged by grazing and, especially, by trampling; the spruce is very sensitive to damage to its bark (Heikinheimo, 1938; Lampimäki, 1939).

Romell (1964) estimates the loss in wood production from pasturing at roughly 50 percent. A heavily grazed birch forest looks like a park where a lawn mower has been in frequent use and where there is generally little understory in which the birds may seek cover and nest sites (locally, though, the juniper may be abundant).

In 1938, nearly half of the privately owned forests and uncultivated areas in south and central Finland were still used for pasturing (Jäntti, 1945) (Figure 11); the ground not used for this purpose was presumably largely barren. The decline in the use of forests as pastures cannot be traced in detail. In 1920, it had already decreased strongly (especially on larger farms) in south and southwest Finland, but in other parts of the country it was still common.

Ahlén (1965, e.g., p. 314–315) has shown that the wild herbivores of Sweden (Red Deer, *Cervus elaphus*, Roe Deer, *Capreolus capreolus*, and elk, *Alces alces*) have profited from the cessation of forest pasturing: "Until the last century the nonagricultural land in Scandinavia was producing a large biomass of domestic stock and very little forest and game. Today the same land produces a considerable biomass of wild herbivores and large amounts of timber." In Finland, the change came even later and the wild herbivore profiting thereby is the elk. Kalela (1948) has attributed the expansion of the Roe Deer to the amelioration of the winter climate; according to Ahlén, this factor has been of minor importance.

Not only the large herbivores, but also the birds seem to have profited from the change in the forests. At Lemsjöholm, soon after cattle grazing was abandoned after 1956, the Blackcap (*Sylvia atricapilla*)

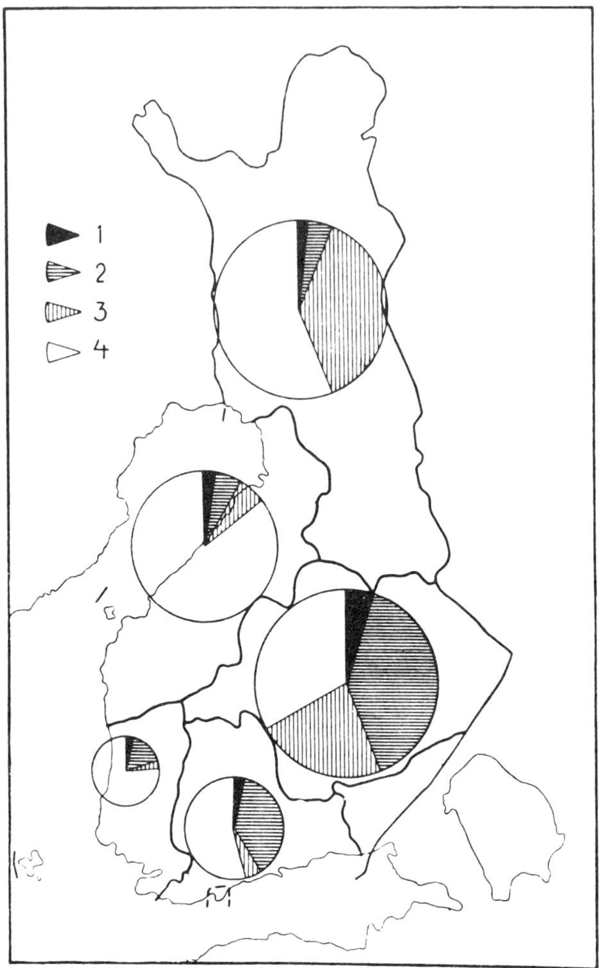

FIGURE 11   Pasturing of forests on privately owned ground in
Finland in 1938 (different shades) versus nonpastured forests (clean
sectors). From Jäntti (1945); redrawn by Valle (1951).

and the Dunnock (*Prunella modularis*) started to increase and within
a short time became noticeably frequent (Figure 3). The very recent
increase of the Bullfinch may perhaps also be attributed to this
change.

   The Dunnock occurs mainly where young spruce grow, either as
an understory in aging birchwoods or as spruce plantations. The
Dunnock's increase has taken place since World War II. The species
has long been distributed all over Finland, but has been exceedingly

rare. Its recent increase in Finland may be characterized as intercalation rather than areal expansion, but in Sweden it has conquered large areas in the eastern part of the country. The causes of the spread have so far been obscure. It is true that the increase in numbers of the Dunnock has been several times larger than the increase of its habitats, but (as was long ago pointed out by the botanist Alvar Palmgren) this is by no means a rare phenomenon in the living world.

An increase of the Blackcap has so far not been noticed elsewhere in Finland, but I assume that it has been overlooked. In Norway and Sweden some increase has been noticed (Haftorn, 1971, Curry-Lindahl, 1959–1963). The censuses of Palmgren (1930) have shown that the Blackcap is absent from heavily pastured woods, whereas it occurs in other forests of grass–herb type, and especially in birch–spruce forest. The Blackcap frequently nests (von Haartman, 1969) in large ferns (*Athyrium filix femina*), which, although not grazed in pastured woods, are heavily damaged by trampling.

CHANGES IN THE TROPHIC CONDITIONS OF THE LAKES

The expansion of the avifauna of the lakes has been one of the most striking features of the recent augmentation of the avifauna in north Europe (see p. 455). Generally, the expansion has come in two waves; an earlier wave, perhaps already weakly discernible around the mid-nineteenth century, and mainly consisting of swimming birds (*Aythya ferina, Fulica atra, Podiceps cristatus*) and a more recent one, dating from the 1920's, consisting of songbirds of the reeds and the bushy vegetation of the shores (the genus *Acrocephalus* being most typical). All these species are inhabitants of the eutrophic type of lakes and their shores. The avifauna of the oligotrophic lakes to some extent have suffered a simultaneous reduction.

Eutrophic lakes are almost exclusively found in the plains. As a rule, they are at least locally surrounded by cultivated fields and human settlement. The nutrients washed into the water from the surroundings are most important for the trophic status of the lake. Two factors connected with human settlement have caused a strong increase in the amount of nutrients received by the lakes—the discharge to their waters of effluents and domestic sewage and the use of commercial fertilizer.

It is not possible to follow these processes in any detail, but a few milestones can be given. The industrialization of Finland began in the 1860's. In 1880, the number of factory workers was 137,875; in 1940, it was 776,668 (Kovero, 1951). In England, where the

industrial revolution first started, Darwin's contemporary Carlyle
(see Selander, 1955, p. 186) was protesting furiously against the
pollution of waters by industrial waste. The history of water
closets has still to be written, but some statistics from Sweden may
suffice to show that they represent a recent development in hydro-
biology: In 1955, they amounted to 1,200,000 and this figure was
expected to at least double before 1965 (Selander, 1955, p. 185).

Commercial fertilizer appeared in Sweden in the middle of the
nineteenth century. By about 1890, it had reached a certain importance, being about a tenth of what it was in 1950. Between 1938
and 1956–1957, the use of nitrogenous fertilizers roughly trebled
and that of phosphates doubled (Stålfelt, 1969); in Finland both
trebled (Salonem, 1968). The recent addition of phosphates to
washing compounds has further increased the amount received by
the lakes.

These nutrients strongly influence the trophic conditions of the
lakes. Under the influence of commercial fertilizers, dystrophic lakes
may develop eutrophic features ("mixotrophic" lakes, Järnefelt,
1936) and eutrophic lakes become much more pronouncedly eutrophic. Selander (1955, p. 171) is probably right in stating that extremely eutrophic lakes are a very recent element in the landscape,
being the result of human activity. The lowering of the water level
of lakes has occasionally resulted in the development of first-class
"bird lakes," since shallow lakes generally become more eutrophic
than deep ones. The regulation of the water level of lakes became
common in Sweden about the middle of the nineteenth century.
Äyräpäänjärvi on the Carelian Isthmus, no longer in Finnish territory,
but once the most famous Finnish bird lake, had its level lowered in
the 1850's (Merikallio, 1929). Lakes of this type have probably been
the centers of expansion of many of the new bird species of north
Europe.

Thus the hypothesis that climatic changes provide the main explanation of the changes in the bird fauna of lakes has been modified
more and more. In 1955, Selander wrote (p. 173) that changing climate and increasing eutrophy had together caused the expansion of
the birds of the lakes; in 1966, Ahlén (p. 91) considered the change
in the nature of the lakes more important.

LAND USE AND SHORE VEGETATION

The recent alterations in the shore vegetation caused by changing land
use have been perhaps even more profound.

In earlier times livestock was allowed to graze on the shores, thereby
changing their vegetation even more drastically than that of the forests.
A grazed shore meadow by a eutrophic lake is cropped quite short. In
places where the ground is not too soft, cattle graze the aquatic vegetation to a depth of about one meter, which results in the development of a zone of clear water between the shoreline and the inner
border of the reed (*Phragmites*) vegetation. As soon as the shore is
protected from grazing, the shore meadows are invaded by tall herbs
(among them *Spirea ulmaria, Lythrum, Lysimachia, Angelica*).
Phragmites moves in and bushes (especially *Salix*) appear. Luxuriant
groves, among the most luxuriant to be found in Finland, may develop
and locally reach the waterline. [For a further description see Ahlén
(1966) and Larsson (1969).]

These effects of the cessation of grazing are enhanced by the fact
that the old-fashioned practice of cutting reeds for fodder has also
been discontinued.

Except for a single bay, which is still used as a pasture, the shores
at Lemsjöholm have been protected from grazing since 1957. Consequently, the Reed Bunting (*Emberiza schoeniclus*), which before 1957
occurred only occasionally (although it nested in the vicinity in
some bays with dense clumps of reeds), has steadily increased in numbers. The same is true of the Reed Warbler (*Acrocephalus scirpaceus*),
although its increase is not as marked as that of the Reed Bunting.
The single bay at Lemsjöholm where these species have not appeared
is that still grazed by cattle. The vegetation succession on the shores
of Lemsjöholm is still in an early stage, and I expect that, for example,
the Sedge Warbler (*Acrocephalus schoenobaenus*), which is now a
rarity in the area, will become a permanent member of this fauna.
A species that has probably suffered from the change of the shore
meadows is the Wigeon (*Anas penelope*), which has not been found
nesting since 1960.

In some other parts of south Finland a moderate increase of the
Reed Bunting has been noticed (von Haartman *et al.*, 1963–1972).
In the Norwegian lowland area (Haftorn, 1971) and in south and
central Sweden (Larsson, 1969), the increase and spread of this species have been very noticeable. The increase in Sweden has taken
place much earlier than in Finland, as might be expected from the
earlier cessation of grazing.

In an important study of grazed and ungrazed lake shores in south
Sweden, Larsson (1969) has demonstrated the effect of grazing on
the avifauna (Table 5).

The members of the species group at the top of Table 5 are inhabitants of wet ground that are favored by protection from grazing. To these species Larsson adds the Marsh Harrier (*Circus aeruginosus*). The Grasshopper Warbler (*Locustella naevia*) is not only a bird of shores but also has profited from developing bushy vegetation on waste areas such as roadsides (Ahlén, 1966).

The Thrush Nightingale (*Erithacus luscinia*) is a typical inhabitant of shore groves. It has clearly (like a number of other species, not listed above) profited from the cessation of pasturing.

Finally, the lowermost group in Table 5 has been adversely affected by the increase in the vegetation of the shores. To this group also belong, among others, the Redshank (*Tringa totanus*), the Yellow Wagtail (*Motacilla flava*), the Lapwing *(Vanellus vanellus)*, and the Sky Lark (*Alauda arvensis*); these species, however, also frequent other habitats.

The species favored by the cessation of grazing represent the majority of the expanding birds of eutrophic lakes (see p. 455).

Among the species suffering from the recent land use, Ahlén also lists the White Stork. Its best foraging grounds, along rivers and other shores, are protected at present from grazing by wire fences, which are often electrified, these being most easy to erect. This allows a dense vegetation to develop, which makes the area inaccessible to the White Stork.

CONCLUDING REMARKS

The effect of human settlement on the avifauna is not restricted to cultivated fields. All habitats have directly or indirectly been remodeled by man. The above sketch is intended to illustrate some aspects of his influence on the birds of forests, lakes, and shores.

Many points remain uncertain, however. It is true that the amelioration of the climate cannot explain the spectacular northward penetration of, for example, the Lapwing, but it is not easy to see how recent human influences can explain it either.

Continuous and detailed studies will have to be carried out in the field before we can evaluate the role of the different factors in the augmentation of the bird fauna.

INTRINSIC CHANGES IN BIRD POPULATIONS

In no other species has the northward extension of the range been of the same magnitude as in the Collared Turtle Dove, *Streptopelia*

TABLE 5 Census Averages from Two Grazed and Two Ungrazed Lake Areas[a]

| | Lake Vombsjön | | Lake Egeside | |
| | Grazed | Ungrazed | Grazed | Ungrazed |
Species	2.3 ha	3.8 ha	38.9 ha	17.5 ha
Podiceps cristatus	—	5.5	—	—
Fulica atra	—	10	+	—
Rallus aquaticus	—	+	—	+
Gallinula chloropus	—	+	—	—
Acrocephalus scirpaceus	—	6	—	4
Acrocephalus schoenobaenus	—	5.5	—	21
Locustella naevia	—	—	—	2
Emberiza schoeniclus	1	7	—	12
Erithacus luscinia	—	5.5	—	4
Capella gallinago	2	—	2	3
Calidris alpina schinzii	1	—	4	—

[a]Data from Larsson (1969).

decaocto (Nowak, 1971) and the Serin, *Serinus serinus* (Mayr, 1926).

The Collared Turtle Dove is a typical species of suburbs, and the recent expansion of the urban areas must have facilitated its northward penetration, although it can hardly be the sole explanation. In the Serin, no external factor of major importance has been found that could adequately explain the northward penetration. "In such cases one must suspect that some genotypic changes occurred in a peripheral population which permitted the precipitant expansion. The genotypic change, obviously, resulted in a change in climatic or habitat tolerance that permitted the subtropical species to enter the temperate zone" (Mayr, 1966, p. 564).

When a species enters a new area and becomes abundant a process begins (von Haartman, 1971) that has already been described by Mayr with respect to the Serin. Soon the surplus birds are forced by the territoriality-and-avoidance system into suboptimal or, at any rate, new habitats. To take a fresh example, the Mute Swan was originally an inhabitant of the reeds of eutrophic lakes and bays. In Sweden, after becoming numerous in this habitat, it began to occupy islands in the archipelago of the Baltic, even nesting at relatively barren rocky islets near the open sea. The Mute Swan began to nest in Finland in 1934 (see p. 460) and, long before it had become numerous in its original habitat, it started to nest in the Finnish archipelago. It seems likely that the Finnish archipelago swans are descended from the Swedish archipelago population, some of which were induced to follow the Finnish Mute Swans on their migration.

In what respect do the archipelago swans differ from the swans nesting in reeds? It seems possible that the preference for the one or the other habitat is just phenotypic, a "modification"; the Mute Swans may have learned to prefer the archipelago for nesting.

The processes of adaptation and modification probably always take place when a bird becomes more numerous in an old area or enters a new one. The extent of the role played by these processes, however, is difficult to evaluate. A further example of modification may suffice. In recent times, the Herring Gull (*Larus argentatus*) and the Common Gull (*L. canus*) have increased enormously in many parts of the Finnish archipelago. The increase has partly depended on these gulls' ability to acquire new food habits. For instance, both species have learned (in Helsinki from about 1935 on) by following the Black-headed Gulls (*L. ridibundus*) to search for earthworms on newly ploughed fields (Bergman, 1960).

The synchronism of the northward extension of range or increase in numbers in numerous southerly species in the last 100 years makes it difficult, however, to assume intrinsic processes as its general cause. Adaptation and modification may be indispensable processes in the expansion, but other factors must be responsible for the synchronization of the expansion of the different species. These factors seem to be the change in climate and the influence exerted by man, both directly and indirectly, on the habitats.

REFERENCES

Ahlén, I. 1965. Studies on the Red Deer, *Cervus elaphus* L., in Scandinavia. I. History of distribution. III. Ecological investigations. Viltrevy 3(1):1–88, 3(3):177–376.

Ahlén, I. 1966. Landskapets utnyttjande och faunan. Sveriges Nat. Årsbok 1966:73–99.

Ångström, A. 1939. Temperaturkilmatets ändring i nuvarande tid och dess orsaker. Ymer 59:62–76.

Ångström, A. 1958. Sveriges klimat. Generalstabens litografiska anstalts förlag, Stockholm. 154 p.

Atlas öfver Finland. 1899. Sällskapet för Finlands Geografi. Helsingfors.

Bergman, G. 1957. Das Porkalegebiet als biologisches Experimentalfeld. Acta Soc. Fauna Flora Fenn. 74:3:1–52.

Bergman, G. 1960. Über neue Futtergewohnheiten der Möwen an den Küsten Finnlands. Ornis Fenn. 37:11–28.

Bergström, U. 1940. Näktergalen (*Luscinia luscinia* L.) som häckfågel i Mellansverige. Fauna Flora 1940:253–274.

Bjärvall, A. 1965. En inventering av näktergalens (*Luscinia luscinia*) förekomst
i Stockholmstrakten. Vår Fågelvärld 24:294–300.

Buchinsky, I. E. 1963. Climatic fluctuations in the arid zone of the Ukraine,
p. 91–95. *In* Changes of climate (proceedings of a symposium). UNESCO,
Liège, Belgium.

Curry-Lindahl, K. 1959–1963. Våra Fåglar i norden. Natur och Kultur, Stockholm. 2294 p.

Dansgaard, W., S. J. Johnesen, and H. B. Clausen. 1971. Hvad gør naturen ved
klimaet Naturens Verden, p. 134–142.

Durango, S. 1946. Blåkråkan i Sverige. Vår Fågelvärld 5:145–190.

Durango, S. 1950. Om klimatets inverkan på törnskatans (*Lanius collurio* L.)
utbredning och levnadsmöjligheter. Fauna Flora 45:49–78.

Eriksson, K. 1969a. On the occurrence of the Grasshopper Warbler (*Locustella
naevia*) and the River Warbler (*L. fluviatilis*) in Finland related to bird
watching activity. Ornis Fenn. 46:113–125.

Eriksson, K. 1969b. On the occurrence and ecology of Blyth's Reed Warbler
(*Acrocephalus dumetorum*) and Marsh Warbler (*A. palustris*) in Finland.
Ornis Fenn. 46:157–170.

Haapanen, A. 1965–1966. Bird fauna of the Finnish forests in relation to forest
succession. Ann. Zool. Fenn. 2:153–196, 3:176–200.

Haftorn, S. 1971. Norges fugler. Universitetsforlaget, Oslo. 862 p.

Heikinheimo, O. 1915. Kaskiviljelyksen vaikutus Suomen metsiin. Acta For.
Fenn. 4:1–264, 1–149, 1–59.

Heikinheimo, O. 1938. Lauiduntamisen puuntaimistolle aiheuttamista tuhoista.
Metsätaloudellinen aikakausilehti, p. 16–20.

Hildén, O. 1966. Changes in the bird fauna of Valassaaret, Gulf of Bothnia,
during recent decades. Ann. Zool. Fenn. 3:245–269.

Hortling, I. 1929–1931. Ornitologisk handbok. J. Simelii Arvingar, Helsingfors.
1142 p.

Hyytiä, K. 1970. Lintuharrastusaktiivisuus ja havaintomäärät. Ornis Fenn.
47:83–86.

Jägerskiöld, L. A. 1919. Om förändringar i Sveriges fågelvärld under de senaste
75 åren. Sveriges natur, p. 47–73.

Jäntti, A. 1945. Suomen laidunolot. PhD thesis, Privately printed. 255 p.

Järnefelt, H. 1936 Suomen järvityyppien alueellinen levinneisyys. Terra 48:1–10.

Jozefik, M. 1960. Modifications of the south-west border of the range of
Erythrina e. erythrina during the last two hundred years. Acta Ornithol.
Mus. Zool. Pol. 5:307–324.

Kaisila, J. 1962. Immigration und Expansion der Lepidopteren Finnlands in den
Jahren 1869–1960. Acta Entomol. Fenn. 18:1–452.

Kalela, O. 1938. Über die regionale Verteilung der Brutvogelfauna im Flussgebiet
des Kokemäenjoki. Ann. Zool. Soc. Zool. Bot. Fenn. Vanamo. 5(9)1–15,
1–297.

Kalela, O. 1940. Zur Frage der neuzeitlichen Anreicherung der Brutvogelfauna
in Fennoskandien mit besonderer Berücksichtigung der Austrocknung in den
früheren Wohngebieten der Arten. Ornis Fenn. 17:41–59.

Kalela, O. 1948. Metsäkauriin esiintymisestä Suomessa ja sen levinneisyyden
muutoksista lähialueilla. Suom. Riista 3:34–56.

Kalela, O. 1949. Changes in geographic ranges in the avifauna of northern and

central Europe in relation to recent changes in climate. Bird-Banding 20:77–
103.

Kalela, O. 1955. Die neuzeitliche Ausbreitung des Kiebitzes, *Vanellus vanellus*
(L.), in Finnland. Ann. Zool. Soc. Zool. Bot. Fenn. Vanamo 16(11):1–80.

Källander, H. 1970. Förekomsten av gräshoppsångare *Locustella naevia* i Sverige
1968. Vår Fågelvärld 29:6–10.

Kalliola, R. 1950. Suomen kaunis luonto. Werner Söderström, Porvoo. 466 p.

Kolkki, O. 1966. Tables and maps of temperature in Finland during 1931–1960.
Meteor. Yearb. Finland 1965 Suppl. 65:1–8, 1–42.

Kolthoff, G. 1896. Vårt villebråd 1–2. Fr. Skoglunds förlag, Stockholm. 826 p.

Kolthoff, G. 1907. Om förändringer i svenska fågelfaunan under det sist
förflutna halfseklet, p. 155–171. *In* Zoologiska studier tillägnade. T. Tallberg,
Uppsala.

Komonen, A. 1963. Töyhtöhyypän, *Vanellus vanellus* (L.), esiintymisestä
Pohjois-Suomessa. Luonnon Tutkija 67:96–101.

Kovero, M. 1951. Teollisuus, p. 188–210. *In* Suomen maantieteen käsikirja.
Otava, Helsinki.

Lampimäki, T. 1939. Nautakarjan laiduntamisesta metsämailla. Silva Fenn.
50:1–106.

Larsson, T. 1969. Land use and bird fauna on shores in southern Sweden. Oikos
20:136–155.

Leivo, O. 1946. Neue Funde des Sumpfrohsängers, *Acrocephalus palustris*
(Bechst.), in Finnland nebst einigen Bemerkugen zu seiner Ausbreitung in
jüngster Zeit. Ornis Fenn. 23:65–77.

Lönnberg, E. 1924. Ett bidrag till den svenska faunans invandringshistoria.
Fauna Flora 1924:97–119.

Løppenthin, B. 1967. Danske ynglefugle i fortid og nutid. Odense Universitets-
forlag, Odense. 609 p.

Mayr, E. 1926. Die Ausbreitung der Girlitz. J. Ornithol. 74:571–671.

Mayr, E. 1966. Animal species and evolution. Belknap Press of Harvard University
Press, Cambridge, Mass. 797 p.

Merikallio, E. 1929. Äyräpäänjärvi. Otava, Helsinki. 202 p.

Merikallio, E. 1951. Der Einfluss der letzen Wärmeperiode (1930–49) auf die
Vogelfauna Nordfinnlands, p. 484–493. *In* S. Hörstadius [ed.] Proceedings
of the Xth International Ornithological Congress. Almqvist & Wiksells
Boktryckeri, Uppsala.

Nordström, G. 1956. Über die Expansion des Karmingimpels, *Carpodacus
erythrinus* Pall., während der letzten Jahre in Finnland. Ornis Fenn. 33:19–28.

Norris, C. A. 1945. Summary of a report on the distribution and status of the
Corncrake (*Crex crex*). Brit. Birds 38:142–148, 162–168.

Nowak, E. 1971. The range expansion of animals and its causes. Polska Akad.
Nauk Inst. Ekol. Zesz. Nauk. 3:1–255.

Otterlind, G. 1954. Flyttning och utbredning. Vår Fågelvärld 13:1–31, 83–113,
147–167, 245–261.

Palmén, J. A. 1874. Om foglarnes flyttningsvägar. J. C. Frenckell & Son,
Helsingfors. 200 p.

Palmgren, P. 1930. Quantitative Untersuchungen über die Vogelfauna in den
Wäldern Südfinnlands mit besonderer Berücksichtigung Ålands. Acta Zool.
Fenn. 7:1–218.

Palmgren, P. 1936. Über den Massenwechsel bei *Regulus r. regulus* (L). Ornis
Fenn. 13:159–164.

Reinikainen, A. 1939. Punavarpusen, *Carpodacus e. erythrinus* (Pallas),
pesimisekologiasta. Ornis Fenn. 16:73–95.

Risberg, E. L. 1970. Rosenfinkens *Carpodacus erythrinus* invandring till Sverige
samt studier av dess häckningsbiologi. Vår Fågelvärld 29:77–89.

Romell, L.-G. 1964. Skog och odling i svensk "natur". Sveriges Nat. Årsbok
1964:110–124.

Salomonsen, F. 1967. Systematisk översikt över Nordens fåglar, p. 1–509. *In*
N. Blaedel [ed.] Nordens fåglar i färg. Vol. VII. Malmö, Sweden.

Salonen, K. 1968. Lannoitteiden käyttö maa-ja metsätaloudessa. Soumen Luonto
27:109–111.

Selander, S. 1955. Det levande landskapet i Sverige. Albert Bonniers Förlag,
Stockholm. 485 p.

Siivonen, L. 1956. The correlation between the fluctuations of Partridge and
European Hare populations and the climatic conditions of winters in southwest Finland during the last thirty years. Papers Game Res. 17:1–30.

Siivonen, L., and O. Kalela. 1937. Über die Veränderungen in der Vogelfauna
Finnlands während der letzten Jahrzehnte und die darauf einwirkenden
Faktoren (Vorläufige Mitteilung). Acta Soc. Fauna Flora Fenn. 60:606–634.

Sirén, G. 1955. Skogen vandrar åter mot norr. Nordenskiöld-Samfundets Tidskr.
18:101–108.

Soveri, J. 1940. Die Vogelfauna von Lammi, ihre regionale Verbreitung und
Abhängigkeit von den ökologischen Faktoren. Acta Zool. Fenn. 27:1–176.

Stålfelt, M. G. 1969, Växtekologi. Svenska Bokförlaget, 444 p.

Udvardy, M. D. F. 1969. Dynamic zoogeography. Van Nostrand Reinhold Co.,
New York, 455 p.

Välikangas, I. 1951. The expansion of the Greenish Warbler (*Phylloscopus
trochiloides viridanus* Blyth) in the Baltic area, especially in Finland, towards
north and northwest, and its causes, p. 527–531. *In* S. Hörstadius [ed.]
Proceedings of the Xth International Ornithological Congress. Almquist &
Wiksells, Boktryckeri, Uppsala.

Valle, O. 1951. Maatalous, p. 404–427. *In* Suomen maantieteen käsikirja. Otava,
Helsinki.

Vepsäläinen, K. 1968. The effect of the cold spring 1966 upon the Lapwing
(*Vanellus vanellus*) in Finland. Ornis Fenn. 45:33–47.

von Haartman, L. 1958. The decrease of the Corncrake (*Crex crex*). Commentat.
Biol. Soc. Sci. Fenn. 18:2:1–29.

von Haartman, L. 1971. Population dynamics, p. 391–459. *In* D. S. Farner and
R. King [ed.] Avian biology. Vol. I. Academic Press, New York and London.

von Haartman, L., O. Hildén, P. Linkola, P. Suomalainen, and R. Tenovuo.
1963–1972. Pohjolan linnut värikuvin. Otava, Helsinki. 1092 + 192 p.

von Rudloff, H. 1967. Die Schwankungen und Pendelungen des Klimas in Europa
seit dem Beginn der regelmässigen Instrumenten-Beobachtungen (1670). Friedr.
Vieweg & Sohn, Braunschweig. 370 p.

Wallengren, H. D. J. 1854–1856. Die Brütezonen der Vögel innerhalb Skandi-
navien. Naumannia 4–6.

Disparate Sex Ratios in Waterfowl *

JOHN W. ALDRICH

The issue of disparate sex ratios was singled out for special discussion because of its interest from the standpoint of application to wildlife management, and because it brings to bear a multidisciplinary scrutiny of the reproductive biology of birds. The reason for current interest is the growing sentiment favoring special hunting regulations of wild ducks that encourage the shooting of more males than females. The rationale here is that there is a disproportionately large number of males in wild duck populations and that it would increase hunting opportunities and at the same time tend to reduce the excess males, which are considered either of no value or even detrimental to reproduction. But the assumption that the extra male ducks have no value in reproduction has not been adequately tested and considerable effort will be required to do this. Some hypotheses requiring testing follow:

1. An excess of drakes may be damaging to production because they harass the hens, either causing them to fail to nest or to disperse too far from water to nest during drought periods, with resulting hazard to ducklings.

*Based in part on a general discussion by participants.

482

2. An excess of drakes in group courtship is necessary to stimulate behavior—endocrine reactions leading to pair formation and synchrony of breeding condition in pairs.

3. An excess of drakes is necessary to ensure a mate for renesting of hens that lose their first clutches of eggs at a time after their first mates have departed.

4. An excess of drakes is necessary to ensure that there will be at least one available male to pair with each female in species the males of which mature at an older age than females.

Harvey K. Nelson commented on the work of a committee of the Central Flyway Technical Section that was formed in 1970 to investigate the function of excess drakes in various waterfowl populations. In particular, he summarized a draft of a paper by Alex Dzubin, titled "Biological Implications of Spring Drake Removal on Productivity of Hen Waterfowl" that was prepared for this committee. Dzubin raised seven questions that are still pertinent to the overall subject as far as ducks are concerned:

1. What is the impact of unnatural sex ratios upon the continued survival and growth of a breeding population?

2. Is a sex ratio distorted above 53:47 (112 ♂♂:100 ♀♀) biologically meaningful in terms of providing renesting hens with readily available males after their primary drakes have abandoned the home ranges at midincubation?

3. Does the presence of the male with the hen in a pair-bond confer some positive advantage to successful laying of a fertilized clutch when we consider that, in semitame flocks, one drake can successfully fertilize 6–10 hens?

4. Does a hen with an accompanying male have some advantage in siting her home range, activity center, and waiting area in an optimum location (i.e., for survival of herself and her progeny)?

5. Are unmated drakes in the spring mostly yearling or adult birds? Are they capable of fertilizing ova?

6. Most Mallards (*Anas platyrhynchos*) now breed in artifically altered habitats under greatly increased spring grain food resources but under restricted acreages of nesting cover. Mallard populations are now subject to "unnatural" hunting mortality. We must ask the unanswerable questions: What were sex ratios in the pristine breeding habitats before the coming of the white man? Are the present day distorted ratios favoring males an "unnatural" condition due to in-

creased hen mortality from hunting or from higher loss to predators incurred nesting in suboptimal habitats?

7. Can we, in fact, do as Crissey (1969) proposes; i.e., can we distort the sex ratios to favor hens and thus decrease pair interaction and pair spacing and increase production?

Nelson suggested that the following additional questions need to be answered:

1. When does the greatest female mortality occur and does this vary for different geographic areas and species?

2. What is the difference in waterfowl sex ratios in the northern portions of the breeding range as against the southern portion of the breeding range?

3. Since there is some evidence that drakes, particularly Mallards, winter farther north, is there any indication of a greater excess of hens in the more southern wintering populations?

4. What is the actual potential to harvest excess drakes, in view of some of the experimental seasons that have been tried?

5. Do we really understand the role of male waterfowl as they influence breeding biology and behavior?

In attempting to determine what is really known about waterfowl sex ratios, Dzubin's paper is one of the most useful sources. He presented the following summary information based on his own field experience and an intensive literature review:

Sex ratios can be separated into four categories: primary at fertilization, secondary at hatching, tertiary in immatures, and quaternary in adults. Spring sex ratios of most adult ducks, divers or dabblers, are distorted and favor males (Bellrose *et al.*, 1961; Hochbaum, 1944; and Sowls, 1955). Generally, diver sex ratios are more heavily distorted than dabblers. Secondary sex ratios of incubator-hatched wild Mallard eggs show a 53:47 ratio ($N = 630$) (Hochbaum, 1944, p. 51).

The tertiary sex ratio of the juvenile Mallard component is apparently close to 1:1. Low (1957), summarizing bandings from prairie Canada, showed a sex ratio of 50.9:49.1 (1,138:1,097) for Alberta-banded locals and 51.6:48.4 (4,796:4,505) for Saskatchewan locals. Using only ducklings banded in the years 1950–1961, Lensink (1964, p. 93–97) reported a sex ratio of 51.2:48.8 (16,102:15,227) for the western interior of Canada and 50.5:49.5 (12,040:11,721) for the northern plains. In the Kindersley area of Saskatchewan, for the period 1952–1959, Gollop (1965) gives a sex ratio of 50.4:49.6 (6523: 6408) for locally banded young. In all, of some 79,726 Mallard ducklings sexed (an unknown proportion duplicated in above three studies), 50.92 percent were drakes.

The sex ratio in the harvest (1964–1967) shows a preponderance of adult drakes per female harvested, i.e., Mississippi Flyway: 2.04, 2.17, 1.74; Central Flyway: 2.02, 2.08, 1.72, 1.85 ♂♂/♀♀ (Croft and Carney, 1968). The comparable figures for immatures are Mississippi Flyway–1.20, 1.23, 1.12, 1.22 immature ♂♂/immature ♀♀; Central Flyway–1.46, 1.35, 1.41, 1.46 immature ♂♂/immature ♀♀. Male:female relative recovery rates indicated that in the United States males were more likely to be shot than females (Anderson, 1968). Anderson also points out that both the Arkansas and Washington banding data corroborate the general pattern of males being taken in a relatively greater extent in the United States and hens being taken to a greater extent in Canada.

The Canadian species composition survey showed the following sex ratios in the harvest for 1967, 1968, and 1969: Alberta–Ad ♂♂/Ad ♀♀ 1.13, 1.61, 1.52; Saskatchewan–2.41, 2.44, 1.75 (N + 500 + per year) Kaiser *et al.*, 1970). The sex ratio in the harvest need not reflect the true "wild" ratio, as there may be an active choice of drakes by hunters.

As reported by Geis *et al.* (1969, p. 852), Mallard hens banded in winter through North America generally show a higher first season band recovery rate and show higher average annual mortality rates than do drakes. The processes leading to distorted sex ratios are fairly evident.

Bellrose *et al.* (1961) concluded that "the effect of hunting on the sex ratios of the entire North American duck population is probably insignificant" while other authors (in Olson, 1965) have argued that hunting may be a major contributory factor to the disproportionate sex ratios of ducks. Olson (1965) concluded that differences in vulnerability to hunting, especially the high vulnerability of adult hens and juveniles, may be the cause of an unbalanced sex ratio in Canvasbacks (*Aythya valisneria*). The same may be happening with Mallards. Early opening dates for hunting on breeding or molting areas may lead to more of the vulnerable adult female component being killed. To date, we have no extensive evidence to show that natural mortality rates of hens during the breeding season are substantially higher than drakes, except for the preliminary results provided by Keith

(1961) and Dzubin and Gollop (in press), who reported approximately 5 percent mortality in drakes and 8–10 percent in hens, May through July. Differential sex mortality on molting lakes is undocumented. Postmolting hens may be more available to Canadian hunters.

There is a differential distribution of sexes and ages on the wintering grounds. Generally, females and juveniles winter further south than do drakes and adults (Bellrose *et al.*, 1961; Gollop, 1965). In the Central Flyway states, wintering Mallard flocks at the northern edge of the wintering range are predominantly drakes (60–75 percent). Since most display and pairing activity occurs during December through February and since all hens arriving on the breeding grounds in April are associated with drakes, do these extra drakes join into breeding activities; that is, are they biologically meaningful?

Dzubin's conclusions were that the Mallard pair-bond is a necessary prerequisite for successful breeding in a wild population of Mallards and that there is insufficient behavioral evidence to suggest that a higher proportion of unmated Mallard hens could nest successfully and produce more progeny if the sex ratio were artifically altered to favor hens.

Nelson stressed that it is obvious that the relationship of waterfowl sex ratios to breeding behavior and production rates is not fully understood and that future research on this subject should be directed specifically at determining these relationships for each of the principal species presently involved in sex-oriented regulations. There is ample evidence to indicate the need for caution in encouraging specific sex-oriented waterfowl hunting regulations directed at the harvest of excess males and assumed protection of females on a *broad-scale basis*, until there are better answers to the questions raised.

John P. Rogers commented on seasonal change in sex ratios in Mallards. His remarks follow:

For more than three decades ornithologists and wildlife biologists have been puzzling over the significance of uneven sex ratios among ducks. In reviewing this subject today it is apparent that while we now have more data to look at, we have not progressed very far in resolving many of the questions raised years ago. John Aldrich and Harvey Nelson have discussed some of these, and their importance to waterfowl management, especially as they relate to recent interest in increasing the harvest of drake Mallards. My remarks will be concerned with briefly describing a recent evaluation of sex ratios in the North American Mallard population which we believe is pertinent to an understanding of Mallard population dynamics.

The general approach used in this evaluation was described in an unpublished

administrative report by Anderson *et al.* (1970). The figures presented below
were developed by W. F. Crissey for a presentation to the Mississippi Flyway
Council Technical Committee at Biloxi, Mississippi, in 1972.

In the past, information on sex ratios usually has been obtained by direct field
observations. The methods used and the biases caused by the mobility of the
birds, as well as differences in seasonal and geographic distribution of sex and age
groups have been discussed by Bellrose *et al.* (1961). These factors make it extremely difficult to obtain a measure of sex ratio representative of the population.

The method used for estimating sex ratios presented here begins with a determination of the size and sex composition of the kill by hunters during the fall
and winter hunting season. This is used in conjunction with direct recovery rates
from prehunting season banding which, when adjusted to account for unreported
bands, yields an estimate of harvest rate by age and sex. With an estimate of harvest and harvest rate it is possible to estimate the size of the population by age
and sex from which the kill was taken. Subtraction of the harvest by age and sex
from the preseason population yields an estimate of the sex ratio for adults and
immatures immediately following the hunting season.

The following table summarizes Mallard sex ratios calculated for different
times of the year for three years beginning with the hunting season of 1967–68.

	1967–68		1968–69		1969–70	
	Adults	Immature	Adults	Immature	Adults	Immature
Prehunting season population	1.60	0.88	1.33	1.13	1.62	0.99
Hunting season harvest	1.87	1.23	1.94	1.27	2.00	1.29
Posthunting season population	1.08	[a]	1.16	[a]	1.12	[a]

[a]Immatures combined with adults in postseason period.

These figures show that in early fall, just prior to the hunting season, the sex
ratio among adult Mallards is strongly unbalanced in favor of males, while that
for immatures is nearly even. It is probable that a sex ratio of 1.6 males per female among adults, and 1.0 males per female among immatures closely approximates the true situation, and that differences between years appearing in the table reflect deficiencies in the data.

Sex ratios in the harvest (second line) show that, for both age groups, hunting
takes a higher proportion of the males than of the females present in the population. This is due partly to hunter selectivity but other factors, related to males
being more accessible than females to hunters, are believed to play a role, also.
Regardless of the reasons, the differential harvest of males and females is sufficiently great that by the end of the hunting season the sex ratio is nearly even,
and there is only a relatively slight preponderance of males.

At present, there is no evidence that population losses between the end of the

hunting season and the beginning of the breeding season affect one sex more
than the other. Thus, it is believed that the sex ratio in the posthunting season
population is practically identical to that at the beginning of the breeding season.

The following conclusions are suggested by this analysis: In the annual cycle
of the Mallard, large and significant changes in sex ratio occur (1) during the
hunting season and (2) during the breeding season. During the hunting season,
losses of males (due to hunting) change the sex ratio from a strongly unbalanced
to a nearly balanced condition. During the breeding season, losses of females
(due to natural mortality) are sufficiently great to cause a return to a strongly
unbalanced sex ratio favoring males.

It appears that there is (1) no great excess of males in the continental Mallard
population in terms of the population available to begin breeding; and (2) natural
mortality of females, associated with breeding, is greater than previously sus-
pected.

Gordon H. Orians of the University of Washington presented a
number of points concerning natural selection for sex ratios in birds.

REFERENCES

Anderson, D. R. 1968. Summary of first hunting season recovery rates of Mal-
    lards banded during winter, summer and early fall of 1967. Administrative
    Report 165. Bureau of Sport Fisheries and Wildlife, U.S. Department of the
    Interior. 4 p. (Unpublished)
Anderson, D. R., R. S. Pospahala, H. M. Reeves, J. P. Rogers, and W. F. Crissey.
    1970. Estimates of the sex composition of the 1967–1969 Mallard popula-
    tion. Administrative Report 182. Bureau of Sport Fisheries and Wildlife,
    U.S. Department of the Interior. 8 p. (Unpublished)
Bellrose, F. C., Jr., T. G. Scott, A. S. Hawkins, and J. B. Low. 1961. Sex ratios
    and age ratios in North American ducks. Ill. Nat. Hist. Sur. Bull. 27:391–474.
Crissey, W. S. 1969. Prairie potholes from a continental viewpoint. Saskatoon
    wetlands seminar, p. 161–177. *In* Canadian Wildlife Service Report Series.
    No. 6. Canadian Wildlife Service, Ottawa.
Croft, R. L., and S. M. Carney. 1968. Sex ratios of ducks in the hunting kill,
    1966–1967. Administrative Report 153. Bureau of Sport Fisheries and Wild-
    life, U.S. Department of the Interior. 43 p. (Unpublished)
Dzubin, A., and J. B. Gollop. In press. Population ecology and spacing of Mal-
    lards on a Canadian grassland and parkland area. American Institute of Bio-
    logical Sciences and Bureau of Sport Fisheries and Wildlife, U.S. Department
    of the Interior. U.S. Government Printing Office, Washington, D.C.
Geis, A. D., R. K. Martinson, and D. R. Anderson. 1969. Establishing hunting
    regulations and allowable harvest of Mallards in the United States. J. Wildlife
    Manage. 38:848–859.
Gollop, J. B. 1965. Dispersal and annual survival of the Mallard. Canadian Wild-
    life Service, Saskatoon, Saskatchewan. 104 p. (Unpublished)

Hochbaum, H. A. 1944. The Canvasback on a prairie marsh. American Wildlife Institute, Washington, D.C. 201 p.

Kaiser, G. W., L. Wight, and F. G. Cooch. 1970. Hand tally—1969 species composition survey. Canadian Wildlife Service, Ottawa, Ontario. (Unpublished)

Keith, L. B. 1961. A study of waterfowl ecology on small impoundments in southeastern Alberta. Wildlife Monogr. 6:1–88.

Lensink, C. J. 1964. Distribution of recoveries from bandings of ducklings. Special Scientific Report—Wildlife No. 89. Bureau of Sport Fisheries and Wildlife, U.S. Department of the Interior, Washington, D.C. 146 p.

Low, S. H. 1957. Waterfowl banding in the Canadian prairie provinces. Special Scientific Report—Wildlife No. 36. Bureau of Sport Fisheries and Wildlife, U.S. Department of the Interior, Washington, D.C. 30 p.

Olson, D. P. 1965. Differential vulnerability of male and female Canvasbacks to hunting. Trans. Thirtieth N. Am. Wildlife Nat. Res. Conf. 30:121–135.

Sowls, L. K. 1955. Prairie ducks. Stackpole Co., Harrisburg, Pa. 193 p.

Commentary

THOMAS S. BASKETT

I am aware of the underlying thought that research by avian ethologists, physiologists, and nutritionists and by wildlife biologists has progressed to a point that the need for interactions among these scientists is crucial. Because of the knowledge explosion, better avenues are needed for exchange of information and ideas among workers in these diverse but intricately related fields.

The wildlife field is not a separate discipline. The people in it represent an assemblage of disciplines; they are bound together by a common interest in conservation and management of wild animals, and by organizational responsibilities within agencies and universities. The field includes many broadly oriented people, best termed wildlife ecologists, who are generalists interested in interdisciplinary approaches. It also includes physiologists, nutritionists, behaviorists, plant ecologists, and members of several other scientific disciplines.

The wildlife field, its responsibilities, and its expertise are both broad and deep. Wildlife biologists are frequently regarded as biologists who turned into land-managers, or as manipulators of wild animal populations for the benefit of the hunter or the bird watcher, or as persons concerned with control of damage to agriculture by birds and mammals. All of these are important facets of the wildlife field and will remain so, but they are only a part of it.

As the field has developed, wildlife scientists have become deeply involved in research on relationships between environmental pollutants and wildlife. Many wildlife scientists, including university researchers and scientists of state and provincial conservation departments, the Canadian national government and the Denver and Patuxent Centers of the U.S. Bureau of Sport Fisheries and Wildlife, have developed information that has been very useful in identifying environmental challenges to wild animals. Insults to wild animal populations from chlorinated hydrocarbons, for example, have been assessed, and the testimony of wildlife scientists at hearings before congressional committees and governmental regulatory agencies has been a factor in slowing the rate of application of these materials. I submit that influencing regulatory procedures governing these compounds by providing solid data is just as important a form of wildlife conservation as any other.

In addition, wildlife scientists have been charged with responsibility for studying endangered species of wildlife to develop ways of improving their habitats, of preserving their gene pools, and of restocking them in their native range when conditions are once more suitable. Propagation is one measure employed to benefit endangered species, both in North America and abroad (for references see IUCN, Red Data Book, 1966–1971). Several "prototype" species, whose propagation and intensive husbandry might provide useful information for handling their endangered relatives, have been brought into captivity in the United States (Erickson, 1968; Goodwin and Denson, 1971). These responsibilities have stimulated the wildlife field to marshal a considerable expertise in the areas of nutrition, physiology, behavior, and pathology.

In controlling animal damage, wildlife biologists employed by federal and state natural resource agencies in this and many other countries are charged with finding and applying means of lessening or redistributing damage to crops, livestock, and human dwellings and their environs. Although lethal chemicals are sometimes involved, research in this country now emphasizes other solutions to vertebrate damage. Better means of assessing the damage and putting it in perspective are inherent in the new thrust. Substantial research in avian physiology, behavior, and population dynamics is being carried out to provide background information for work in this area. Recent publications documenting facets of this work include those of Thompson *et al.* (1968); Kadlec and Drury (1968); Lefebre and Seubert (1970), and Murton (1971).

When this symposium was first suggested, many phases of our wildlife research programs in the United States were being guided closer to the basic areas that tie in very well with the interests of ethologists, avian physiologists, and nutritionists. Such ties persist in many phases of wildlife research—the influence of pesticides and other pollutants upon wildlife populations, for example. In many areas of interest, thorough research into physiological matters is essential and, if people in the wildlife field are unable to perform it, they must be in contact with those who can.

More recently, a great wave of environmental concern has swept the North American continent, and we are redirecting some of our research programs to anticipate environmental challenges while there is still time to prevent or correct serious degradation. Consequently, we must commit some of our research funds and energies to provide wildlife population and habitat data quickly when environmental upheavals, such as large-scale engineering feats, are in prospect. This demand on time and resources explains why some wildlife agencies have not sponsored more research in such experimental fields as avian physiology, nutrition, and ethology in recent years.

Having established that wildlife scientists are a very heterogeneous lot, we can examine more closely the relationships between them and scientists in the three fields just mentioned.

ETHOECOLOGICAL ASPECTS OF REPRODUCTION

One research area in which rapport has been good was brought out by McKinney. Ethologists, particularly those interested in behavior of waterfowl, have certainly worked closely with wildlife biologists on this continent and to some degree the world over.

In Canada and the United States, McKinney has contributed greatly to that good rapport. Several of the studies to which he referred in his paper, and other studies related to them, were conducted by people who are identified with the wildlife field. (See Dzubin, 1955; Weller, 1959; and Smith, 1968, for example.) The quality of the research reported by these authors was strongly influenced by McKinney, who had effective communication with them and helped them as their research progressed.

For want of better information, wildlife managers responsible for land purchase and manipulation in favor of waterfowl have often assumed that the amount of space observed to be used by breeding waterfowl, and the kinds and degrees of dispersion of vegetation and water that attracted them, were necessary for successful breeding. This may be an expensive assumption, for certain studies have indicated ways to achieve remarkably efficient natural waterfowl production without resorting to game farm techniques. A classic study illustrating this idea was published by Duebbert (1966). He found Gadwalls (*Anas strepera*) nesting in densities exceeding 10 per acre (25 per ha) on a 7-acre (2.8 ha) island, isolated from most terrestrial predators, in a wildlife refuge in North Dakota. Nesting success was unusually high for a ground-nesting bird: 86 and 92 percent in successive years. Although the conditions may have been exceptional and the Gadwall particularly capable of congregating to breed, the implications are clear.

There are many excellent reasons other than waterfowl production to preserve Canadian and northern United States prairie wetlands, but, where production of aquatic birds is an important consideration, input by ethologists is essential. Wildlife managers need to know which species can be compressed and still retain biological dignity, and they need interpretive help from ethologists in this matter. McKinney's point that breeding behavior must be examined species by species, without prior assumptions that the "rules" are the same even for closely related forms, is relevant here. On their side, the ethologists will be provided a flood of information and questions certain to prove stimulating.

DISPARATE SEX RATIOS IN WATERFOWL: BEHAVIORAL-NEUROENDOCRINE REACTIONS

The possible consequences of manipulations that produce disparate sex ratios in wild waterfowl are of great interest to wildlife managers. In fact, the opportunity to discuss this subject with ethologists and avian physiologists was one of the reasons that the wildlife ecologists were interested in participating in this symposium. Yet in the discussion on this subject, although numerous miscellaneous observations on sex ratios were brought out, possible effects of altered sex ratios on breeding behavior and physiology of waterfowl were touched only lightly.

Beer, however, did give a number of examples of behavioral–neuroendocrine reactions, which suggested the existence of a great unexplored field of wildlife reproductive behavior that we should know more about in order to guide wildlife management practices.

For the record, it is possible to formulate hunting regulations that result in selective harvest by sex in a waterfowl species—the Mallard, particularly—with marked sexual dimorphism. If such regulations were applied on a national scale (a practice not now anticipated, to my knowledge), it is conceivable that the sex ratio of the continental Mallard population would be altered. What would be the effect of changing an unbalanced ratio favoring males to an equal ratio, or to one favoring females? Do extra drakes stimulate earlier and more effective breeding? Are they needed to fertilize hens for renesting attempts?

Results of pen studies might not be applicable to unconfined Mallards. Field studies of wild Mallards equipped with radio transmitters are being conducted at the Northern Prairie Wildlife Research Center. Although these studies may give direct evidence in answer to some of these questions, a behavioral–physiological rationale will be needed to explain field results.

NUTRITION IN REPRODUCTION

Scott's presentation about nutritional requirements of several game species has obvious relevance to the wildlife field. The traditional reasons for propagation, such as for release of game birds on preserves, still hold to some extent. But beyond that, wildlife research has followed avenues that require successful breeding in captivity of many wild birds for experimental purposes. These birds are used in pesticide experiments or as "pilot" animals in endangered species programs, and their propagation and maintenance in captivity are often quite difficult.

ENERGY ASPECTS OF REPRODUCTION

King's paper on the energy aspects of reproduction and the ensuing discussion had several points of direct concern to the wildlife manager and to the researcher. King pointed out that the energy cost for laying a clutch of eggs is very high for Anseriformes—much higher than for gallinaceous birds. In this case, the wildlife manager intuitively did the right thing. As Hickey remarked, practices on national wildlife refuges in the United States where geese overwinter have been designed to send the geese back to the breeding grounds with the maximum amounts of fat, and Canada Goose (*Branta canadensis*) populations are probably now at an all-time high.

Another point of special relevance for wildlife researchers is King's remark that no matter what his special interest in avian biology, a researcher can contribute significant information leading to the understanding of the caloric cost of reproduction. Information on such subjects as the fresh weights of gonads, rates of growth of ovarian follicles, rates of egg-laying, and weights of fresh eggs are needed for many species of birds. Some of these are subjects upon which wildlife ecologists routinely gather data for other reasons, and the opportunity for useful interchange is obvious.

EFFECTS OF ENVIRONMENTAL MANIPULATIONS ON AVIFAUNA: ENVIRONMENTAL MANAGEMENT STRATEGY

Two presentations were particularly relevant to problems the resource manager now faces. In the first, Orians pointed out that traditional wildlife investigations have often been long range, intensive studies of individual species. Although these have their value, they are time consuming and have little general applicability. He proposed that one way to respond to today's press for quick answers in development of environmental management strategy is to construct more general models, using data already available even at the sacrifice of some accuracy. His remark about the need for heterogeneity in the environment was hardly new but it struck a responsive chord with wildlife researchers who must "solve" a vertebrate damage problem whose fundamental cause is corn monoculture.

Second, von Haartman elucidated changes in distribution of certain Scandinavian songbirds. A northward spread was correlated in part with a warming trend starting in the 1930's and in part with a 30-year burning cycle employed in coniferous forests. In discussions that followed, it was apparent that many kinds of activity by man affect songbird populations and distribution both in Europe and in North America. For example, increases in numbers of the Great Tit, *Parus major*, in Europe (as with the Cardinal, *Richmondena cardinalis*, in the United States) may be attributed to increased feeding at suburban houses. Decreases of the Pileated Woodpecker (*Dryocopus pileatus*), according to Hickey, are related to the fact that few large trees are allowed to persist in the United States.

Studies like those of von Haartman will become increasingly important on this continent, and his work may often be used as a pattern by wildlife ecologists as interest rises in many aspects of environmental quality.

During the last 2 years, resource managers in the United States, including wildlife biologists, have found themselves deeply involved in preparation or review of "environmental impact statements." By federal law, these statements must be submitted before important environmental challenges are made. An interesting observation, related by Norton, was that gravel roads and pads near Prudhoe Bay on the Beaufort Sea in Alaska, cause premature melting of ice in spring and that this, in turn, triggers early breeding by shorebirds before the seasonal emergence of food organisms needed by the young. Such observations, backed by solid data, are precisely the type of information needed for impact statements prepared before other similar environmental changes are made.

REPRODUCTIVE ENDOCRINOLOGY

Although the connections with wildlife conservation and management seem remote, progress in the great field of reproductive endocrinology may well prove valuable in applied fields. Dane pointed out the possibility of using knowledge of the chemistry of neurohormonal control to inhibit reproduction in birds causing damage, or to stimulate it in species that are difficult to breed in captivity. (Farner has suggested brain lesions as a possible research approach to the latter problem.) Dane also mentioned the need for field and laboratory studies on the termination of breeding and on means of shortening refractory periods; until now, most research has centered on the initiation of breeding.

DEMOGRAPHIC ASPECTS OF REPRODUCTION

In discussing demographic aspects of avian reproduction, Ricklefs demonstrated that there are indeed some areas in which communication between wildlife scientists and others interested in similar biological problems is poor. Contrary opinions notwithstanding, there is a tremendous body of published data, well analyzed, about demography of game birds. It is based largely on mark and recovery tech-

niques, on immature–adult ratios determined by morphological
differences among bagged birds or on nesting studies. These subjects
were reviewed comprehensively by Farner and by Hickey in their
chapters of the book by Wolfson (1955). Some outstanding studies
have appeared in state, provincial, or federal publication series rather
than in standard journals, and this fact may have limited their circulation somewhat. Hickey's studies (1952) may be an example; it is a
landmark contribution that has served as a reference for many ornithologists for 20 years. The review papers in Wolfson (1955) and
Hickey's (1952) paper stimulated many more recent studies by wildlife scientists who used refined techniques for gathering and analyzing
data on population dynamics in birds.

I see no easy way to ensure that all researchers interested in avian
population dynamics are fully aware of each other's work; surely, however, thorough search of the literature is equally incumbent on all
scientists. One suggestion may prove helpful: Regardless of their
source many North American papers about avian demography (and
most of those concerning game species) are abstracted in *Wildlife
Review* and *Wildlife Abstracts*, both published by the U.S. Bureau of
Sport Fisheries and Wildlife.

BEHAVIORAL AND PHYSIOLOGICAL ASPECTS OF WILDLIFE-POLLUTANT PROBLEMS

At several points discussions touched on one of the most important
types of wildlife research now being conducted—the relationship of
pesticides and other pollutants to wildlife. Until now, research in this
area has been concerned principally with effects of pollutants on
mortality or natality (see Dahlgren and Linder, 1971; Fimreite and
Karstad, 1971; Longcore *et al.*, 1971). A growing number of studies
is concerned with other effects, as on behavior. The latter subject was
discussed briefly. McKinney has used motion pictures successfully to
quantify behavior, and these techniques seem applicable in pesticide
experiments.

One important area that was not covered here is the precise effects
of pesticides, particularly chlorinated hydrocarbons, upon the reproductive physiology of birds. Based on good experimental and field
evidence, we know that many species of birds are adversely affected
by eggshell thinning. Physiological mechanisms involved are not well
understood [although induction of hyperthyroidism has recently

been suggested as a contributing factor; see Jeffries and French (1972)]. Several researchers are at work on this problem. For example, one project being conducted at the Denver Wildlife Research Center bears on this, but will not produce all the answers we need. Perhaps, if we knew the physiological mechanisms involved in eggshell thinning, it would be possible for us to look for several subtle manifestations that we do not now recognize as abnormalities induced by pollutants.

CONCLUSION

I think that if another meeting of this sort is held, certain areas not treated in this symposium would interest the wildlife ecologists greatly. Avian population phenomena, with emphasis on natural regulation of populations, should be treated in much greater depth than was possible here. A related topic of value to the wildlife ecologist is the relationship of environmental stress to adrenal physiology and to survival and reproductive performance of birds.

REFERENCES

Dahlgren, R. B., and R. L. Linder. 1971. Effects of polychlorinated biphenyls on pheasant reproduction, behavior, and survival. J. Wildlife Manage. 35:315–319.

Duebbert, H. F. 1966. Island nesting of the Gadwall in North Dakota. Wilson Bull. 78:12–25.

Dzubin, A. 1955. Some evidences of home range in waterfowl, p. 278–298. *In* Transactions of the 20th North American Wildlife Conference. Wildlife Management Institute, Washington, D.C.

Erickson, R. C. 1968. A research program for endangered wildlife, p. 418–433. In Transactions of the 33rd North American Wildlife and Natural Resources Conference. Wildlife Management Institute, Washington, D.C.

Fimreite, M., and L. Karstad. 1971. Effects of dietary methyl mercury on Red-tailed Hawks. J. Wildlife Manage. 35:293–300.

Goodwin, H. A., and E. P. Denson. 1971. Status of endangered species program, p. 331–343. *In* Transactions of the 36th North American Wildlife and Natural Resources Conference. Wildlife Management Institute, Washington, D.C.

Hickey, J. J. 1952. Survival studies of banded birds. Special Scientific Report No. 15. Fish and Wildlife Service, U.S. Department of the Interior. U.S. Government Printing Office, Washington, D.C. 177 p.

IUCN [International Union for the Conservation of Nature and Natural Resources]. 1966–1971. Red data book. Morges, Switzerland.

Jeffries, D. J., and M. C. French. 1972. Changes induced in the pigeon thyroid by p, p'-DDE and dieldrin. J. Wildlife Manage. 36:24–30.

Kadlec, J. A., and W. B. Drury, Jr. 1968. Structure of the New England Herring Gull population. Ecology 49:644–676.

Lefebre, P. W., and J. L. Seubert. 1970. Surfacants as blackbird stressing agents. Proc. Fourth Vertebr. Pest Conf. 4:156–161.

Longcore, J. R., F. B. Samson, and T. W. Whittendale, Jr. 1971. DDE thins eggshells and lowers reproductive success of captive Black Ducks. Bull. Environ. Contam. Toxicol. 6:485–490.

Murton, R. K. 1971. Man and birds. Collins, London. 364 p.

Smith, R. I. 1968. The social aspects of reproductive behavior in the Pintail. Auk 85:381–396.

Thompson, R. D., C. V. Grant, E. W. Pearson, and G. W. Corner. 1968. Differential heart rate response of Starlings to sound stimuli of biological origin. J. Wildlife Manage. 32:888–893.

Weller, M. W. 1959. Parasite egg-laying in the Redhead (*Aythya americana*) and other North American Anatidae. Ecol. Monogr. 29:333–365.

Wolfson, A. 1955. Recent studies in avian biology. American Ornithologists, Union. University of Illinois Press, Urbana. 479 p.

Summary

A. VAN TIENHOVEN

In Professor Immelmann's extremely clear exposé he pointed out the phylogenetic aspects involved in the regulation of breeding activity and of the adaptations that are found for the "fine" regulation of the onset of the final phases of reproduction. He particularly stressed the predictive value of photoperiod for birds breeding in the North Temperate zone. It is clear from the evidence he presented that light may function "simply" as a cue to keep circennial rhythms of testicular recrudesence and regression, of fat deposition and of molt, in phase. On the other hand, changes in photoperiod can stimulate the hypothalamo–hypophysio–gonadal system because the light period starts to impinge on this system at times during which the system is sensitive to such stimulation. This aspect was also illustrated by Dr. Follett and by Professor Lofts.

For birds breeding in the tropics and for opportunistic breeders, particularly in Australia, cues other than photoperiod are used. Examples are rainfall and temperature changes. Dr. Frith illustrated the importance of water levels for the breeding of various species of Australian waterfowl.

The question as to how photoperiodic and other information is processed, resulting in gonadal stimulation, has been illuminated by the presentations by Doctors Assenmacher, Follett, and Oksche. It is evident that, in general, there is anatomical evidence that there may be as many as four neurosecretory systems in the avian hypothalamus. It would be vain for me to try to summarize these exquisite studies, which were so well illustrated by excellent slides.

In general, and conceptually, insights derived from the anatomical studies and the physiological studies in which lesions were made in different parts of the hypothalamus are in agreement. There appear to be some differences between results obtained by different experimenters working with different species. It is possible that these differences can be reduced to technical differences; for example, it may be that in one species a lesion in the anterior hypothalamus impinges also on some connections with the tuberal nuclei, whereas in another species a slightly different anatomical arrangement may prevent such an involvement.

To me, one of the most important advances has been the development of a reliable radioimmunoassay for avian luteinizing hormone (LH) by Follett *et al.* (1971). In the past, one had to rely on measurements of gonadotropin (GTH) concentrations and contents of the anterior pituitary and on the cytology, histochemistry, and fine structure of this gland in order to evaluate secretion of these hormones. However, there were always difficulties of interpretation because a high GTH content may mean either storage of hormone, with little release, or high level of secretion. Similarly, a low GTH content may reflect a lack of secretion or a high secretion, with rapid release of the hormones.

To interpret the GTH data, one needs to assess the state of gonads, as a type of "internal bioassay." But, if gonadal data are required for the interpretation of the pituitary GTH data, one might just as well rely on the gonadal data in the first place.

Several participants pointed out that there is a great deal of variability among the LH concentrations in the blood of the same species kept in similar, or identical, conditions. With the new radioimmunoassays and with the new protein-binding assays for steroid hormones that allow one to determine nanograms of progesterone (Peterson and Common, 1971) and picograms of estrogens (Laguë, 1972) in avian blood, it seems to me that simultaneous determinations of LH, progesterone, estrogen, and androgens in the same blood sample may give considerable insight into the quantitative aspects of the hypophysial–gonadal relationships. Such investigations may also establish whether the variations in LH concentrations among birds kept in similar conditions is the result of different "set points" of the servo mechanism.

In spite of the many advances that have been made, there are some inadequacies in our concepts of the control of the pituitary by the hypothalamus. It would be convenient if there were a separate FSH-releasing hormone (FSH-RH) and a separate LH-RH, because this would account for variations in FSH:LH ratios in the blood. However, the pure ovine LH-RH (Amoss *et al.*, 1971) and the pure porcine LH-RH (Schally *et al.*, 1971a,b) not only release LH but also FSH. The porcine LH-RH/FSH-RH has the following structure:

pyroGlu-His-Try-Ser-Tyr-Gly-Leu-Arg-Pro-Gly NH_2

(Baba *et al.*, 1971; Matsuo *et al.*, 1971a) and has been synthesized (Matsuo *et al.*, 1971b).

A molecule consisting of four amino acids that releases LH, but not FSH, from rat pituitaries has been synthesized, but it has not been found in hypothalamic extracts (Bowers *et al.*, 1971; Chang *et al.*, 1971).

Recent investigations by Jackson (1971a,b) suggest that mammalian and avian gonadotropin-releasing hormones may differ chemically, yet show biological overlap. Previous experiments by Clark and Fraps (1967) and Opel and Lepore (1967) have shown that crude bovine and avian hypothalamic extracts infused into the anterior pituitary caused premature ovulations, an indicator of LH release.

Dr. Schally recently made some synthetic LH-RH/FSH-RH available to us. After intravenous injections of 10 μg of this material, we obtained about 50 percent premature ovulations; after injection of 20 μg, we obtained 100 percent premature ovulations in laying domestic fowl.

Many important questions still remain unanswered with respect to the photosexual response of birds:

What is the location of the extraretinal receptor?
What is the structure of the extraretinal receptor?
To what extent do the eyes provide information about the photoperiod with respect to the hypothalamo–hypophysio–gonadal system. On the one hand, the experiments by Menaker *et al.* (1970), in which the amount of light reaching the extra retinal receptor was decreased by injections of India ink under the skin of the skull of House Sparrows (*Passer domesticas*), indicate that information about the photoperiod available via the retina is ignored. On the other hand, Dr. Homma has cited evidence, with Japanese Quail, suggesting that

upon switching the birds from long to short days, the eyes perform inhibitory functions.

What is the location of the circadian clock for the photoperiodic response? Experimental evidence suggest that removal of the pineal causes a rhythmic locomotor activity (Gaston and Menaker, 1968) and abolished the normal circadian rhythm of body temperature (Binkley *et al.*, 1971) of House Sparrows. However, Follett has mentioned during our discussions that in experiments requiring an involvement of the biological clock, because "unnatural" light regimes are used, pinealectomy did not affect the photosexual response.

What is the role of prolactin in different species with respect to testicular and ovarian stimulation? Professor Farner has pointed out that, in view of experiments by Meier and his associates, the entire subject needs reinvestigation. Meier *et al.* (1971) found that prolactin injections, made either 4 or 8 hours after corticosterone administration, were effective in stimulating testicular growth of photorefractory House Sparrows exposed to continuous light (LL). Prolactin injections at other intervals after corticosterone administration were ineffective. The testes of photosensitive White-throated Sparrows (*Zonotrichia albicollis*) under LL were stimulated when prolactin was injected 12 hours after corticosterone administration but were inhibited by daily prolactin injections 8 hours after corticosterone injections.

Dr. Murton gave an account of the relationship between photoperiod and date of egg-laying of geese and swans at Slimbridge, on the one hand, and the time of breeding of these birds at their normal breeding latitude, on the other hand. This significant relationship provides evidence for a fundamental mechanism that regulates the onset of breeding and for the adaptations of different species of waterfowl that may have involved modifications in the threshold of day length at which they respond.

Dr. Murton also challenged many of us when he indicated that a feedback mechanism was not involved, that the feedback mechanism and the photorefractory period were possibly artifacts. However, it may be worth exploring to what extent variations in circulating gonadotropins are the result of feedback mechanisms and to what extent they occur as the result of daily rhythms. There is evidence that the daily variation in corticotropin-releasing factor in the hypothalamus of rats is independent of the adrenal (Hiroshige and Sakakura, 1971). With the new radioimmunoassays of avian LH, it should be of considerable interest to determine whether daily variations in LH concentrations occur in castrated birds and in birds with regressed testes.

The entire question of refractoriness of the hypothalamo–hypophysio–gonadal system is an extremely difficult one and many enigmas remain. Professor Lofts' contributions provide us with a number of new insights. First, he showed that the regressed testes of photosensitive and photorefractory Mallards differ in some important histological and histochemical aspects. This may mean that part of the refractoriness of the system is the result of a refractoriness at the level of the gonad. There is evidence in pullets that the post-juvenile refractoriness of the hypothalamo–hypophysio–ovarian system is the result of the inability of the ovary to respond. Massive doses of mammalian gonadotropin and of crude chicken pituitary extracts (15 rooster anterior pituitary equivalents/day for 20 days) failed to stimulate the deposition of yolk (Phillips, 1959).

Second, Professor Lofts provided evidence that the testicular tubules and, to be more specific, the Sertoli cells may secrete steroid hormones that, in turn, may be involved in a feedback mechanism causing a refractoriness at the level of the hypothalamus. Dr. F. E. Wilson's experiments with implants of androgens and antiandrogens into the hypothalamus strongly suggest that such a mechanism could account for refractoriness at this site. The new techniques for determining small amounts of steroids and LH should aid in solving this aspect of the problem.

We have been interested in the possible role of the thyroid in determining the refractory period in Starlings (*Sturnus vulgaris*). Woitkewitsch (1940) was the first to report that thyroidectomy of Starlings prevented the normal regression of the testes at the end of the breeding season. Wieselthier (1969), working in my laboratory, confirmed Woitkewitsch's observation but also established that thyroidectomy did not affect the length of the photorefractory period. It appears that 6–8 weeks after thyroidectomy, there is an increase in gonadotropin secretion, as indicated by testicular size, pituitary histology and the fact that thyroidectomy does not affect the testicular response to exogenous gonadotropin injections.

In addition to the effect of photoperiod on hypothalamo–hypophysio–gonadal system, other inputs can affect this system. An example of inputs via the retina is the reciprocal influence of male and female pigeons (Matthews, 1939) and the experiments reviewed

by Professor Beer for Ring Doves and canaries. The question was
raised during the discussion whether data obtained with such domesticated species as Ring Doves and canaries might be valid for non-
domesticated birds. One of the difficulties encountered in attempting
to investigate such interrelationships between male and female, on
the one hand, and the neuroendocrine system, on the other, is that
females of many avian species do not breed in captivity. However,
an example of the importance of visual inputs in regulating gonadal
activity can be found in the field experiments by Neal Smith (1966)
with *Larus* gulls. Smith found that the contrast between eye ring and
head feathers is the most important isolating mechanism for the
sympatric species of *Larus* he was studying. The female selects her
mate on the basis of this contrast in eye-ring feathers. After the pair
is formed, the male uses the same criterion for maintenance of the
pair bond. After experimentally changing the ring color of mated
females, the pair-bond broke and the testes did not recrudesce,
whereas in control pairs the testes increased in size.

These experiments seem to provide a link between the physiological considerations and Professor McKinney's account of the comparative ethology of the behavior of Steller's (*Polysticta stelleri*) and
Common (*Somateria mollissima*) Eiders and the Shoveler (*Anas
clypeata*) and Pintail (*Anas acuta*). Such studies will give a clearer
understanding of social behavior and its evolution. The better our
understanding of these aspects of behavior, the more it may be possible to study the physiologic mechanisms involved that, in turn, will
allow a study of the evolution of the particular physiologic adaptations. Such considerations are important to the "laboratory physiologist" and the "wildlife biologist."

It is clear, of course, that spermatogenesis, ovulation, copulation,
and fertilization are only parts of the aspects of the entire reproductive process.

The incubation of eggs requires not only behavioral adaptations
but in many species morphological adaptations such as the formation
of the brood-patch. Professor Beer cited evidence on how behavior
and circulating hormones interact to provide the optimal development and use of such a brood-patch. Dr. Drent gave us a full account
of the physical and behavioral aspects involved in keeping the eggs
warm and of conserving the incubating parents' calories. Professor
Farner pointed out to me that the breakthrough by Dr. Drent occurred because he considered the eggs as part of the incubating birds.
Indeed, the entire problem of thermoregulation considered in this

manner again involves the hypothalamus. There is considerable evidence that the biogenic amines produced and released in the hypothalamus play a vital role in temperature regulation of mammals (Feldberg, 1970a,b; Myers, 1970). Experiments involving intraventricular injections of such amines in domestic fowl (Grunden and Marley, 1970) suggest that the same mechanisms may be involved in birds.

In all these aspects of reproduction, nutrition plays an important role. Professor Scott has shown that carcass analysis can give one a close approximation of the requirements for specific nutrients of an animal. To determine in what manner an animal acquires these nutrients is a matter for ecological studies.

We have heard about experiments in which the effect of reduced energy intake on reproduction was studied by simply reducing feed intake. I want to point out that it takes rather sophisticated techniques to study the effect of reduced energy intake per se. The reduction of food intake also reduces the intake of all nutrients so one does not know what is responsible for the observed effect— energy deficiency or deficiency of some nutrient(s). Similarly, if one wishes to study the effect of a specific nutrient deficiency, one has to make sure that such a deficiency is not accompanied by reduced total food intake. (This is true for many vitamin, mineral, and amino acid deficiencies.) There are several experimental set-ups one can employ to avoid such pitfalls. One can use a pair-feeding technique, but this may result in a stress phenomenon because the control birds are hungry most of the time. One can manipulate the diets so that the nutrient intake for controls and experimentals are the same except for the particular nutrient under consideration. In general, it takes fairly sophisticated manipulation of diets to accomplish this.

Professor King and Professor Kendeigh have given us estimates of the energy budgets required for different aspects of the reproductive process.

Dr. von Haartman gave a thorough account of the effects of changes in the temperature in Scandinavia over the centuries upon the avifauna. His account of how land use and forestry and animal husbandry practices can change the avifauna over a short time should serve as a warning. The question is, can we as avian biologists predict to agriculturists what will be the consequences on the avifauna of changes in farm management practices?

During this symposium, we all have debated the worthwhileness of the various models that were presented. However, it seems to me

that the greatest advances in our understanding of avian breeding
biology will come through such modeling. I was particularly impressed with Professor Orians' "rough and ready" modeling that
could readily be tested in the field. The data obtained in such experiments can then, in turn, be used to refine the model. Another value
of a model is, of course, that it can be of great value in guiding the
experimenter in designing the experiments crucial from a conceptual
view, or required to obtain quantitative insights in very complex
relationships.

It is, in my judgment, the quantitation of circulating hormones
that will bring physiologists and behaviorists closer together.

Neither models nor detailed quantitative data, however, can give
us by themselves the wisdom to handle the results ideally. Obviously,
wildlife biology and wildlife management are going to be under
pressures involving political, social, and economic considerations. To
what extent we can guide decision-makers by symposia such as these
and to what extent we can convince such decision-makers that there
is a real relationship between the exquisite anatomy of the
hypothalamo–hypophysial system and public policy needs to be
determined.

REFERENCES

Amoss, M., R. Burgus, R. Blackwell, W. Vale, R. Fellows, and R. Guillemin.
1971. Purification, amino acid composition and N terminus of the
hypothalamic luteinizing hormone releasing factor (LRF) of ovine origin.
Biochem. Biophys. Res. Commun. 44:205–210.
Baba, Y., H. Matsuo, and A. V. Schally. 1971. Structure of the porcine LH-
and FSH-releasing hormone. II. Confirmation of the proposed structure by
conventional sequential analysis. Biochem. Biophys. Res. Commun.
44:459–463.
Binkley, S., E. Kluth, and M. Menaker. 1971. Pineal function in sparrows:
Circadian rhythms and body temperature. Science 174:311–314.
Bowers, C. Y., J-K. Chang, H. Sievertsson, C. Bogentoft, B. L. Currie, and K.
Folkers. 1971. Activity of a new synthetic tetrapeptide in hypothalamic
luteinizing and follicle-stimulating releasing hormone assay systems. Biochem.
Biophys. Res. Commun. 44:414–421.
Chang, J-K., H. Sievertsson, C. Bogentoft, B. L. Currie, K. Folkers, and C. Y.
Bowers. 1971. Discovery of a new synthetic tetrapeptide having luteinizing
releasing hormone (LRH) activity. Biochem. Biophys. Res. Commun.
44:409–413.
Clark, C. E., and R. M. Fraps. 1967. Induction of ovulation in the chicken with
median eminence extracts. Poultry Sci. 46:1245–1246.

Feldberg, W. S. 1970a. The monoamines of the hypothalamus as mediators of temperature responses, p. 493–506. *In* J. D. Hardy, A. P. Gagge, and J. A. J. Stolwjk [ed.] Physiological and behavioral temperature regulation. Charles C Thomas, Springfield, Ill.

Feldberg, W. S. 1970b. Monoamines of the hypothalamus as mediators of temperature response, p. 213–232. *In* L. Martini, M. Motta, and F. Fraschini [ed.] The hypothalamus. Academic Press, New York.

Follett, B. K., C. G. Scanes, and F. J. Cunningham. 1971. A radioimmunoassay for avian luteinizing hormone. J. Endocrinol. 51:v–vi.

Gaston, S., and M. Menaker. 1968. Pineal function: The biological clock in the sparrow. Science 160:1125–1127.

Grunden, L. R., and E. Marley. 1970. Effects of sympathomimetic amines injected into the third ventricle in adult chickens. Neuropharmacology 9:119–128.

Hiroshige, T., and M. Sakakura. 1971. Circadian rhythm of corticotropin releasing activity in the hypothalamus of normal and adrenalectomized rats. Neuroendocrinology 7:25–36.

Jackson, G. L. 1971a. Avian luteinizing hormone releasing factor. Endocrinology 89:1454–1459.

Jackson, G. L. 1971b. Comparison of rat and chicken luteinizing-hormone releasing factors. Endocrinology 89:1460–1463.

Laguë, P. C. 1972. Plasma estrone and estradiol concentrations in the laying hen in relation to the ovulatory cycle. PhD thesis, Cornell University, Ithaca, New York. 114 p.

Matsuo, H., Y. Baba, R. M. G. Nair, A. Arimura, and A. V. Schally. 1971a. Structure of the porcine LH- and FSH-releasing hormone. I. The proposed amino acid sequence. Biochem. Biophys. Res. Commun. 43:1334–1339.

Matsuo, H., A. Arimura, R. M. G. Nair, and A. V. Schally. 1971b. Synthesis of the porcine LH- and FSH-releasing hormone by the solid-phase method. Biochem. Biophys. Res. Commun. 45:822–826.

Matthews, L. H. 1939. Visual stimulation and ovulation in pigeons. Proc. Roy. Soc. London Ser. B 126:557–560.

Meier, A. H., D. D. Martin, and R. MacGregor, III. 1971. Temporal synergism of corticosterone and prolactin controlling gonadal growth in sparrows. Science 173:1240–1242.

Menaker, M., R. Roberts, J. Elliott, and H. Underwood. 1970. Extraretinal high perception in the sparrow. III. The eyes do not participate in photoperiodic photoreception. Proc. Natl. Acad. Sci. U.S.A. 67:320–325.

Myers, R. D. 1970. The role of hypothalamic transmittor factors in the control of body temperature, p. 648–665. *In* J. D. Hardy, A. P. Gagge, and J. A. J. Stolwjk [ed.] Physiological and behavioral temperature regulation. Charles C Thomas, Springfield, Ill.

Opel, H., and P. D. Lepore. 1967. Ovulating hormone releasing factor in the chicken hypothalamus. Poultry Sci. 46:1302.

Peterson, A. J., and R. H. Common. 1971. Progesterone concentration in peripheral plasma of laying hens as determined by competitive protein-binding assay. Can. J. Zool. 49:599–605.

Phillips, R. E. 1959. Endocrine mechanisms of the failure of Pintails (*Anas acuta*) to reproduce in captivity. PhD thesis, Cornell University, Ithaca, New York. 76 p.

Schally, A. V., A. Arimura, A. J. Kastin, H. Matsuo, Y. Baba, T. W. Redding, R. M. G. Nair, and L. Debeljuk. 1971a. Gonadotropin-releasing hormone: One polypeptide regulates secretion of luteinizing and follicle stimulating hormones. Science 173:1036–1038.

Schally, A. V., A. Arimura, Y. Baba, R. M. G. Nair, H. Matsuo, T. W. Redding, L. Debeljuk, and W. F. White. 1971b. Isolation and properties of the FSH and LH-releasing hormone. Biochem. Biophys. Res. Commun. 43:393–399.

Smith, N. G. 1966. Evolution of some Artic Gulls *Larus*: An experimental study of isolating mechanisms. Ornothological Monographs No. 4. American Ornothologists' Union, Lawrence, Kansas. 99 p.

Wieselthier, A. S. 1969. The stimulation and prolongation of testicular activity in the European Starling (*Sturnus vulgaris*) by thyroidectomy and reduced photoperiod. PhD thesis, Cornell University, Ithaca, New York. 120 p.

Woitkewitsch, A. A. 1940. Dependence of seasonal periodicity in gonadal changes on the thyroid gland in *Sturnus vulgaris* L. C.R. (Dokl.) Akad. Sci. U.R.S.S. 27:741–747.

Participants

ABBOTT, URSULA K., Department of Avian Sciences, University of California, Davis, California 95616

ABEL, JOHN H. JR., Department of Physiology and Biophysics, Colorado State University, Ft. Collins, Colorado 80521

ALDRICH, JOHN W., Bird and Mammal Laboratories, Fish and Wildlife Service, Museum of Natural History, Washington, D.C. 20560

ASSENMACHER, I., Laboratoire de Physiologie animale, Universitè de Montpellier, Montpellier, France

BARTHOLOMEW, GEORGE, Department of Zoology, University of California, Los Angeles 90024

BASKETT, THOMAS S., Fish and Wildlife Service, Bureau of Sport Fisheries and Wildlife, U.S. Department of the Interior, Washington, D.C. 20240

BAXTER, WILLIAM L., Nebraska Game and Parks Commission, Lincoln, Nebraska 68503

BEER, COLIN G., Institute for Animal Behavior, Rutgers University, Newark, New Jersey 07102

BESSER, JEROME F., Denver Wildlife Research Center. Denver, Colorado 80226

BIELLIER, HAROLD V., Department of Poultry Husbandry, University of Missouri, Columbia, Missouri 65201

BOAG, D. A., Department of Zoology, University of Alberta, Edmonton 7, Alberta, Canada

BRAUN, CLAIT E., Division of Game, Fish and Parks, Fort Collins, Colorado 80521

BROWN, R. G. B., Canadian Wildlife Service, Marine Ecology Laboratory, Bedford Institute, Dartmouth, Nova Scotia, Canada

CAIN, J. RICHARD, Poultry Science Department, Texas A&M University, College Station, Texas 77843

CHIASSON, ROBERT B., Department of Biological Sciences, University of Arizona, Tucson, Arizona 85721

CONLEY, WALT, Department of Fisheries and Wildlife, Michigan State University, East Lansing, Michigan 48823

DANE, CHARLES W., Northern Prairie Wildlife Research Center, U.S. Department of the Interior, Jamestown, North Dakota 58401

DARDEN, TOM, Department of Zoology, University of Washington, Seattle, Washington 98105

DAVIES, WILLIAM, Department of Avian Sciences, University of California, Davis, California 95616

DIETER, MICHAEL, Patuxent Wildlife Research Center, Laurel, Maryland 20810

DRENT, RUDOLF, Zoölogisch Laboratorium der Rijksuniversiteit te Groningen, Haren Groningen, The Netherlands

DUSTMAN, E. H., Patuxent Wildlife Research Center, Laurel, Maryland 20810

DYER, M. I., Fish and Wildlife Service, Ohio Field Station, Sandusky, Ohio 44870

ELDER, WILLIAM H., Department of Zoology, University of Missouri, Columbia, Missouri 65201

ELLIS, JAMES E., Natural Resource Ecology Laboratory, Colorado State University, Fort Collins, Colorado 80521

ENG, ROBERT L., Department of Zoology and Entomology, Montana State University, Bozeman, Montana 59715

ERSKINE, A. J., Canadian Wildlife Service, Ottawa 4, Ontario, Canada

FARNER, DONALD S., Department of Zoology, University of Washington, Seattle, Washington 98195

FOLLETT, BRIAN K., Department of Zoology, University College of North Wales, Bangor, Caernarvonshire, U.K.

FREDRICKSON, LEIGH H., Gaylord Memorial Laboratory, University of Missouri, Puxico, Missouri 63960

FRETWELL, S. D., Division of Biology, Kansas State University, Manhattan, Kansas 66502

FRIEND, MILTON, Denver Wildlife Research Center, Denver, Colorado 80225

FRITH, HARRY, CSIRO Division of Wildlife Research, Lyneham, A.C.T. 2602, Australia

GARRETT, R. L., Department of Avian Sciences, University of California, Davis, California 95616

GEE, GEORGE F., Endangered Wildlife Research, Patuxent Wildlife Research Center, Laurel, Maryland 20810

GESSAMAN, JAMES A., Department of Zoology, Utah State University, Logan, Utah 84321

GOFORTH, W. REID, Missouri Cooperative Wildlife Research Unit, University of Missouri, Columbia, Missouri 65201

GRANT, C. VAL, Department of Zoology, Utah State University, Logan, Utah 84321

GUARINO, JOSEPH T., Bureau of Sport Fisheries and Wildlife, Denver Wildlife Research Center, Denver, Colorado 80225

HAILMAN, JACK P., Department of Zoology, University of Wisconsin, Madison, Wisconsin 53706

HEINZ, GARY, Patuxent Wildlife Research Center, Laurel, Maryland 20810

HELMS, CARL W., Department of Zoology, University of Georgia, Athens, Georgia 30601

HICKEY, J. J., Department of Wildlife Ecology, University of Wisconsin, Madison, Wisconsin 53706

HOMMA, K., Department of Veterinary Medicine and Animal Sciences, Faculty of Agriculture, University of Tokyo, Bukyo-ku, Tokyo, Japan

IMMELMANN, KLAUS, Zoologisches Institut der Technischen Universität, 33 Braunschweig, Federal Republic of Germany

JENNI, DONALD A., Department of Zoology, University of Montana, Missoula, Montana 58901

JOHNSGARD, PAUL A., Department of Zoology and Physiology, University of Nebraska, Lincoln, Nebraska 68508

KENDEIGH, S. CHARLES, Department of Zoology, University of Illinois, Urbana, Illinois 61820

KENNELLY, JAMES J., Bureau of Sport Fisheries and Wildlife, Denver, Colorado 80225

KERN, MICHAEL DON, Department of Biological Sciences, Fordham University, Bronx, New York 10458

KING, JAMES R., Laboratories of Zoophysiology, Washington State University, Pullman, Washington 99163

KRAPU, GARY, Northern Prairie Wildlife Research Center, Jamestown, North Dakota 58401

LINDER, RAYMOND L., South Dakota Cooperative Wildlife Research Unit, South Dakota State University, Brookings, South Dakota 57006

LOCK, ROSS A., Nebraska Game and Parks Commission, Lincoln, Nebraska 68503

LOFTS, BRIAN, Department of Zoology, University of Hong Kong, Hong Kong

MATHER, F. BEN, Department of Poultry Science, University of Nebraska, Lincoln, Nebraska 68503

MATSCHKE, GEORGE H., Denver Wildlife Research Center, Denver, Colorado 80225

MAY, TERRY A., Institute of Arctic and Alpine Research, University of Colorado, Boulder, Colorado 80302

McKINNEY, FRANK, Museum of Natural History, University of Minnesota, Minneapolis, Minnesota 55455

MENDALL, HOWARD L., Maine Cooperative Wildlife Research Unit, University of Maine at Orono, Orono, Maine 04473

MODAFFERI, RONALD D., Alaska Cooperative Wildlife Research Unit, University of Alaska, College, Alaska 99701

MULLER, H. D., Extension Poultry Scientist, University of Georgia, Athens, Georgia 30601

MURTON, R. K., The Nature Conservancy, Monks Wood Experimental Station, Abbots Ripton, Huntingdon, U.K.

NELSON, HARVEY K., Northern Prairie Wildlife Research Center, Jamestown, North Dakota 58401

NISH, DARRELL H., Utah Division of Wildlife Resources, Salt Lake City, Utah 84116

NORMAN, JAMES A., Kansas Forestry, Fish and Game Commission, Hays, Kansas 67601

NORTON, DAVID W., Insitute of Arctic Biology, University of Alaska, College, Alaska 99701

OKSCHE, A., Anatomisches Institut, Friedrichstrasse 24, 63 Giessen, Federal Republic of Germany

OPEL, HOWARD L., JR., Animal Science Research Division, Agricultural Research Center, Beltsville, Maryland 20705

ORIANS, GORDON, Department of Zoology, University of Washington, Seattle, Washington 98195

ORING, LEWIS W., Department of Biology, University of North Dakota, Grand Forks, North Dakota 58201

PALMER, ROBERT A., Department of Zoology, Utah State University, Logan, Utah 84321

PITELKA, FRANK A., Department of Zoology, University of California, Berkeley, California 94720

PRINCE, HAROLD H., Department of Fish and Wildlife, Michigan State University, East Lansing, Michigan 48823

RAVELING, DENNIS G., Department of Animal Physiology, University of California, Davis, California 95616

REED, AUSTIN, Canadian Wildlife Service, Saint Foy, Quebec, Canada

RICKLEFS, ROBERT E., The College Department of Biology, University of Pennsylvania, Philadelphia, Pennsylvania 19104

ROBEL, R. J., Division of Biology, Kansas State University, Manhattan, Kansas 66502

ROGERS, JOHN P., Migratory Bird Population Station, Bureau of Sport Fisheries and Wildlife, U.S. Department of the Interior, Laurel, Maryland 20810

RUDD, ROBERT L., Department of Zoology, University of California, Davis, 95616

RUSHING, JUNE, Department of Zoology, Utah State University, Logan, Utah 84321

RUSSELL, KENNETH R., Iowa Cooperative Wildlife Research Unit, Iowa State University, Ames, Iowa 50010

RUTLEDGE, JAMES, Department of Avian Sciences, University of California, Davis, California 95616

SCHAFER, EDWARD W., JR., Denver Wildlife Research Center, Denver, Colorado 80225

SCHWAB, ROBERT G., Department of Animal Physiology, University of California, Davis, California 95616

SCHWILLING, MARVIN D., Kansas Forestry, Fish and Game Commission, Hays, Kansas 67601

SCOTT, M. L., Department of Poultry Science, Cornell University, Ithaca, New York 14850

SEUBERT, JOHN, Patuxent Wildlife Research Center, Laurel, Maryland 20810

SHUMAKE, STEPHEN A., Denver Wildlife Research Center, Denver, Colorado 80225

SIMPSON, RONALD, Georgia Game and Fish Commission, Albany, Georgia 31701

STETSON, MILTON H., Department of Zoology, University of Texas, Austin, Texas 78712

STEVENS, RUSSELL B., Division of Biology and Agriculture, National Research Council, Washington, D.C. 20418

STONE, CHARLES P., Denver Wildlife Research Center, Denver, Colorado 80225

SUGDEN, LAWSON G., Prairie Migratory Bird Research Centre, University of Saskatchewan Campus, Saskatoon, Saskatchewan, Canada

TESTER, JOHN R., Department of Ecology and Behavioral Biology, University of Minnesota, Minneapolis, Minnesota 55455

VAN TIENHOVEN, ARI, Department of Poultry Science, Cornell University, Ithaca, New York 14850

VON HAARTMAN, LARS, Universitetets Zoologiska Institut, N. Järnvägsgatan 13, Helsinki, Finland

WATSON, ADAM,* The Nature Conservancy, Blackhall Banchory Kincardineshire AB3 3PS, U.K.

WIENS, JOHN A., Department of Zoology, Oregon State University, Corvallis, Oregon 97331

WEISE, CHARLES M., Zoology Department, University of Wisconsin, Milwaukee, Wisconsin 53211

WIGHT, HOWARD M., Oregon Cooperative Wildlife Research Unit, Oregon State University, Corvallis, Oregon 97330

WILSON, FRED E., Division of Biology, Kansas State University, Manhattan, Kansas 66502

WILSON, W. O., Department of Avian Sciences, University of California, Davis, California 95616

ZIMMERMAN, JOHN L., Division of Biology, Kansas State University, Manhattan, Kansas 66502

ZWICKEL, FRED, Department of Zoology, University of Alberta, Edmonton 7, Alberta, Canada

*Unable to attend symposium.